Algorithms and Computation in Mathematics

Volume 32

With this forward-thinking series Springer recognizes that the prevailing trend in mathematical research towards algorithmic and constructive processes is one of long-term importance. This series is intended to further the development of computational and algorithmic mathematics. In particular, Algorithms and Computation in Mathematics emphasizes the computational aspects of algebraic geometry, number theory, combinatorics, commutative, non-commutative and differential algebra, geometric and algebraic topology, group theory, optimization, dynamical systems and Lie theory. Proposals or manuscripts that center on content in non-computational aspects of one of these fields will also be regarded if the presentation gives consideration to the contents' usefulness in algorithmic processes.

Wolfram Decker · Christian Eder · Claus Fieker ·
Max Horn · Michael Joswig
Editors

The Computer Algebra System OSCAR

Algorithms and Examples

Springer

Editors
Wolfram Decker
Department of Mathematics
RPTU Kaiserslautern-Landau
Kaiserslautern, Rheinland-Pfalz, Germany

Christian Eder
Department of Mathematics
RPTU Kaiserslautern-Landau
Kaiserslautern, Rheinland-Pfalz, Germany

Claus Fieker
Department of Mathematics
RPTU Kaiserslautern-Landau
Kaiserslautern, Rheinland-Pfalz, Germany

Max Horn
Department of Mathematics
RPTU Kaiserslautern-Landau
Kaiserslautern, Rheinland-Pfalz, Germany

Michael Joswig
Institute of Mathematics
Technical University of Berlin
Berlin, Germany

ISSN 1431-1550
Algorithms and Computation in Mathematics
ISBN 978-3-031-62126-0 ISBN 978-3-031-62127-7 (eBook)
https://doi.org/10.1007/978-3-031-62127-7

This Springer imprint is published by the registered company Springer Nature Switzerland AG
The registered company address is: Gewerbestrasse 11, 6330 Cham, Switzerland

Preface

This book describes the new Computer Algebra System OSCAR in its first complete version 1.0. OSCAR is free open source software written in Julia. OSCAR builds on and vastly extends four *cornerstone systems*: ANTIC for number theory, GAP for group and representation theory, polymake for polyhedral and tropical geometry, and Singular for commutative algebra and algebraic geometry. These topics and their interaction describe the scope and focus of OSCAR.

This book is organized into three parts. In the first part we discuss technical aspects, including the installation, the programming language Julia, as well as the overall design. The second part is dedicated to the four mathematical directions of OSCAR, which come from its four cornerstones. These chapters are accessible on the graduate student level and serve as a basic introduction, explaining how the four directions are reflected in the software. The third part comprises a great variety of more specialized topics, some of which are addressed to more expert audiences.

The OSCAR project has been supported from the start, in 2017, by the Deutsche Forschungsgemeinschaft (DFG, German Research Foundation) through "Symbolic Tools in Mathematics and their Application" (TRR 195, project-ID 286237555)[1]. Since 2019, it has also been supported by the "Research Initiative of the Federal State of Rhineland-Palatinate".

Berlin and Kaiserslautern,
March 2024,

Wolfram Decker
Christian Eder
Claus Fieker
Max Horn
Michael Joswig

[1] Gefördert durch die Deutsche Forschungsgemeinschaft (DFG) – Projektnummer 286237555 – TRR 195.

Acknowledgements

A big software project like `OSCAR` requires the support of a large number of developers. While many of them are also authors of chapters of this book, some are not, so we give here the complete list of contributors to the code. More precisely, the following programmers contributed to the master branch of `OSCAR`, excluding merge commits, on or before 1st March 2024:

John Abbott, Mohamed Barakat, Mara Belotti, Martin Bies, Tobias Boege, Janko Böhm, Alessio Borzi, Simon Brandhorst, Cameron Braunstein, Thomas Breuer, Taylor Brysiewicz, Andrei Comaneci, Daniel Corey, Firoozeh Dastur, Giovanni DeFranceschi, Wolfram Decker, Antony Della Vecchia, Alexander Dinges, Christian Eder, Raul Epure, Kamillo Ferry, Claus Fieker, Ghislain Fourier, Rafael Fourquet, Anne Frühbis-Krüger, Alheydis Geiger, Zoe Geiselmann, Fedor Glazov, Lars Göttgens, William Hart, Tommy Hofmann, Max Horn, Alexej Jordan, Michael Joswig, Marek Kaluba, Lars Kastner, Aaruni Kaushik, Lukas Kühne, Markus Kurtz, Benjamin Lorenz, Dante Luber, Yingbo Ma, Rafael Mohr, Stevell Muller, Giosue Muratore, Erik Paemurru, Marta Panizzut, Syx Pek, Helena Petri, Nicolas Faroß, Delphine Pol, Daniel Rademacher, Lakshmi Ramesh, Yue Ren, Friedrich Rober, Morgan Rodgers, Felix Röhrich, Kemal Rose, Daniel Schaefer, Victoria Schleis, Johannes Schmitt, Benjamin Schröter, Daniel Schultz, Alexander Seiler, Carlo Sircana, Jieao Song, Justus Springer, Ulrich Thiel, Jan P. Thiele, Andrew P. Turner, Zlatan Vasović, Laura Voggesberger, Marcel Wack, Martin Wagner, Lena Weis, Claudia Yun, Oguzhan Yürük, Matthias Zach.

Contents

List of Contributors

Martin Bies
RPTU Kaiserslautern-Landau, Kaiserslautern, Germany, e-mail: bies@mathematik.uni-kl.de

Janko Böhm
RPTU Kaiserslautern-Landau, Kaiserslautern, Germany, e-mail: boehm@mathematik.uni-kl.de

Simon Brandhorst
Universität des Saarlandes Fakultät für Mathematik und Informatik, Campus E2.4 66123 Saarbrücken, Germany, e-mail: brandhorst@math.uni-sb.de

Thomas Breuer
RWTH Aachen University, Aachen, Germany, e-mail: thomas.breuer@math.rwth-aachen.de

Taylor Brysiewicz
University of Western Ontario, Department of Mathematics, London ON, Canada. e-mail: tbrysiew@uwo.ca

Daniel Corey
University of Nevada, Department of Mathematical Sciences, Las Vegas, USA, e-mail: daniel.corey@unlv.edu

Firozeeh Dastur
RPTU Kaiserslautern-Landau, Kaiserslautern, Germany, e-mail: firoozeh.aga@edu.rptu.de

Wolfram Decker
RPTU Kaiserslautern-Landau, Kaiserslautern, Germany, e-mail: decker@mathematik.uni-kl.de

Christian Eder
RPTU Kaiserslautern-Landau, Kaiserslautern, Germany, e-mail: ederc@mathematik.uni-kl.de

Xin Fang
RWTH Aachen University, Aachen, Germany, e-mail: fang@art.rwth-aachen.de

Claus Fieker
RPTU Kaiserslautern-Landau, Kaiserslautern, Germany, e-mail: claus.fieker@rptu.de

Ghislain Fourier
RWTH Aachen University, Aachen, Germany, e-mail: fourier@art.rwth-aachen.de

Alain Hoffmann
RPTU Kaiserslautern-Landau, Kaiserslautern, Germany, e-mail: hoffmann@mathematik.uni-kl.de

Tommy Hofmann
Universität Siegen, Naturwissenschaftlich-Technische Fakultät, 57068 Siegen, Germany, e-mail: tommy.hofmann@uni-siegen.de

Isaac Holt
Durham University, Department of Mathematical Sciences, Durham DH1 3LE, United Kingdom, e-mail: isaac.a.holt@durham.ac.uk

Max Horn
RPTU Kaiserslautern-Landau, Kaiserslautern, Germany, e-mail: max.horn@rptu.de

Michael Joswig
Technische Universität Berlin, Chair of Discrete Mathematics/Geometry, 10623 Berlin, Germany, e-mail: joswig@math.tu-berlin.de
Max-Planck Institute for Mathematics in the Sciences, 04103 Leipzig, Germany

Lars Kastner
Technische Universität Berlin, Chair of Discrete Mathematics/Geometry, 10623 Berlin, Germany, e-mail: lkastner@math.tu-berlin.de

Lukas Kühne
Universität Bielefeld, Fakultät für Mathematik, Germany, e-mail: lukas.kuehne@math.uni-bielefeld.de

Benjamin Lorenz
Technische Universität Berlin, Chair of Discrete Mathematics/Geometry, 10623 Berlin, Germany, e-mail: lorenz@math.tu-berlin.de

Hannah Markwig
Universität Tübingen, Fachbereich Mathematik, Auf der Morgenstelle 10, 72076 Tübingen, e-mail: hannah@math.uni-tuebingen.de

Rafael Mohr
RPTU Kaiserslautern-Landau, Kaiserslautern, Germany
Sorbonne Université, CNRS, Paris, France, e-mail: Rafael.Mohr@lip6.fr

Gabriele Nebe
RWTH Aachen University, Aachen, Germany, e-mail: gabriele.nebe@rwth-aachen.de

Richard Parker
Cambridge, e-mail: richpark7920@gmail.com

Lakshmi Ramesh
RPTU Kaiserslautern-Landau, Kaiserslautern, Germany, e-mail: ramesh@mathematik.uni-kl.de

Yue Ren
Durham University, Department of Mathematical Sciences, Durham DH1 3LE, United Kingdom, e-mail: yue.ren2@durham.ac.uk

Lukas Ristau
Fraunhofer-Institut für Techno- und Wirtschaftsmathematik ITWM, Fraunhofer-Platz 1, 67663 Kaiserslautern, e-mail: lukas.ristau@itwm.fraunhofer.de

Kemal Rose
Max-Planck Institute for Mathematics in the Sciences, 04103 Leipzig, Germany, e-mail: krose@mis.mpg.de

Mohab Safey El Din
Sorbonne Université, CNRS, Paris, France, e-mail: Mohab.Safey@lip6.fr

Victoria Schleis
Universität Tübingen, Fachbereich Mathematik, Auf der Morgenstelle 10, 72076 Tübingen, e-mail: victoria.schleis@student.uni-tuebingen.de

Johannes Schmitt
Universität Siegen, Naturwissenschaftlich-Technische Fakultät, 57068 Siegen, Germany, e-mail: johannes2.schmitt@uni-siegen.de

Frank-Olaf Schreyer
Universität des Saarlandes Fakultät für Mathematik und Informatik, 66123 Saarbrücken, Germany, e-mail: schreyer@math.uni-sb.de

Benjamin Schröter
KTH Royal Institute of Technology, Department of Mathematics, 10044 Stockholm, Sweden, e-mail: schrot@kth.se

Bernd Sturmfels
Max-Planck Institute for Mathematics in the Sciences, 04103 Leipzig, Germany, e-mail: bernd@mis.mpg.de

Simon Telen
Max-Planck Institute for Mathematics in the Sciences, 04103 Leipzig, Germany, e-mail: simon.telen@mis.mpg.de

Ali Traore
RPTU Kaiserslautern-Landau, Kaiserslautern, Germany, e-mail: atraore@rptu.de

Andrew P. Turner
University of Pennsylvania, Department of Physics and Astronomy, Philadelphia 19146, USA, e-mail: turnerap@sas.upenn.edu

Matthias Zach
RPTU Kaiserslautern-Landau, Kaiserslautern, Germany, e-mail: zach@mathematik.uni-kl.de

Part I
OSCAR

What Is OSCAR?

Claus Fieker and Michael Joswig

Abstract We give a general overview of the software and discuss some design decisions. This basic sketch of OSCAR's inner workings is meant to support users in making best use of our system.

1 Introduction

OSCAR is a computer algebra system focusing on exact algebraic computations. The acronym stands for

Open **S**ource **C**omputer **A**lgebra **R**esearch

and is software specifically developed for and by research mathematicians. Its full documentation is available online[1].

The current use of research software in essentially all areas of mathematics is a success story. The breadth of this success is based on myriad mathematical software systems, which come in numerous flavors. For instance, the web service `swmath.org` lists 44594 systems in total[2]. This census covers comprehensive frameworks (such as `Maple`, `Mathematica`, `Matlab` or `SageMath`), large but more specialized systems (such as `Macaulay2` for algebraic geometry and commutative algebra, or `Pari/GP` for number theory) as well as individual libraries (such as `cddlib` for convex hull computations and linear optimization,

Claus Fieker
RPTU Kaiserslautern-Landau, e-mail: `fieker@rptu.de`

Michael Joswig
TU Berlin, Chair of Discrete Mathematics/Geometry, Berlin, Germany; Max-Planck-Institut für Mathematik in den Naturwissenschaften, Leipzig, Germany, e-mail: `joswig@math.tu-berlin.de`

[1] `https://docs.oscar-system.org/stable/`

[2] Counted on February 29, 2024.

W. Decker et al. (eds.), *The Computer Algebra System OSCAR*,
Algorithms and Computation in Mathematics 32,
https://doi.org/10.1007/978-3-031-62127-7_1

or FLINT for arithmetic and number theory). Where does OSCAR fit in this classification? OSCAR is oriented towards combinatorics, commutative algebra, number theory, algebraic, polyhedral and tropical geometry, with particular emphasis on the interaction of these fields and their applications. An important point is that OSCAR explicitly excludes non-exact methods. This distinguishes OSCAR from Maple, Mathematica or SageMath, which aim for comprising a wider range of topics, including numerical algorithms. On a different account, OSCAR is a free open source computer algebra system; this is a distinction from Magma, Maple, Mathematica or Matlab. Many users may not care about the intricacies of the inner workings of a modern computer algebra system. Making the source code open for everyone to read and scrutinize is a key first step to making any results obtained by such systems reproducible and thus reliable. We believe that this combination of exact computation and open source makes OSCAR a particularly well suited tool for research in mathematics.

Most of the systems listed on swmath.org fulfill one or more specific purposes. For any new mathematical software system this raises the question of how to deal with the mathematical and algorithmic knowledge that those systems are imbued with. Due to the sheer size of the combined code base, it is impossible to reimplement all the relevant methods from scratch, even for a narrow area such as commutative algebra alone. This asks for ways to build upon existing systems, which has been tried successfully in the past. A prominent example is SageMath, which features interfaces to more than 100 standard and optional packages. Such an approach, to build upon other software, comes with challenges of its own. Firstly, software maintenance, stability and installation is an issue, which becomes delicate if the number of moving parts becomes large. A second aspect is more subtle. Each software system imposes a semantic view on the objects that it deals with. Combining existing software requires some way of dealing with these semantic choices, which may or may not be aligned. Either the combining system interfaces the underlying packages "as is", in which case the user needs to provide any translation across package boundaries. Or the combining system imposes a global semantic layer on top, hiding the semantic views of the packages. OSCAR adheres to the latter principle, building on four cornerstone systems: ANTIC (for number theory), GAP (for representation and group theory), polymake (for polyhedral and tropical geometry) and Singular (for commutative algebra and algebraic geometry). We emphasize that OSCAR provides a semantic layer of its own, which may differ from the semantic views of the cornerstones. For instance, polymake uses homogeneous coordinates for polyhedra throughout, whereas OSCAR employs a more traditional affine geometric view of polyhedra. It is worth noting that each cornerstone is a fairly large software system by itself, with numerous other packages and libraries underneath. On a low level, there is a large intersection between the code bases of SageMath and OSCAR. Libraries which are in common include, e.g., FLINT and cddlib. Even more, substantial capabilities of SageMath rely on our cornerstones GAP and Singular; and polymake is an optional package. Compared with

SageMath, OSCAR provides a deeper integration of the cornerstones and a more coherent semantics. A crucial difference is that SageMath treats GAP, Singular, and polymake as third-party software which can be used, whereas the development of OSCAR and that of the cornerstones go hand in hand. In fact, leading developers of OSCAR are at the same time leading developers of the cornerstones. In this way, the cornerstones can be adjusted so as to suit OSCAR best, which often means that the cornerstones themselves are improved. This pattern is meant to stay, which means that OSCAR is not supposed to replace any cornerstone, not even in the long run. We hope that many mathematicians will like to work with this rich software environment, which allows for many different ways of usage.

2 Why Julia?

The key challenge in designing mathematical software is the necessity to bridge the gap between the ideal world of theoretical mathematics and the reality of current hard- and software. Often the team of developers is small compared to the wealth of existing mathematical knowledge. This fact requires building such bridges efficiently, keeping an eye on future maintenance. In our opinion these requirements favor standard over custom made programming languages for mathematical software. The obvious question is: which programming language?

A unique selling point of OSCAR among current computer algebra systems is the choice of Julia. One reason is that the Julia ecosystem allows for a particularly easy delivery of the code to the user. After installing Julia as described at

https://julialang.org/downloads,

within the Julia interactive command-line (REPL) you type

```
julia> import Pkg; Pkg.add("Oscar");
```

This will take a moment to automatically download and install the latest version of the software, which is then ready to be used:

```
julia> using Oscar

  ___   ____   ____    _    ____
 / _ \ / ___| / ___|  / \  |  _ \   |  Combining ANTIC, GAP, Polymake,...
| | | |\___ \| |     / _ \ | |_) |  |  Type "?Oscar" for more information
| |_| | ___) | |___ / ___ \|  _ <   |  Manual: https://docs.oscar-syst...
 \___/ |____/ \____/_/   \_\_| \_\  |  Version 1.0.0
```

More detailed instructions are under

https://www.oscar-system.org/install/.

There are many features of Julia which are often promoted in advertisements, and we will discuss some of them below. For example from a software maintenance and stability point of view, the built-in package manager is Julia's number one asset which is most useful to OSCAR. To explain this, it is useful to describe how OSCAR's code base is organized on https://github.com/oscar-system. Three of the four cornerstones, GAP, polymake and Singular, have a considerable pre-history before OSCAR, in the realm of C and C++, with a bit of Perl and other individual, specialized languages in the mix. GAP, polymake and Singular are made available to the Julia universe via separate interface layers, GAP.jl, Polymake.jl and Singular.jl, which can be used separately and independently from OSCAR. The purpose of these interfaces is to make the full functionality of each cornerstone accessible in Julia. In this way, e.g., Singular.jl works as a replacement for Singular, but with a Julia interface. The situation for ANTIC is easier, as its components AbstractAlgebra, Nemo and Hecke are Julia natives. Crucially, even though some of the original cornerstone code are not written in Julia, they too are available through Julia as Artifacts, which are distributed via Julia's Yggdrasil. These artifacts are automatically installed from the internet by the package manger, too. So the user does not need to know where the various parts of the code actually come from. It should be noted that an expert user who does know and care can easily override Julia's versions with their own.

It would require great effort to measure the total size of OSCAR's code base. The cornerstones combined with the Julia code of our own easily adds up to a few million lines of code, distributed over many libraries in several programming languages. That does not count further dependencies such as, e.g., FLINT, which provides long integer arithmetic, linear algebra and much more. The main task of Julia's package manager is to keep track of all of this, including the compatibility requirements among the various versions of the individual pieces. That kind of information is impossible to maintain by hand, and Julia's capabilities in this direction set the state of the art. Package managers for older programming languages like Perl or Python are not part of the respective language and are thus considerably more restricted. The full list of OSCAR dependencies is listed in its Project.toml file; see Figure 1. Further dependencies are implicit via imports of imports etc.

There are many criteria by which programming languages can be compared. One such aspect is energy efficiency, which is difficult to assess precisely, and which also comes in more than one flavor. Pereira et al. carried out a comprehensive study of several dozen programming languages in [Per+21]. For instance, Pereira et al. found that Julia 1.3.1 loses to Rust 1.40.0 a factor of 1.8 with respect to energy efficiency and a factor of 3.39 with respect to time (normalized global results). This should be compared with Python 3.5.2 which looses to C (gcc 6.2.0) a factor of 75.88 in terms of energy and 71.9 with respect to time. For completeness, the older version 1.16.0 of Rust was measured to be almost on par with C (gcc 6.2.0), the energy factor reads

```
name = "Oscar"
uuid = "f1435218-dba5-11e9-1e4d-f1a5fab5fc13"
authors = ["The OSCAR Team <oscar@mathematik.uni-kl.de>"]
version = "1.0.0"

[deps]
AbstractAlgebra = "c3fe647b-3220-5bb0-a1ea-a7954cac585d"
AlgebraicSolving = "66b61cbe-0446-4d5d-9090-1ff510639f9d"
Distributed = "8ba89e20-285c-5b6f-9357-94700520ee1b"
DocStringExtensions = "ffbed154-4ef7-542d-bbb7-c09d3a79fcae"
GAP = "c863536a-3901-11e9-33e7-d5cd0df7b904"
Hecke = "3e1990a7-5d81-5526-99ce-9ba3ff248f21"
JSON = "682c06a0-de6a-54ab-a142-c8b1cf79cde6"
JSON3 = "0f8b85d8-7281-11e9-16c2-39a750bddbf1"
LazyArtifacts = "4af54fe1-eca0-43a8-85a7-787d91b784e3"
Nemo = "2edaba10-b0f1-5616-af89-8c11ac63239a"
Pkg = "44cfe95a-1eb2-52ea-b672-e2afdf69b78f"
Polymake = "d720cf60-89b5-51f5-aff5-213f193123e7"
Preferences = "21216c6a-2e73-6563-6e65-726566657250"
Random = "9a3f8284-a2c9-5f02-9a11-845980a1fd5c"
RandomExtensions = "fb686558-2515-59ef-acaa-46db3789a887"
Serialization = "9e88b42a-f829-5b0c-bbe9-9e923198166b"
Singular = "bcd08a7b-43d2-5ff7-b6d4-c458787f915c"
TOPCOM_jll = "36f60fef-b880-50dc-9289-4aaecee93cc3"
UUIDs = "cf7118a7-6976-5b1a-9a39-7adc72f591a4"
cohomCalg_jll = "5558cf25-a90e-53b0-b813-cadaa3ae7ade"

[compat]
AbstractAlgebra = "0.40.1"
AlgebraicSolving = "0.4.11"
Distributed = "1.6"
DocStringExtensions = "0.8, 0.9"
GAP = "0.10.2"
Hecke = "0.30.0"
JSON = "^0.20, ^0.21"
JSON3 = "1.13.2"
LazyArtifacts = "1.6"
Nemo = "0.43.0"
Pkg = "1.6"
Polymake = "0.11.14"
Preferences = "1"
Random = "1.6"
RandomExtensions = "0.4.3"
Serialization = "1.6"
Singular = "0.22.4"
TOPCOM_jll = "0.17.8"
UUIDs = "1.6"
cohomCalg_jll = "0.32.0"
julia = "1.6"
```

Fig. 1: Project.toml file of `Oscar.jl`.

1.03, while the time factor is 1.04. Of course, all of these languages continue to evolve, and so these comparisons are to be considered as snapshots. Yet, based on this data, it looks like `Julia` is about 20 times more efficient (with respect to energy or time) than, e.g., `Python`. Note that `Julia` was not tested in the published version of [Per+21]; but it was added on a web page with more recent updates; its URL is given in [Per+21].

`OSCAR` benefits from many more features of the `Julia` language and ecosystem. In addition to the package manager, the following are particularly useful for `OSCAR`:

- parametric polymorphism and multiple dispatch
- static types
- just-in-time compilation
- garbage collection

In mathematics everything is precisely defined, building upon logic and set theory. However, mathematical practice, in lectures or written texts, requires a good deal of natural language to be able to communicate mathematics. Consequently, some words have very different definitions in different contexts. For example, irreducibility for polynomials has little to do (at face value) with irreducibility of linear representations of groups. Thus a language where the code that is run when a function is called is depending on the type of the argument is very natural for a mathematician. This is precisely the multiple dispatch mechanism. In `Julia` a function is selected based on the types of *all* arguments. In comparison, in classical `C` the dispatch is on the name alone, in `C++` it implicitly depends on the type of the first argument only.

Such mechanisms are also useful for algorithm selection. Consider a typical algebraic task, such as the computation of a determinant. For a matrix over a field, we probably want to use Gaussian elimination or one of its variants. On the other hand, for matrices over Euclidean rings like the integers or the ring of univariate polynomials with field coefficients, this does not work. Here we need the Hermite normal form, or, in the absence of zero-divisors, division free versions of Gauss. `Julia` supports this in a range of ways using parametric polymorphism and multiple dispatch. The data type for a matrix is parametrized by the type of the entries. Thus multiple dispatch allows us to give the same name to several distinct implementations, which are tailored to arguments of different types.

3 Basic OSCAR Usage

Here we want to give a first few examples for using `OSCAR`, highlighting some of our design decisions. On the way, we will explain how `Julia`'s features mentioned above fit into the picture. Basic familiarity with the `Julia` language is assumed. In case of need, readers are referred to the excellent tutorials, books and further documentation found at

https://julialang.org/learning.

Basic Number Types

Julia is designed to be efficient in (numerical) mathematical applications and to act "naturally" in this context. "Naturally" here means that the expected behavior is similar to Matlab or R when applicable. This also produces

```
julia> typeof(2)
Int64

julia> 2^100
0
```

as, for efficiency, 2 is of machine type, directly supported by the CPU. Consequently, 2^{100} is computed modulo 2^{64} hence yields 0. Doing computer algebra, often we mean to compute something else. The command

```
julia> ZZ(2)^100
1267650600228229401496703205376
```

employs "long integers", providing the correct value for 2^{100}. The expression ZZ(2) constructs the natural number 2 as an element of the ring of integers, which has exact arithmetic in OSCAR. Some users will consider this behavior a disadvantage, because mathematically there is only one version of the integers. However, this is an unavoidable consequence, as we want to make use of the rich set of features of a modern and mature general programming language. Furthermore, the distinction between machine supported integers of fixed size and their arbitrarily long siblings contributes to the overall efficiency of OSCAR. For a quick introduction to integer arithmetic see [GG03, §2.1]. Particular care is needed for fractions:

```
julia> 1/2
0.5
```

behaves like expected in the context of floating–point computations. To obtain an exact fraction:

```
julia> 1//2
1//2
```

By default / performs the division in $\mathbb{R}$ (as double or Float64 in Julia), while // creates an exact object, a rational number. As a rule-of-thumb, Julia will revert to floating–point arithmetic as a fallback. In contrast, in OSCAR, operations will either complete in the same domain as the input (or fail), or, in rare cases, the domain will be extended to represent the result correctly. For most of the normal applications the types are invisible, they just make the correct things happening automatically and fast.

Lists

Julia supports a range of natural list operations, which are convenient to write mathematical expressions concisely:

```
julia> prod(1-1/a^2 for a = 1:100 if is_prime(a))
0.6090337253995164
```

will compute

$$\prod_p 1 - \frac{1}{p^2}$$

where the product runs over the primes up to 100. The product will converge to $1/\zeta(2) = 6/\pi^2$, but very slowly. The computation above shows that a random integer has about 60% chance of being square-free.

Static Types

Julia is statically typed, which means that the type of an object will never change, it is fixed on creation. To illustrate this feature, the command

```
julia> R, p = residue_ring(ZZ, 3);
```

creates a finite quotient ring of the integers, a "generic" $\mathbb{Z}/n\mathbb{Z}$, together with the canonical projection $\mathbb{Z} \to \mathbb{Z}/n\mathbb{Z}$. Calling

```
julia> typeof(R)
zzModRing
```

reveals that the type is a ring. This type does not "know" that 3 is a prime, and thus that R is actually a field. Functions that expect a field as input will not work with R. This can be remedied as follows.

```
julia> F, p = residue_field(ZZ, 3);

julia> typeof(F)
FqField
```

The type of F is now a field, and its elements are recognized as field elements. Once assigned, the type cannot change. The advantage is that usually the compiler can deduce the types of expressions. This allows us to rely on multiple dispatch for calling the most appropriate functions directly. Typically the latter functions are also the most efficient ones.

Polynomials

There is more than one way to implement polynomials and their arithmetic operations; see [GG03, §2.2]. Julia, or better OSCAR, does not support symbolic polynomials. This means that, in order to work with polynomials, you have to specify both the coefficient ring and the variables. Moreover, note that the dispatch is done on Julia-types and not on implied mathematics:

```
julia> QQx, x = polynomial_ring(QQ, "x");

julia> typeof(x)
QQPolyRingElem

julia> Ry, y = polynomial_ring(QQ, ["y"]);

julia> typeof(y)
Vector{QQMPolyRingElem} (alias for Array{QQMPolyRingElem, 1})
```

The first example gives a univariate polynomial ring with rational coefficients, while the second one returns a multivariate polynomial ring (with a vector of variables), also with rational coefficients. While mathematically there is no difference, the algorithms for univariate and multivariate polynomials are completely independent, with different performance characteristics. There are also some subtle differences in the semantics, as can be seen from the following simple example. The derivative of x is unambiguous, while the one of y_1 is not uniquely defined: as the ring is multivariate in the latter case, the variable for the differentiation has to be specified.

OSCAR provides shortcuts to allow for code which looks naturally to mathematicians. For instance, the construction of the univariate polynomial ring above would also work via

```
julia> QQx, x = QQ["x"];
```

Just-In-Time Compilation

Julia is an interpreted language which features just-in-time compilation. This means that a function is compiled on the fly when the interpreter sees it for the first time. Calling the same function again, with the same argument types, is much faster. You can see that by trying, in a fresh OSCAR session:

```
julia> @time class_group(quadratic_field(7)[1]);
  0.659554 seconds (916.73 k allocations: 60.414 MiB, 4.44% gc time,
  ↪  94.30% compilation time)

julia> @time class_group(quadratic_field(11)[1]);
  0.003981 seconds (82.12 k allocations: 4.215 MiB)
```

The first command takes some time, which depends on your setup. The second command is almost immediate.

Traditionally, languages are either interpreted (and mostly interactive to use) or compiled, in which case one has to use an editor, compile and link, and then is able to run. The advantage of the first category is ease of use, while the second allows for faster execution. Julia is in-between: one can work interactively in a shell (REPL: **r**ead, **e**valuate, **p**rint **l**oop), however the evaluation is always prefixed by a compiling step. For a user it means a (small) delay when running a function for the first time, but in exchange, the speed of execution is often comparable to C (when coded carefully) while still retaining the spontaneity of interactive use.

A further key ingredient is the availability of a garbage collector as it frees developers from worrying about allocating and releasing objects by hand, something that is notoriously error prone. On the other hand, by using the direct interface to C, one can also use C-style memory management for critical code sections.

Parents Versus Types

In OSCAR the relationship between types and the domains for computation is more involved that it may seem at first sight. A basic example are the rings $\mathbb{Z}/n\mathbb{Z}$ that we saw before; all of them share the same type `Nemo.zzModRing`. However, we never want to add an element in $\mathbb{Z}/2\mathbb{Z}$ to an element in $\mathbb{Z}/3\mathbb{Z}$. In OSCAR, this issue is resolved via *parents*: each object in the OSCAR-system has both a type and a parent. In general, objects from different parents cannot be combined — even though they share the same type:

```
julia> R, mR = quo(ZZ, 2)
(Integers modulo 2, Map: ZZ -> ZZ/(2))

julia> S, mS = quo(ZZ, 3)
(Integers modulo 3, Map: ZZ -> ZZ/(3))

julia> typeof(R(1))
zzModRingElem

julia> typeof(S(1))
zzModRingElem

julia> S(1) + R(1)
ERROR: Operations on distinct residue rings not supported
Stacktrace:
 [1] error(s::String)
   @ Base ./error.jl:35
 [2] check_parent
   @ ~/.julia/packages/Nemo/F0iQ2/src/flint/nmod.jl:22 [inlined]
 [3] +(x::zzModRingElem, y::zzModRingElem)
   @ Nemo ~/.julia/packages/Nemo/F0iQ2/src/flint/nmod.jl:138
[...]
```

Caching

The parent-type difference also affects other operations, e.g., with polynomials rings: What happens if we try to create the same ring twice?

```
julia> Qx,x = QQ[:x]
(Univariate polynomial ring in x over QQ, x)

julia> QQx, xx = QQ[:x]
(Univariate polynomial ring in x over QQ, x)

julia> Qx === QQx
true

julia> x == xx
true
```

Here we used the same (since we used symbols, actually identical) input twice and, indeed, obtained identical rings. In order to achieve this, OSCAR maintains a dictionary of such rings, together with the information how they were constructed. So when a user asks for a polynomial ring, the system first checks if the same ring was created before, and if so returns it. Otherwise a new ring is built, returned — and added to that global dictionary.

This default behavior is chosen mainly for the user's convenience. The same commands can be run repeatedly, obtaining compatible results. Of course, this choice comes with the price of maintaining that dictionary, which can become a burden with respect to memory usage. For instance, consider a typical modular gcd-calculation where the gcd of rational polynomials is computed by mapping the problem to (many) finite fields and then combining results. Then caching would result in a huge number of rings stored throughout the entire session. In that case, the default behavior is undesirable; a ring should be deleted (i.e., garbage collected) when it is not needed any more. To this end, the `polynomial_ring` constructor allows to *disable* caching:

```
julia> Qx, x = polynomial_ring(QQ, :x, cached = false)
(Univariate polynomial ring in x over QQ, x)

julia> Qx == QQx
false

julia> x + xx
ERROR: Incompatible polynomial rings in polynomial operation
[...]
```

In fact, within the OSCAR system, in the internal implementation, all (polynomial) rings are constructed without caching. The same mechanism is available for many other data types which occur as parents.

Matrices and Linear Algebra

The distinction between types and parents also implies that `Julia`'s linear algebra, and in particular many of the matrix operations cannot be used. The reason is that `Julia` expects that `one` applied to the type is available:

```
julia> one(Int)
1

julia> one(Float64)
1.0
```

That cannot work in OSCAR as the unit matrix needs to know the unit in the respective ring of coefficients (which depends on the type *and* the parent). Therefore OSCAR provides its own matrix types. These types keep track of the coefficient rings, via parents. This mechanism also prevents users from creating matrices mixing coefficients from various rings:

```
julia> [S(1) R(1)]
1×2 Matrix{zzModRingElem}:
 1  1

julia> dot(ans, ans)
ERROR: Operations on distinct residue rings not supported
Stacktrace:
[...]
```

The following code works.

```
julia> matrix(S, 1, 2, [1, 2])
[1   2]

julia> dot(ans, ans)
2
```

Note, that in particular for $\mathbb{Z}/n\mathbb{Z}$ with $n < 2^{64}$, it would theoretically be possible to add the n to the type and use `Julia`'s matrices. But now the just-in-time-compilation interferes. The effect can be seen, e.g., when we want to compute determinants of integer matrices using the Chinese remainder theorem; see [GG03, §5.4]. This amounts to computing the determinant modulo p for several primes p and then combining this to get the determinant in $\mathbb{Z}$. But when p is part of the type, `Julia` will compile the "entire" linear algebra code for each prime! This would dramatically inflate both the memory consumption and the running time. Finally, since `Julia` allows only bit-types as type parameters, this cannot even work for large n.

It should be noted that most of `Julia`'s interface for matrices is also available for OSCAR matrices. However, due to different data structures, not all functions can be supported. Where it makes sense, some OSCAR functions also allow the use of `Julia` matrices and vectors as input. This can be tricky due to different interpretations:

```
julia> det([1 2; 3 4])
-2.0
```

computes the determinant of an `Int`-matrix as a `Float64`, while

```
julia> det(BigInt[1 2; 3 4])
-2
```

keeps it as a `BigInt`, which is `Julia`'s built-in type for long integers. So determinants are already tricky, but let us look at inverses:

```
julia> inv(BigInt[1 2; 3 4])
2×2 Matrix{BigFloat}:
 -2.0   1.0
  1.5  -0.5
```

Clearly there is no integral inverse, but the conversion to `BigFloat` would have happened also for invertible matrices:

```
julia> inv(BigInt[1 2; 0 1])
2×2 Matrix{BigFloat}:
 1.0  -2.0
 0.0   1.0
```

On the other hand, in `OSCAR` the behavior is different:

```
julia> inv(ZZ[1 2; 3 4])
ERROR: Matrix not invertible
[...]
```

No automatic ring change is performed! In order to invert we either have to have a matrix with determinant ± 1, or create the matrix over a field (or ring with enough inverses).

Calling `C` Code

`Julia` also integrates very well with `C`. As long as structures and arrays are constructed using only machine types, the memory layout is consistent with `C`. The following example illustrates this feature, even if the computation could be done as efficiently in `Julia` directly.

```
julia> ccall(:sin, Float64, (Float64,), 1.0)
0.8414709848078965

julia> u = Int64[1,2,3,4,5,6];

julia> ccall(:memset, Cvoid, (Ptr{Int}, Int, UInt), u, 0,
↪  length(u)*sizeof(Int64))
julia> show(u)
[0, 0, 0, 0, 0, 0]
```

Similarly, a `Julia` function taking machine type input can be compiled into machine code and the function then be called using a pointer. While not important for most users, this allows to easily re-use existing high-performance `C` or `fortran` libraries.

More Details on Types

For most "casual" use, the exact types do not matter, in order to use the multiple dispatch they do:

```
julia> function next(a)
           return a+1
       end;

julia> next(2)
3

julia> next(x)
x + 1

julia> next('a')
'b': ASCII/Unicode U+0062 (category Ll: Letter, lowercase)

julia> next(sin)
ERROR: MethodError: no method matching +(::typeof(sin), ::Int64)
[...]
```

To make it work, we'll have to write a new function:

```
julia> function next(a::typeof(sin))
         return x->a(x)+1
       end;

julia> next(sin)
#55 (generic function with 1 method)

julia> ans(1)
1.8414709848078965

julia> sin(1)+1
1.8414709848078965

julia> next(cos)
ERROR: MethodError: no method matching +(::typeof(cos), ::Int64)
[...]
```

As to not have to write for each possible function, we can look at the graph (or partially ordered set) of types:

```
julia> supertype(typeof(sin))
Function

julia> supertype(typeof(cos))
Function

julia> function next(a::Function)
         return x->a(x)+1
       end;
```

This will now also work for a wide range of functions, which can be combined, e.g., with `next`:

```
julia> a = next(next(tan))
#57 (generic function with 1 method)

julia> a(1)
3.5574077246549023

julia> tan(1)+2
3.5574077246549023
```

In order to get an idea how to use OSCAR, please consult the next four chapters, they provide step-by-step introductions into the four core areas represented in OSCAR: number theory, group theory, algebraic geometry and polyhedral geometry. Afterwards the later chapters introduce more specialized aspects spanning a broad range of mathematics.

4 This Book

There are numerous code examples spread over the entire book. We will maintain a web site

`https://book.oscar-system.org`

with that code and keep it up to date. In this way these code examples will remain useful, even for future versions of OSCAR. More information is given in the chapter on confirmable workflows [JKL24] (cf. page 473). Detailed information about the relevant software versions is listed on page 484.

References

[BJMSS23] M. Belotti, M. Joswig, C. Meroni, V. Schleis, and J. Schmitt. "Algebraic and geometric computations in OSCAR". *SIAM News* 56.07 (2023), pp. 9–10.

[GG03] J. von zur Gathen and J. Gerhard. *Modern computer algebra.* Second. Cambridge University Press, 2003, pp. xiv+785.

[JKL24] M. Joswig, L. Kastner, and B. Lorenz. "Confirmable Workflows". In: *The Computer Algebra System OSCAR. Algorithms and Examples.* Ed. by W. Decker, C. Eder, C. Fieker, M. Horn, and M. Joswig. Springer, 2024, pp. 473–491.

[Per+21] R. Pereira et al. "Ranking programming languages by energy efficiency". *Science of Computer Programming* 205 (2021). Updates and extensions at `https://sites.google.com/view/energy-efficiency-languages`, p. 102609. DOI: `10.1016/j.scico.2021.102609`.

Part II
The Cornerstones

Commutative Algebra and Algebraic Geometry

Janko Böhm, Wolfram Decker, and Frank-Olaf Schreyer

Abstract We give illustrative examples of how the computer algebra system `OSCAR` can support research in commutative algebra and algebraic geometry. We start with a thorough introduction to Gröbner basis techniques, with particular emphasis on the computation of syzygies, then apply these techniques to deal with ideal and ring theoretic concepts such as primary decomposition and normalization, and finally use them for geometric case studies which concern curves and surfaces, both from a local and global point of view.

Key words: Computational commutative algebra, computational algebraic geometry, Gröbner bases, syzygies, free resolutions, cohomology, parametrizations, deformations, algebraic surfaces

1 Introduction

Most of mathematics is concerned at some level with setting up and solving various types of equations. Algebraic geometry is the mathematical discipline handling solution sets of systems of polynomial equations. By making use of a correspondence which relates the solution sets to ideals in polynomial rings, algebraic geometers translate problems concerning the geometry of the

Janko Böhm
Fachbereich Mathematik, RPTU Kaiserslautern-Landau, e-mail: `boehm@mathematik.uni-kl.de`

Wofram Decker
Fachbereich Mathematik, RPTU Kaiserslautern-Landau, e-mail: `decker@mathematik.uni-kl.de`

Frank-Olaf Schreyer
Fakultät für Mathematik und Informatik, Universität des Saarlandes, e-mail: `schreyer@math.uni-sb.de`

W. Decker et al. (eds.), *The Computer Algebra System OSCAR*,
Algorithms and Computation in Mathematics 32,
https://doi.org/10.1007/978-3-031-62127-7_2

solution sets into algebra. Commutative algebra has developed to help solve these problems as well as problems originating from number theory.

In modern times, support comes from computer algebra techniques, notably from the concept of Gröbner bases, where the main workhorse is Buchberger's algorithm for computing them. We will discuss a number of techniques relying on Buchberger's algorithm, with particular emphasis on the computation of syzygies. And we will show these techniques at work in `OSCAR`, mainly focusing on geometrically motivated examples.

The integration of the commutative algebra and algebraic geometry functionality of `OSCAR` with techniques such as those presented in the other cornerstone chapters allows for extended algorithmic applications. See, for example, the chapter on algorithms for computing GIT fans [BB24] (cf. page 213), the chapter on fibration hopping for elliptically fibered K3 surfaces [BZ24] (cf. page 225), the chapter on invariant theory [DRS24] (cf. page 297), and the chapter on topics in F-Theory [BT24] (cf. page 451).

The textbooks [GP08; CLO15] provide excellent introductions to both the geometry–algebra dictionary and the use of Gröbner bases in the study of geometric questions by algebraic means. We refer to them for basic details and proofs not given here.

The Role of the Ground Field. Throughout this chapter, K will be an algebraically closed field, and $k \subset K$ a subfield. Our geometric objects of study will be algebraic sets in affine or projective space over K which are defined by polynomial equations with coefficients in k. When it comes to explicit computations, we suppose that the arithmetic of k can be handled by `OSCAR`. As a typical example, consider $K = \mathbb{C}$ and $k = \mathbb{Q}$. In explicit examples, unless otherwise mentioned, we tacitly assume that we work with this pair. Note that on the computational side, most questions considered can be checked using Gröbner bases. A crucial point then is that Buchberger's algorithm for finding such bases does not extend the given ground field. So most problems considered can be handled by computations over k. For example, based on Hilbert's Nullstellensatz, one can decide whether an algebraic system of equations defined over k has a solution over K. On the other hand, testing whether an ideal is radical cannot be decided using Gröbner bases alone. Note, however, that if k is perfect, and $I \subset k[x_1, \ldots, x_n]$ is a radical ideal[1], then also the extended ideal[2] $IK[x_1, \ldots, x_n]$ is a radical ideal (see [ZS60, VII, §11]). We therefore tacitly assume that k is perfect whenever the radical condition comes into play. The concept of primary decomposition depends on the choice of k. See Example 17.

---

[1] If R is a commutative ring, the *radical* of an ideal $I \subset R$ is the ideal $\mathrm{rad}(I) := \{f \in R \mid \text{ there exists } N \geq 0 \text{ such that } f^N \in I\}$. I is a *radical ideal* if $I = \mathrm{rad}(I)$.

[2] The extended ideal $IK[x_1, \ldots, x_n]$ is the ideal generated by the elements of I in the larger polynomial ring obtained by extending the coefficient field from k to K.

2 Gröbner Basics and the Affine Geometry–Algebra Dictionary

We write $\mathbb{A}^n = \mathbb{A}^n(K)$ for the *affine* n*-space* over K. Given a multivariate polynomial $f \in k[x_1, \ldots, x_n]$, we call the set

$$V(f) = \{(a_1, \ldots, a_n) \in K^n \mid f(a_1, \ldots, a_n) = 0\} \subset \mathbb{A}^n$$

the *vanishing locus* of f (over K). The *vanishing locus* of finitely many polynomials $f_1, \ldots, f_r \in k[x_1, \ldots, x_n]$ is the simultaneous solution set

$$V(f_1, \ldots, f_r) \;=\; \bigcap_{i=1}^{r} V(f_i) \subset \mathbb{A}^n.$$

The *ideal* generated by $f_1, \ldots, f_r \in k[x_1, \ldots, x_n]$ is written as

$$(f_1, \ldots, f_r) := \{g_1 f_1 + \ldots + g_r f_r \mid g_i \in k[x_1, \ldots, x_r]\}.$$

If I is this ideal, we write $V(I) = V(f_1, \ldots, f_r)$.

Theorem 1 (Hilbert's Nullstellensatz) *With notation as above,*

$$V(f_1, \ldots, f_r) = \emptyset \iff 1 \in (f_1, \ldots, f_r).$$

Thus, in our setting, the existence question for solutions over K can be decided by solving the more general problem below:

Ideal Membership Problem Given $f, f_1, \ldots, f_r \in k[x_1, \ldots, x_n]$, decide whether

$$f \in (f_1, \ldots, f_r).$$

Taking our cue from the case of one variable, in which Euclidean division with remainder provides a solution to the ideal membership problem, we extend the division algorithm to polynomials in more than one variable, allowing at the same time more than one divisor. The extended algorithm, however, does not provide an immediate solution to the multivariate ideal membership problem. Gröbner bases are designed to remedy this problem.

Univariate monomials are usually sorted according to their degree. The starting point for the concepts of multivariate division with remainder and Gröbner bases is that we have ways of sorting multivariate monomials.

Definition 2 A *monomial* in $k[x_1, \ldots, x_n]$ is a power product

$$x^\alpha = x_1^{\alpha_1} \cdots x_n^{\alpha_n}, \quad \text{where} \quad \alpha = (\alpha_1, \ldots, \alpha_n) \in \mathbb{N}^n.$$

A *monomial ideal* in $k[x_1, \ldots, x_n]$ is an ideal generated by monomials. A *monomial ordering* $>$ on $k[x_1, \ldots, x_n]$ is a total ordering $>$ on the set of monomials in $k[x_1, \dots, x_n]$ which is compatible with multiplication:

$$x^\alpha > x^\beta \implies x^\alpha x^\gamma > x^\beta x^\gamma \text{ for any triple } x^\alpha, x^\beta, x^\gamma \text{ of monomials.}$$

A *term* in $k[x_1 \dots, x_n]$ is the product ax^α of a scalar $a \in k$ and a monomial x^α. If $0 \neq f = \sum_\alpha f_\alpha x^\alpha \in k[x_1 \dots, x_n]$ is a polynomial written as the finite sum of its nonzero terms, then the *leading term* of f with respect to $>$ is

$$\mathrm{L}(f) := \mathrm{L}_>(f) := f_\beta x^\beta, \text{ where } x^\beta = \max\{x^\alpha \mid f_\alpha \neq 0\}.$$

A monomial ordering $>$ is called *global* if $x_i > 1$ for all i. In contrast, a *local* monomial ordering satisfies $1 > x_i$ for all i. Note that $>$ is global iff it is a well-ordering. The word global is used to indicate that Gröbner basis computations with respect to such orderings are used to study vanishing loci in their entirety. Rather than taking this global point of view, however, one may be interested in examining the behaviour of such a set "near" one of its points as we do in the next section. Here, local orderings are used. **For the rest of this section**, $>$ will be a global monomial ordering on $k[x_1, \dots, x_n]$. Unless otherwise indicated, all our considerations will be relative to this ordering.

Theorem 3 (Division With Remainder) *Let $f_1, \dots, f_r \in k[x_1, \dots, x_n]$ be nonzero. For each $f \in k[x_1, \dots, x_n]$, there is a unique expression $f = g_1 f_1 + \dots + g_r f_r + h$, with $g_1, \dots, g_r, h \in k[x_1, \dots, x_n]$, and such that*

1. *for $i > j$, no term of $g_i \,\mathrm{L}(f_i)$ is divisible by $\mathrm{L}(f_j)$;*
2. *for all i, no term of h is divisible by $\mathrm{L}(f_i)$.*

Proof (Existence) The statement is obvious if $f_1, \dots, f_r$ are terms. Applying this to f and the $\mathrm{L}(f_i)$, we get an expression

$$f = g_1^{(0)}\,\mathrm{L}(f_1) + \dots + g_r^{(0)}\,\mathrm{L}(f_r) + h^{(0)},$$

where $h^{(0)}, g_1^{(0)}, \dots, g_r^{(0)}$ satisfy conditions 1 and 2 above. Setting

$$f^{(1)} = f - (g_1^{(0)} f_1 + \dots + g_r^{(0)} f_r + h^{(0)}),$$

either $f^{(1)}$ is zero, and we are done, or $\mathrm{L}(f) > \mathrm{L}(f^{(1)})$. Iterating this, the resulting process must terminate since our underlying monomial ordering is assumed to be global and, thus, a well-ordering. □

The uniqueness part above corresponds to the fact that the division algorithm in the proof is *determinate*[3] in that it does not make any choices in the process. We call h the *remainder upon determinate division of f by $f_1, \dots, f_r$*. Note, however, that due to condition 1, changing the order in which $f_1, \dots, f_r$ are listed may lead to a different remainder. Even worse,

[3] For many applications we may allow some indeterminacy of the division algorithm. See the OSCAR user manual.

$f \in I = (f_1, \dots, f_r)$ does not necessarily imply that h is zero. As we will see, the latter is guaranteed if $f_1, \dots, f_r$ form a Gröbner basis of $(f_1, \dots, f_r)$.

Definition 4 The *leading ideal* of an ideal $I \subset k[x_1, \dots, x_n]$ is the monomial ideal

$$\mathrm{L}(I) := \mathrm{L}_>(f) := (\mathrm{L}(f) \mid f \in I).$$

A collection of elements $g_1, \dots, g_s \in I$ is called a *Gröbner basis* of I if

$$\mathrm{L}(I) = (\mathrm{L}(g_1), \dots, \mathrm{L}(g_s)).$$

Proposition 5 *Let $I \subset k[x_1, \dots, x_n]$ be an ideal.*

1. *If $g_1, \dots, g_s$ is a Gröbner basis of an ideal I, then a polynomial $f \in k[x_1, \dots, x_n]$ is contained in I iff the remainder upon determinate division of f by $g_1, \dots, g_s$ is zero.*
2. *The monomials x^α with $x^\alpha \notin \mathrm{L}(I)$ represent a k-vector space basis of the quotient $k[x_1, \dots, x_n]/I$. We refer to these monomials as* standard monomials *(for I, with respect to $>$).*

Remark 6 If $g_1, \dots, g_s$ is a Gröbner basis of an ideal $I \subset k[x_1, \dots, x_n]$, and $f \in I$, then the remainder upon determinate division of f by $g_1, \dots, g_s$ is determined by f, I, and $>$ (and does not depend on the choice of Gröbner basis). It represents the residue class $f + I \in k[x_1, \dots, x_n]/I$ in terms of the standard monomials. We speak of the *normal form* of f mod I, with respect to $>$.

If $I \subset k[x_1, \dots, x_n]$ is an ideal, a Gröbner basis of I can be computed from any given set of generators using *Buchberger's algorithm*. This algorithm is a generalization of both Gaussian elimination for systems of linear equations and Euclidean division with remainder in the univariate case. In Section 5, we will make a few comments on how the algorithm works.

Remark 7 A Gröbner basis $g_1, \dots, g_s$ computed by `OSCAR` is *minimal* in the sense that $\mathrm{L}(g_i)$ is not divisible by $\mathrm{L}(g_j)$, for $i \neq j$. A Gröbner basis is *reduced* if it is minimal and no term of g_i is divisible by $\mathrm{L}(g_j)$, for $i \neq j$. Note that these definitions of minimal and reduced deviate from those in textbooks as we do not ask that the leading coefficients of the Gröbner basis elements are one. A reduced Gröbner basis fulfilling this last property is unique.

Remark 8 Gröbner bases were introduced by Gordan [Gor] who used them to give his own proof of Hilbert's basis theorem. In fact, Gordan deduced the theorem for arbitrary ideals $I \subset k[x_1, \dots, x_n]$ from the combinatorial statement that monomial ideals are finitely generated (Dixon's Lemma).

There is a multitude of monomial orderings. Of particular importance are the *lexicographic ordering* $>_{\mathrm{lex}}$ and the *degree reverse lexicographic ordering* $>_{\mathrm{degrevlex}}$:

$$x^\alpha >_{\text{lex}} x^\beta \quad :\Longleftrightarrow \text{ the first nonzero entry of } \alpha - \beta \in \mathbb{Z}^n \text{ is positive.}$$

$$x^\alpha >_{\text{degrevlex}} x^\beta \quad :\Longleftrightarrow \deg x^\alpha > \deg x^\beta \text{ or } (\deg x^\alpha = \deg x^\beta \text{ and the last nonzero entry of } \alpha - \beta \in \mathbb{Z}^n \text{ is negative}).$$

Note that the above definitions depend on the ordering $x_1 > \cdots > x_n$ of the variables. The difference between $>_{\text{lex}}$ and $>_{\text{degrevlex}}$ is subtle but crucial. For example, we have

$$x_1^2 >_{\text{lex}} x_1x_2 >_{\text{lex}} x_1x_3 >_{\text{lex}} x_2^2$$

but

$$x_1^2 >_{\text{degrevlex}} x_1x_2 >_{\text{degrevlex}} x_2^2 >_{\text{degrevlex}} x_1x_3.$$

Using $>_{\text{degrevlex}}$ aims at obtaining leading terms involving as few variables as possible. This typically has the effect that using Buchberger's algorithm with $>_{\text{degrevlex}}$ results in a better performance concerning CPU time and memory usage.

The lexicographic ordering, on the other hand, can be used to eliminate any initial set of variables. In fact, the *elimination property*

$$\mathrm{L}(f) \in k[x_i, \ldots, x_n] \implies f \in k[x_i, \ldots, x_n]$$

holds for each subring $k[x_i, \ldots, x_n] \subset k[x_1, \ldots, x_n]$. See Example 10 below.

Remark 9 So a single Gröbner basis computation with respect to $>_{\text{lex}}$ yields the whole flag of *elimination ideals* $I_m = I \cap k[x_{m+1}, \ldots, x_n]$ for $0 \leq m < n$. If only one of the elimination ideals is needed, other monomial orders are usually more efficient. The `OSCAR` function `eliminate` takes this into account.

Example 10 Consider the ideal

$$I = \left(x^2 + y^2 + 2\,z^2 - 8,\; x^2 - y^2 - z^2 + 1,\; x - y + z\right) \subset \mathbb{Q}[x, y, z].$$

Its vanishing locus $V(I) = \bigcap_{f \in I} V(f) \subset \mathbb{A}^3(\mathbb{C})$ is the intersection of an ellipsoid, a hyperboloid, and a plane, see Figure 1.

We use `OSCAR` to compute a reduced lexicographic Gröbner basis of I:

```
julia> R, (x, y, z) = polynomial_ring(QQ, ["x", "y", "z"]);

julia> I = ideal(R, [x^2+y^2+2*z^2-8, x^2-y^2-z^2+1, x-y+z]);

julia> groebner_basis(I, ordering = lex(R), complete_reduction = true)
Gröbner basis with elements
1 -> 6*z^4 - 18*z^2 + 1
2 -> y + 3*z^3 - 9*z
3 -> x + 3*z^3 - 8*z
with respect to the ordering
lex([x, y, z])
```

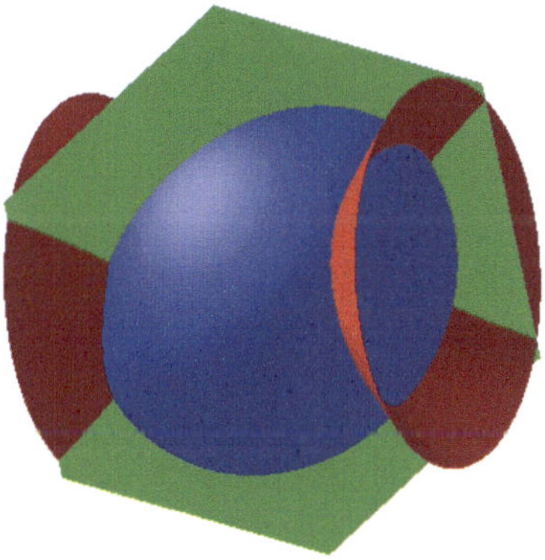

Fig. 1: Intersection of an ellipsoid, a hyperboloid, and a plane.

Solving the first equation $z^4 - 3\,z^2 + \frac{1}{6} = 0$, we find the four values $z = \pm\sqrt{\frac{3}{2} \pm \sqrt{\frac{9}{4} - \frac{1}{6}}}$. Substituting these into the other two equations, we get four points with coordinates $(-3z^3 + 8z, -3z^3 + 9z, z)$. To determine the distance of each point to the origin, we use division with remainder:

```
julia> normal_form(x^2+y^2+z^2, I, ordering = lex(R))
-z^2 + 8
```

Thus, the four points are grouped into two pairs of points with equal distance.

Remark 11 OSCAR functions such as `groebner_basis` and `normal_form` depend on the choice of a monomial ordering which is entered as a keyword argument as shown above. If no ordering is entered, the default ordering of the underlying ring R is used. This ordering is `degrevlex` except if R is $\mathbb{Z}$-graded with positive weights. Then the corresponding `wdegrevlex` ordering is used:

```
julia> S, (x, y) = graded_polynomial_ring(QQ, ["x", "y"], [1, 2]);

julia> default_ordering(S)
wdegrevlex([x, y], [1, 2])
```

Remark 12 Each Gröbner basis computed in OSCAR is cached and associated to the corresponding ideal for later use. In our example above, the cached lexicographic Gröbner basis of `I` is used in the normal form computation. If subsequently another monomial ordering is specified when entering the command `normal_form`, the required Gröbner basis will be computed and cached in addition to the lexicographic Gröbner basis.

The triangular structure of the reduced lexicographic Gröbner basis in the example above is typical in the following sense:

Proposition 13 *Let $I \subset k[x_1, \ldots, x_n]$ be a radical ideal such that $V(I) \subset \mathbb{A}^n$ is a finite set consisting of, say, d points which project to different points on the x_n-axis. Then a reduced lexicographic Gröbner basis of I has n elements. One element is a polynomial in $k[x_n]$ of degree d, and for each $1 \leq i \leq n-1$, there is an element contained in $k[x_i, x_n]$ which depends linearly on x_i.*

More generally, if $>$ is any global monomial ordering, then we have:

Proposition 14 *Let $I \subset k[x_1 \ldots, x_n]$ be an ideal. Then $V(I) \subset \mathbb{A}^n$ is finite iff only finitely many monomials x^α are not contained in $\mathrm{L}(I)$. In this case, the number of points in $V(I)$ is bounded by*

$$|V(I)| \leq \left|\{x^\alpha \mid x^\alpha \notin \mathrm{L}(I)\}\right|,$$

with quality holding if the extended ideal $IK[x_1, \ldots, x_n]$ is the vanishing ideal of $V(I)$ as defined below.

If $A \subset \mathbb{A}^n$ is any subset, its *vanishing ideal* is the ideal

$$\mathrm{I}(A) = \{f \in K[x_1, \ldots, x_n] \mid f(a) = 0 \text{ for all } a \in A\}.$$

The residue ring $K[A] = K[x_1, \ldots, x_n]/\mathrm{I}(A)$ can be identified with a K-subalgebra of the K-algebra $K^A = \{f : A \to K\}$ of all K-valued functions on A, namely the subalgebra which is generated by the restrictions $x_i|_A$ of the coordinate functions x_i to A. It is therefore called the *coordinate ring* of A.

Note that vanishing ideals are radical ideals.

Theorem 15 (Hilbert's Strong Nullstellensatz) *Let $I \subset K[x_1, \ldots, x_n]$ be an ideal. Then*

$$\mathrm{I}(V(I)) = \mathrm{rad}(I).$$

A subset $A \subset \mathbb{A}^n$ is called an *algebraic subset*, or simply an *algebraic set*, if $A = V(I)$ for some ideal $I \subset K[x_1, \ldots, x_n]$. If I is generated by polynomials $f_1, \ldots, f_r \in k[x_1, \ldots, x_n]$, then $A = V(f_1, \ldots, f_r)$, and we call k a *field of definition* of A. Note that the algebraic subsets of $\mathbb{A}^n$ form the closed sets of a topology on $\mathbb{A}^n$. This topology is called the *Zariski topology* on $\mathbb{A}^n$. An algebraic set A is called *irreducible* if

$$A = A_1 \cup A_2 \implies A = A_1 \text{ or } A = A_2$$

holds for every pair of nonempty algebraic subsets A_1, A_2 of $\mathbb{A}^n$. Note that A is irreducible iff $\mathrm{I}(A)$ is a prime ideal iff $K[A]$ is an integral domain. An *affine variety* is an irreducible algebraic subset of some $\mathbb{A}^n$.

Every algebraic set A is a finite union of irreducible algebraic sets C_i. If

$$A = C_1 \cup \ldots \cup C_r \text{ with } C_i \not\subset C_j \text{ for } i \neq j,$$

then each C_i is called a *component* of A.

Example 16

Consider the monomial ideal

$$(xy, yz) \subset \mathbb{Q}[x, y, z].$$

Its vanishing locus has two components:

$$V(xy, yz) = V(y) \cup V(x, z) \subset \mathbb{A}^3.$$

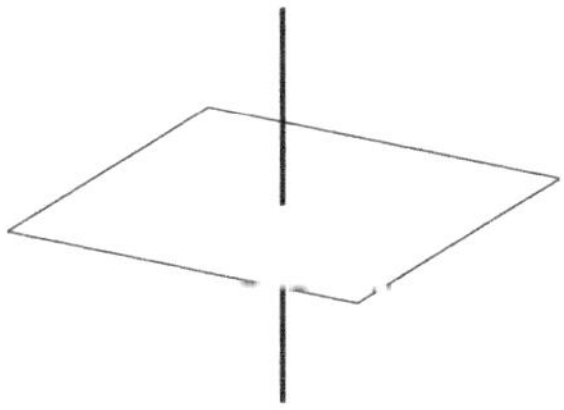

Indeed, both $V(y)$ and $V(x, z)$ are irreducible since $K[x, y, z]/(y) \cong K[x, z]$ and $K[x, y, z]/(x, z) \cong K[y]$ are integral domains.

Example 17 On the algebraic side, decomposing an algebraic set is related to the concept of primary decomposition. Here is OSCAR code illustrating this:

```
julia> R, (x, y, z) = polynomial_ring(QQ, ["x", "y", "z"]);

julia> I = ideal(R, [x*y, y*z]);

julia> minimal_primes(I)
2-element Vector{MPolyIdeal{QQMPolyRingElem}}:
 Ideal (z, x)
 Ideal (y)

julia> J = ideal(R, [x^2+1])^2;

julia> primary_decomposition(J)
1-element Vector{Tuple{MPolyIdeal{QQMPolyRingElem},
↪   MPolyIdeal{QQMPolyRingElem}}}:
 (Ideal (x^4 + 2*x^2 + 1), Ideal (x^2 + 1))

julia> L = absolute_primary_decomposition(J)
1-element Vector{Tuple{MPolyIdeal{QQMPolyRingElem},
↪   MPolyIdeal{QQMPolyRingElem},
↪   MPolyIdeal{AbstractAlgebra.Generic.MPoly{AbsSimpleNumFieldElem}},
↪   Int64}}:
 (Ideal (x^4 + 2*x^2 + 1), Ideal (x^2 + 1), Ideal (x - _a), 2)

julia> base_ring(L[1][3])
Multivariate polynomial ring in 3 variables x, y, z
  over number field of degree 2 over QQ
```

Note that there is a single primary component over $\mathbb{Q}$. Furthermore, there is one corresponding class of conjugated absolute associated primes. This is defined over a number field of degee two and is represented by `L[1][3]`.

Theorem 18 *The correspondences* V *and* I *induce bijections*

$$\begin{array}{ccc}\{\textit{radical ideals of } K[x_1,\dots,x_n]\} & \longleftrightarrow & \{\textit{algebraic subsets of } \mathbb{A}^n\}, \\ \cup & & \cup \\ \{\textit{prime ideals of } K[x_1,\dots,x_n]\} & \longleftrightarrow & \{\textit{irreducible alg. subsets of } \mathbb{A}^n\}, \\ \cup & & \cup \\ \{\textit{maximal ideals of } K[x_1,\dots,x_n]\} & \longleftrightarrow & \{\textit{points of } \mathbb{A}^n\}.\end{array}$$

If a system of algebraic equations has infinitely many solutions, we might ask for the dimension of the solution set. Equivalently, if $A \subset \mathbb{A}^n$ is this set, we might ask for the Krull dimension of $K[A]$. Note that A is finite iff its dimension is zero. We say that A is *equidimensional* of dimension d if every component of A has the same dimension d. If A is equidimensional of dimension one (resp. two), we speak of a *curve* (resp. *surface*).

Theorem 19 (Gröbner Basis Criterion for Dimension) *Let an ideal $I = (f_1, \dots, f_r) \subset k[x_1, \dots, x_n]$ be given. If $\mathrm{rad}(\mathrm{L}_>(I)) = (x_1, \dots, x_c)$ for some c, then $V(I)$ has dimension $n - c$ and the projection*

$$\mathbb{A}^n \to \mathbb{A}^{n-c}, (a_1, \dots, a_n) \mapsto (a_{c+1}, \dots, a_n),$$

restricts to a finite surjective map $V(I) \to \mathbb{A}^{n-c}$. If, moreover, $\mathrm{L}_>(I)$ is generated by monomials in the subring $k[x_1, \dots, x_c]$, then I is unmixed[4] *of dimension $n - c$, and the residue ring $k[x_1, \dots, x_n]/I$ is Cohen–Macaulay (see [BH93] for Cohen–Macaulay rings). In particular, A is equidimensional of dimension $n - c$.*

Example 20 Consider the twisted cubic curve $C = V(x_2 - x_1^2, x_3 - x_1x_2) \subset \mathbb{A}^3$:

```
julia> R, (x1, x2, x3) = polynomial_ring(QQ, ["x1", "x2", "x3"]);

julia> I = ideal(R, [x2 - x1^2, x3 - x1*x2]);

julia> radical(leading_ideal(I))
Ideal generated by
  x2
  x1
```

If k is infinite, and $>$ is $>_{\mathrm{degrevlex}}$, then the assumption

$$\mathrm{rad}(\mathrm{L}_>(I)) = (x_1, \dots, x_c) \text{ for some } c$$

is satisfied after a general linear change of coordinates. However, it is typically not a good idea to apply a change of coordinates as this may destroy any kind of sparseness of the given data. A more promising approach here is to use the equality $\dim V(I) = \dim V(\mathrm{L}(I))$ in order to reduce the computation of dimension to the purely combinatorial case of monomial ideals. The `OSCAR` function `dim` is based on this idea. In Example 20:

```
julia> dim(I)
1
```

[4] That is, all associated primes of I have Krull dimension $n - c$.

To study a given variety of dimension ≥ 1, one may wish to parametrize the variety, for example by rational functions.

Example 21 The circle $x^2 + y^2 = 1$ can be parametrized by rational functions. To find these functions, we intersect the circle $x^2 + y^2 = 1$ with the one-parameter family of lines $y = tx + 1$ through $(0, 1)$, obtaining one fixed and one variable intersection point. See Figure 2.

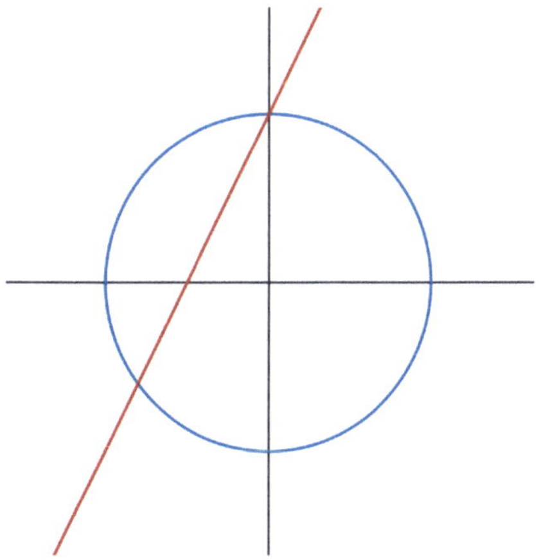

Fig. 2: Parametrization of a circle.

The following computation, which eliminates y, shows that either $x = 0$ (the x-coordinate of the fixed point) or $x = \frac{-2t}{1+t^2}$ (the x-coordinate of the variable point).

```
julia> R, (x, y, t) = polynomial_ring(QQ, ["x", "y", "t"]);

julia> I = ideal(R, [x^2 + y^2 - 1, y - t*x - 1]);

julia> Gy = groebner_basis(I, ordering = lex([y, x, t]),
↪   complete_reduction=true)
Gröbner basis with elements
1 -> x^2*t^2 + x^2 + 2*x*t
2 -> y - x*t - 1
with respect to the ordering
lex([y, x, t])

julia> factor(Gy[1])
1 * (x*t^2 + x + 2*t) * x
```

Eliminating x, we observe that either $y = 1$ or $y = \frac{1-t^2}{1+t^2}$.

```
julia> Gx = groebner_basis(I, ordering = lex([x, y, t]),
↪   complete_reduction=true)
Gröbner basis with elements
```

```
1 -> y^2*t^2 + y^2 - 2*y - t^2 + 1
2 -> x*t - y + 1
3 -> x*y - x + y^2*t - t
4 -> x^2 + y^2 - 1
with respect to the ordering
lex([x, y, t])

julia> factor(Gx[1])
1 * (y - 1) * (y*t^2 + y + t^2 - 1)
```

Remark 22 In general, rational parametrizations do only rarely exist. The question of whether a variety of dimension 1 or 2 is rationally parametrizable is well understood, and for the affirmative case, we have algorithms to compute such a parametrization. For varieties of dimension ≥ 3, this question is at the frontier of current research in algebraic geometry.

The answers given in dimension 1 and 2 can be better understood in the setting of projective algebraic geometry. Before we come to this, we include a few words on local studies.

3 Local Studies

The local point of view will, for example, be taken when discussing the intersection multiplicity of two plane curves at a point. In general, for local studies, we may assume that the given point is the origin $o = (0, \ldots, 0)$ of $\mathbb{A}^n$. The case of an arbitrary point $p \in \mathbb{A}^n$ can be dealt with by translating p to o. This requires that we extend k by adjoining each coordinate of p not contained in k.

Algebraically, local studies lead us to enlarge $k[x_1, \ldots, x_n]$ by considering the ring extension $k[x_1, \ldots, x_n] \hookrightarrow \mathcal{O}_o$, where $\mathcal{O}_o$ is the localization of $k[x_1, \ldots, x_n]$ at the maximal ideal corresponding to o. That is, $\mathcal{O}_o$ is the ring of fractions

$$\mathcal{O}_o := \left\{ \frac{g}{h} \;\middle|\; g, h \in k[x_1, \ldots, x_n], h(o) \neq 0 \right\}.$$

Here, g/h stands for the equivalence class under the equivalence relation given by $(g, h) \sim (g', h') \iff gh' = hg'$.

We now focus on plane curves.

Definition 23 Let $f, g \in k[x, y]$ be non-constant square-free polynomials without a common factor. Let $C, D \subset \mathbb{A}^2$ be the plane curves defined by f, g, and let $p \in \mathbb{A}^2$ be a point. Then the *intersection multiplicity* of C and D at p is defined as follows. If $p = o$, set

$$i(C, D; o) = \dim_k(\mathcal{O}_o/(f, g)\mathcal{O}_o).$$

If p is different from o, translate p to o, extending k if needed, and use the formula above over the extended field.

Intersection multiplicities can be found using Mora division with remainder to compute a Gröbner basis with respect to a local monomial ordering as implicitly done in the example below.

Example 24 We compute the intersection multiplicity of the tangential cusps $C = V(y^2 - x^3)$ and $D = V(2y^2 - x^3)$ at the origin:

```
julia> R, (x, y) = polynomial_ring(QQ, ["x", "y"]);

julia> f = y^2-x^3;

julia> g = 2*y^2-x^3;

julia> U = complement_of_point_ideal(R, [0 ,0]);

julia> Rloc, _ = localization(R, U);

julia> I = ideal(Rloc, [f, g]);

julia> A, _ = quo(Rloc, I);

julia> vector_space_dimension(A)
6
```

Alternatively, we may use the built-in function `intersection_multiplicity`:

```
julia> C = plane_curve(f)
Affine plane curve
  defined by 0 = x^3 - y^2

julia> D = plane_curve(g);

julia> P = D([0, 0])
Rational point
  of scheme(x^3 - 2*y^2)
with coordinates (0, 0)

julia> intersection_multiplicity(C, D, P)
6
```

Our next definition introduces a local invariant which is 0 at every smooth point of a plane curve, and measures the deviation of a singular point from being a smooth point, otherwise. For this, we encourage the reader to recall the concept of normalization of rings. Note that OSCAR provides a number of highly efficient algorithms for computing normalization. See [GLS10; Böh+13; BDLP22].

Definition 25 Let $f \in k[x,y]$ be non-constant and square-free, let $C = V(f) \subset \mathbb{A}^2$, and let $p \in C$. The *delta invariant* $\delta_p(C)$ of C at p is defined as follows. If $p = o$, set

$$\delta_o(C) = \dim_k \left(\overline{A_o}/A_o\right),$$

where $A_o = \mathcal{O}_o/f\mathcal{O}_o$ and $\overline{A_o}$ is the normalization of A_o. If p is different from o, proceed as in Definition 23.

The *total delta invariant* $\delta(C) = \sum_{p \in C} \delta_p(C)$ of an affine plane curve can be computed from the normalization of its coordinate ring as shown below:

Example 26

```
julia> R, (x, y) = polynomial_ring(QQ, ["x", "y"]);

julia> f = -3*x^5 - 2*x^4*y - 3*x^3*y^2 + x*y^4 + 3*y^5 + 6*x^4 + 7*x^3*y
↪   + 3*x^2*y^2 - 2*x*y^3 - 6*y^4 - 3*x^3 - 5*x^2*y + x*y^2 + 3*y^3;

julia> A, _ = quo(R, ideal(R, [f]));

julia> L = normalization_with_delta(A);

julia> L[3]
6
```

We will come back to this example in the next section.

Different types of singularities of plane curves can often be distinguished by considering the tangent lines at these points. To define tangent lines to plane curves at singular points, note that each non-constant homogeneous polynomial $f \in k[x, y]$ decomposes over K into linear factors.

Definition 27 Let $C = V(f) \subset \mathbb{A}^2$ be as above. Let $f = f_0 + f_1 + f_2 + \ldots + f_d$ be the Taylor expansion of f at o, where, for each i, the polynomial f_i collects the degree-i terms of f. The *multiplicity* $\mathrm{mult}(C, o)$ *of* C *at* o is defined to be the least m such that $f_m \neq 0$. The *tangent lines* to C at o are the lines defined by the linear factors of $f_{\mathrm{mult}(f,o)}$ over K. We say that o is an *ordinary multiple point* of C if $\mathrm{mult}(C, o) \geq 2$ and all tangent lines at o are distinct. For an arbitrary $p \in \mathbb{A}^2$, the definitions carry over as in Definition 23.

Note that the delta invariant and the multiplicity m at an ordinary multiple point p of C are related as follows: $\delta_p(C) = \frac{m(m-1)}{2}$.

4 Projective Algebraic Geometry

Typically, two different lines in $\mathbb{A}^2$ intersect in precisely one point. This is not true, however, for parallel lines. To treat pairs of lines on equal footing, one has to add points at infinity, one for each class of parallel lines. This idea goes back to Renaissance artists who considered such points to introduce perspective in their paintings. In any dimension n, the *projective n-space* over K is obtained as the union[5]

[5] In case $K = \mathbb{C}$, the projective space $\mathbb{P}^n$ carries the structure of a complex manifold which is a compactification of $\mathbb{A}^n$.

$$\mathbb{P}^n = \mathbb{A}^n \cup H$$

of $\mathbb{A}^n$ with a *hyperplane at infinity* $H \cong \mathbb{P}^{n-1}$. To treat points of $\mathbb{P}^n$ on equal footing, the formal definition of $\mathbb{P}^n$ introduces *homogeneous coordinates*. The *homogeneous coordinate ring* of $\mathbb{P}^n$ is the graded polynomial ring $S = K[x_0, \ldots, x_n]$, where each variable x_i has degree 1. *Projective algebraic subsets* of $\mathbb{P}^n$ are defined by homogeneous equations and correspond, thus, to homogeneous ideals $I \subset S$. In this way, we get a projective geometry–algebra dictionary. In particular, if $A \subset \mathbb{P}^n$ is a projective algebraic set, we can speak of its *homogeneous vanishing ideal* $\mathrm{I}(A) \subset S$ and its *homogeneous coordinate ring* $K[A] := S/\mathrm{I}(A)$. Moreover, the formal definition of $\mathbb{P}^n$ gives rise to many ways of writing $\mathbb{P}^n$ as the union of an "affine chart" $\mathbb{A}^n$ and a hyperplane at infinity. Local concepts can, then, be extended from the affine to the projective case by considering a covering of $\mathbb{P}^n$ by affine charts. For example,

$$\mathbb{P}^n = \bigcup_{i=0}^{n} U_i,$$

with *coordinate charts* $\mathbb{A}^n \cong U_i := \{(a_0 : \cdots : a_n) \in \mathbb{P}^n \mid a_i \neq 0\}$ and hyperplanes at infinity $H_i = V(x_i) \cong \mathbb{P}^{n-1}$. A local question can then be studied by considering the question in an appropriate chart. For example, the notion of intersection multiplicity of two plane curves at a point as introduced in the previous section carries over, independently of the choice of chart in which the point under consideration lies. The *projective closure* $\overline{A} \subset \mathbb{P}^n$ of an algebraic set $A = V(I) \subset \mathbb{A}^n \cong U_0$ is realized by homogenizing the elements of a Gröbner basis of I with respect to a degree refining monomial ordering. Note that if A is defined over k, then $\overline{A}$ has k as a field of definition as well since Gröbner basis computations do not extend the ground field.

Example 28 Consider the twisted cubic curve

$$C = V(x_2 - x_1^2, x_3 - x_1 x_2) \subset \mathbb{A}^3.$$

Its projective closure $\overline{C}$ is realized as follows:

```
julia> R, (x1, x2, x3) = polynomial_ring(QQ, ["x1", "x2", "x3"]);

julia> I = ideal(R, [x2 - x1^2, x3 - x1*x2]);

julia> H = homogenizer(R, "x0"; pos=1);

julia> J = H(I)
Ideal generated by
  -x1*x3 + x2^2
  -x0*x3 + x1*x2
  -x0*x2 + x1^2
```

Note that it is not enough to homogenize the original two generators:

```
julia> J1 = ideal([H(I[1]), H(I[2])])
Ideal generated by
  x0*x2 - x1^2
  x0*x3 - x1*x2

julia> MP = minimal_primes(J1)
2-element Vector{MPolyIdeal{MPolyDecRingElem{QQFieldElem,
↪  QQMPolyRingElem}}}:
 Ideal (x1*x3 - x2^2, x0*x3 - x1*x2, x0*x2 - x1^2)
 Ideal (x1, x0)

julia> J in MP
true
```

This shows that $V(J_1) \subset \mathbb{P}^2$ contains the additional line $L = V(x_0, x_1)$ which is completely contained in the hyperplane $H = V(x_0)$ at infinity. Moreover:

```
julia> pt = radical(J+ideal([base_ring(J)[1]]))
Ideal generated by
  x2
  x1
  x0
```

Thus, the projective closure of C intersects H in the single point

$$\overline{C} \cap H = \{(0:0:0:1)\}$$

which is the limit of the points $(1:t:t^2:t^3) = (\frac{1}{t^3} : \frac{1}{t^2} : \frac{1}{t} : 1)$ for $t \to \infty$.

In $\mathbb{P}^2$, we have the line at infinity. Any other line in $\mathbb{P}^2$ is a line $L \subset \mathbb{A}^2$ together with the common point at infinity of all lines parallel to L. Now, two different lines meet in precisely one point. In fact, much more is true:

Theorem 29 (Bézout) *Let C and D be projective plane curves of degrees c and d, respectively. Suppose that the two curves do not have a common component. Then, counted with multiplicity, C and D intersect in precisely $c \cdot d$ points:*

$$\sum_{p \in C \cap D} i(C, D; p) = c \cdot d.$$

Here, the *degree* of a projective plane curve is the degree of a square-free homogeneous polynomial defining it.

Example 30 Consider the projective closures C and D of the affine plane curves defined by $f = (x-y)((x+y)^2 - (x-y)^3) - (x+y)^4$ respectively $g = y^2 - x^2 + 3x^3$:

We show that $i(C, D; o) = 8$. In this example, as an alternative to the recipe from Example 24, we can also proceed as follows:

```
julia> R, (x, y) = polynomial_ring(QQ, ["x", "y"]);

julia> f = (x-y)*((x+y)^2-(x-y)^3)-(x+y)^4;

julia> g = y^2-x^2+3*x^3;

julia> H = homogenizer(R, "z");

julia> F = H(f);

julia> G = H(g);

julia> I = ideal([F, G]);

julia> degree(radical(I))
5

julia> cI = primary_decomposition(I);

julia> degrees = map(comp -> degree(comp[1]), cI)
2-element Vector{ZZRingElem}:
 4
 8
```

The OSCAR implementation of the primary decomposition algorithm applied above depends on some random choices. As a result, another run of the algorithm may return the primary components in a different order.

```
julia> pos = findfirst(deg -> deg == 8, degrees)
2

julia> cI[pos][2]
Ideal generated by
  y
  x
```

There is a conjugated group of four more intersection points, two of which are visible and real, as one can see from the minimal polynomial of the component in the absolute primary decomposition of I:

```
julia> cIabs = absolute_primary_decomposition(I);

julia> pos_abs = findfirst(cp -> degree(cp[1]) == 8, cIabs)
2

julia> degree(cIabs[pos_abs][1])
8

julia> other_pos_abs = pos_abs == 1 ? 2 : 1
1

julia> cI2 = cIabs[other_pos_abs][3]
Ideal generated by
  648*y + (-160*_a^3 - 1269*_a^2 + 22446*_a + 972)*z
  184147758075888*x + (2850969000960*_a^3 + 22611747888864*_a^2 -
  ↪   399955313722176*_a - 17319636680832)*y + (2884707374400*_a^3 +
  ↪   22782606070410*_a^2 - 405172045313820*_a - 57843867366864)*z

julia> R2 = base_ring(cI2)
Multivariate polynomial ring in 3 variables over number field graded by
  x -> [1]
  y -> [1]
  z -> [1]

julia> K = coefficient_ring(R2)
Number field with defining polynomial 160*x^4 + 1269*x^3 - 22446*x^2 -
↪   972*x - 648
  over rational field

julia> a = gen(K);

julia> mp =  minpoly(a)
x^4 + 1269//160*x^3 - 11223//80*x^2 - 243//40*x - 81//20

julia> S, (x,) = polynomial_ring(QQ, ["x"]);

julia> g = mp(x);

julia> sols, _ = real_solutions(ideal([g]));

julia> length(sols)
2
```

Since $8 + 4 = 4 \cdot 3$, there are no intersection points on the line at infinity.

Most curves do not have a rational parametrization. In the case $K = \mathbb{C}$, there is a topological obstruction. Let $C \subset \mathbb{P}^n$ be a smooth projective curve defined over $\mathbb{C}$. Then $V(C)$ carries the structure of a 1-dimensional complex manifold, that is, a compact Riemann surface. The underlying real 2-dimensional differential manifolds of Riemann surfaces are oriented and classified by their genus g, that is, by their number of handles, see Figure 3.

A rational parametrization gives rise to a differentiable surjective finite branched covering $\mathbb{P}^1(\mathbb{C}) \to C(\mathbb{C})$. Hence such a map cannot exist if $g =$

Fig. 3: Illustration of Riemann surfaces of genus zero, one, and two.

$g(C) > 0$ since the preimage of a non-contractible loop in C would be contractible in $\mathbb{P}^1(\mathbb{C})$.

Theorem 31 *An irreducible projective plane curve $C \subset \mathbb{P}^2$ has a rational parametrization iff its geometric genus (see below) is zero.*

In the smooth case over $\mathbb{C}$, the geometric genus coincides with the topological genus considered above. We do not recall the definition of the geometric genus in full generality. Instead, we recall a formula which relates the geometric genus of a projective plane curve to the *total delta invariant* of the curve (compare with the previous section):

Proposition 32 *The geometric genus $p_g(C)$ of a curve C of degree d with isolated singularities is*

$$p_g(C) = \frac{(d-1)(d-2)}{2} - \sum_{p \in C} \delta_p(C).$$

Note that $\delta_p(C) > 0$ only for the finitely many singularities p of C. The first term on the right hand side is the *arithmetic genus* $p_a(C)$ of C (see Section 5 for the general definition of the arithmetic genus).

Example 33 Consider the plane quintic curve with defining polynomial

$$\begin{aligned} f = & -3x^5 - 2x^4y - 3x^3y^2 + xy^4 + 3y^5 + 6x^4 + 7x^3y \\ & + 3x^2y^2 - 2xy^3 - 6y^4 - 3x^3 - 5x^2y + xy^2 + 3y^3 \end{aligned}$$

as in Example 26. See Figure 4 for a real picture. The projective closure $C \subset \mathbb{P}^2$ has a triple point at the origin, and double points at the points with coordinates $(0,1), (1,0), (1,1)$ as its only singularities. So its geometric genus is (see Proposition 32 and Example 26):

$$p_g(C) = \binom{4}{2} - \binom{3}{2} - 3\binom{2}{2} = \binom{4}{2} - 6 = 0.$$

It follows that C has a rational parametrization. Using Bézout's theorem, we can find a parametrization as follows: Consider the pencil (one parameter family) of conics through the four singular points,

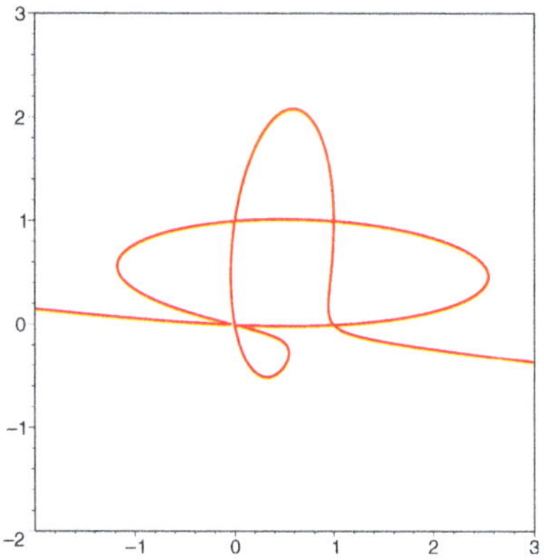

Fig. 4: Degree 5 plane curve with three double points and one triple point.

$$D_t = V(t(x^2 - x) + y^2 - y).$$

Each curve D_t intersects C with intersection multiplicity 3 at the origin and

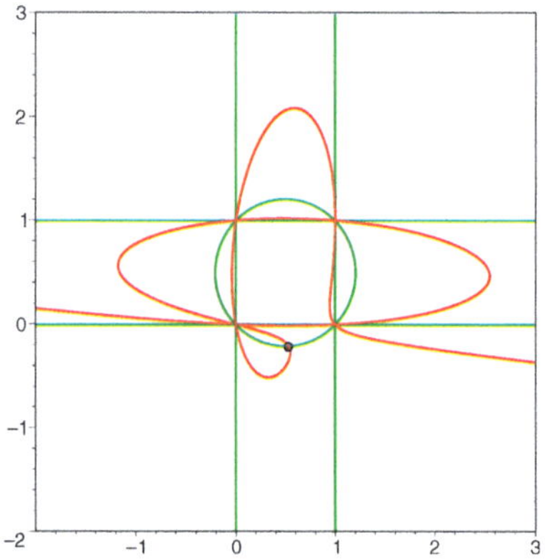

Fig. 5: Three curves in the pencil of quadrics through the singular points of C.

intersection multiplicity 2 at the double points. Three curves in the pencil (corresponding to $t = 0, \infty, 1$) are depicted in Figure 5. Since

$$2 \cdot 5 - 3 - 2 - 2 - 2 = 1,$$

Bézout's theorem implies that there is exactly one moving intersection point $p(t)$. In the figure, for $t = 1$, this intersection point is shown. Computing the coordinates of this point gives a rational parametrization. We observe that for both t and C we have to consider points at infinity. This motivates us to consider parametrizations as maps in the projective setting.

```
julia> P, (x, y, z) = graded_polynomial_ring(QQ, ["x", "y", "z"]);

julia> f = x^5 + 10*x^4*y + 20*x^3*y^2 + 130*x^2*y^3 - 20*x*y^4 + 20*y^5 -
↪   2*x^4*z - 40*x^3*y*z - 150*x^2*y^2*z - 90*x*y^3*z - 40*y^4*z +
↪   x^3*z^2 + 30*x^2*y*z^2 + 110*x*y^2*z^2 + 20*y^3*z^2;

julia> C = plane_curve(f)
Projective plane curve
  defined by 0 = x^5 + 10*x^4*y - 2*x^4*z + 20*x^3*y^2 - 40*x^3*y*z +
  ↪   x^3*z^2 + 130*x^2*y^3 - 150*x^2*y^2*z + 30*x^2*y*z^2 - 20*x*y^4 -
  ↪   90*x*y^3*z + 110*x*y^2*z^2 + 20*y^5 - 40*y^4*z + 20*y^3*z^2

julia> conics = [x^2-x*z, y^2-y*z];

julia> BM = invert_birational_map(conics, C);

julia> phi = BM["inverse"]
3-element Vector{QQMPolyRingElem}:
 -10*y(1)^5 - 430*y(1)^4*y(2) - 5020*y(1)^3*y(2)^2 - 15100*y(1)^2*y(2)^3 -
  ↪   4800*y(1)*y(2)^4 - 400*y(2)^5
 y(1)^5 + 50*y(1)^4*y(2) + 690*y(1)^3*y(2)^2 + 1620*y(1)^2*y(2)^3 -
  ↪   1800*y(1)*y(2)^4 - 400*y(2)^5
 y(1)^5 - 60*y(1)^4*y(2) - 2240*y(1)^3*y(2)^2 - 18100*y(1)^2*y(2)^3 -
  ↪   4800*y(1)*y(2)^4 - 400*y(2)^5

julia> evaluate(defining_equation(C), phi)
0
```

For applications it would be nice to have an algorithm for computing rational parametrizations which, ideally, are defined over the given field of definition of C. The proof of the theorem below yields the basis of such an algorithm:

Theorem 34 (Hilbert and Hurwitz) *Let $C \subset \mathbb{P}^2 = \mathbb{P}^2(K)$ be an irreducible curve of degree d with defining equation over k. If d is odd, then there exists a rational parametrization of C defined over k. If d is even, then there exists a rational parametrization defined over an algebraic extension field $k(a)$ of k of degree $[k(a) : k] \leq 2$.*

Note that the quadratic field extension is not needed if a particular conic computed by the algorithm has a k-rational point. A considerably enhanced variant of the algorithm is used by `OSCAR`.

Example 35 The idea of Theorem 34 is to determine the adjoint ideal of C, see [BDLP17]. The degree $d-2$ part of this ideal determines a rational map to a rational normal curve in $\mathbb{P}^{d-2}$. We then iteratively use the anti-canonical map to arrive either at $\mathbb{P}^1$ or a plane conic. These computations do not leave the field of definition. In the latter case, one can then decide, whether the conic has a rational point. Note that in the above example, this step is not required since the curve has odd degree. The corresponding algorithms are implemented in OSCAR in the commands `adjoint_ideal`, and `rational_point_conic`:

```
julia> I = adjoint_ideal(C)
Ideal generated by
  y^3 - y^2*z
  x*y^2 - x*y*z
  x^2*y - x*y*z
  x^3 - x^2*z

julia> D = Oscar.map_to_rational_normal_curve(C)
Projective curve
  in projective 3-space over QQ with coordinates [y(1), y(2), y(3), y(4)]
defined by ideal with 3 generators

julia> betti(free_resolution(defining_ideal(D)))
       0  1
-----------
2    : 3  2
-----------
total: 3  2

julia> Oscar.rat_normal_curve_anticanonical_map(D)
2-element Vector{MPolyDecRingElem{QQFieldElem, QQMPolyRingElem}}:
 2*y(2) + 13*y(4)
 y(4)

julia> rational_point_conic(plane_curve(y^2 - x*z))
3-element Vector{QQMPolyRingElem}:
 0
 0
 -1
```

The command `parametrization` combines all steps for finding a rational parametrization into one command.

```
julia> I = parametrization(C)
3-element Vector{QQMPolyRingElem}:
 25*s^5 - 1025*s^4*t + 14825*s^3*t^2 - 85565*s^2*t^3 + 146420*s*t^4 -
 ↪  20780*t^5
 25*s^5 - 1400*s^4*t + 30145*s^3*t^2 - 305650*s^2*t^3 + 1396410*s*t^4 -
 ↪  2023972*t^5
 25*s^5 - 1025*s^4*t + 15575*s^3*t^2 - 116205*s^2*t^3 + 562440*s*t^4 -
 ↪  1898652*t^5
```

5 Syzygies

Write $R = k[x_1, \dots, x_n]$, and let $F = R^s$ be the free R-module with basis $e_1, \dots, e_s$. Elements of F are simply column vectors

$$\begin{pmatrix} p_1 \\ \vdots \\ p_s \end{pmatrix} = \sum_{i=1}^{s} p_i e_i,$$

with polynomial entries $p_i \in R$. In the same spirit, we will usually not distinguish between an $s \times r$-matrix A with polynomial entries $a_{i,j} \in R$ and the corresponding R-module homomorphism $\varphi_A \colon R^r \to R^s$.

Remark 36 In `OSCAR`, we have a different convention: Free module elements are thought of as row vectors, and matrices operate by multiplication on the right. This explains the use of `transpose` in Example 38 below.

A *syzygy* on $f_1, \ldots, f_r \in F$ is a tuple $(g_1, \ldots, g_r)^t \in R^r$ such that $\sum_{i=1}^r g_i f_i = 0$. That is, the syzygies are precisely the elements of the kernel of the R-module homomorphism $\varphi_A \colon R^r \to R^s$ defined by the $s \times r$-matrix A with columns $f_1, \ldots, f_r$. This kernel is a finitely generated R-module which is called the *syzygy module* of $f_1, \ldots, f_r$. An $r \times t$ matrix B whose columns generate the kernel is called a *syzygy matrix* of A.

The computation of syzygies has plenty of applications. For instance:

Proposition 37 *Let* $I = (f_1, \ldots, f_r)$ *and* $J = (g_1, \ldots, g_s) \subset R$ *be two ideals. Consider the* $2 \times (r + s + 1)$ *matrix*

$$A = \begin{pmatrix} 1 & f_1 & \ldots & f_r & 0 & \ldots & 0 \\ 1 & 0 & \ldots & 0 & g_1 & \ldots & g_r \end{pmatrix}.$$

Let $B = (b_{ij})$ *be a syzygy matrix of* A *of size* $(r + s + 1) \times t$, *say. Then the ideal* $I \cap J \subset R$ *is generated by the entries of the first row of* B.

We leave it to the reader to formulate a recipe for computing *colon ideals*

$$I : J = \{f \in k[x_1, \ldots, x_n] \mid fg \in I \text{ for all } g \in J\}.$$

Iteratively, this allows us to compute the *saturation*

$$I : J^\infty := \{f \in k[x_1, \ldots, x_n] \mid fJ^m \subset I \text{ for some } m \geq 1\} = \bigcup_{m=1}^{\infty} (I : J^m)$$

Example 38 We have

$$(x_0, x_1) \cap (x_2, x_3) = (x_0x_2, x_1x_2, x_0x_3, x_1x_3) \subset \mathbb{Q}[x_0, \ldots, x_3].$$

Indeed, setting

$$A = \begin{pmatrix} 1 & x_0 & x_1 & 0 & 0 \\ 1 & 0 & 0 & x_2 & x_3 \end{pmatrix},$$

we get:

```
julia> R, (x0, x1, x2, x3) = polynomial_ring(QQ, ["x0", "x1", "x2",
↪ "x3"]);

julia> I = ideal([x0, x1]);

julia> J = ideal([x2, x3]);

julia> A = matrix(R, [1 x0 x1 0 0; 1 0 0 x2 x3])
[1   x0   x1    0    0]
[1    0    0   x2   x3]

julia> B = transpose(syz(transpose(A)))
[  0     0   x1*x3   x0*x3   x1*x2   x0*x2]
[  0   -x1       0     -x3       0     -x2]
[  0    x0     -x3       0     -x2       0]
[-x3     0       0       0     -x1     -x0]
[ x2     0     -x1     -x0       0       0]

julia> intersect(I,J)
Ideal generated by
  x1*x3
  x0*x3
  x1*x2
  x0*x2
```

Syzygies play a particular important role when it comes to Buchberger's algorithm. We discuss this algorithm in the general context of submodules of free modules. We extend some of our terminology from ideals to this case.

Definition 39 A *monomial* in F is the product $x^\alpha e_i$ of a monomial $x^\alpha \in R$ and a basis vector e_i of F, a *term* in F is a scalar times a monomial. A *monomial ordering* $>$ on F is a total ordering $>$ on the set of monomials satisfying

$$x^\alpha e_i > x^\beta e_j \implies x^{\alpha+\gamma} e_i > x^{\beta+\gamma} e_j$$

for any triple consisting of two monomials $x^\alpha e_i$, $x^\beta e_j$ in F and a monomial x^γ in R. For simplicity, we require in addition that

$$x^\alpha e_i > x^\beta e_i \iff x^\alpha e_j > x^\beta e_j \text{ for all } i, j.$$

Then $>$ induces a unique monomial ordering on R in the obvious way, and we say that $>$ is *global* if the induced ordering on R is global.

For the rest of this section, $>$ will be a global monomial ordering on F. Unless otherwise stated, all our considerations will be relative to this ordering.

If $0 \neq f = \sum f_{\alpha,i} x^\alpha e_i \in F$ is written as the finite sum of its nonzero terms, then the *leading term* of f with respect to $>$ is

$$\mathrm{L}(f) := \mathrm{L}_>(f) := f_{\beta,j} x^\beta e_j, \text{ where } x^\beta e_j = \max\{x^\alpha e_i \mid f_{\alpha,i} \neq 0\}.$$

The notions *leading module* and *Gröbner basis* of a submodule $I \subset F$ extend similarly from the ideal case.

Theorem 40 (Division With Remainder in Free Modules) *Let elements $f_1, \ldots, f_r \in F \setminus \{0\}$ be given. For each $f \in F$, there is a unique expression $f = g_1 f_1 + \ldots g_r f_r + h$, with $g_1, \ldots, g_r \in R$ and a* remainder $h \in F$, *and such such that:*

1. *for $i > j$, no term of $g_i \, \mathrm{L}(f_i)$ is a multiple of $\mathrm{L}(f_j)$,*
2. *for all i, no term of h is a multiple of $\mathrm{L}(f_i)$.*

Let now $I = (f_1, \ldots, f_r) \subset F$ be a submodule with given generators f_i. Consider the *induced monomial ordering* $>_1$ on the free R-module $F_1 := R^r$ with basis $e_1^{(1)}, \ldots, e_r^{(1)}$:

$$x^\alpha e_i^{(1)} >_1 x^\beta e_j^{(1)} \;:\Longleftrightarrow\; x^\alpha f_i >_0 x^\beta f_j \text{ or}$$
$$(x^\alpha f_i = x^\beta f_j \text{ up to a scalar}^6 \text{ and } i > j).$$

Then $>_1$ is global since $>$ is global.

Starting from some set of generators for a given submodule of a free module, the basic idea of Buchberger's algorithm for computing Gröbner bases is to consider at each step the remainders of differences of monomial multiples of two elements of the intermediate generating set, chosen so that the original leading terms cancel. If a remainder is nonzero, it is added to the generating set. To make this an algorithm, a criterion for termination is required.

To avoid superfluous *Buchberger tests*, we consider the following monomial ideals: For a given set $f_1, \ldots, f_r$ of generators of I, set

$$M_i = (\mathrm{L}(f_1), \ldots, \mathrm{L}(f_{i-1})) : \mathrm{L}(f_i), \text{ for } i = 2, \ldots, r.$$

Here, $I : J = \{g \in R \mid gJ \subset I\}$ denotes the colon ideal of two submodules I, J of F. Notice that if $x^\alpha \in M_i$, then the element $x^\alpha f_i$ is not allowed in a division expression by condition 1 of Theorem 40 above.

Theorem 41 (Buchberger's Criterion) *Let $I = (f_1, \ldots, f_r) \subset F$ be a submodule. Then $f_1, \ldots, f_r$ is a Gröbner basis of I iff for each i and each minimal generator x^α of the monomial ideal M_i above, the remainder upon division of $x^\alpha f_i$ by $f_1, \ldots, f_r$ is zero.*

Proof The condition is clearly necessary. To prove that it is sufficient, we consider a division expression $x^\alpha f_i = \sum_{j=1}^r g_j^{(i,\alpha)} f_j$ with remainder zero which satisfies condition 1 of Theorem 40. Then

$$G^{(i,\alpha)} = (-g_1^{(i,\alpha)}, \ldots, x^\alpha - g_i^{(i,\alpha)}, \ldots, -g_r^{(i,\alpha)})^t \in F_1 = R^r$$

is a syzygy on $f_1, \ldots, f_r$. With respect to the induced monomial ordering $>_1$ on F_1, we have $\mathrm{L}_{>_1}(G^{(i,\alpha)}) = x^\alpha e_i^{(1)}$. Indeed, precisely one leading term

[6] Note that the phrase "up to a scalar" would be superfluous, if we assume that the leading coefficients of the f_i are all equal to 1.

$\mathrm{L}_>(g_j^{(i,\alpha)})f_j$ coincides with $x^\alpha \mathrm{L}_>(f_i)$ and this j satisfies $i > j$ since condition 1 of Theorem 40 is satisfied for the $g_j^{(i,\alpha)}$. All other terms of a product $g_\ell^{(i,\alpha)} \mathrm{L}_>(f_\ell)$ are smaller than $x^\alpha \mathrm{L}_>(f_i)$. Now, consider an arbitrary element $f = a_1 f_1 + \ldots, a_r f_r \in I$, and let $\sum_{j=1}^r g_j e_j^{(1)} \in F_1$ be the remainder of $\sum_{j=1}^r a_j e_j$ on division by the $G^{(i,\alpha)}$. Then $f = \sum_{j=1}^r g_j f_j$ since the $G^{(i,\alpha)}$ are syzygies on $f_1, \ldots, f_r$. Hence,

$$\mathrm{L}_>(f) = \max\{\mathrm{L}_>(g_j)\,\mathrm{L}(f_j) \mid j = 1, \ldots, r\} \in (\mathrm{L}_>(f_1), \ldots, \mathrm{L}_>(f_r))$$

since these leading terms have pairwise different monomial parts. □

The $G^{(i,\alpha)}$ in the proof are called *Buchberger's test syzygies.*

Corollary 42 (Schreyer) *Consider the map $\varphi\colon F_1 \to F$ defined by $e_i^{(1)} \mapsto f_i$. If $f_1, \ldots, f_r$ is a Gröbner basis of I with respect to $>$, then Buchberger's test syzygies form a Gröbner basis of* $\ker(\varphi)$ *with respect to $>_1$.*

The corollary is the starting point for Schreyer's syzygy algorithm.

Theorem 43 (Hilbert's Syzygy Theorem) *Every finitely generated $R = k[x_1, \ldots, x_n]$-module M has a finite free resolution*

$$0 \leftarrow M \leftarrow F_0 \leftarrow F_1 \leftarrow \ldots \leftarrow F_c \leftarrow 0$$

of length $c \leq n$.

Proof (Schreyer) Since M is finitely generated, there exists a surjection $\psi\colon F_0 \to M$ of a free module $F_0 = R^{r_0}$ onto M. Choose a global monomial ordering $>_0$ on F_0 and compute a Gröbner basis $f_1, \ldots, f_{r_1}$ of $\ker(\psi)$. By the corollary, Buchberger's test syzygies form a Gröbner basis of $\ker(\varphi_1)$, where $\varphi_1\colon F_1 = R^{r_1} \to F_0$ is defined by $e_i^{(1)} \mapsto f_i$. Next, we can consider the induced monomial ordering $>_1$ on F_1 and the Buchberger test syzygies among the $G^{(i,\alpha)}$ to obtain a surjection $\varphi_2 : F_2 \to \ker(\varphi_1)$. Repeating this process yields a free resolution of M. Note that the result of the process depends, in particular, on how the Gröbner basis elements at each step are sorted. Sort the f_j such that $j < i$ holds if for $\mathrm{L}(f_j) = a x^{\alpha(j)} e_{\ell(j)}$ the following condition is satisfied:

$$\begin{aligned}
&\ell(j) < \ell(i) \text{ or} \\
&(\ell(j) = \ell(i) \text{ and} \quad \deg x^{\alpha(j)} < \deg x^{\alpha(i)} \text{ or} \\
&\qquad\qquad (\deg x^{\alpha(j)} = \deg x^{\alpha(i)} \text{ and} \quad x^{\alpha(j)} >_{\mathrm{drlex}} x^{\alpha(i)}))
\end{aligned}$$

Sort the Buchberger test syzygies at each step similarly. With this specification, the process stops after at most n steps. □

Example 44 Consider the ideal $J \subset S = k[w, x, y, z]$ generated by the entries of the first column in the following table.

$w^2 - xz$	$-x$	y	0	$-z$	0	$-y^2 + wz$
$wx - yz$	w	$-x$	$-y$	0	z	z^2
$x^2 - wy$	$-z$	w	0	$-y$	0	0
$xy - z^2$	0	0	w	x	$-y$	$-yz$
$y^2 - wz$	0	0	$-z$	$-w$	x	w^2
	0	y	$-x$	w	$-z$	1
	$-y^2 + wz$	z^2	$-wy$	yz	$-w^2$	x

The original generators turn out to be a degree reverse lexicographic Gröbner basis of J, and the algorithm produces a free resolution of shape

$$0 \longleftarrow S/J \longleftarrow S \xleftarrow{\varphi_1} S^5 \xleftarrow{\varphi_2} S^6 \xleftarrow{\varphi_3} S^2 \longleftarrow 0\,,$$

with matrices $\begin{array}{|c|c|}\hline \varphi_1^t & \varphi_2 \\ \hline & \varphi_3^t \\ \hline\end{array}$ as in the table.

In the graded case, one keeps track of the grading. Let $S = k[x_0, \dots, x_n]$ be the standard graded polynomial ring. The twisted module $S(e) = \oplus_d S_{d+e}$ is the free S-module with the generator $1 \in S_0 = S(e)_{-e}$ in degree $-e$. Requiring that the matrices in the resolution respect the grading, we obtain in the example above a graded complex

$$0 \leftarrow S/J \leftarrow S \leftarrow S(-2)^5 \leftarrow S(-3)^5 \oplus S(-4) \leftarrow S(-4) \oplus S(-5) \leftarrow 0.$$

We can compute this resolution and access the graded free modules and differentials of it as follows[7]:

```
julia> S, (w, x, y, z) = graded_polynomial_ring(QQ, ["w", "x", "y",
↪ "z"]);

julia> J = ideal([w^2-x*z,w*x-y*z,x^2-w*y,x*y-z^2,y^2-w*z]);

julia> A, _ = quo(S, J);

julia> FA = free_resolution(A)
Free resolution of A
S^1 <---- S^5 <---- S^6 <---- S^2 <---- 0
0         1         2         3         4

julia> FA[1]
Graded free module S^5([-2]) of rank 5 over S
```

[7] There are different strategies for computing free resolutions. In OSCAR, the default way of doing this is Schreyer's algorithm which is based on Corollary 42.

```
julia> FA[2]
Graded free module S^5([-3]) + S^1([-4]) of rank 6 over S

julia> FA[3]
Graded free module S^1([-4]) + S^1([-5]) of rank 2 over S

julia> map(FA,1)
S^5 -> S^1
e[1] -> (-w*z + y^2)*e[1]
e[2] -> (x*y - z^2)*e[1]
e[3] -> (-w*y + x^2)*e[1]
e[4] -> (w*x - y*z)*e[1]
e[5] -> (w^2 - x*z)*e[1]
Homogeneous module homomorphism

julia> map(FA,2)
S^6 -> S^5
e[1] -> -x*e[1] + y*e[2] - z*e[4]
e[2] -> w*e[1] - x*e[2] + y*e[3] + z*e[5]
e[3] -> -w*e[3] + x*e[4] - y*e[5]
e[4] -> z*e[1] - w*e[2] + y*e[4]
e[5] -> z*e[3] - w*e[4] + x*e[5]
e[6] -> (-w^2 + x*z)*e[1] + (-w*z + y^2)*e[5]
Homogeneous module homomorphism

julia> map(FA,3)
S^2 -> S^6
e[1] -> -w*e[2] - y*e[3] + x*e[4] - e[6]
e[2] -> (-w^2 + x*z)*e[1] + y*z*e[2] + z^2*e[3] - w*y*e[4] + (w*z -
↪  y^2)*e[5] + x*e[6]
Homogeneous module homomorphism
```

Hilbert's original motivation for the syzygy theorem was to prove the polynomial nature of the Hilbert function. Let $M = \sum_{d \in \mathbb{Z}} M_d$ be a finitely generated graded S-module. The function

$$H_M \colon \mathbb{Z} \to \mathbb{Z}, d \mapsto \dim_k M_d,$$

is called the *Hilbert function* of M. If

$$0 \leftarrow M \leftarrow F_0 \leftarrow F_1 \leftarrow \ldots \leftarrow F_c \leftarrow 0$$

is a finite graded free resolution with, say, $F_i = \bigoplus_j S(-j)^{\beta_{ij}}$, then

$$H_M(d) = \sum_{i=0}^{c} (-1)^i \sum_j \beta_{ij} \binom{d+n-j}{n} \tag{1}$$

since $\dim_k S_d = \binom{n+d}{n}$. Interpreting the binomial coefficients as a polynomial $\binom{t+n-j}{n} = \frac{1}{n!} \prod_{\ell=1}^{n} (t+\ell-j)$, the same formula defines a polynomial $P_M(t) \in \mathbb{Q}[t]$ such that $H_M(d) = P_M(d)$ for all $d \gg 0$. We call $P_M(t)$ the *Hilbert polynomial* of M.

Theorem 45 *Let $I \subset S$ be a homogeneous ideal. Then $V(I) \subset \mathbb{P}^n$ has dimension r iff the Hilbert polynomial $P_{S/I}(t) \in \mathbb{Q}[t]$ has degree r. In that case, $P_{S/I}(t)$ has the form*

$$P_{S/I}(t) = d\frac{t^r}{r!} + \text{ lower degree terms },$$

for some integer $d > 0$.

If $A \subset \mathbb{P}^n$ is a projective algebraic set, we apply this to the homogeneous coordinate ring $K[A] = S/\mathrm{I}(A)$, calling $P_A(t) := P_{K[A]}(t)$ the *Hilbert polynomial* of A. The integer d is then called the *degree* of A. This degree has a geometric interpretation: Bertini's theorem (see [Har77]) implies that the degree of an r-dimensional algebraic set coincides with the number of intersection points of A with a general linear subspace $\mathbb{P}^{n-r}$ of complementary dimension. The *arithmetic genus* of A is defined to be $p_a(A) = (-1)^r(P_{K[A]}(0) - 1)$.

The geometric genus p_g of a smooth irreducible projective curve $C \subset \mathbb{P}^n$ can also be read off its Hilbert polynomial:

$$P_C(t) = dt + 1 - p_g(C),$$

where $d = \deg C$. This is an easy consequence of the Riemann–Roch theorem.

Example 46 A smooth complete intersection C of two quadrics in $\mathbb{P}^3$ is a curve of degree $d = 4$ and geometric genus $p_g(C) = 1$. Indeed, the homogeneous coordinate ring $K[C] = S/\mathrm{I}(C)$ has the free resolution

$$0 \leftarrow S/\mathrm{I}(C) \leftarrow S \leftarrow S(-2)^2 \leftarrow S(-4) \leftarrow 0,$$

so that

$$p_C(t) = \binom{t+3}{3} - 2\binom{t+1}{3} + \binom{t-1}{3} = 4t.$$

In `OSCAR`, we get:

```
julia> S, (w,x,y,z) = graded_polynomial_ring(QQ, ["w", "x", "y", "z"]);

julia> I = ideal(S, [x*w-y*z, y*w-(x-z)*(x-2*z)]);

julia> Q = projective_scheme(I);

julia> is_smooth(Q)
true

julia> hilbert_polynomial(Q)
4*t

julia> degree(Q)
4
```

```
julia> arithmetic_genus(Q)
1
```

A finitely generated graded module M has a minimal graded free resolution which is obtained by choosing a minimal set of homogeneous generators of M and all its syzygy modules $\ker(\varphi_i)$. A resolution with differentials $\varphi_i \colon F_i \to F_{i-1}$ for $i \geq 1$ is minimal iff the entries of the matrices φ_i are contained in the homogeneous maximal ideal $(x_0, \ldots, x_n)$. Such a minimal resolution for M is unique up to isomorphism and the numbers β_{ij} appearing in the resolution are called the *graded Betti numbers of* M. These numbers are numerical invariants of M which refine the Hilbert function and the Hilbert polynomial as described in Equation 1.

More generally, we may speak of the graded Betti numbers of an arbitrary graded free resolution. It is convenient to display these numbers in a *Betti table* (b_{ji}), with $b_{ij} = \beta_{i,i+j}$.

Example 47 The resolution of the ideal J found in Example 44 is not minimal[8]: Note that 1 is an entry of the matrix giving φ_3. The OSCAR computation below shows the corresponding Betti table as well as the table of minimal Betti numbers of J (the symbol - refers to a zero entry):

```
julia> betti_table(FA)
       0  1  2  3
-----------------
0    : 1  -  -  -
1    : -  5  5  1
2    : -  -  1  1
-----------------
total: 1  5  6  2

julia> minimal_betti_table(FA)
       0  1  2  3
-----------------
0    : 1  -  -  -
1    : -  5  5  -
2    : -  -  -  1
-----------------
total: 1  5  5  1
```

Note that $\beta_{2,4} = b_{2,2}$ cancels against $\beta_{3,4} = b_{3,1}$.

Currently, the fastest method to compute the minimal Betti numbers of a graded module M is to compute a not necessarily minimal resolution as in Example 44 above, and then use this resolution to compute the dimensions of the Tor-groups

$$\beta_{ij} = \dim_k \operatorname{Tor}_i^S(M, k)_j = \dim_k (H_i(F_* \otimes_S k))_j.$$

[8] Computing graded free resolutions with Schreyer's algorithm is typically quite efficient, but the result may be far from being minimal.

To find free resolutions, Schreyer's algorithm first computes a Gröbner basis of the given submodule (ideal), starting from the given generators. In particular, the algorithm computes the syzygies on the Gröbner basis elements. For many applications, however, it is necessary to compute the syzygies on the given set of generators. This requires some bookkeeping when performing Buchberger's algorithm:

Algorithm (Computation of Syzygies)
Input. Vectors $f_1, \ldots, f_r \in F$.
Output. A matrix $\psi \in R^{r \times t}$ whose columns generate the kernel of the R-module homomorphism

$$\varphi : R^r \to F, e_i \mapsto f_i.$$

1. Choose a global monomial ordering on F and compute a Gröbner basis

$$f_1, \ldots, f_r, f_{r+1}, \ldots, f_{r'}$$

   of $(f_1, \ldots, f_r)$, keeping track of the Buchberger test syzygies $G^{(i,\alpha)}$.
2. Sort the $G^{(i,\alpha)}$ such that the test syzygies which produced new Gröbner basis elements come first.
3. The matrix with columns $G^{(i,\alpha)}$ has now shape

$$\psi' = \begin{pmatrix} A & B \\ C & D \end{pmatrix}, \text{ where } C = \begin{pmatrix} 1 & & * \\ & \ddots & \\ 0 & & 1 \end{pmatrix}$$

   is an $(r' - r) \times (r' - r)$ upper triangular square matrix with all diagonal entries being 1. Return

$$\psi = B - AC^{-1}D.$$

Note that one can compute C^{-1} by applying row operations to the matrix $(E|C)$ to obtain $(C'|E)$. The inverse matrix $C' = C^{-1}$ has entries in R.

The compution of kernels of morphisms between modules is central to many homological constructions in `OSCAR`.

Algorithm (Computation of Kernels)
Input. An R-module homomorphism $\varphi\colon M \to N$ between finitely presented R-modules given by a commutative diagram of type

$$\begin{array}{ccccccc} R^{r_1} & \xrightarrow{\phi} & R^{r_0} & \longrightarrow & M & \longrightarrow & 0 \\ \downarrow{\scriptstyle \varphi_1} & & \downarrow{\scriptstyle \varphi_0} & & \downarrow{\scriptstyle \varphi} & & \\ R^{s_1} & \xrightarrow{\psi} & R^{s_0} & \longrightarrow & N & \longrightarrow & 0\,. \end{array}$$

Output. A presentation matrix C of $\ker(\varphi)$.

1. Compute the syzygy matrix $\begin{pmatrix} A \\ B \end{pmatrix}$ of $(\varphi_0|\psi)$.
2. Compute the syzygy matrix $\begin{pmatrix} C \\ D \end{pmatrix}$ of $(A|\phi)$.
3. Then C is a presentation matrix of $\ker(\varphi)$:

$$R^{t_1} \xrightarrow{C} R^{t_0} \longrightarrow \ker(\varphi) \longrightarrow 0.$$

It is an important design feature of OSCAR that modules are represented as subquotients rather than being given by free representations.

Definition 48 Let $A\colon R^a \to R^s$ and $B\colon R^b \to R^s$ be two morphisms between free R-modules with the same codomain. The *subquotient* defined by A and B is

$$\operatorname{subquo}(A,B) = \frac{\operatorname{im}(A)+\operatorname{im}(B)}{\operatorname{im}(B)} \cong \frac{\operatorname{im}(A)}{\operatorname{im}(A)\cap\operatorname{im}(B)}.$$

In the OSCAR manual, we refer to

- the common codomain R^s as the *ambient free module* of M,
- the images of the basis vectors of R^a in R^s as the *ambient representatives of the generators* of M, and
- the images of the basis vectors of R^b in R^s as the *relations* of M.

Example 49 (Touching Surfaces) Consider the polynomials

$$f = \det\begin{pmatrix} x_3 & x_2 \\ x_2 & x_1 \end{pmatrix} \text{ and } g = \det\begin{pmatrix} x_3 & x_2 & x_1 \\ x_2 & x_1 & x_0 \\ x_1 & x_0 & 0 \end{pmatrix} \in S = K[x_0,\ldots,x_3].$$

See Figure 6 for an illustration of the corresponding surfaces.

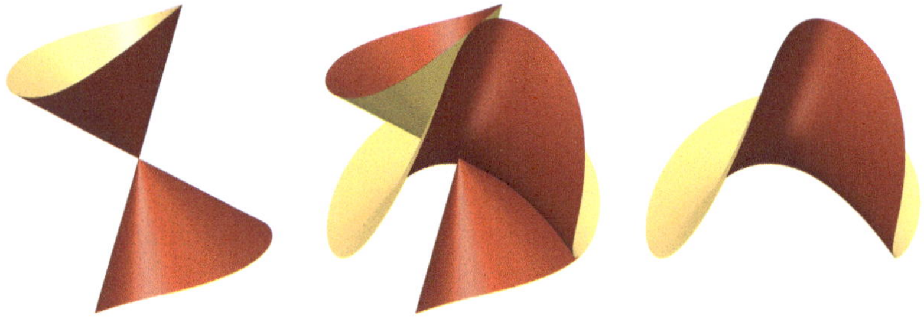

Fig. 6: Touching surfaces.

With respect to $>_{\text{degrevlex}}$, we have $\mathrm{L}(f) = -x_2^2$ and $\mathrm{L}(g) = -x_1^3$:

```
julia> S, x = graded_polynomial_ring(QQ, ["x_0", "x_1", "x_2", "x_3"]);

julia> m3x3 = matrix(S, 3, 3, [(i + j == 4) ? 0 : x[4 - (i + j)] for i in
↪  0:2, j in 0:2])
[x_3   x_2   x_1]
[x_2   x_1   x_0]
[x_1   x_0     0]

julia> f = det(m3x3[1:2, 1:2])
x_1*x_3 - x_2^2

julia> g = det(m3x3)
-x_0^2*x_3 + 2*x_0*x_1*x_2 - x_1^3

julia> leading_term(f)
-x_2^2

julia> leading_term(g)
-x_1^3
```

Thus, f, g form a Gröbner basis of $J = (f, g)$, J is unmixed of dimension one by Theorem 19, and $A = V(J) \subset \mathbb{P}^3$ is a curve. The twisted cubic curve $C \subset \mathbb{P}^3$, which is defined by the 2×2-minors of the matrix

$$\begin{pmatrix} x_3 & x_2 & x_1 \\ x_2 & x_1 & x_0 \end{pmatrix},$$

is a component of A. Thus, $\mathfrak{p}_1 = (x_1^2 - x_0x_2, x_1x_2 - x_0x_3, x_2^2 - x_1x_3)$ is one of the associated primes of J. Actually, $\mathrm{rad}(J) = \mathfrak{p}_1 = \mathrm{I}(C)$:

```
julia> p1 = radical(J);

julia> p1 == ideal(minors(m3x3[1:3, 1:2],2))
true
```

The intersection multiplicity of the two surfaces along C is computed by counting the number of times $\mathfrak{p}_1$ occurs in a filtration of the module $M = S/J$ via primes (see [Har77, Ch. I, Sec. 7]). To compute such a filtration, we first construct the graded S-module $\mathrm{Hom}(S/\mathfrak{p}_1, S/J)$:

```
julia> mp1 = ideal_as_module(p1)
Graded submodule of S^1
1 -> (-x_1*x_3 + x_2^2)*e[1]
2 -> (-x_0*x_3 + x_1*x_2)*e[1]
3 -> (-x_0*x_2 + x_1^2)*e[1]
represented as subquotient with no relations

julia> M1, _ = quo(ambient_free_module(mp1), mp1);

julia> M1
Graded subquotient of submodule of S^1 generated by
1 -> e[1]
by submodule of S^1 generated by
1 -> (-x_1*x_3 + x_2^2)*e[1]
```

```
2 -> (-x_0*x_3 + x_1*x_2)*e[1]
3 -> (-x_0*x_2 + x_1^2)*e[1]

julia> mJ = ideal_as_module(J);

julia> M, _ = quo(ambient_free_module(mJ),mJ);

julia> homM1M, psi = hom(M1, M);

julia> hom1, tohomM1M = prune_with_map(homM1M);

julia> hom1
Graded subquotient of submodule of S^2 generated by
1 -> e[1]
2 -> e[2]
by submodule of S^2 generated by
1 -> -x_2*e[1] + x_3*e[2]
2 -> x_0*e[1] - x_1*e[2]
3 -> -x_1*e[1] + x_2*e[2]
```

Actually, $\mathrm{Hom}(S/\mathfrak{p}_1, S/J)$ is generated in degree 2 with generators corresponding to multiplication by $x_0x_2 - x_1^2$ and $x_0x_3 - x_1x_2$, respectively:

```
julia> degrees_of_generators(hom1)
2-element Vector{FinGenAbGroupElem}:
 [2]
 [2]

julia> phi1 = psi(tohomM1M(hom1[1]))
M1 -> M
e[1] -> (-x_0*x_3 + x_1*x_2)*e[1]
Graded module homomorphism of degree [2]

julia> phi2 = psi(tohomM1M(hom1[2]))
M1 -> M
e[1] -> (-x_0*x_2 + x_1^2)*e[1]
Graded module homomorphism of degree [2]
```

As the first step towards the desired filtration, we may consider

$$M_1 = S/\mathfrak{p}_1(-2) \hookrightarrow M = S/J,$$

where the inclusion is induced by multiplication with $x_0x_2 - x_1^2$:

```
julia> iszero(kerphi2)
true
```

The annihilator of the image of $x_1x_2 - x_0x_3$ in M/M_1 is the prime ideal $\mathfrak{p}_2 = (x_0, x_1, x_2)$:

```
julia> MmodM1 = cokernel(phi2)
Graded subquotient of submodule of S^1 generated by
1 -> e[1]
by submodule of S^1 generated by
1 -> (x_1*x_3 - x_2^2)*e[1]
```

```
2 -> (-x_0^2*x_3 + 2*x_0*x_1*x_2 - x_1^3)*e[1]
3 -> (-x_0*x_2 + x_1^2)*e[1]

julia> p2 = ideal([x[1],x[2],x[3]])
Ideal generated by
  x_0
  x_1
  x_2

julia> f = x[2]*x[3]-x[1]*x[4]
-x_0*x_3 + x_1*x_2

julia> v = f*MmodM1[1];

julia> U, inclU = sub(MmodM1, [v]);

julia> annihilator(U) == p2
true
```

Now, we may take $M_2 \subset M$ as the preimage under the projection map $M \to M/M_1$ of the image of the map $S/\mathfrak{p}_2(-2) \hookrightarrow M/M_1$ and $M_3 = M$ because $M/M_2 \cong S/\mathfrak{p}_1$:

```
julia> Smodp2, _ = quo(ambient_free_module(mp2), mp2);

julia> homp2M1, tau = hom(Smodp2, MmodM1);

julia> hom2, tohomp2M1 = prune_with_map(homp2M1);

julia> psi = tau(tohomp2M1(hom2[1]));

julia> iszero(kernel(psi)[1])
true

julia> annihilator(cokernel(psi))
Ideal generated by
  -x_1*x_3 + x_2^2
  -x_0*x_3 + x_1*x_2
  -x_0*x_2 + x_1^2

julia> annihilator(cokernel(psi)) == p1
true
```

So in the filtration

$$0 \subset M_1 \subset M_2 \subset M_3 = M,$$

the prime ideal $\mathfrak{p}_1$ occurs twice and $\mathfrak{p}_2$ occurs once. We conclude: The surfaces $Q = V(f)$ and $H = V(g)$ intersect with multiplicity 2 along C, that is, they touch each other tangentially along C. There are no further intersections since $\deg Q \cdot \deg H = 2 \cdot 3$ and $2 \deg C = 2 \cdot 3$ coincide. Note that $\mathfrak{p}_2$ is not an associated prime ideal of M, since J is unmixed. In this case, it is not possible to find a filtration where only associated primes of M occur.

6 Applications to Surfaces in $\mathbb{P}^4$

Ellingsrud and Peskine [EP89] proved that there are only finitely many components of the Hilbert scheme of $\mathbb{P}^4$ whose general points correspond to smooth surfaces of Kodaira dimension < 2, that is, to surfaces of non-general type. In particular, this confirms a conjecture of Hartshorne about smooth rational surfaces in $\mathbb{P}^4$. Referring to [DES93; DS00] for a unified approach to constructing non-general type surfaces in $\mathbb{P}^4$, we illustrate this approach by giving two examples of rational surfaces in $\mathbb{P}^4$. Typically, the approach requires that we find graded modules with particular syzygies. In the first example below, the required module was found by a random computer algebra search in small characteristic. In Section 9, we will discuss how to infer from this that smooth surfaces with the same invariants exist over $\mathbb{C}$. For the general theory of compact complex surfaces see [BPV84].

Example 50 (A Rational Surface in $\mathbb{P}^4$ [Sch96]) Consider the ideal `m` defined below and its minimal free resolution:

```
julia> S, (x0, x1, x2, x3, x4) = graded_polynomial_ring(GF(3), ["x0",
↪   "x1", "x2", "x3", "x4"]);

julia> m = ideal(S, [x1^2+(-x1+x2+x3-x4)*x0, x1*x2+(x1-x3+x4)*x0,
↪   x1*x3+(-x1+x4+x0)*x0, x1*x4+(-x1+x3+x4-x0)*x0, x2^2+(x1-x2-x4-x0)*x0,
↪   x2*x3+(x1-x2+x3+x4-x0)*x0, x2*x4+(x1+x2-x3-x4-x0)*x0,
↪   x3^2+(x3+x4-x0)*x0,x3*x4+(-x3-x4+x0)*x0, x4^2+(x1+x3-x4-x0)*x0]);

julia> Qm, _ = quo(S, m);

julia> FQm = free_resolution(Qm, algorithm = :mres);

julia> betti_table(FQm)
       0  1  2  3  4 5
---------------------------
0    : 1  -  -  -  - -
1    : -  10 15  2  - -
2    : -  -   7 26 20 5
---------------------------
total: 1  10 22 28 20 5
```

The minimal free resolution gives rise to a subcomplex

$$G : 0 \leftarrow S \leftarrow S^{10}(-2) \leftarrow S^{10}(-3) \leftarrow S^2(-4) \leftarrow 0$$

whose homology $H_1(G) \cong I_X(2)$ is isomorphic to the twist of the homogeneous ideal I_X of a smooth surface X of degree $d = 11$ and sectional genus $\pi = 11$, with geometric genus and irregularity $p_g(X) = q(X) = 0$. Here, the *irregularity* of X is the difference $q(X) = p_g(X) - p_a(X)$.

We create the surface in OSCAR:

```
julia> phi = map(FQm, 3);
```

```
julia> M = transpose(matrix(phi)[1:2, 1:15]);

julia> DM = [map_entries(x->derivative(x,i), M) for i = 1:5];

julia> MM = transpose(hcat(DM...));

julia> size(MM)
(10, 15)

julia> NN = matrix(map(FQm, 2))[1:15, 1:10];

julia> size(NN)
(15, 10)

julia> D = graded_cokernel(transpose(MM*NN));

julia> FD = free_resolution(D, algorithm = :mres);

julia> betti_table(FD)
        0   1  2
----------------
0     : 10  10 -
1     : -   -  1
2     : -   -  -
3     : -   -  1
----------------
total: 10  10  2

julia> P = cokernel(transpose(matrix(map(FD, 2))));

julia> I = annihilator(P);

julia> QI, _ = quo(S, I);

julia> FQI = free_resolution(QI, algorithm = :mres);

julia> betti_table(FQI)
       0   1   2   3  4
-----------------------
0     : 1  -   -   -  -
1     : -  -   -   -  -
2     : -  -   -   -  -
3     : -  -   -   -  -
4     : -  5   -   -  -
5     : -  7  26  20  5
-----------------------
total: 1  12  26  20  5

julia> dim(I)
3

julia> degree(I)
11
```

So the resulting ideal is generated by 5 quintics and 7 sextics. It defines a surface of the desired degree. See the example of the Alexander surface below for how to check smoothness and for how to compute the sectional genus.

We would now like to determine how this surface fits into the Enriques-Kodaira classification of smooth projective surfaces. For this, we apply the adjunction process of Van de Ven and Sommese [SV87]. Recall that a smooth projective surface arises from a minimal smooth projective surface $X_{\min}$ by repeatedly blowing-up a point. The exceptional curve E of the blow-up is a so called (-1)-*curve*. Such a curve is rational, $E \cong \mathbb{P}^1$, with self-intersection number $E^2 = -1$. A surface is *minimal* iff it contains no (-1)-curves.

Let $X \subset \mathbb{P}^n$ be a smooth projective surface over $\mathbb{C}$ of codimension c. Let S and S_X denote the homogeneous coordinate rings of $\mathbb{P}^n$ and X, respectively. Consider $\omega_X = \operatorname{Ext}^c_S(S_X, S(-\dim S))$, the graded *dualizing module* of S_X. A basis of $(\omega_X)_1$ corresponds to the linear system $|K_X + H|$, where K_X is a canonical divisor on X and H is the hyperplane class. This linear system defines a birational morphism $\varphi_{|K_X+H|}\colon X \to X'$ onto another smooth projective surface X' which blows down precisely all (-1)-lines on X unless

1. $X \subset \mathbb{P}^n$ is a linearly or quadratically embedded $\mathbb{P}^2$ or X is ruled by lines; hence $|K_X + H| = \emptyset$ holds,
2. $X \subset \mathbb{P}^n$ is an anti-canonically embedded del Pezzo surface, in which case $\varphi_{|K_X+H|}$ maps X to a point,
3. $X \subset \mathbb{P}^n$ is a conic bundle, in which case $\varphi_{|K_X+H|}\colon X \to B$ maps X to a curve B and the fibers of $\varphi_{|K_X+H|}$ are the conics, or
4. $X \subset \mathbb{P}^n$ is a surface in one of four families identified by Sommese and Van de Ven [SV87], and $\varphi_{|K_X+H|}\colon X \to X'$ is not birational, but finite to one.

Unless we are in one of the exceptional cases, a (-1)-conic C in X is mapped to a (-1)-line in X' because $(K_X + H) \,.\, C = -1 + 2 = 1$. Thus, continuing this *adjunction process*, we eventually arrive at a minimal model $X_{\min}$ of X unless X has negative Kodaira dimension. In the rational case, we finally arrive at $\mathbb{P}^2$, the Veronese surface, a Hirzebruch surface, a Del Pezzo surface, a conic bundle, or, with bad luck, in one of the exceptional cases of part (4) above.

For the surface under consideration, the adjunction process produces a sequence of birational maps

$$\begin{array}{ccccccccc} \mathbb{P}^4 & & \mathbb{P}^9 & & \mathbb{P}^{10} & & \mathbb{P}^8 & & \mathbb{P}^5 \\ \cup & & \cup & & \cup & & \cup & & \cup \\ X^{11}_{10} & \stackrel{3}{\to} & X^{19}_{11} & \stackrel{2}{\to} & X^{18}_{9} & \stackrel{0}{\to} & X^{13}_{6} & \stackrel{3}{\to} & X^{6}_{2} \end{array}$$

Here, X^d_π denotes a surface of degree d with sectional genus π, and $\stackrel{\ell}{\to}$ denotes a birational map which blows down ℓ (-1)-curves. The final surface $X^6_2 \subset \mathbb{P}^5$ is the complete intersection of the Segre product $\mathbb{P}^1 \times \mathbb{P}^2$ with a quadric Q. Thus X^6_2 is a conic bundle which has 6 singular fibers. Blowing down one of the two (-1)-curves in each singular fiber, we arrive at a Hirzebruch surface. In general, the choice of curves in such a process is not canonical, and we might need to extend our field of definition to define the maps and, consequently, a rational parametrization of X. Luckily, in the example under consideration, no field extension is needed. The resulting parametrization is by forms of bi-degree $(10, 11)$ on $\mathbb{P}^1 \times \mathbb{P}^1$ with 14 base points of multiplicities $(5^6, 4^4, 2^2, 1^3)$. For a tutorial containing the complete `OSCAR` code for this example and further explanations, we refer to the `OSCAR` webpage.

Example 51 (The Alexander Surface [Ale88]) Consider a module M with minimal free presentation $S^3(-1) \oplus S^{15}(-2) \to S \oplus S^3(-1) \to M \to 0$ over $S = \mathbb{Q}[x_0, \ldots, x_4]$. The minimal Betti table of M is

	0	1	2	3	4	5
0 :	1	3	3	1	–	–
1 :	3	15	26	15	–	–
2 :	–	–	–	6	10	3
total :	4	18	29	22	10	3

and the dual of the linear strand $0 \to S^6(-5) \to S^{10}(-6) \to S^3(-7) \to 0$ is a complex G whose homology $H_1(G) \cong I_X(9)$ is the twisted homogeneous ideal of a smooth projective surface X. Rather than showing this construction, we demonstrate how to load the surface as a projective scheme from the corresponding `OSCAR` database which offers preconstructions of all surfaces listed in [DES93] as well as some of the other surfaces found later[9]:

```
julia> X = rational_d9_pi6();

julia> is_smooth(X)
true

julia> degree(X)
9

julia> S = ambient_coordinate_ring(X)
Multivariate polynomial ring in 5 variables over GF(31991) graded by
  x -> [1]
  y -> [1]
  z -> [1]
  u -> [1]
  v -> [1]

julia> B, _ = quo(S, ideal(S, [gens(S)[1]]));

julia> Y = proj(B)
Projective scheme
```

[9] To ease subsequent computations, the surfaces are constructed over finite fields. Note, however, that in contrast to the surface in Example 50, which was found by a random search in small characteristic, the surfaces in the data base were constructed using recipes which also work in characteristic zero. So all computations can be confirmed in characteristic zero, although this may be time consuming.

```
  over finite field of characteristic 31991
defined by ideal (x)

julia> C = intersect(X, Y);

julia> arithmetic_genus(C)
6

julia> A = homogeneous_coordinate_ring(X);

julia> FA = free_resolution(A);

julia> minimal_betti_table(FA)
       0   1   2   3  4
-----------------------
0    : 1   -   -   -  -
1    : -   -   -   -  -
2    : -   -   -   -  -
3    : -   -   -   -  -
4    : -  15  26  15  3
5    : -   1   3   3  1
-----------------------
total: 1  16  29  18  4
```

The sextic generator of the ideal is needed because the quintics alone define the union of X with a 6-secant line:

```
julia> I = defining_ideal(X);

julia> IQ = ideal([x for x in gens(I) if degree(x)[1] == 5]);

julia> J = saturation(IQ, I);

julia> degree(J)
1

julia> M = I + J;

julia> degree(M)
6
```

The astute reader will notice that the linear strand

$$0 \to S^6(-9) \to S^3(-8) \to S^3(-7) \to S(-6) \to 0$$

arising from the resolution `FA` is (up to twist) the Koszul complex of a line in $\mathbb{P}^4$. It should not come as a surprise that this line is the 6-secant above.

This time, the adjunction process yields

$$\begin{array}{ccccccccc} \mathbb{P}^4 & & \mathbb{P}^5 & & \mathbb{P}^5 & & \mathbb{P}^4 & & \mathbb{P}^2 \\ \cup & & \cup & & \cup & & \cup & & \| \\ X_6^9 & \xrightarrow{0} & X_6^{10} & \xrightarrow{0} & X_5^9 & \xrightarrow{0} & X_3^6 & \xrightarrow{10} & X_0^1 \end{array}.$$

Recursively substituting the parametrizations of the intermediate surfaces in the adjunction matrices and computing the syzygies of the transposed matrices yields a rational parametrization of X. In the example, X is parametrized by forms of degree 13 which vanish to order four at ten base points.

The module M has Hilbert function $(1, 5, 3, 0, \ldots)$. A general module with this Hilbert function has syzygies of type

	0	1	2	3	4	5
0 :	1	–	–	–	–	–
1 :	–	12	25	15	–	–
2 :	–	–	–	6	10	3
total :	1	12	25	21	10	3

and a surface obtained as above from such a module has syzygies of type

	0	1	2	3	4
0 :	1	–	–	–	–
1 :	–	–	–	–	–
2 :	–	–	–	–	–
3 :	–	–	–	–	–
4 :	–	15	25	12	–
5 :	–	–	–	–	1
total :	1	15	25	12	1

This time, the resulting surface has no 6-secant line, instead it has a (-1)-line. The adjunction process

$$\begin{array}{ccccccc} \mathbb{P}^4 & & \mathbb{P}^5 & & \mathbb{P}^5 & & \mathbb{P}^5 \\ \cup & & \cup & & \cup & & \cup \\ Y_6^9 & \xrightarrow{1} & Y_6^{10} & \xrightarrow{0} & Y_6^{10} & \xrightarrow{0} & Y_6^{10} \end{array}$$

becomes periodic, because $Y' = Y_6^{10}$ is a minimal Enriques surface and $2K_{Y'} \sim 0$.

Notice that the twelve quadric generators of the ideal $\widetilde{m}$ with $M = S/\widetilde{m}$ are annihilated by three quadrics in $k[\partial_0, \ldots, \partial_4]$, where $\partial_i = \frac{\partial}{\partial x_i}$. The three quadrics generate the homogeneous ideal to a canonical curve of genus 5 in $\check{\mathbb{P}}^4$. Conversely, the curve determines the ideal $\widetilde{m}$ and hence M and the non-minimal Enriques surface Y. This implies the curious fact that an open part of the universal family $\mathcal{U} \to \mathcal{M}_E$ of Fano polarized Enriques surfaces is isomorphic to an open part of the moduli space $\mathcal{M}_5$ of genus 5 curves. This was first observed in [DES], but awaits a geometric explanation till today.

7 Cohomology of Coherent Sheaves

The OSCAR command `sheaf_cohomology` offers two algorithms for computing the cohomology of coherent sheaves over projective n-space. The algorithms are based on local cohomology (see [EGSS02]) and on Tate resolutions via the Bernstein–Gelfand–Gelfand correspondence as introduced in [EFS03], respectively. The first algorithm makes use of syzygy computations over the symmetric algebra, and the second algorithm is based on syzygy computations over the exterior algebra (see [DE02] for a tutorial). Thus, in most examples, the second algorithm is much faster. The command `sheaf_cohomology` takes as input a finitely generated graded module whose sheafification gives the desired coherent sheaf as well as numerical information on what cohomology groups should be considered. It returns the vector space dimensions of these groups in the form of a cohomology table.

Example 52 For the ideal sheaf $\mathcal{I}_X$ of the surface $X \subset \mathbb{P}^4$ from Example 50, the dimensions $h^i(\mathbb{P}^4, \mathcal{I}_X(j))$ in the range $j = -2, \ldots 8$ are:

```
julia> MI = ideal_as_module(I);

julia> sheaf_cohomology(MI, -2, 8, algorithm = :loccoh)
twist:   -2   -1    0    1    2    3    4    5    6    7    8
------------------------------------------------------------
4:        -    -    -    -    -    -    -    -    -    -    -
3:       30   10    -    -    -    -    -    -    -    -    -
2:        -    -    -    2    -    -    -    -    -    -    -
1:        -    -    -    -    1    5    5    -    -    -    -
0:        -    -    -    -    -    -    -    5   32   84  170
------------------------------------------------------------
chi:     30   10    -    2    1    5    5    5   32   84  170
```

Here, as for Betti tables, the symbol - refers to a zero entry. For example, the $h^1(\mathbb{P}^4, \mathcal{I}_X(j))$ in the range $j = 1, \ldots 5$ are $0, 1, 5, 5, 0$.

Now we come to the Tate resolution. Let $\mathcal{F} = \widetilde{M}$ be the coherent sheaf on $\mathbb{P}^n$ associated to a finitely generated graded module $M = \sum M_d$ over $S = k[x_0, \ldots, x_n]$. We describe how to compute the dimensions $h^i(\mathbb{P}^n, \mathcal{F}(j))$ via exterior syzygies over the algebra $E = k[e_0, \ldots, e_n]$ dual to S.

Note that M defines a complex of graded free E-modules

$$R(M) : \ldots \to \mathrm{Hom}_k(E, M_{d-1}) \to \mathrm{Hom}_k(E, M_d) \to \mathrm{Hom}_k(E, M_{d+1}) \to \ldots.$$

The complex $R(M_{\geq r})$ is acyclic if r is greater than the Castelnuovo-Mumford regularity of M. For such an r we extend $R(M_{\geq r})$ by computing a minimal free resolution

$$\ldots \to T^{r-2} \to T^{r-1}$$

of the E-module $P = \ker(\mathrm{Hom}_k(E, M_r) \to \mathrm{Hom}_k(E, M_{r+1}))$. This yields a doubly infinite exact complex

$$\mathbb{T} = \mathbb{T}(\mathcal{F}) : \ldots \to T^{r-2} \to T^{r-1} \to T^r \to T^{r+1} \to \ldots$$

whose isomorphism class depends only on the sheaf $\mathcal{F}$. We call $\mathbb{T}(\mathcal{F})$ the *Tate resolution* of $\mathcal{F}$.

Theorem 53 (Eisenbud–Fløystad–Schreyer) *With notation as above,*

$$\mathbb{T}^e(\mathcal{F}) = T^e = \sum_{i=0}^{n} \mathrm{Hom}_k(E, H^i(\mathbb{P}^n, \mathcal{F}(e-i))).$$

Thus each cohomology group $H^i(\mathbb{P}^n, \mathcal{F}(j))$ occurs precisely once in the complex.

Example 54 For the ideal sheaf from Example 52, we get:

```
julia> sheaf_cohomology(MI, -2, 8)
twist:   -2   -1    0    1    2    3    4    5    6    7    8
-------------------------------------------------------------
4:        -    -    -    -    -    -    -    *    *    *    *
3:        *   10    -    -    -    -    -    -    *    *    *
2:        *    *    -    2    -    -    -    -    -    *    *
1:        *    *    *    -    1    5    5    -    -    -    *
0:        *    *    *    *    -    -    -    5   32   84  170
-------------------------------------------------------------
chi:      *    *    *    *    1    5    5    *    *    *    *
```

With no algorithm specified, `OSCAR` applies the default option `algorithm = :bgg`. Due to the shape of the Tate resolution, the function does then not compute all values in a given range of twists $l < h$. The missing values are indicated by a `*`. To determine all values in the range $l < h$, enter `sheaf_cohomology(MI, l-ngens(base_ring(M)), h+ngens(base_ring(M)))`.

8 Plane Curves, their Duals, and Riemann–Roch Spaces

Let $C = V(f) \subset \mathbb{P}^2$ be an irreducible plane curve, given by a square-free homogeneous polynomial f. The space of tangent lines $\check{C} \subset \check{\mathbb{P}}^2$ of C is a curve in the dual projective space.

Example 55 We demonstrate how to write an `OSCAR` function which computes the dual of a plane curve specified by a homogeneous polynomial:

```
function dual_curve(f::MPolyRingElem, P_dual::MPolyRing)
  P = parent(f)
  vars_P = gens(P)
  nvars_P = ngens(P)
  vars_P_dual = gens(P_dual)
  # Extend the original polynomial ring to include the variables of
  ↪  P_dual
```

```
  P_ext, vars_ext = polynomial_ring(base_ring(P), [[string(v) for v in
  ↪  vars_P]; [string(v) for v in vars_P_dual]])
  inc = hom(P, P_ext, vars_ext[1:nvars_P])
  f_ext = inc(f)
  # Compute the Jacobian matrix with respect to the original variables
  jf = transpose(jacobian_matrix(f_ext)[1:nvars_P, 1:1])
  # Form the matrix with the last 'ngens(P_dual)' variables of P_ext
  A = matrix([vars_ext[(end-ngens(P_dual)+1):end]])
  # Stack the Jacobian matrix and the matrix A
  m2x3 = vcat(jf, A)
  # Compute minors and saturate
  I = ideal(minors(m2x3, 2))
  J = ideal([jf[1, i] for i in 1:ncols(jf)])
  Isat = saturation(I + ideal([f_ext]), J)
  # Project to the dual space
  proj_dual_images = vcat([zero(P_dual) for _ in 1:nvars_P],
  ↪  gens(P_dual))
  proj = hom(P_ext, P_dual, proj_dual_images)
  dual_curve = groebner_basis(proj(Isat))
  return dual_curve[1]
end
```

We apply the function to a smooth plane quartic curve:

```
julia> P, (x, y, z) = graded_polynomial_ring(QQ, ["x", "y", "z"]);

julia> f = 8*x^4+20*x^2*y^2+8*y^4-48*x^2*z^2-48*y^2*z^2+65*z^4;

julia> P_dual, (u, v, w) = graded_polynomial_ring(QQ, ["u", "v", "w"]);

julia> f_dual = dual_curve(f, P_dual)
101920*u^12 - 283920*u^10*v^2 - 424704*u^10*w^2 - 329160*u^8*v^4 -
↪  420192*u^8*v^2*w^2 + 701152*u^8*w^4 + 1211860*u^6*v^6 -
↪  200976*u^6*v^4*w^2 + 1603016*u^6*v^2*w^4 - 585600*u^6*w^6 -
↪  329160*u^4*v^8 - 200976*u^4*v^6*w^2 + 1873041*u^4*v^4*w^4 -
↪  1405488*u^4*v^2*w^6 + 261000*u^4*w^8 - 283920*u^2*v^10 -
↪  420192*u^2*v^8*w^2 + 1603016*u^2*v^6*w^4 - 1405488*u^2*v^4*w^6 +
↪  489600*u^2*v^2*w^8 - 58752*u^2*w^10 + 101920*v^12 - 424704*v^10*w^2 +
↪  701152*v^8*w^4 - 585600*v^6*w^6 + 261000*v^4*w^8 - 58752*v^2*w^10 +
↪  5184*w^12
```

The curve C and its dual are depicted in Figure 7.

By the Plücker formulas from the theorem below the curve C in the example has 24 flexes and 28 bi-tangents. These correspond to 24 ordinary cusps and 28 ordinary double points of the dual curve.

Theorem 56 (Plücker Formulas) *Let $C \subset \mathbb{P}^2$ be an irreducible plane curve of degree d defined over a field of characteristic 0, and such that C and the dual curve $\check{C} \subset \check{\mathbb{P}}^2$ have only ordinary nodes and cusps as singularities. Let f, b, κ, δ denote the number of flexes, bi-tangents, cusps and nodes of C, respectively. Let g denote the geometric genus of C, and let $\check{d}$ be the degree of $\check{C}$. Then*

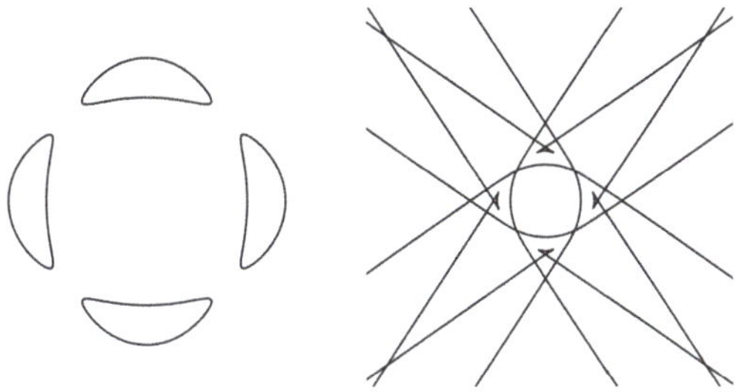

Fig. 7: Smooth plane quartic and its dual curve.

$$g = \frac{1}{2}(d-1)(d-2) - \delta - \kappa = \frac{1}{2}(\check{d}-1),(\check{d}-2) - b - f,$$
$$\check{d} = d(d-1) - 2\delta - 3\kappa, \qquad d = \check{d}(\check{d}-1) - 2b - 3f,$$
$$f = 3d(d-2) - 6\delta - 8\kappa, \qquad \kappa = 3\check{d}(\check{d}-2) - 6b - 8f,$$

and $b = \frac{1}{2}d(d-2)(d-3)(d+3) - (4d^2-4d-20)\delta - (6d^2-6d-27)\kappa + (2\delta+3\kappa)^2$.

In Example 55, eight of the flexes/cups are real and visible in Figure 7. All 28 bi-tangents are real. The 24 flexes are precisely the $g^3 - g$ Weierstrass points on this curve of genus $g = 3$.

The only other possible gap sequence[10] of a Weierstrass point on a non-hyperelliptic curve of genus 3 is $1, 2, 5$. In that case, the plane quartic has a tangent line which intersects the curve with multiplicity 4 at the Weierstrass points and the dual curve becomes more singular than the singularities of curves allowed in the Plücker formulas.

Note that in the characteristic zero context of Theorem 56, the double dual $\check{\check{C}}$ is identical to C. This explains the symmetry of the formulas. In positive characteristic, $\check{\check{C}} = C$ might not hold.

What kind of singularities can occur in the dual curve of a smooth plane quartic with a non-ordinary Weierstrass point?

Example 57 Consider the quartic curves defined by $f_1 = x^4 + y^3 - y$ and $f_2 = x^4 + y^3 - \frac{2}{9}x^2 - y + \frac{1}{81}$. The curves are depicted in Figure 8. Their projective closures are smooth and the dual curves have the following affine equations:

```
julia> f1 = x^4+y^3*z-y*z^3;

julia> f1_dual = dual_curve(f1, P_dual);

julia> f1_dual(u,1,w)
```

[10] The gap sequence at a point p on a curve C of genus g is the sequence of g integers that are not pole orders at p of regular functions on $C \setminus \{p\}$. The point p is a Weierstrass point if the sequence is different from $\{1, 2, \ldots, g\}$.

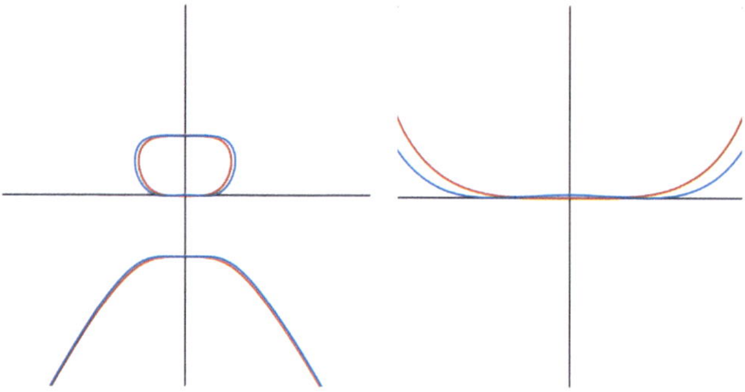

Fig. 8: Smooth quartic curves.

```
4*u^12 - 48*u^8*w^3 + 48*u^8*w - 27*u^4*w^8 + 84*u^4*w^6 - 546*u^4*w^4 +
↪  84*u^4*w^2 - 27*u^4 - 256*w^9 + 768*w^7 - 768*w^5 + 256*w^3

julia> f2 = x^4+y^3*z-y*z^3-2//9*x^2*z^2+1//81*z^4;

julia> f2_dual = dual_curve(f2, P_dual);

julia> f2_dual(u,1,w)
235953*u^12 + 8748*u^10*w^2 + 314928*u^10*w + 2916*u^10 - 118098*u^8*w^4 -
↪  2834352*u^8*w^3 + 8748*u^8*w^2 + 2831760*u^8*w - 1458*u^8 +
↪  708588*u^6*w^6 - 866052*u^6*w^4 + 90720*u^6*w^3 + 3700404*u^6*w^2 +
↪  33696*u^6*w + 235940*u^6 - 1594323*u^4*w^8 + 4960116*u^4*w^6 -
↪  1189728*u^4*w^5 - 32240754*u^4*w^4 + 46656*u^4*w^3 + 4967028*u^4*w^2 +
↪  303264*u^4*w - 1589715*u^4 + 6928416*u^2*w^7 - 8188128*u^2*w^5 -
↪  62208*u^2*w^4 + 3149280*u^2*w^3 - 1889568*u^2*w - 20736*u^2 -
↪  15116544*w^9 + 45349632*w^7 + 186624*w^6 - 45349632*w^5 - 373248*w^4 +
↪  15116544*w^3 + 186624*w^2
```

Figure 9 depicts the real points of the affine curves and a magnification of the plot around the point $p' = [0 : 1 : 0] \in \check{\mathbb{P}}^2$.

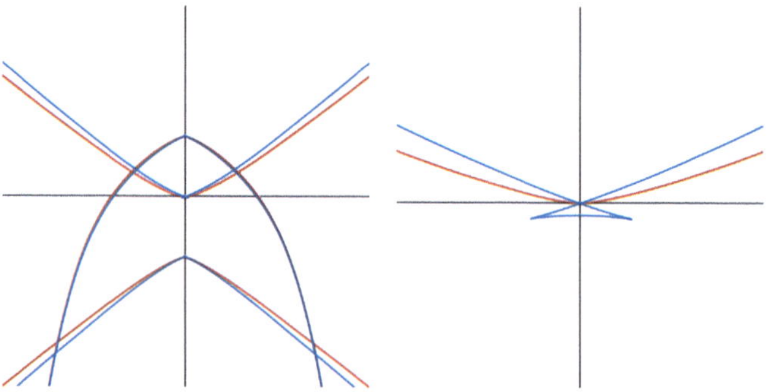

Fig. 9: Plots of the real points of the curves and a magnification of the plot around the point $p' = [0 : 1 : 0] \in \check{\mathbb{P}}^2$.

The dual curve of $C_1 = V(f_1)$ has an E_6-singularity at p' which deforms to a node and two cusps in the dual curve of $C_2 = V(f_2)$. The corresponding point $p = [0 : 0 : 1] \in \mathbb{P}^2$ is a Weierstrass point with gap sequence $1, 2, 5$. Actually C_1 has 12 non-ordinary Weierstrass points. There are no other Weierstrass points since $2 \cdot 12 = 3^3 - 3$. Furthermore, C_1 has 16 bi-tangents apart from the tangents at the Weierstrass points, of which 4 are real.

Example 58 It is known that for arbitrary genus g a general curve $C \in \mathcal{M}_g$ has $g^3 - g$ simple Weierstrass points and that the monodromy group over $\mathcal{M}_g \setminus B$, where B is the set of curves with a non-ordinary Weierstrass point or a non-trivial automorphism, is the full symmetric group on $g^3 - g$ points, see [EH87].

In case $g = 3$ we can verify this fact via a Galois group computation. Consider the family of plane curves defined by

$$V(s(8x^4+20x^2y^2+8y^4-48x^2z^2-48y^2z^2+65z^4)+t(x^3y+y^3z+z^3x)) \subset \mathbb{P}^1 \times \mathbb{P}^2.$$

The set of Weierstrass points in the fibers of this family is a curve X with a $24:1$ map $X \to \mathbb{P}^1$. One can verify that the monodromy group of $X(\mathbb{C}) \to \mathbb{P}^1(\mathbb{C})$ is S_{24} by Galois group computations over $\mathbb{Q}$ as follows. We consider two fibers, and commence showing that the corresponding two extension fields have Galois group S_{24} over $\mathbb{Q}$.

```
julia> P2, (x, y, z) = polynomial_ring(QQ,["x","y","z"]);

julia> f1 = 8*x^4+20*x^2*y^2+8*y^4-48*x^2*z^2-48*y^2*z^2+65*z^4+x*y^3;

julia> f2 = x^3*y+y^3*z+z^3*x+x^4;

julia> g1 = 1771*f1 - 1317*f2;

julia> hess1 = hessian(g1);

julia> I1 = ideal([hess1,g1]);

julia> g1yz = eliminate(I1,[x]);

julia> Q, t = polynomial_ring(QQ,"y");

julia> phi = hom(P2, Q, [0,t,1]);

julia> g1t = phi(g1yz[1]);

julia> G1,_ = galois_group(g1t);

julia> G1
Sym(24)

julia> g2 = 7713*f1 - 1313*f2;

julia> g2t = phi(eliminate(ideal([hessian(g2),g2]),[x])[1]);
```

```
julia> G2,_ = galois_group(g2t);

julia> G2==G1
true
```

Furthermore, we show that the extension fields do not have a common subfield other than $\mathbb{Q}$: The only possible ramified primes are divisors of both polynomial discriminants. We observe that the gcd of the polynomial discriminants of the fibers is a power of 2.

```
julia> dg1 = discriminant(g1t);

julia> dg2 = discriminant(g2t);

julia> ggT = gcd(dg1,dg2);

julia> factor(ZZ(ggT))
1 * 2^4
```

Since the field discriminant (which contains exactly the ramified primes) divides the polynomial discriminant, only 2 is a possible common prime divisor of the two field discriminants under consideration. Computing a maximal order over 2 for one of the fibers shows that 2 is not ramified in the normalization:

```
julia> KK, _ = number_field(g2t);

julia> degree(KK)
24

julia> O = any_order(KK);

julia> OO = pmaximal_overorder(O, 2);

julia> d = discriminant(OO);

julia> gcd(d,2)
1
```

Hence, 2 cannot divide the discriminant of the intersection of the extension fields, which thus has discriminant one, and thus is $\mathbb{Q}$.

Thus Galois group computations provide an alternative approach to the commonly used homotopy continuation technique.

Let $C \subset \mathbb{P}^2$ be an irreducible plane curve of degree d with only ordinary singularities $p_1, \ldots, p_s$ of multiplicities $r_1, \ldots, r_s$. Using the completeness of the adjoint systems, we can compute for a divisor D on the normalisation $X = \widetilde{C}$ the Riemann–Roch space

$$H^0(X, \mathcal{O}(D)) = L(D) = \{f \in K(X)^* \mid (f) + D \geq 0\} \cup \{0\}$$

as follows.

Algorithm (Riemann–Roch Space)
Input. An irreducible plane curve $C' = V(f)$ with only ordinary singularities given by a square-free homogeneous polynomial $f \in K[x_0, x_1, x_2]$.
A divisor $D = \sum n_i p_i$ with support disjoint from the singularities of C'.
Output. $\ell(D) = \dim L(D)$ and a basis of $L(D)$.

1. If the divisor is given by a list of pairs of multiplicities and points $\{(n_i, p_i)\}$ compute
$$\mathrm{I}(D_1) = \bigcap_{n_i>0} (\mathrm{I}(p_i)^{n_i} + (f)) \text{ and } \mathrm{I}(D_2) = \bigcap_{n_i<0} (\mathrm{I}(p_i)^{-n_i} + (f)).$$
2. Compute the adjoint ideal J_{adj}.
3. Verify that C has only ordinary singularities, if necessary.
4. Compute $I = \mathrm{I}(D_1) \cap J_{\mathrm{adj}}$.
5. Choose $e > 0$ such that $I_e \supsetneq (f)_e$ and a form $h \in I_e \setminus (f)_e$.
6. Compute the residual ideal $I' = (f, h) : I$.
7. If $V(I' + \mathrm{I}(D_2)) = \emptyset$ then $J = I' \cap \mathrm{I}(D_2)$ else $J = (I' \cdot \mathrm{I}(D_2) + (f)) : (x_0, x_1, x_2)^\infty$.
8. Compute $\ell = \ell(D) = \dim J_e/(f)_e$ and, if $\ell > 0$, forms $h_1, \dots h_\ell$, which represent a basis of $J_e/(f)_e$.
9. Return $\ell(D)$ and if $\ell(D) > 0$ the rational functions $h_1/h, \dots, h_\ell/h$.

Example 59 (Taken from [Sch]). Consider the family of sextic curves $C_t = V(f_t) \subset \mathbb{P}^2$ with affine parts defined by the polynomials

$$\begin{aligned} f_t =& xy(x+y) - 3(x^5+y^5) - 2(x^6+y^6) \\ &+ t[-2xy + 12(x^4+y^4) + x^3y + xy^3 - 20x^2y^2 + 8(x^5+y^5) - 12(x^4y+xy^4) \\ &+ 6(x^3y^2+x^2y^3) - 5(x^5y+xy^5) - 2(x^4y^2+x^2y^4) + 14x^3y^3] \\ &+ t^2[-12(x^3+y^3) - 2(x^2y+xy^2) - 8(x^4+y^4) + 24(x^3y+xy^3) \\ &- 44x^2y^2 + 10(x^4y+xy^4) + 20(x^3y^2+x^2y^3) + 10(x^4y^2+x^2y^4) + 24x^3y^3]. \end{aligned}$$

The curve C_t has four ordinary double points $p_0 = (0,0), p_1 = (2t, 0), p_2 = (0, 2t)$, and $p_3 = (-1,-1)$, and no further singularities for general values for t. However, C_0 has an ordinary triple point at p_0, see Figure 10 for a plot of the curves $C_{1/10}$ and C_0. Thus this is a family of curves of genus 6.

The curve C_0 is trigonal. The projection from the triple point induces a $3:1$ map $C \to \mathbb{P}^1$. From Petri's Theorem [Pet23], the homogeneous ideal of the canonical model of this curve in $\mathbb{P}^5$ needs cubic generators. We check this computationally:

```
julia> Rt, (x,y,t) = polynomial_ring(QQ, ["x","y","t"]);
```

Fig. 10: Plots of the curves $C_{1/10}$ and C_0.

```
julia> ft = -2*x^6 - 5*x^5*y*t + 8*x^5*t - 3*x^5 + 10*x^4*y^2*t^2 -
↪  2*x^4*y^2*t + 10*x^4*y*t^2 - 12*x^4*y*t - 8*x^4*t^2 + 12*x^4*t +
↪  24*x^3*y^3*t^2 + 14*x^3*y^3*t + 20*x^3*y^2*t^2 + 6*x^3*y^2*t +
↪  24*x^3*y*t^2 + x^3*y*t - 12*x^3*t^2 + 10*x^2*y^4*t^2 - 2*x^2*y^4*t +
↪  20*x^2*y^3*t^2 + 6*x^2*y^3*t - 44*x^2*y^2*t^2 - 20*x^2*y^2*t -
↪  2*x^2*y*t^2 + x^2*y - 5*x*y^5*t + 10*x*y^4*t^2 - 12*x*y^4*t +
↪  24*x*y^3*t^2 + x*y^3*t - 2*x*y^2*t^2 + x*y^2 - 2*x*y*t - 2*y^6 +
↪  8*y^5*t - 3*y^5 - 8*y^4*t^2 + 12*y^4*t - 12*y^3*t^2;

julia> R, (x,y) = polynomial_ring(QQ, ["x","y"]);

julia> phi0 = hom(Rt,R,[x,y,0]);

julia> f = phi0(ft);

julia> H = homogenizer(R, "z");

julia> F = H(f);

julia> S = parent(F);

julia> C = plane_curve(F);

julia> I = adjoint_ideal(C)
Ideal generated by
  x*y - y^2
  x^2 - y^2
  y^3 + y^2*z

julia> m3 = ideal(gens(S))^3;

julia> K = intersect(I, m3)
Ideal generated by
  x*y*z - y^2*z
  x^2*z - y^2*z
  y^3 + y^2*z
  x*y^2 + y^2*z
```

```
  x^2*y + y^2*z
  x^3 + y^2*z

julia> P5, _ = graded_polynomial_ring(QQ, ["x_0", "x_1", "x_2", "x_3",
↪  "x_4", "x_5"]);

julia> PC, pr = quo(S, ideal([F]));

julia> psi = hom(P5, PC, [pr(K[i]) for i = 1:6]);

julia> J = kernel(psi);

julia> JJ = ideal(minimal_generating_set(J));

julia> Q , _ = quo(P5, JJ);

julia> re = free_resolution(Q);

julia> minimal_betti_table(re)
       0  1   2  3  4
---------------------
0     : 1  -   -  -  -
1     : -  6   8  3  -
2     : -  3   8  6  -
3     : -  -   -  -  1
---------------------
total: 1  9  16  9  1
```

The curves C_t for $t \neq 0$, on the other hand, are 4-gonal and their canonical images are generated by quadrics alone. We verify this for the curve $C_{1/10}$.

```
julia> phi1 = hom(Rt,R,[x,y,R(1//10)]);

julia> f = phi1(ft);

julia> F = H(f);

julia> C = plane_curve(F);

julia> I = adjoint_ideal(C)
Ideal generated by
  6*x*y - 5*y^2 + y*z
  5*x^2 - x*z - 5*y^2 + y*z

julia> m3 = ideal(gens(S))^3;

julia> K = intersect(I, m3)
Ideal generated by
  6*x*y*z - 5*y^2*z + y*z^2
  5*x^2*z - x*z^2 - 5*y^2*z + y*z^2
  5*y^3 + 4*y^2*z - y*z^2
  6*x*y^2 + 5*y^2*z - y*z^2
  6*x^2*y + 5*y^2*z - y*z^2
  25*x^3 - x*z^2 + 20*y^2*z - 4*y*z^2
```

```
julia> PC, pr = quo(S, ideal([F]));

julia> psi = hom(P5, PC, [pr(K[i]) for i = 1:6]);

julia> J = kernel(psi);

julia> JJ = ideal(minimal_generating_set(J));

julia> Q , _ = quo(P5, JJ);

julia> re = free_resolution(Q);

julia> minimal_betti_table(re)
       0 1  2 3 4
---------------------
0    : 1 -  - - -
1    : - 6  5 - -
2    : - -  5 6 -
3    : - -  - - 1
---------------------
total: 1 6 10 6 1
```

Remark 60 The phenomenon above has a wide generalization. According to Green's Conjecture [Gre84] for smooth projective curves defined over $\mathbb{C}$ one can read off the gonality (more precisely the Clifford index) of the curve from the Betti table of its canonical model. This was proved in landmark papers by Voisin [Voi02; Voi05] for the general curve using $K3$ surfaces and in [AF11] for curves on arbitrary $K3$ surfaces. The conjecture is known to be wrong for some smooth curves over some fields of positive characteristic [Sch86; BS21]. Green's conjecture is open in its full generality.

9 Deformations

Example 61 The surface from Example 50 is defined over the field $\mathbb{F}_3$. Actually, we would like to prove the existence of an example over $\mathbb{C}$. The key point in the construction is the module S/m with Hilbert function $(1, 5, 5, 0, \ldots)$. A general module of this Hilbert function has Betti numbers

	0	1	2	3	4	5
0 :	1	–	–	–	–	–
1 :	–	10	15	–	–	–
2 :	–	–	5	26	20	5
total :	1	10	20	26	20	5

instead of

	0	1	2	3	4	5
0 :	1	–	–	–	–	–
1 :	–	10	15	2	–	–
2 :	–	–	7	26	20	5
total :	1	10	22	28	20	5

The family of modules

$$M = \{\text{good } m's\} \subset \mathbb{G}(10, H^0(\mathbb{P}^4, \mathcal{O}(2))$$

has at the given point m from Example 50 at most codimension $7 \cdot 2 = 14$. We will show by computation that equality holds and that M is smooth of dimension $10 \cdot 5 - 14 = 36$ at our given m. We start out by determining the free resolution:

```
julia> K = GF(3);

julia> S, (x0, x1, x2, x3, x4) = graded_polynomial_ring(K, ["x0", "x1",
↪  "x2", "x3", "x4"]);

julia> m = ideal(S, [x1^2+(-x1+x2+x3-x4)*x0, x1*x2+(x1-x3+x4)*x0,
↪  x1*x3+(-x1+x4+x0)*x0, x1*x4+(-x1+x3+x4-x0)*x0, x2^2+(x1-x2-x4-x0)*x0,
↪  x2*x3+(x1-x2+x3+x4-x0)*x0, x2*x4+(x1+x2-x3-x4-x0)*x0,
↪  x3^2+(x3+x4-x0)*x0,x3*x4+(-x3-x4+x0)*x0, x4^2+(x1+x3-x4-x0)*x0]);

julia> R, _ = quo(S, m);

julia> FR = free_resolution(R, algorithm = :mres);
```

Let $\varphi_1, \ldots, \varphi_3$ denote the first three maps in the resolution of S/m. We add a 1×10 deformation matrix A_1 to φ_1 and solve recursively the equations

$$A_1 \cdot \varphi_2 = -\varphi_1 \cdot A_2, \quad A_2 \cdot \varphi_3 = -\varphi_2 \cdot A_3.$$

We write a helper function to solve this problem, finding the normal space generator-by-generator:

```
julia> L = monomial_basis(R, 2)
5-element Vector{MPolyDecRingElem{FqFieldElem, FqMPolyRingElem}}:
 x4^2
 x3*x4
 x2*x4
 x2*x3
 x1*x4

julia> versal_unfolding = [[i == div((j-1), 5) + 1 ? S(L[(j-1) % 5 + 1]) :
↪  S(0) for i in 1:10] for j in 1:50];

julia> function normal_space_generator(FR, A1t)
         phi1 = map(FR, 1)
         e1 = gen(codomain(phi1),1)
         phi2 = map(FR, 2)
         phi3 = map(FR, 3)
         A1 = hom(domain(phi1), codomain(phi1), [p*e1 for p in A1t])
         A1phi2 = phi2 * A1
         A2 = lift(A1phi2, phi1)
         A2phi3 = phi3 * A2
         A3 = lift(A2phi3, phi2)
         A3m = matrix(A3)
         return A3m[1:2, 16:22]
       end;
```

```
julia> nlist = [normal_space_generator(FR, v) for v in versal_unfolding];

julia> TS, t = graded_polynomial_ring(K, "t"=>(1:50));

julia> nlist_t = [map_entries(x -> x * t[i],
↪   map_entries(constant_coefficient, nlist[i])) for i in 1:50];

julia> B = sum(nlist_t);
```

The complex with differentials

$$\varphi_1 + A_1, \varphi_2 + A_2 \text{ and } \varphi_3 + A_3$$

is a flat first order deformation defined over $\mathbb{F}_3[t_1, \ldots, t_{50}]/(t_1, \ldots, t_{50})^2$. The entries of the 7×2 submatrix B of A_3 returned by the function define the tangent space

$$T_m M \subset T_m \mathbb{G}(10, H^0(\mathbb{P}^4, \mathcal{O}(2)) = \operatorname{Hom}(\mathbb{F}_3^{10}, \mathbb{F}_3^5).$$

```
julia> tangent_space = ideal(vec(collect(B)));

julia> ngens(leading_ideal(tangent_space))
14
```

Now, M is defined over the integers. Taking linear equations $\ell_1, \ldots, \ell_{14}$ on $\mathbb{P}(\Lambda^5 H^0(\mathbb{P}^4, \mathcal{O}(2))^*)$ defined over $\mathbb{Z}$ which cut mod 3 the variety $M(\mathbb{F}_3)$ transversal at m, we get a scheme $V(\ell_1, \ldots, \ell_{14}) \cap M$ over $\operatorname{Spec} \mathbb{Z}$ which has as one of its components an open part of $\operatorname{Spec} \mathcal{O}_L$ of a number field L and a prime ideal $\mathfrak{p}$ over (3) with $\mathcal{O}_{L,\mathfrak{p}}/\mathfrak{p} \cong \mathbb{F}_3$. The other steps in the construction of the surface work over $\operatorname{Spec} \mathcal{O}_{L,\mathfrak{p}}$, since the conditions that the construction works are open conditions. In particular, the generic fiber of the family is a smooth surface defined over L with the same adjunction behavior as the surface over the finite field.

Remark 62 Notice that this technique allows one to prove the existence of interesting objects in algebraic geometry with a search using computer algebra. The example above was found by randomly picking points $p \in \mathbb{G}(10, 15)(\mathbb{F}_3)$ and testing for 2 extra syzygies. The probability for a random p to lie in $M(\mathbb{F}_3)$ is roughly $1 : 3^{14}$ because the codimension M in $\mathbb{G}(10, 15)$ is 14. We can improve our search by searching in the unirational subvariety

$$M_1 = \{m \in \mathbb{G}(10, H^0(\mathbb{P}^4, \mathcal{O}(2))) \mid \dim \operatorname{Tor}_3^S(S/m, \mathbb{F}_3)_4 \geq 1\}$$

in which M has codimension at most 7. Four families of smooth non-general type surfaces with $p_g = q = 0$, degree $d = 11$, and sectional genus $\pi = 10$ were found with this method [Sch96]. This is interesting because of the aforementioned result of Ellingsrud and Peskine [EP89] which states that there are only finitely many components of the Hilbert scheme of $\mathbb{P}^4$ whose general points correspond to smooth surfaces of Kodaira dimension < 2, confirming a conjecture of Hartshorne about smooth rational surfaces in $\mathbb{P}^4$. Currently there are 24 families of smooth rational surfaces known in $\mathbb{P}^4$. The largest degree of a known smooth rational surface in $\mathbb{P}^4$ is 12, see [AR06].

A problem on which the cornerstone system **Singular** grew is the search by Greuel and Pfister for a counterexample to the Zariski Conjecture [GP21]:
Is the multiplicity of an isolated singularity of a hypersurface $V(f) \subset \mathbb{C}^n$ a topological invariant?
Let $f \in \mathbb{C}\{x, y, z\}$ be a convergent power series with an isolated singularity at 0. Recall that the *Milnor fiber*

$$M_f = V(f - t) \cap S^5_\varepsilon \subset \mathbb{C}^3$$

is homotopic to a bouquet of μ spheres S^2, where μ is the *Milnor number*

$$\mu = \dim_{\mathbb{C}} \mathbb{C}\{x, y, z\}/(\frac{\partial f}{\partial x}, \frac{\partial f}{\partial y}, \frac{\partial f}{\partial z}).$$

The *Tjurina number*

$$\tau = \dim_{\mathbb{C}} \mathbb{C}\{x, y, z\}/(f, \frac{\partial f}{\partial x}, \frac{\partial f}{\partial y}, \frac{\partial f}{\partial z})$$

is the dimension of the (smooth) base space B of the semi-universal deformation of f. Zariski asked whether the multiplicity is constant along the μ-constant stratum $\subset B$.

The speed with which **Singular** operates comes from the desire of Greuel and Pfister to treat examples with five digit Milnor numbers.

Example 63 Consider

$$f_t = x^a + y^b + z^{3c} + x^{c+2}y^{c-1} + x^{c-1}y^{c-1}z^3 + x^{c-2}y^c(y^2 + tx)^2$$

with $(a, b, c) = (40, 30, 8)$. In **OSCAR**, multiplicities and Milnor numbers can be found in the spirit of Section 3 using local monomial orderings. We get:

$$m(f_0) = 17,\ m(f_t) = 16,\ \mu(f_0) = 10661,\ \mu(f_t) = 10655.$$

For example[11]:

```
julia> k = GF(31991);

julia> S, (x, y, z, t) = polynomial_ring(k, ["x", "y", "z", "t"]);
```

[11] In experiments which aim at finding a particular example in characteristic zero, it is often convenient to first work over a finite field to speed up computations. Once a promising example has been found, and no theoretical argument is known for deducing the existence of an example in characteristic zero, one has to verify the computation in characteristic zero. In the example under consideration here, the latter is feasible due an algorithm given in [GPS22].

```
julia> a, b, c = 40, 30, 8
(40, 30, 8)

julia> ft=x^a+y^b+z^(3*c)+x^(c+2)*y^(c-1)+x^(c-1)*y^(c-1)*z^3+x^(c-2)*y^
↪   c*(y^2+t*x)^2;

julia> R, (x, y, z) = polynomial_ring(k, ["x", "y", "z"]);

julia> f0 = hom(S, R, [x, y, z, 0])(ft);

julia> f1 = hom(S, R, [x, y, z, 1])(ft);

julia> MI0 = jacobian_ideal(f0);    MI1 = jacobian_ideal(f1);

julia> U = complement_of_point_ideal(R, [0 ,0, 0]);

julia> Rloc, phi = localization(R, U);

julia> A0, _ = quo(Rloc, phi(MI0));  A1, _ = quo(Rloc, phi(MI1));

julia> vector_space_dimension(A0)
10661

julia> vector_space_dimension(A1)
10655
```

The multiplicity goes down for $t \neq 0$. Unfortunately (or luckily), the Milnor number goes down as well. Thus f_t is not a counterexample to the Zariski Conjecture. However, $\frac{6}{10661} \approx .00056$ is rather small compared to $\frac{1}{17} \approx .059$.

Acknowledgements Funded by the Deutsche Forschungsgemeinschaft (DFG, German Research Foundation) – Project-ID 286237555 – TRR 195. Janko Böhm and Wolfram Decker also acknowledge support of the Potentialbereich *SymbTools – Symbolic Tools in Mathematics and their Application* of the Forschungsinitiative Rheinland-Pfalz.

References

[AR06] H. Abo and K. Ranestad. "Construction of rational surfaces of degree 12 in projective fourspace". *J. Algebr. Geom.* 15.2 (2006), pp. 323–338. DOI: 10.1090/S1056-3911-06-00424-3.

[Ale88] J. Alexander. "Surfaces rationelles non-spéciales dans $\mathbb{P}^4$. (Non special rational surfaces in $\mathbb{P}^4$)". *Math. Z.* 200.1 (1988), pp. 87–110.

[AF11] M. Aprodu and G. Farkas. "Green's conjecture for curves on arbitrary $K3$ surfaces". *Compos. Math.* 147.3 (2011), pp. 839–851. DOI: 10.1112/S0010437X10005099.

[BPV84] W. Barth, C. Peters, and A. Van de Ven. *Compact complex surfaces.* Vol. 4. Ergeb. Math. Grenzgeb., 3. Folge. Springer, Cham, 1984.

[BT24] M. Bies and A. P. Turner. "F-Theory Applications". In: *The Computer Algebra System OSCAR. Algorithms and Examples.* Ed. by W. Decker, C. Eder, C. Fieker, M. Horn, and M. Joswig. Springer, 2024, pp. 451–472.

[BB24] J. Böhm and T. Breuer. "Algorithms for GIT-Fans of Affine Torus Actions". In: *The Computer Algebra System OSCAR. Algorithms and Examples.* Ed. by W. Decker, C. Eder, C. Fieker, M. Horn, and M. Joswig. Springer, 2024, pp. 213–224.

[BDLP22] J. Böhm, W. Decker, S. Laplagne, and G. Pfister. "Computing integral bases via localization and Hensel lifting". *J. Symb. Comput.* 109 (2022), pp. 283–324. DOI: 10.1016/j.jsc.2020.07.007.

[BDLP17] J. Böhm, W. Decker, S. Laplagne, and G. Pfister. "Local to global algorithms for the Gorenstein adjoint ideal of a curve". In: *Algorithmic and Experimental Methods in Algebra, Geometry, and Number Theory.* Cham: Springer, 2017, pp. 51–96. DOI: 10.1007/978-3-319-70566-8_3.

[Böh+13] J. Böhm et al. "Parallel algorithms for normalization". *J. Symb. Comput.* 51 (2013), pp. 99–114. DOI: 10.1016/j.jsc.2012.07.002.

[BS21] C. Bopp and F.-O. Schreyer. "A version of Green's conjecture in positive characteristic". *Exp. Math.* 30.4 (2021), pp. 475–480. DOI: 10.1080/10586458.2019.1576082.

[BZ24] S. Brandhorst and M. Zach. "Elliptic Fibrations on Vinberg's Most Algebraic K3 Surface". In: *The Computer Algebra System OSCAR. Algorithms and Examples.* Ed. by W. Decker, C. Eder, C. Fieker, M. Horn, and M. Joswig. Springer, 2024, pp. 225–241.

[BH93] W. Bruns and J. Herzog. *Cohen-Macaulay rings.* Vol. 39. Cambridge Studies in Advanced Mathematics. Cambridge University Press, 1993, pp. xii+403.

[CLO15] D. A. Cox, J. Little, and D. O'Shea. *Ideals, varieties, and algorithms.* Fourth. Undergraduate Texts in Mathematics. Springer, Cham, 2015, pp. xvi+646. DOI: 10.1007/978-3-319-16721-3.

[DES93] W. Decker, L. Ein, and F.-O. Schreyer. "Construction of Surfaces in $\mathbb{P}^4$". *J. Algebr. Geom.* 2 (1993), pp. 185–237.

[DE02] W. Decker and D. Eisenbud. "Sheaf algorithms using the exterior algebra". In: *Computations in algebraic geometry with Macaulay 2.* Berlin: Springer, 2002, pp. 215–249.

[DRS24] W. Decker, L. Ramesh, and J. Schmitt. "Invariant Theory". In: *The Computer Algebra System OSCAR. Algorithms and Examples.* Ed. by W. Decker, C. Eder, C. Fieker, M. Horn, and M. Joswig. Springer, 2024, pp. 297–331.

[DS00] W. Decker and F.-O. Schreyer. "Non-general type surfaces in $\mathbb{P}^4$: Some remarks on bounds and constructions". *J. Symb. Comput.* 29.4-5 (2000), pp. 545–582. DOI: 10.1006/jsco.1999.0323.

[EFS03] D. Eisenbud, G. Fløystad, and F.-O. Schreyer. "Sheaf cohomology and free resolutions over exterior algebras". *Transactions of the American Mathematical Society* 355.11 (2003), pp. 4397–4426. DOI: 10.1090/S0002-9947-03-03291-4.

[EGSS02] D. Eisenbud, D. R. Grayson, M. Stillman, and B. Sturmfels, eds. *Computations in algebraic geometry with Macaulay 2.* Vol. 8. Algorithms and Computation in Mathematics. Springer-Verlag, Berlin, 2002, pp. xvi+329. DOI: 10.1007/978-3-662-04851-1.

[EH87] D. Eisenbud and J. Harris. "Irreducibility and monodromy of some families of linear series". *Annales scientifiques de l'École Normale Supérieure* Ser. 4, 20.1 (1987), pp. 65–87. DOI: 10.24033/asens.1524.

[EP89] G. Ellingsrud and C. Peskine. "Sur les surfaces lisses de $\mathbf{P}_4$". *Invent. Math.* 95.1 (1989), pp. 1–11. DOI: 10.1007/BF01394141.

[Gor] P. Gordan. "Neuer Beweis des Hilbert'schen Satzes über homogene Funktionen." *Nachr. Ges. Wiss. Göttingen, Math.-Phys. Kl.* 1899, pp. 240–242.

[Gre84] M. L. Green. "Koszul cohomology and the geometry of projective varieties. Appendix: The nonvanishing of certain Koszul cohomology groups (by Mark Green and Robert Lazarsfeld)". *J. Differ. Geom.* 19 (1984), pp. 125–167, 168–171. DOI: 10.1007/s100970200042.

[GLS10] G.-M. Greuel, S. Laplagne, and F. Seelisch. "Normalization of rings". *J. Symb. Comput.* 45.9 (2010), pp. 887–901. DOI: 10.1016/j.jsc.2010.04.002.

[GP08] G.-M. Greuel and G. Pfister. *A Singular introduction to commutative algebra.* Second, extended edition. With contributions by Olaf Bachmann, Christoph Lossen and Hans Schönemann. Berlin: Springer, 2008.

[GP21] G.-M. Greuel and G. Pfister. *History of Singular and its relation to Zariski's multiplicty conjecture.* 2021. arXiv: 2103.00525.

[GPS22] G.-M. Greuel, G. Pfister, and H. Schönemann. "Using semicontinuity for standard bases computations". *Math. Comput. Sci.* 16.4 (2022). Id/No 21, p. 11. DOI: 10.1007/s11786-022-00539-2.

[Har77] R. Hartshorne. *Algebraic geometry.* Vol. No. 52. Graduate Texts in Mathematics. Springer-Verlag, New York-Heidelberg, 1977, pp. xvi+496.

[Pet23] K. Petri. "Über die invariante Darstellung algebraischer Funktionen einer Veränderlichen." *Math. Ann.* 88 (1923), pp. 242–289. DOI: 10.1007/BF01579181.

[Sch] F.-O. Schreyer. *Introduction to Algebraic Geometry with a Computational Approach.* Book manuscript, submitted.

[Sch96] F.-O. Schreyer. "Small fields in constructive algebraic geometry". In: *Moduli of vector bundles. Papers of the 35th Taniguchi symposium, Sanda, Japan and a symposium held in Kyoto, Japan, 1994*. New York, NY: Marcel Dekker, 1996, pp. 221–228.

[Sch86] F.-O. Schreyer. "Syzygies of canonical curves and special linear series". *Math. Ann.* 275 (1986), pp. 105–137. DOI: 10.1007/BF01458587.

[SV87] A. J. Sommese and A. Van de Ven. "On the adjunction mapping". *Math. Ann.* 278 (1987), pp. 593–603. DOI: 10.1007/BF01458083.

[Voi05] C. Voisin. "Green's canonical syzygy conjecture for generic curves of odd genus". *Compos. Math.* 141.5 (2005), pp. 1163–1190. DOI: 10.1112/S0010437X05001387.

[Voi02] C. Voisin. "Green's generic syzygy conjecture for curves of even genus lying on a $K3$ surface". *J. Eur. Math. Soc. (JEMS)* 4.4 (2002), pp. 363–404. DOI: 10.1007/s100970200042.

[ZS60] O. Zariski and P. Samuel. *Commutative Algebra*. Vol. II. Springer-Verlag, 1960.

Number Theory

Claus Fieker and Tommy Hofmann

Abstract We give a brief introduction to computational algebraic number theory in OSCAR. Our main focus is on number fields, rings of integers and their invariants. After recalling some classical results and their constructive counterparts, we showcase the functionality in two examples related to the investigation of the Cohen–Lenstra heuristic for quadratic fields and the Galois module structure of rings of integers.

Key words: Algorithmic number theory, class groups, Galois module theory

1 Introduction

The area of number theory is one of the oldest mathematical disciplines and comprises different streams of research. Many developments have been inspired by the study of Diophantine equations, which is the investigation of integer solutions of polynomial equations with integer coefficients. One of the most famous examples is the equation $x^n + y^n = z^n$ with $n \in \mathbb{N}$, $n \geq 3$. The statement that this equation has no non-trivial integer solution is commonly known as Fermat's last theorem, whose proof by Wiles [Wil95] is regarded as one of the pinnacles of the development of modern number theory in the 20th century. Algebraic number theory, which is concerned with the investigation of number theoretic questions using concepts from abstract algebra and counts Gauß [Gau86] among its founding fathers, forms one of the pillars of modern number theory. The beauty of this subject lies both in the depth of the theory

Claus Fieker
RPTU Kaiserslautern–Landau, Kaiserslautern, Germany, e-mail: `claus.fieker@rptu.de`

Tommy Hofmann
Universität Siegen, Siegen, Germany e-mail: `tommy.hofmann@uni-siegen.de`

W. Decker et al. (eds.), *The Computer Algebra System OSCAR*,
Algorithms and Computation in Mathematics 32,
https://doi.org/10.1007/978-3-031-62127-7_3

and the possibility to do concrete computations and experiments. In recent years, this fruitful interplay between theory and practical questions has been intensified further by the connection to the theory of secure communication, that is, cryptography. In this chapter, we give a brief introduction to some of the main characters from algebraic number theory and show how we can investigate them using the algorithmic tools provided by OSCAR.

2 Number Fields and Rings of Integers

One of the fundamental notions in algebraic number theory is that of a *number field*, which is a finite extension field of the rational numbers. In this section we will explain the basics for working with these objects in OSCAR (see also [FHHJ17] on some background for the design choices of the number theory in OSCAR). We will freely use theoretical results from algebraic number theory, which can be found, for example, in [Lan94; Neu99; Mar18]. An algorithmic treatment of the subject can be found in [Coh93; PZ97; Coh00].

2.1 Construction of Number Fields and First Properties

From theory we know that any number field K, that is, a finite extension of $\mathbb{Q}$, can be generated by a primitive element, say $K = \mathbb{Q}(\alpha)$ for some $\alpha \in K$. Since we are interested at this point only in the algebraic properties of K, we can realize K also as $K \cong \mathbb{Q}[x]/(f)$, where $f \in \mathbb{Q}[x]$ is the minimal polynomial of α (which is necessarily irreducible). Thus specifying a number field K is as simple as specifying an irreducible rational polynomial, a *defining polynomial* for K. As an example, we want to construct the quadratic number field $K = \mathbb{Q}[x]/(x^2 - 235) \cong \mathbb{Q}(\sqrt{5 \cdot 47})$.

```
julia> Qx, x = QQ["x"];

julia> K, a = number_field(x^2 - 235, "a")
(Number field of degree 2 over QQ, a)

julia> a^2 - 235 # confirm that a is a root of x^2 - 235
0
```

Note that the construction of the number field returns as a second value a primitive element α satisfying $\alpha^2 - 235 = 0$. Since $\{1, \alpha\}$ is a $\mathbb{Q}$-basis of K, we can now express arbitrary elements as linear combinations of 1 and α. Vice versa we can obtain the coordinates of any element. Another way of constructing elements of K is by mapping elements of $\mathbb{Q}[x]$ under the canonical surjection $\mathbb{Q}[x] \to \mathbb{Q}[x]/(x^2 - 235) = K$.

```
julia> b = 2 + 1//3 * a
1//3*a + 2

julia> coordinates(b)
2-element Vector{QQFieldElem}:
 2
 1//3

julia> K(x^3 + x + 2) # map the polynomial x^3 + x + 2 to K
236*a + 2
```

Of course, since a number field K is foremost an extension of $\mathbb{Q}$, we can ask for basic field theoretic properties of this extension as well as of the elements of K.

```
julia> degree(K)
2

julia> trace(a)
0

julia> norm(a)
-235
```

2.2 Three Theorems From Algebraic Number Theory

For arithmetic purposes, one is often interested in the *ring of integers* $\mathcal{O}_K$ of a number field K, which is the subring comprised of all elements of K with integral minimal polynomial. Equivalently, $\mathcal{O}_K$ is the integral closure of $\mathbb{Z}$ in K. Because of this maximality property, it is also known as the *maximal order* of K. The structure of $\mathcal{O}_K$ is described by three classical theorems from algebraic number theory. Before stating these theorems, we first have to introduce some notation. The set I_K of invertible fractional ideals of $\mathcal{O}_K$ forms an abelian group with respect to multiplication. It contains the subgroup P_K of principal fractional ideals and the quotient $\mathrm{Cl}(K) = I_K/P_K$ is the *class group* of K. The order $h_K = |\mathrm{Cl}(K)|$ is the *class number* of K. The *signature* of K is the tuple (r, s) of non-negative integers, where r denotes the number of real embeddings $K \to \mathbb{R}$ and $2s$ the number of non-real embedding $K \to \mathbb{C}$. With this at hand, the ring, ideal and unit structure of $\mathcal{O}_K$ can be described as follows:

1. As an abelian group, $\mathcal{O}_K$ is a free $\mathbb{Z}$-module of rank $d = [K : \mathbb{Q}]$. Thus there exists an *integral basis* $(\omega_i)_{1 \le i \le d}$ satisfying $\mathcal{O}_K = \bigoplus_{i=1}^d \mathbb{Z}\omega_i$.
2. The ring $\mathcal{O}_K$ is a Dedekind domain with finite class group $\mathrm{Cl}(K)$.
3. The unit group $\mathcal{O}_K^\times$ is finitely generated of rank $r + s - 1$ with finite cyclic torsion subgroup, where (r, s) is the signature of K. Thus we have $\mathcal{O}_K^\times \cong \mathbb{Z}^{r+s-1} \times \mathbb{Z}/m\mathbb{Z}$, where $m \in \mathbb{Z}_{\ge 1}$ is the number of roots of unity of K.

Since we approach this subject from a constructive perspective, these statements beg the questions: Can we compute an integral basis, the class group or generators of the unit group? We will now discuss how these questions can be answered with the help of OSCAR.

The Ring of Integers

For statement 1, note that already asking for the ring of integers of K in OSCAR will implicitly compute an integral basis, which can be queried for afterwards.

```
julia> OK = ring_of_integers(K);

julia> basis(OK)
2-element Vector{AbsSimpleNumFieldOrderElem}:
 1
 a

julia> discriminant(OK)
940
```

Continuing with our example $K = \mathbb{Q}[x]/(x^2 - 235)$, we get as a result $\{1, \alpha\}$, which means that $\mathcal{O}_K = \mathbb{Z} \cdot 1 \oplus \mathbb{Z} \cdot \alpha$. From this we can also derive the discriminant $\operatorname{disc}(K)$ of K, which is equal to $\det((\operatorname{Tr}_{K/\mathbb{Q}}(\omega_i \omega_j))_{1 \le i,j \le d}))$ and a measure for the "size" of the ring of integers.

We want to mention that this result has a simple geometric interpretation in case K is an imaginary quadratic field. In this case we can identify K with a subset of $\mathbb{C}$, such that the set $\mathcal{O}_K$ becomes a lattice (a free $\mathbb{Z}$-submodule of rank 2). For example, if $L = \mathbb{Q}(\alpha)$, $\alpha^2 + 3 = 0$, then $\mathcal{O}_L = \mathbb{Z}[(1+\alpha)/2]$ has integral basis $(\omega_1, \omega_2) = (1, (1+\alpha)/2)$. We illustrate this using the Julia package Plots.jl

```
using Plots;
w1, w2 = Complex(1),(sqrt(Complex(-3)) + 1)/2
pts = filter(z -> max(abs(imag(z)), abs(real(z))) <= 3,
             [ a*w1 + b*w2 for a in -5:5 for b in -5:5]);
x = range(0, real(w1) + real(w2), 100);
y1 = (x -> min(sqrt(3)*x, imag(w2))).(x);
y2 = (x -> max(0, sqrt(3)*x - sqrt(3))).(x);
p = scatter(real.(pts), imag.(pts), framestyle=:origin,
            legend = false, mc = :red)
plot!(p, x, y1, fillrange = y2, fillalpha = 0.25, color = :blue)
```

The result is displayed in Figure 1. The red dots are the points of $\mathcal{O}_K$ with bounded real and imaginary part. The blue area is a fundamental domain for this lattice with area equal to $\sqrt{3}/2 = \sqrt{|\operatorname{disc}(K)|}/2$ (the factor of 2 is introduced by an implicit choice of measure on $\mathbb{C}$), thus giving a precise meaning to the aforementioned claim on the relation between discriminant and "size" of the ring of integers.

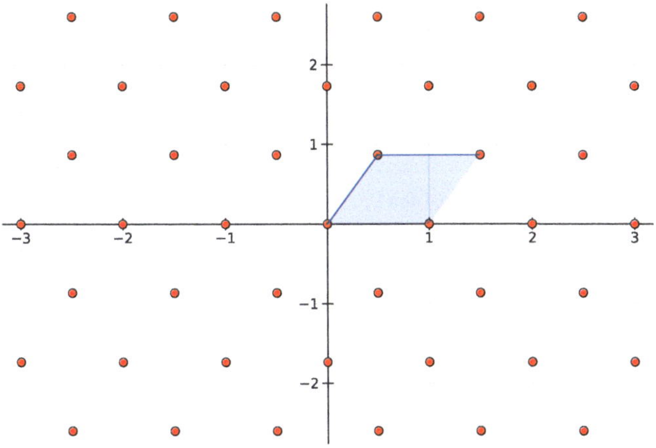

Fig. 1: The ring $\mathbb{Z}[(1+\sqrt{-3})/2]$ as a lattice of rank 2 in $\mathbb{C}$ and a fundamental domain.

Unique Factorization of Ideals and the Class Group

Let us now turn to statement 2, which describes the ideal structure of the ring of integers. That this ring is a Dedekind domain is equivalent to the unique factorization of non-trivial ideals into prime ideals. Before giving an example of this, we first consider the structure of prime ideals and highlight the celebrated theorem of Dedekind–Kummer, which, for almost all prime ideals, gives an explicit description in terms of properties of the defining polynomial of the number field. Since $\mathcal{O}_K$ is an integral extension of $\mathbb{Z}$, the canonical map $\operatorname{Spec}(\mathcal{O}_K) \to \operatorname{Spec}(\mathbb{Z})$ is surjective. Thus we obtain all prime ideals of $\mathcal{O}_K$ by determining the prime ideals lying over the rational primes in $\mathbb{Z}$.

```
julia> prime_ideals_over(OK, 7)
2-element Vector{AbsSimpleNumFieldOrderIdeal}:
 <7, a + 5>
 <7, a + 2>
```

While this a priori only shows that $\mathfrak{p}_1 = (7, \alpha+5)$ and $\mathfrak{p}_2 = (7, \alpha+2)$ are the prime ideals lying above 7, we have in fact the stronger statement that $7\mathcal{O}_K = (7, \alpha+4) \cdot (7, \alpha+3)$. This is the factorization of $7\mathcal{O}_K$ into prime ideals. We verify this by applying the theorem of Dedekind–Kummer, which states that the prime ideal decomposition of $7\mathcal{O}_K$ is in shape equal to the unique factorization of $\overline{x^2 - 235} \in \mathbb{F}_7[x]$ into irreducible polynomials, and that generators of the prime ideals are obtained by evaluating the lifted factors at α.

```
julia> factor(change_coefficient_ring(GF(7), x^2 - 235))
1 * (x + 5) * (x + 2)
```

The factorization into prime ideals is not limited to prime numbers. We can also factor arbitrary ideals, for example, $\alpha\mathcal{O}_K$.

```
julia> factor(a * OK)
Dict{AbsSimpleNumFieldOrderIdeal, Int64} with 2 entries:
  <47, a> => 1
  <5, a>  => 1
```

We now turn our attention to the class group $\mathrm{Cl}(K)$. Let us first ask OSCAR for this invariant and then see how this should be interpreted.

```
julia> A, m = class_group(OK);

julia> A
Z/6
```

This tells us that the class group $\mathrm{Cl}(K)$ is, as an abelian group, isomorphic to the cyclic group $A \cong \mathbb{Z}/6\mathbb{Z}$. If A is not trivial, as in our case, this does not give a complete description of the group structure of $\mathrm{Cl}(K) = I_K/P_K$, since it does not tell us how the classes $[I], [J]$ of two (fractional) ideals I, J of $\mathcal{O}_K$ behave under multiplication. Here we encounter one of OSCAR's design principles—the concept of *interpretation maps*. If the result of a computation carries the structure of, say, an abelian group, we will return an abstract abelian group together with an appropriate map (the *interpretation map*), which is an isomorphism or can be used to reconstruct the isomorphism. In our case, the returned map $m\colon A \to I_K$ is such that the composition

$$A \xrightarrow{m} I_K \longrightarrow I_K/P_K = \mathrm{Cl}(K)$$

with the natural projection is an isomorphism. We can evaluate m and also ask for preimages of arbitrary ideals, respectively, their classes.

```
julia> m(zero(A)) # apply m to the neutral element of A
<1, 1>
[...]

julia> P = prime_ideals_over(OK, 2)[1]
<2, a + 1>
[...]

julia> preimage(m, P)
Abelian group element [3]
```

Thus we have shown that the ideal $\mathfrak{p} = (2, \alpha+1)$ corresponds to the element $\overline{3} \in \mathbb{Z}/6\mathbb{Z}$ and thus has order 2. In particular this implies that $\mathfrak{p}^2 \in P_K$, that is, $\mathfrak{p}^2$ is principal. The additional information about a possible generator of $\mathfrak{p}^2$ is not part of the map m, but we can ask for it separately.

```
julia> is_principal_with_data(P^2)
(true, 2)

julia> P^2 == 2*OK
true
```

The Unit Group

Finally we turn to statement 3, which describes the structure of the unit group $\mathcal{O}_K^\times$. First note that if the signature of $K = \mathbb{Q}[x]/(f)$ is (r, s), then the rational polynomial f has r real roots and $2s$ non-real complex roots. Hence the rank of $\mathcal{O}_K^\times$, which is equal to $r + s - 1$, can be computed very easily. But as in the case of class groups, we also want to know an explicit isomorphism describing the structure of $\mathcal{O}_K^\times$ as an abelian group.

```
julia> U, mU = unit_group(OK);

julia> U
Z/2 x Z
```

Thus $\mathcal{O}_K^\times \cong U = \mathbb{Z} \times \mathbb{Z}/2\mathbb{Z}$ as expected, since here $(r, s) = (2, 0)$. The interpretation map $U \to \mathcal{O}_K^\times$ is in this case an isomorphism. We can use it to determine explicit generators of the unit group, as well as, to map units to U, thus expressing elements in terms of the generators.

```
julia> mU(U[1]), mU(U[2])
(-1, 3*a + 46)

julia> preimage(mU, -214841715*a - 3293461126)
Abelian group element [1, 5]

julia> -214841715*a - 3293461126 == (-1)^1 * (3a + 46)^5
true
```

Finally we want to mention how the unit group can be interpreted geometrically. For the sake of simplicity we consider the number field $K = \mathbb{Q}(\alpha)$ with $\alpha^2 - 3 = 0$, whose ring of integers is equal to $\mathbb{Z}[\alpha] = \{a + b\alpha \mid a, b \in \mathbb{Z}\}$. Since this a quadratic extension, we can embed K into $\mathbb{R}^2$ via the $\mathbb{Q}$-linear map

$$\varphi \colon K \to \mathbb{R}^2,\ a + b\alpha \mapsto (a, b).$$

Now for any number field, the unit group is equal to $\mathcal{O}_K^\times = \{x \in \mathcal{O}_K \mid \mathrm{N}_{K/\mathbb{Q}}(x) = \pm 1\}$. For the field under consideration this implies

$$\varphi(\mathbb{Z}[\alpha]^\times) = \{(a, b) \in \mathbb{Z}^2 \mid a^2 - 3b^2 = \pm 1\}.$$

Thus $\mathbb{Z}[\alpha]^\times$ corresponds to the points of $H = \{(a, b) \in \mathbb{R}^2 \mid a = \pm\sqrt{3b^2 \pm 1}\}$ with integer coordinates. We can plot this as follows:

```
pts = [ (a, b) for a in -8:8 for b in -5:5]
x = range(-7.5, 7.5, 500)
y1 = (x -> x^2 - 1 < 0 ? missing : sqrt((x^2 - 1)/3)).(x)
y2 = (x -> sqrt((x^2 + 1)/3)).(x)
scatter(pts, framestyle=:origin, leg = false, mc = :blue, alpha = 0.25)
plot!(x, [y1 -y1 y2 -y2], color = :black)
units = [(1, 0), (-1, 0), (2, 1), (-2, 1), (-2, -1),
         (2, -1), (7, 4), (7, -4), (-7, 4), (-7, -4)];
scatter!(units, color = :red)
```

The result is depicted in Figure 2. The blue dots correspond to elements of $\mathbb{Z}[\alpha]$ with red dots corresponding to points lying in H, thus to units of $\mathbb{Z}[\alpha]$. The points $(\pm 1, 0)$ correspond to the torsion units $\pm 1 \in \mathbb{Z}[\alpha]^\times$, and $(\pm 2, \pm 1)$ correspond to the units $\pm 2 \pm \alpha$, which are in fact generators of the torsion-free part of $\mathbb{Z}[\alpha]^\times$. The prominent symmetries of $\varphi(\mathbb{Z}[\alpha])$ (and of H) are explained by the existence of the two automorphisms $x \mapsto -x$ and $x \mapsto x^{-1}$ of $\mathbb{Z}[\alpha]^\times$.

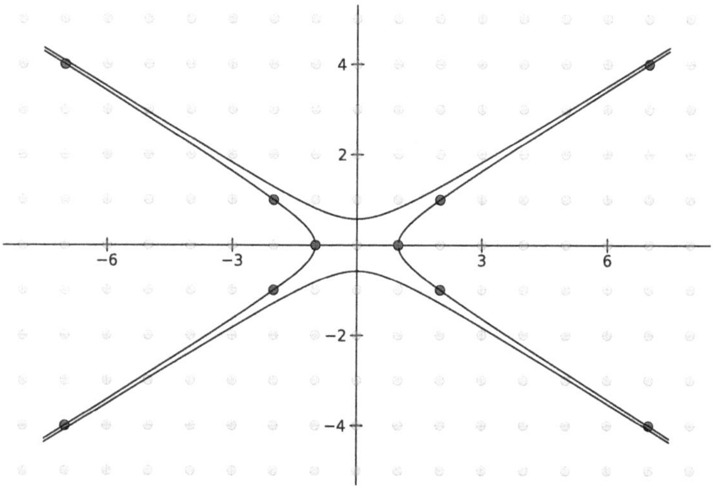

Fig. 2: The unit group $\mathbb{Z}[\alpha]^\times$ interpreted as integer points lying on the union of two algebraic curves.

Note that while Figure 2 illustrates that units are quite likely to be "sparse" within the ring of integers, it does not depict the group structure of the unit group. To visualize the latter, we will make use of the so-called logarithmic embedding (note that this historical name is slightly misleading, since this map is *not* injective):

$$\ell\colon K^\times \to \mathbb{R}^2,\ a + b\alpha \mapsto (\log|a + b\sqrt{3}|, \log|a - b\sqrt{3}|).$$

The image $\ell(\mathbb{Z}[\alpha]^\times)$ is a free $\mathbb{Z}$-submodule of $\mathbb{R}^2$ contained in the hyperplane $H = \{(a,b) \in \mathbb{R}^2 \mid a+b=0\}$, generated by $\ell(2+\alpha) = (1.316..., -1.316...)$ We can plot this with the following commands:

```
l(a, b) = (log(abs(a + sqrt(3)*b)), log(abs(a - sqrt(3)*b)))
pts = filter!(z -> maximum(abs.(z)) < 3,
              [l(a, b) for a in -50:50 for b in -50:50])
scatter(pts, framestyle=:origin, leg = false, mc = :blue, alpha = 0.25)
x = range(-3, 3, 100); y = -x; plot!(x, y, color = :black)
units = [ i .* l(2, 1) for i in -2:2]
scatter!(units, color = :red)
```

The result is depicted in Figure 3. As before, the blue dots correspond to elements of $\mathbb{Z}[\alpha]$ with red dots corresponding to points lying on H, that is, corresponding to units of $\mathbb{Z}[\alpha]$. The abelian structure of $\mathbb{Z}[\alpha]$ is now clearly visible. Note that, in comparison to Figure 2, we have "lost" one symmetry. This is explained by the fact that $-1 \in \ker(\ell)$, and hence the automorphism $x \mapsto -x$ of $\mathbb{Z}[\alpha]^\times$ becomes the identity.

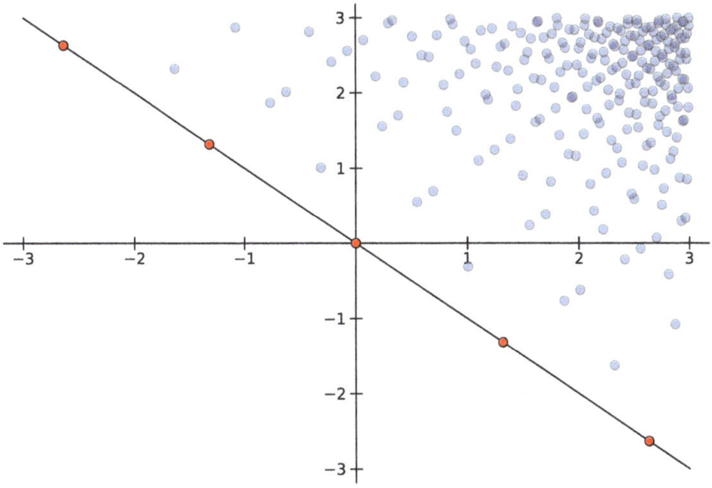

Fig. 3: The unit group $\mathbb{Z}[\alpha]^\times$ under the logarithmic embedding ℓ.

3 General Number Fields and Morphisms

So far we have constructed number fields only as simple extensions of $\mathbb{Q}$, where we have provided the minimal polynomial of a generator. Although due to the primitive element theorem, any number field is of this kind, in practice, number fields are not always given in this form. Often, in the context of the specific problem under consideration, there is a description as a non-simple extension of $\mathbb{Q}$ or of other number fields. In this regard, OSCAR is quite flexible in that a number field is allowed to be either

- a simple extension $\mathbb{Q}(\alpha)$ or a non-simple extension $\mathbb{Q}(\alpha_1, \ldots, \alpha_n)$ of $\mathbb{Q}$ (referred to as an *absolute field*); or
- a simple extension $K(\alpha)$ or a non-simple extension $K(\alpha_1, \ldots, \alpha_n)$, where K is a previously defined number field (referred to as *a relative field*).

Let us illustrate this by constructing the biquadratic non-simple extension $K = \mathbb{Q}(\sqrt{2}, \sqrt{3})$. This is as simple as providing the defining polynomials of the generators, which in our case are $x^2 - 2$ and $x^2 - 3$.

```
julia> K, a = number_field([x^2 - 2, x^2 - 3]);

julia> a[1]^2 == 2 && a[2]^2 == 3
true
```

To illustrate the relative fields, we now construct $\mathbb{Q}(\sqrt{2}, \sqrt{3})$ as the quadratic extension $L = k(\sqrt{3})$, where $k = \mathbb{Q}(\sqrt{2})$. Note that although $t^2 - 2 \in k[t]$ has rational coefficients, it is constructed as a polynomial over k. As a consequence, when we use it to define a number field, we obtain an extension of k.

```
julia> k, b = quadratic_field(3);

julia> kt, t = k["t"];

julia> L, c = number_field(t^2 - 2);
```

Note that although L and K are isomorphic as extensions of $\mathbb{Q}$, they behave very differently. Since L was created as an extension of k, all the field theoretic properties of L and its elements are relative to the base field k.

```
julia> base_field(L) == k
true

julia> base_field(K) == QQ
true

julia> degree(K)
4

julia> degree(L)
2
```

We can still ask the system for information of the number field L, regarded as an absolute field, that is, as a field extension of $\mathbb{Q}$.

```
julia> absolute_degree(K)
4

julia> absolute_norm(c)
4
```

The flexible framework to construct number fields as arbitrary extensions of number fields is accompanied by a framework for working with maps between number fields of similar flexibility. Specifying a map $L \to M$ from a number field L with base field k is done by providing the image(s) of the generator(s) and an optional map $k \to M$. We illustrate this by constructing the map defined by $\sqrt{2} \mapsto -\sqrt{2}$ and $\sqrt{3} \mapsto -\sqrt{3}$ for both fields K and L encountered before.

```
julia> f = hom(K, K, [-a[1], -a[2]]);

julia> f(a[1]) == -a[1] && f(a[2]) == -a[2]
true

julia> g = hom(L, L, hom(k, k, -b), -c);

julia> g(c) == -c && g(L(b)) == L(-b)
true
```

As we have seen, the construction of number fields is quite flexible and allows for arbitrary towers of number fields. On the other hand, it is often necessary to collapse a given number field L into an absolute simple extension, that is, into the form $\mathbb{Q}(\alpha)$ for some α. This need arises, for example, when a certain functionality is provided only for absolute simple number fields. We illustrate this by computing the class group of $L = \mathbb{Q}(\sqrt{3})(\sqrt{2})$ and proving that $h_L = 1$.

```
julia> Lprime, LprimetoL = absolute_simple_field(L);

julia> defining_polynomial(Lprime)
x^4 - 10*x^2 + 1
```

The first return value L' is a simple extension of $\mathbb{Q}$ and the second return value is a $\mathbb{Q}$-linear isomorphism $L' \to L$. We now determine the class number of L via L':

```
julia> C, _ = class_group(ring_of_integers(Lprime)); order(C)
1
```

Note that similar functionality exists to collapse a non-simple extension into a simple extension (`simple_extension`) and to collapse a relative extension $L/K/k$ into an extension L'/k (`collapse_top_layer`).

4 Embeddings of Number Fields

We now want to stress an important point about computational aspects of number fields, which requires some adjustment to the conventional way of working with number fields. If we define the number field K as a field extension $\mathbb{Q}[x]/(f)$ of $\mathbb{Q}$ for some irreducible polynomial $f \in \mathbb{Q}[x]$, then we do not consider K to be embedded into $\mathbb{C}$. For example, if $K = \mathbb{Q}[x]/(x^2 - 2)$, then the primitive element $\alpha = \overline{x} \in K$ satisfies $\alpha^2 - 2 = 0$, but we cannot tell whether α is equal to $\sqrt{2}$ or $-\sqrt{2}$. On the theoretical side, this ambiguity comes from the fact that for a number field K, there is in general more than one morphism $K \to \mathbb{C}$.

4.1 Embeddings

From the standpoint of embeddings and places, the presence of more than one morphism $K \to \mathbb{R}$ is seen as additional arithmetic information of the number field. A *complex embedding* of a number field K is a morphism $\iota \colon K \to \mathbb{C}$. If $\iota(K) \subseteq \mathbb{R}$, the embedding is called *real*, and *imaginary* otherwise. If $K = \mathbb{Q}(\alpha) = \mathbb{Q}[x]/(f)$ is a simple absolute extension, imaginary embeddings correspond to the roots of f in $\mathbb{C} \setminus \mathbb{R}$, with the real embeddings corresponding to real roots.

```
julia> K, a = number_field(x^3 - 2, "a");

julia> signature(K)
(1, 1)

julia> real_embeddings(K)
1-element Vector{AbsSimpleNumFieldEmbedding}:
 Complex embedding corresponding to 1.26 of K

julia> embs = complex_embeddings(K) # printed with limited precision
3-element Vector{AbsSimpleNumFieldEmbedding}:
 Complex embedding corresponding to 1.26 of K
 Complex embedding corresponding to -0.63 + 1.09 * i of K
 Complex embedding corresponding to -0.63 - 1.09 * i of K

julia> embs[1](a - 1)
[0.2599210499 +/- 8.44e-11]
```

4.2 Embedded Number Fields

We continue with the number field $K = \mathbb{Q}[x]/(x^3 - 2)$ and primitive element α satisfying $\alpha^3 - 2 = 0$. To obtain an ordered field, where α is in fact equal to $\sqrt[3]{2}$, we can construct an embedded number field, where we additionally supply an approximation of the root during the construction.

```
julia> K, a = embedded_number_field(x^3 - 2, 1.26);

julia> 0 < a < 2
true
```

Note that such an embedded number field does not support number theoretic operations, but is useful when there is a need for an ordered field, for example, in polyhedral geometry [BJ24] (cf. page 139).

5 Symmetries

The symmetries of a number field arise as a purely field-theoretic concept and play a fundamental role in algebraic number theory. On the one hand, these symmetries are an important tool when studying arithmetic objects connected to or defined over number fields. On the other hand, various deep open problems in algebraic number theory are related to the study of the symmetries of number fields themselves, one of the most prominent being the Inverse Galois Problem. In this section we explain how to compute and work with symmetries of number fields, which come in two different flavors: automorphism groups and Galois groups.

5.1 Automorphisms

Let $K/\mathbb{Q}$ be a number field. The automorphism group $\mathrm{Aut}_{\mathbb{Q}}(K)$ is by definition equal to the set of all ring isomorphisms $K \to K$. Let us illustrate this using the number field $K/\mathbb{Q}$ with defining polynomial $x^4 - 13x^2 + 16$.

We can easily compute this with OSCAR as follows. After defining the field, we tell the system to compute the automorphism group as a permutation group.

```
julia> K, a = number_field(x^4 - 13*x^2 + 16, "a");

julia> G, m = automorphism_group(PermGroup, K); G # we only print G
Permutation group of degree 4
```

(Note that the abstract permutation group G representing the automorphism group is not unique. Due to the randomized nature of the algorithm, the permutations generating the group can differ.) Since the two permutations commute and are of order 2, we see that G is isomorphic to $\mathrm{C}_2 \times \mathrm{C}_2$, which we can also confirm as follows:

```
julia> describe(G)
"C2 x C2"
```

So far we have not seen any automorphism of K, but only some abstract description of the group of all automorphisms. This additional information is provided by the map m, which is an isomorphism $G \to \mathrm{Aut}_{\mathbb{Q}}(K)$.

```
julia> m(G[1])
Map
  from number field of degree 4 over QQ
  to number field of degree 4 over QQ

julia> m(G[1])(a)
-1//4*a^3 + 13//4*a
```

These automorphisms are just special cases of morphisms that we already encountered in Section 3. For example, we can construct the automorphism of K defined by $a \mapsto -a$ and ask for its preimage in G:

```
julia> preimage(m, hom(K, K, -a))
(1,2)(3,4)
```

In agreement with the overall design principles for number fields, the automorphism group respects also the (possible) relative structure of a number field. For example, consider $k = \mathbb{Q}(\alpha)$ with defining polynomial $x^2 - 18$ and $K = k(\beta)$ with defining polynomial $x^4 + (\alpha + 6)x^2 + 2\alpha + 9$.

```
julia> k, a = number_field(x^2 - 18, "a"); kt, t = k["t"];

julia> K, b = number_field(t^4 + (a + 6)*t^2 + 2a + 9, "b");

julia> G, m = automorphism_group(PermGroup, K); describe(G)
"C4"
```

Thus $\mathrm{Aut}_k(K) \cong G \cong \mathrm{C}_4$ is cyclic of order 4. While any k-linear automorphism is necessarily $\mathbb{Q}$-linear, not every $\mathbb{Q}$-linear automorphism must be k-linear, and in fact, it might not even stabilize k. Thus, in general we have $\mathrm{Aut}_k(K) \subsetneq \mathrm{Aut}_{\mathbb{Q}}(K)$. As we will see now, the field K has this property. The group $\mathrm{Aut}_{\mathbb{Q}}(K)$, which we refer to as the *absolute automorphism group* can be computed as follows.

```
julia> G, m = absolute_automorphism_group(PermGroup, K); describe(G)
"Q8"
```

5.2 Galois Groups

The second kind of symmetry we will discuss are Galois groups of number fields. We will illustrate this using the number field $K = \mathbb{Q}(\alpha)$ with defining polynomial $x^4 - 2$.

```
julia> K, a = number_field(x^4 - 2);

julia> G, mA = automorphism_group(PermGroup, K);

julia> describe(G)
"C2"
```

Given that this field is not normal, the automorphism group of K is not "large enough". Here we only have one non-trivial automorphism, sending α to $-\alpha$. One way to overcome this apparent lack of automorphisms is to pass to the Galois group, which is by definition the automorphism group of a splitting field of a defining polynomial f of K. In OSCAR we can directly ask for the Galois group.

```
julia> G, C = galois_group(K); describe(G)
"D8"
```

We can verify that this is (isomorphic to) the automorphism group of a splitting field of $x^4 - 2$.

```
julia> L = splitting_field(x^4 - 2);

julia> GL, = automorphism_group(PermGroup, L);

julia> fl, = is_isomorphic(G, GL); fl
true
```

The automorphism group of the splitting field of f acts naturally on the $d = \deg(f)$ roots of f and is therefore naturally isomorphic to a subgroup of the symmetric group $\mathfrak{S}_d$ on d letters. For this reason the permutation group G computed is not any permutation group isomorphic to the Galois group, but a group of explicit permutations of the roots of the defining polynomial in a suitable field. The information about the roots can be accessed via the second return value of the galois_group command.

```
julia> rts = roots(C, 2)
4-element Vector{QadicFieldElem}:
 (8*11^0 + O(11^2))*a + 8*11^0 + 9*11^1 + O(11^2)
 (3*11^0 + 10*11^1 + O(11^2))*a + 7*11^0 + 4*11^1 + O(11^2)
 (3*11^0 + 10*11^1 + O(11^2))*a + 3*11^0 + 11^1 + O(11^2)
 (8*11^0 + O(11^2))*a + 4*11^0 + 6*11^1 + O(11^2)

julia> parent(rts[1])
Unramified extension of 11-adic numbers of degree 2
```

We can read off that the roots of $x^4 - 2$ are approximated in a local field, more precisely, the unramified quadratic extension of the field of 11-adic numbers $\mathbb{Q}_{11}$. The approximated roots can be computed with arbitrary precision—for the sake of simplicity we only asked for a precision of 2. An interesting side-effect of Stauduhar's algorithm [Sta73] (see also [FHJ22] for details on the implementation in OSCAR), which is the algorithm underlying the Galois group computation, is that it allows us to determine subfields of the splitting field exactly, still using only approximated roots. As an illustration we consider the field K with defining polynomial $x^9 - 3x^8 + x^6 + 15x^5 - 13x^4 - 3x^3 + 4x - 1$. The Galois group of this field is the group $G = \mathrm{ASL}_2(\mathbb{F}_3)$, the affine special linear group on $\mathbb{F}_3^2$ of order 216. This group has a subgroup H isomorphic to the Heisenberg group He_3 of order 27 with $G/\operatorname{Core}(H) \cong \mathrm{SL}_2(\mathbb{F}_3)$ of order 24. Thus the fixed field $k = K^H$ is non-normal of degree 8 with Galois group $\mathrm{SL}_2(\mathbb{F}_3)$. We can compute a number field (and therefore a defining polynomial) with this Galois group as follows:

```
K, a = number_field(x^9 - 3*x^8 + x^6 + 15*x^5 - 13*x^4 -
                     3*x^3 + 4*x - 1, "a")
G, C = galois_group(K)
subsG = subgroups(G)
H = first(H for H in subsG if order(H) == 27)
k, = simplify(fixed_field(C, H))

julia> defining_polynomial(k)
x^8 + x^7 - 7*x^6 - 23*x^5 + 22*x^4 + 80*x^3 + 99*x^2 + 27*x + 8

julia> describe(galois_group(k)[1])
"SL(2,3)"
```

Thus we have found the field k. Note that the same computation using automorphism groups is prohibitively expensive, since it would require the explicit construction of a normal closure of K, which has degree 216.

We end this section by highlighting a few of the algorithmic differences between Galois group and automorphism group computations of a normal number field K with defining polynomial $f \in \mathbb{Q}[x]$. Although in this situation the automorphism group is isomorphic to the Galois group, the Galois group is still given as a permutation group acting on the (approximated) roots in a local splitting field, while the automorphism group will implicitly use the roots of f in K. As a consequence, the actual size of the Galois group is of much less importance for Galois group computations, while being critical for automorphism group computations. On the other hand, automorphisms can be applied directly to elements and ideals, while elements of the Galois group actually do not act on the field.

In comparison to most older implementations, the Galois group computation in `OSCAR` does not use any tables of precomputed data, hence, subject to runtime and memory constrains, can be applied to fields and polynomials of arbitrary degree, as well as to reducible polynomials.

6 Examples

We end this chapter with some `OSCAR` applications to a few well-known problems from algebraic number theory.

6.1 Class Group Heuristics

One of the striking applications of computational tools for algebraic number theory has been experiments with arithmetic properties in families of number fields or related objects. The tabulation of inequivalent binary quadratic forms in certain discriminant ranges carried out by Gauß in article 303 of his "Disquisitiones" in 1801, which led him to formulate various conjectures about class numbers, is one of the earliest works in this direction.

Here we consider the distribution of class groups in families of quadratic fields, which goes back to seminal work of Cohen–Lenstra [CL84]. More precisely, for $X \in \mathbb{R}$, $X > 0$, we will consider the family

$$\mathcal{F}(X) := \{K/\mathbb{Q} \text{ real quadratic}, \ |\mathrm{disc}(K)| \le X\}$$

of real quadratic fields with bounded discriminant inside a fixed algebraic closure of $\mathbb{Q}$. If p is an odd prime and A a finite p-group, then the Cohen–Lenstra heuristic predicts that

$$\lim_{X\to\infty} \frac{|\{K \in \mathcal{F}(X) \mid \mathrm{Cl}(K)[p^\infty] \cong A\}|}{|\mathcal{F}(X)|} = \frac{w_p}{|A| \cdot |\mathrm{Aut}(A)|},$$

where $w_p = \prod_{k\ge 2}(1 - p^{-k})$ and $\mathrm{Cl}(K)[p^\infty]$ is the Sylow p-subgroup of the class group. One can interpret the left hand side as the probability that the Sylow p-subgroup of the class group of a random real quadratic field is isomorphic to A. Our aim is to numerically investigate this conjecture for $p = 5$ and $X = 10^6$. (Some of the computations described below require a few minutes of runtime. At the expense of working with a smaller data set, one can choose $X = 10^5$ to obtain weaker results more quickly.) We begin by computing the set of quadratic fields $\mathcal{F}(X)$ and their class groups.

```
julia> fields = [d for d in 1:10^6 if
                 d != 1 && is_fundamental_discriminant(d)];

julia> length(fields)
303957

julia> clgrps = [class_group(quadratic_field(d, cached = false)[1])[1]
                 for d in fields];
```

To obtain the statistics, we will make use of the Julia package `Tally`:

```
julia> using Tally;

julia> tally(clgrps,
             by = x -> sylow_subgroup(x, 5)[1],
             equivalence = (x, y) -> is_isomorphic(x, y)[1])
Tally with 303957 items in 4 groups:
[Z/1]      | 291400 | 95.869%
[Z/5]      |  12324 |  4.055%
```

```
[Z/25]     |     230 |  0.076%
[(Z/5)^2] |       3 |  0.001%
```

We now determine the proportions predicted by the conjecture. We first construct the abelian groups (by specifying elementary divisors) and then determine the values.

```
julia> ab_grps = abelian_group.([[1], [5], [25], [5, 5]]);

julia> w = prod(1 - 1/5.0^i for i in 2:1000)
0.9504159948390403

julia> [ 100 * w/(1.0 * (order(A) *
          order(automorphism_group(A)))) for A in ab_grps]
4-element Vector{BigFloat}:
 95.041[...]
  4.752[...]
  0.190[...]
  0.007[...]
```

Thus we see that using the first 300 000 fields, the numbers match approximately. We now want to have a closer look at how the discrepancy between the predicted and actual value evolves with increasing X. For the sake of simplicity, we consider the probability that the class number of a real quadratic field is divisible by 5. The Cohen–Lenstra heuristic predicts this value to be

$$\lim_{X\to\infty} \frac{|\{K \in \mathcal{F}(X) \mid 5 \text{ divides } h_K\}|}{|\mathcal{F}(X)|} = 1 - w_5 = 1 - \prod_{k=2}^{\infty}\left(1 - \frac{1}{5^k}\right).$$

We now collect the predicted and actual proportions for various values of X.

```
julia> clnumbs = order.(clgrps);

julia> x = range(1, length(clnumbs), 1000);

julia> values = [count(c -> is_divisible_by(c, 5),
                        clnumbs[1:Int(ceil(step))])/Int(ceil(step)) for
                 step in x];
```

The plot of this result is presented in Figure 4 for $X \leq 10^7$ (for this bound, the computation of the class groups takes a day). Note that we plot the proportion against $|\mathcal{F}(X)|$ instead of X.

```
julia> pr = 1 - prod(1 - 1/5.0^i for i in 2:1000)
0.04958400516095973

julia> using Plots;
       plot(x, [values fill(pr, length(x))], linewidth = 2,
            xlabel = "Number of fields",
            label = ["Proportion" "Prediction"])
```

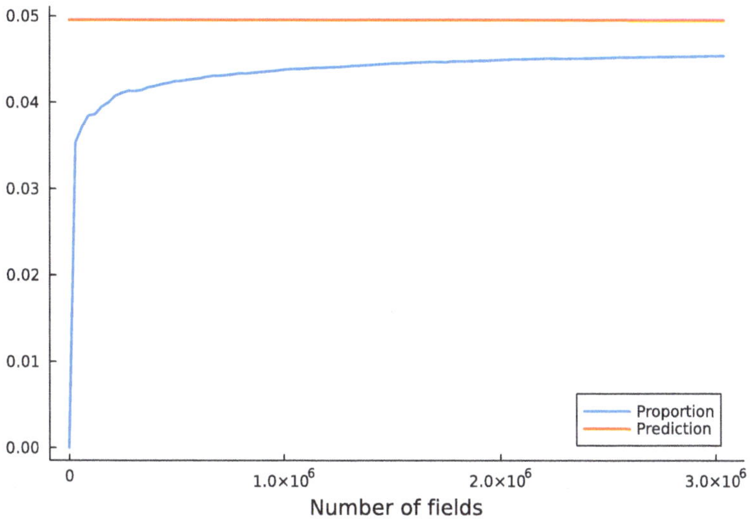

Fig. 4: Predicted and actual proportion of real quadratic fields with class number divisible by 5 with discriminant bounded by 10^7.

Finally we note that the Cohen–Lenstra heuristic has been extended to arbitrary number fields and relative extensions by Cohen–Martinet [CM87] with modifications due to Malle [Mal08; Mal10]; see also [SW23] for some recent progress.

6.2 Galois Module Structure

For a normal extension $K/\mathbb{Q}$, the Galois group $G = \mathrm{Gal}(K/\mathbb{Q})$ (which we interpret simply as the automorphism group $\mathrm{Aut}_{\mathbb{Q}}(K)$) acts naturally on many arithmetic objects derived from or defined over K. The theory of Galois modules is an area of number theory concerned with the study of such actions.

The Normal Basis Theorem

The simplest example of a Galois module is the additive group of the field K itself, on which G acts naturally as automorphism group. In this case the question about the Galois module structure is answered by the celebrated normal basis theorem, which asserts that there exists an element $\beta \in K$ such that $(\sigma(\beta))_{\sigma \in G}$ is a basis of K when considered as a $\mathbb{Q}$-vector space. We refer to such an element β as a *normal basis generator*. There are different ways to find such an element for a given number field. We present a simple randomized approach and illustrate it with $K = \mathbb{Q}(\alpha)$, where $\alpha^4 + 4\alpha^2 + 2 = 0$. Note that $K/\mathbb{Q}$ is a Galois extension. Since there are abundantly many normal basis generators, the idea is to pick a random element $\beta \in K$ and verify the condition of the definition:

```
julia> K, a = number_field(x^4 + 4*x^2 + 2, "a");

julia> b = rand(K, -10:10) # random element with coefficients
                           # between -10 and 10
5*a^3 - 7*a^2 + 8*a - 8

julia> A, mA = automorphism_group(K);
```

Now construct the list $(\sigma(\beta))_{\sigma \in G}$:

```
julia> list = [mA(s)(b) for s in A]
4-element Vector{AbsSimpleNumFieldElem}:
 5*a^3 - 7*a^2 + 8*a - 8
 -7*a^3 + 7*a^2 - 26*a + 20
 -5*a^3 - 7*a^2 - 8*a - 8
 7*a^3 + 7*a^2 + 26*a + 20
```

To verify that these elements are $\mathbb{Q}$-linearly independent, we consider the coordinates with respect to the canonical basis:

```
julia> X = matrix(QQ, coordinates.(list))
[-8     8   -7    5]
[20   -26    7   -7]
[-8    -8   -7   -5]
[20    26    7    7]
```

Thus, we have $X \cdot (1, \alpha, \alpha^2, \alpha^3)^t = (\sigma_1(\beta), \sigma_2(\beta), \sigma_3(\beta), \sigma_4(\beta))^t$ and it remains to check whether X is invertible.

```
julia> det(X) != 0
true
```

This proves that $\beta = -2\alpha^3 + 10\alpha^2 - 2\alpha$ is indeed a normal basis generator of K. Note that not every primitive element is a normal basis generator. For this given quartic field K, the element α is, for example, not a normal basis generator:

```
julia> det(matrix(QQ, [coordinates(mA(s)(a)) for s in A]))
0
```

Finally we mention that the normal basis theorem has an important module theoretic interpretation. It asserts that the number field K is a free $\mathbb{Q}[G]$-module of rank 1 and any element $\beta \in K$ with $K = \mathbb{Q}[G] \cdot \beta$ is a normal basis generator.

Normal Integral Bases

Using the same setup as before, we now consider the integral structure of a normal number field $K/\mathbb{Q}$. Since integrality is preserved under automorphisms, the Galois group $G = \mathrm{Gal}(K/\mathbb{Q})$ acts on $\mathcal{O}_K$, turning the ring of integers $\mathcal{O}_K$ into a $\mathbb{Z}[G]$-module. In view of the normal basis theorem it is only natural to ask whether $\mathcal{O}_K$ is a free $\mathbb{Z}[G]$-module (necessarily of rank 1). An affirmative answer is equivalent to the existence of a *normal integral basis generator*, that is, an element $\beta \in \mathcal{O}_K$ such that $(\sigma(\beta))_{\sigma \in G}$ is a $\mathbb{Z}$-basis of $\mathcal{O}_K$. While it is in general a hard problem to decide freeness of $\mathbb{Z}[G]$-modules, OSCAR provides some functionality to work with such objects. We want to investigate whether the quartic number field $K = \mathbb{Q}(\alpha)$ defined before has a normal integral basis.

To this end, we first construct a $\mathbb{Q}[G]$-linear isomorphism $f\colon K \to V$, where V is the regular $\mathbb{Q}[G]$-module (the existence of which is guaranteed by the normal basis theorem)

```
julia> V, f = galois_module(K);
```

Next, we translate the ring of integers to a $\mathbb{Z}[G]$-submodule M of V and ask whether it is free or not as a $\mathbb{Z}[G]$-module.

```
julia> OK = ring_of_integers(K);

julia> M = f(OK);

julia> is_free(M)
false
```

Note that we could have already seen this by investigating whether M is locally free at 2, that is, whether $M \otimes \mathbb{Z}_2$ is a free $\mathbb{Z}_2[G]$-module.

```
julia> fl = is_locally_free(M, 2)
false
```

In fact, the local non-freeness is equivalent to an arithmetic property of $\mathcal{O}_K$. Noether's criterion asserts that $\mathcal{O}_K$ is locally free at all primes p if and only if $K/\mathbb{Q}$ is tamely ramified. The latter condition means that for all ramified primes p and prime ideals $\mathfrak{p}$ lying above p, the ramification index $e(\mathfrak{p}/p)$ is not divisible by p. This is a criterion that can easily be checked:

```
julia> prime_decomposition_type(OK, 2)
1-element Vector{Tuple{Int64, Int64}}:
 (1, 4)
```

Thus $2 \cdot \mathcal{O}_K = \mathfrak{p}_2^4$ for some prime ideal $\mathfrak{p}_2$, $2 \mid 4 = e(\mathfrak{p}_2/2)$ and the extension is therefore not tamely ramified. If we instead consider the C_4-extension $K = \mathbb{Q}(\alpha)$, with $\alpha^4 - \alpha^3 + \alpha^2 - \alpha + 1 = 0$, then $K/\mathbb{Q}$ is tamely ramified.

```
julia> K, a = number_field(x^4 - x^3 + x^2 - x + 1, "a");

julia> is_tamely_ramified(K)
true
```

We do the same computation as before, but in this case obtain a free generator of the $\mathbb{Z}[G]$-module.

```
julia> V, f = galois_module(K); OK = ring_of_integers(K); M = f(OK);

julia> fl, c = is_free_with_basis(M); # the elements of c form a basis

julia> b = preimage(f, c[1]) # the element may be different per session
a
```

We can verify that the preimage of the element c_1 under the map f is a normal integral basis generator. As for the normal basis generator, we check whether the matrix of coordinates with respect to an integral basis is invertible over $\mathbb{Z}$:

```
julia> A, mA = automorphism_group(K);

julia> X = matrix(ZZ, [coordinates(OK(mA(s)(b))) for s in A])
[0   1   0   0]
[0   0   0   1]
[1  -1   1  -1]
[0   0  -1   0]

julia> det(X)
1
```

One question we can ask is whether not being locally free is the only obstruction for a number field to have a normal integral basis. For abelian extensions of $\mathbb{Q}$, the affirmative answer to this question is the content of the theorem of Hilbert–Speiser, which asserts that an abelian extension $K/\mathbb{Q}$ has a normal integral basis if and only if $K/\mathbb{Q}$ is tamely ramified, or equivalently, there exists $n \in \mathbb{N}$ odd and square-free such that $K \subseteq \mathbb{Q}(\zeta_n)$.

For arbitrary extensions, the answer is in general a (complicated) no. We next want to find a tamely ramified number field $K/\mathbb{Q}$ without normal integral basis. Since in this case $\mathcal{O}_K$ must be a locally free $\mathbb{Z}[G]$-module by Noether's criterion, the obstruction to the existence of a normal integral basis must come from the existence of locally free $\mathbb{Z}[G]$-modules which are not free. An invariant capturing this obstruction (in part) is provided by the locally free class group $\mathrm{Cl}(\mathbb{Z}[G])$, which is a finite abelian group and a generalization of the class group of Dedekind domains to the non commutative setting. A necessary criterion for finding a locally free but non-free $\mathbb{Z}[G]$-module is $\mathrm{Cl}(\mathbb{Z}[G]) \neq 1$. Let us find the first non-abelian group where this is happening.

```
for i in 1:8, j in 1:number_of_small_groups(i)
  G = small_group(i, j)
  if is_abelian(G)
    continue
  end
  QG = QQ[G]
  ZG = integral_group_ring(QG)
```

```
  println(i, " ", j, " ", describe(G), ": ",
  ↪  locally_free_class_group(ZG))
end
```

This yields:

```
6 1 S3: Z/1
8 3 D8: Z/1
8 4 Q8: Z/2
```

We conclude that $G = \mathrm{Q}_8$ is the smallest group with this property. Next we construct a few candidate fields and hope that we find what we are looking for. Since Q_8 has the derived series $1 \subseteq \mathrm{C}_2 \subseteq \mathrm{Q}_8$ with quotient $\mathrm{Q}_8/\mathrm{C}_2 \cong \mathrm{C}_2 \times \mathrm{C}_2$, we can construct Q_8-fields as quadratic extensions of $\mathrm{C}_2 \times \mathrm{C}_2$-fields. As a starting field we take the $\mathrm{C}_2 \times \mathrm{C}_2$-extension $k/\mathbb{Q}$ defined by $x^4 - 13x^2 + 16$. On top of this field, using class field theory, we construct all tamely ramified C_2-extensions K/k which are normal over $\mathbb{Q}$ and satisfy $|\mathrm{disc}(K)| \leq 10^{13}$ as well as $\mathrm{Gal}(K/\mathbb{Q}) \cong \mathrm{Q}_8$.

```
k, = number_field(x^4 - 13*x^2 + 16)
candidates = []
for F in abelian_normal_extensions(k, [2], ZZ(10)^13)
  K, = absolute_simple_field(number_field(F))
  if !is_tamely_ramified(K)
    continue
  end
  if !is_isomorphic(galois_group(K)[1], quaternion_group(8))
    continue
  end
  push!(candidates, K)
end
```

We can now investigate the second candidate:

```
julia> length(candidates)
2

julia> K = candidates[2]; V, f = galois_module(K); OK =
↪  ring_of_integers(K); M = f(OK);

julia> fl = is_free(M)
false

julia> defining_polynomial(K)
x^8 + 105*x^6 + 3465*x^4 + 44100*x^2 + 176400
```

Thus we have found a tamely ramified Q_8-extension $K/\mathbb{Q}$ without normal integral basis. In fact, with a little more effort we could prove that this is in fact the smallest Q_8-extension with this property. For this we would have to consider all biquadratic fields k up to the discriminant bound $10^{\frac{13}{2}}$. For more information on the vast and intriguing subject of Galois module theory of rings of integers, we refer the reader to Fröhlich [Frö83]. Algorithmic aspects of this theory are treated for example in [BJ08; BJ11; HJ20; BHJ22].

Acknowledgements Funded by the Deutsche Forschungsgemeinschaft (DFG, German Research Foundation) – Project-ID 286237555 – TRR 195. The first author also acknowledges funding through MaRDI (Mathematical Research Data Initiative), funded by the Deutsche Forschungsgemeinschaft, project number 460135501, NFDI 29/1 "MaRDI – Mathematische Forschungsdateninitiative" and the state of Rheinland-Pfalz via the Forschungsinitiative and the "SymbTools" project.
The authors are grateful to Henri Johnston for helpful comments and discussions.

References

[BHJ22] W. Bley, T. Hofmann, and H. Johnston. "Computation of lattice isomorphisms and the integral matrix similarity problem". *Forum Math. Sigma* 10 (2022), Paper No. e87, 36. DOI: 10.1017/fms.2022.74.

[BJ08] W. Bley and H. Johnston. "Computing generators of free modules over orders in group algebras". *J. Algebra* 320.2 (2008), pp. 836–852. DOI: 10.1016/j.jalgebra.2008.01.042.

[BJ11] W. Bley and H. Johnston. "Computing generators of free modules over orders in group algebras II". *Math. Comp.* 80.276 (2011), pp. 2411–2434. DOI: 10.1090/S0025-5718-2011-02488-9.

[BJ24] T. Brysiewicz and M. Joswig. "Polyhedral Geometry". In: *The Computer Algebra System OSCAR. Algorithms and Examples.* Ed. by W. Decker, C. Eder, C. Fieker, M. Horn, and M. Joswig. Springer, 2024, pp. 139–164.

[Coh93] H. Cohen. *A course in computational algebraic number theory.* Vol. 138. Graduate Texts in Mathematics. Springer-Verlag, Berlin, 1993, pp. xii+534. DOI: 10.1007/978-3-662-02945-9.

[Coh00] H. Cohen. *Advanced topics in computational number theory.* Vol. 193. Graduate Texts in Mathematics. Springer-Verlag, New York, 2000, pp. xvi+578. DOI: 10.1007/978-1-4419-8489-0.

[CL84] H. Cohen and H. W. Lenstra Jr. "Heuristics on class groups of number fields". In: *Number theory, Noordwijkerhout 1983.* Vol. 1068. Lecture Notes in Math. Springer, Berlin, 1984, pp. 33–62. DOI: 10.1007/BFb0099440.

[CM87] H. Cohen and J. Martinet. "Class groups of number fields: numerical heuristics". *Math. Comp.* 48.177 (1987), pp. 123–137. DOI: 10.2307/2007878.

[FHHJ17] C. Fieker, W. Hart, T. Hofmann, and F. Johansson. "Nemo/Hecke: computer algebra and number theory packages for the Julia programming language". In: *Proceedings of the 2017 ACM International Symposium on Symbolic and Algebraic Computation.* ACM, New York, 2017, pp. 157–164. DOI: 10.1145/3087604.3087611.

[FHJ22] C. Fieker, T. Hofmann, and M. Joswig. "Computing Galois groups of Ehrhart polynomials in OSCAR". *Sém. Lothar. Combin.* 86B (2022), Art. 87, 9.

[Frö83] A. Fröhlich. *Galois module structure of algebraic integers.* Ergebnisse der Mathematik und ihrer Grenzgebiete (3). Springer-Verlag, Berlin, 1983, pp. x+262. DOI: 10.1007/978-3-642-68816-4.

[Gau86] C. F. Gauss. *Disquisitiones arithmeticae.* Translated and with a preface by Arthur A. Clarke, Revised by William C. Waterhouse, Cornelius Greither and A. W. Grootendorst and with a preface by Waterhouse. Springer-Verlag, New York, 1986.

[HJ20] T. Hofmann and H. Johnston. "Computing isomorphisms between lattices". *Math. Comp.* 89.326 (2020), pp. 2931–2963. DOI: 10.1090/mcom/3543.

[Lan94] S. Lang. *Algebraic number theory.* Second. Vol. 110. Graduate Texts in Mathematics. Springer-Verlag, New York, 1994, pp. xiv+357. DOI: 10.1007/978-1-4612-0853-2.

[Mal08] G. Malle. "Cohen–Lenstra heuristic and roots of unity". *J. Number Theory* 128.10 (2008), pp. 2823–2835. DOI: 10.1016/j.jnt.2008.01.002.

[Mal10] G. Malle. "On the distribution of class groups of number fields". *Experiment. Math.* 19.4 (2010), pp. 465–474. DOI: 10.1080/10586458.2010.10390636.

[Mar18] D. Marcus. *Number fields.* Universitext. Springer, Cham, 2018, pp. xviii+203. DOI: 10.1007/978-3-319-90233-3.

[Neu99] J. Neukirch. *Algebraic number theory.* Vol. 322. Grundlehren der mathematischen Wissenschaften. Springer-Verlag, Berlin, 1999, pp. xviii+571. DOI: 10.1007/978-3-662-03983-0.

[PZ97] M. Pohst and H. Zassenhaus. *Algorithmic algebraic number theory.* Vol. 30. Encyclopedia of Mathematics and its Applications. Cambridge University Press, 1997, pp. xiv+499.

[SW23] W. Sawin and M. M. Wood. *Conjectures for distributions of class groups of extensions of number fields containing roots of unity.* 2023. arXiv: 2301.00791 [math.NT].

[Sta73] R. P. Stauduhar. "The determination of Galois groups". *Math. Comp.* 27 (1973), pp. 981–996. DOI: 10.2307/2005536.

[Wil95] A. Wiles. "Modular elliptic curves and Fermat's last theorem". *Ann. of Math. (2)* 141.3 (1995), pp. 443–551. DOI: 10.2307/2118559.

Group Theory

Max Horn and Claus Fieker

Abstract We describe how to do group theory in OSCAR and illustrate this with numerous examples.

Key words: Group theory, character theory, representation theory

1 Introduction

Groups occur everywhere in mathematics, and are thus also central to computer algebra. In fact, groups were among the first objects to be investigated mechanically. The famous Todd-Coxeter algorithm for enumerating cosets in finitely-presented groups [TC36] was implemented in Manchester in the 1950s [Ban56] on an electronic computer (as opposed to human computers, to which the term used to refer), and elsewhere a few years later (see e.g. [Fel61]; for a more comprehensive overview of this early history, see [Can69]).

For a beginner, or in fact any non-specialist, groups are difficult in computer algebra as they come in many different representations depending both on the intended computations as well as the origin of the group. While OSCAR tries to make good choices automatically, it is still important to know what some of the options are. Some standard ways to represent groups include:

1. via transformations of some "space", such as
   - permutations of a finite set, leading to *permutation groups*;
   - linear maps acting on a vector space, leading to *matrix groups*;

Claus Fieker
RPTU Kaiserslautern-Landau, e-mail: `claus.fieker@rptu.de`

Max Horn
RPTU Kaiserslautern-Landau, e-mail: `max.horn@rptu.de`

W. Decker et al. (eds.), *The Computer Algebra System OSCAR*,
Algorithms and Computation in Mathematics 32,
https://doi.org/10.1007/978-3-031-62127-7_4

2. via generators and relations, leading to *finitely presented groups*;
3. via a multiplication table.

To illustrate these different avenues, consider a regular pentagon in the real plane, centered on the origin and with vertices labeled 1 to 5 (see Figure 1). Its symmetries are the reflections σ_i at the symmetry axes through i for

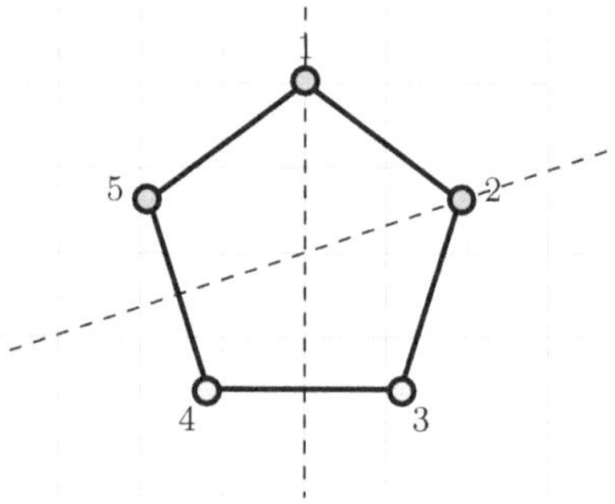

Fig. 1: A regular pentagon.

$i \in \{1, \ldots, 5\}$, as well as rotations by multiples of 72 degrees around the center point. It is well-known that they form a group under composition, which has order 10, with 5 reflections and 5 rotations, and is called the *dihedral group* D_{10}. (We denote the group of symmetries of a regular n-gon as D_{2n}, not as D_n.)

We now show some different ways of constructing this symmetry group.

1.1 Permutation Groups

We can interpret the symmetries in purely combinatorial terms by observing how they permute the vertices of the pentagon. E.g. σ_2 leaves vertex 2 invariant and interchanges 1 with 3, and 4 with 5. We can thus describe our group as a **permutation group**. Here we use that σ_1 and σ_2 already generate it. [1]

```
julia> G_perm = @permutation_group(5, (2,5)(3,4), (1,3)(4,5))
Permutation group of degree 5
```

We can easily verify that it has the correct order and isomorphism type.

```
julia> order(G_perm)
10

julia> describe(G_perm)
"D10"
```

[1] Note the use of `@permutation_group` instead of `permutation_group`, which is a *macro* and allows us to enter the generators conveniently in cycle notation.

The product of two adjacent reflections should give a rotation, and indeed we can multiply the first and second generator (corresponding to our reflections σ_1 and σ_2) to get a 144-degree clockwise rotation. Note that in OSCAR the action is from the right.

```
julia> r = G_perm[1] * G_perm[2]
(1,3,5,2,4)
```

Not every permutation of the vertices corresponds to a symmetry of our pentagon.

```
julia> g = perm([1,2,3,5,4])  # 'tabular' notation
(4,5)

julia> g in G_perm
false
```

There are several other ways to enter permutations, here are some:

```
julia> G_perm([2,1,5,4,3])  # 'tabular' notation with explicit parent
(1,2)(3,5)

julia> cperm([1,2],[4,5])   # cycle decomposition
(1,2)(4,5)

julia> @perm (1,2)(4,5)     # cycle notation via macro
(1,2)(4,5)
```

1.2 Matrix Groups

A completely different approach is to use the embedding of the pentagon into the real plane $\mathbb{R}^2$, and then describe its automorphisms as linear transformations via 2×2 matrices. In Figure 1 the pentagon can be thought of as centered on the origin $(0, 0)$, and we can assign coordinates $(0, 1)$ to vertex 1. Then σ_1 is reflection in the y-axis. To describe a second reflection we would need coordinates for at least one of the other vertices which we don't have. Instead we follow a different approach and exploit that D_{10} is also generated by σ_1 together with a rotation by 72 degree, i.e., $\frac{2\pi}{5}$ radians. We can simply write down the corresponding 2×2 rotation matrix. To do so we work over an algebraic closure of $\mathbb{Q}$.

```
julia> K = algebraic_closure(QQ)
Field of algebraic numbers

julia> e = one(K)
Root 1.00000 of x - 1

julia> s, c = sinpi(2*e/5), cospi(2*e/5)
(Root 0.951057 of 16x^4 - 20x^2 + 5, Root 0.309017 of 4x^2 + 2x - 1)

julia> mat_rot = matrix([ c -s ; s c ]);
```

Together with the matrix for σ_1 we now have generators for our group:

```
julia> mat_sigma1 = matrix(K, [ -1 0 ; 0 1 ]);

julia> G_mat = matrix_group(mat_rot, mat_sigma1)
Matrix group of degree 2
  over field of algebraic numbers
```

This group is isomorphic to the one we constructed in the previous section:

```
julia> is_isomorphic(G_mat, G_perm)
true
```

We can now also compute coordinates for the remaining vertices by computing the orbit of vertex 1 under the group action.

```
julia> p = K.([0,1]);  # coordinates of vertex 1, expressed over K

julia> orb = orbit(G_mat, *, p)
G-set of
  matrix group of degree 2 over QQBar
  with seeds Vector{QQBarFieldElem}[[Root 0 of x, Root 1.00000 of x - 1]]
```

The resulting object is a G-set (more on that later), and at first no actual orbit computation takes place. We can ask OSCAR to actually enumerate the orbit by using the standard Julia function `collect`.

```
julia> pts = collect(orb)
5-element Vector{Vector{QQBarFieldElem}}:
 [Root 0 of x, Root 1.00000 of x - 1]
 [Root 0.951057 of 16x^4 - 20x^2 + 5, Root 0.309017 of 4x^2 + 2x - 1]
 [Root 0.587785 of 16x^4 - 20x^2 + 5, Root -0.809017 of 4x^2 + 2x - 1]
 [Root -0.951057 of 16x^4 - 20x^2 + 5, Root 0.309017 of 4x^2 + 2x - 1]
 [Root -0.587785 of 16x^4 - 20x^2 + 5, Root -0.809017 of 4x^2 + 2x - 1]
```

Often a picture says more than a thousand words (or digits), so we ask OSCAR to draw one for us via `visualize(convex_hull(pts))`, with the result in Figure 2. (This is actually a 3D visualization, which is why the label 5 ends up "behind" the pentagon. Note also that the labels in Figure 2 correspond to the positions in `pts`, which do not coincide with the labels in Figure 1.) All in all we successfully managed to reconstruct the pentagon we started with.

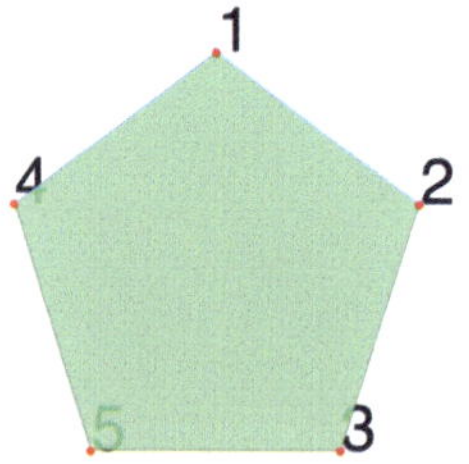

Fig. 2: Visualisation of point orbit.

1.3 Generic Groups

OSCAR also supports **generic groups**, where the user provides arbitrary objects as generators (or group elements) and specifies a binary function implementing the multiplication. The result is represented via a multiplication table – as such this is only suitable for relatively small examples. We again consider our pentagon example, with K the algebraic closure of $\mathbb{Q}$.

```
julia> R2 = vector_space(K, 2); # the "euclidean" plane over K

julia> sigma_1 = hom(R2, R2, [-R2[1], R2[2]])
Module homomorphism
  from vector space of dimension 2 over field of algebraic numbers
  to vector space of dimension 2 over field of algebraic numbers

julia> rot = hom(R2, R2, mat_rot);

julia> G_generic = generic_group(closure([rot, sigma_1], *), *)[1]
Generic group of order 10 with multiplication table
```

While generic groups are limited as explained above, in many basic applications and in teaching, they are perhaps the most natural way of realising a group on a computer. And in any case it is possible to convert them to other representations, e.g. into a permutation group.

```
julia> permutation_group(G_generic)
Permutation group of degree 10

julia> is_isomorphic(ans, G_perm)
true
```

1.4 Finitely Presented Groups

The symmetry group of the pentagon can also be described as a **finitely presented group**, i.e., by generators and relations. One way to get such a group is via conversion of an already known group:

```
julia> G_fp = fp_group(G_perm)
Finitely presented group of order 10

julia> gens(G_fp)
2-element Vector{FPGroupElem}:
 F1
 F2

julia> relators(G_fp)
3-element Vector{FPGroupElem}:
 F1^2
 F1^-1*F2*F1*F2^-4
 F2^5
```

We may also construct finitely presented groups as quotients of free groups. Here we use the well-known presentation for $G \cong D_{10}$ as a reflection group respectively Coxeter group, generated by two elements of order 2, corresponding to two reflections preserving our pentagon.

```
julia> F = free_group(2)
Free group of rank 2

julia> G_cox, _ = quo(F, [F[1]^2, F[2]^2, (F[1]*F[2])^5])
(Finitely presented group, Hom: F -> G_cox)

julia> is_isomorphic(G_cox, G_perm)
true
```

Small changes in a presentation can have great effects. As an example, here are two more Coxeter groups whose presentation differs in only one place, yet they exhibit quite different properties.

```
julia> F = free_group(:a, :b, :c); a,b,c = gens(F);

julia> G, _ = quo(F, [a^2, b^2, c^2, (a*b)^3, (a*c)^2, (b*c)^3])
(Finitely presented group, Hom: F -> G)

julia> H, _ = quo(F, [a^2, b^2, c^2, (a*b)^3, (a*c)^3, (b*c)^3])
(Finitely presented group, Hom: F -> H)

julia> is_finite(G)
true
```

The presentations for G and H are quite similar, and `OSCAR` easily determines that G is finite. But trying to do the same for H does not terminate. In fact H is infinite but the default algorithms are unable to establish this. So how can we do it?

The problem is that while fp-groups are the "most natural" way to present groups in many applications, it is well known, that in general even to test if a given fp-group is trivial, let alone if it is finite, is undecidable [Nov55; Boo58]. On the other hand, for many groups arising from applications, this can be decided quickly. For mathematicians who are not experts in group theory, it can often already be helpful to get *some* information about a given fp-group occurring in their research. The function `describe` we already saw in use before makes a best effort to produce at least some helpful information about a given group, even though this is mainly impossible in general. Unfortunately for the second group it still fails to produce anything helpful:

```
julia> describe(G)
"S4"

julia> describe(H)
"a finitely presented group"
```

One way to deal with this fundamental problem of fp-groups is to restrict to working with special kinds of presentation of groups. For example *polycyclic groups* (pc-groups) have a specific system of relations that allows for fast computation of normal forms of words, in particular, to decide if a group element is trivial or not. On the other hand, polycyclic groups are always solvable, and hence are not suitable for all applications.

Going back to our group H, we can deal with it by finding a subgroup that is "obviously" infinite, namely its second derived subgroup H''.

```
julia> H1, _ = derived_subgroup(H)
(Finitely presented group, Hom: H1 -> H)

julia> H2, _ = derived_subgroup(H1)
(Finitely presented group, Hom: H2 -> H1)

julia> describe(H2)
"Z x Z"
```

This also establishes that H is actually solvable. The quotient modulo H'' is easily found as

```
julia> describe(quo(H, H2)[1])
"(C3 x C3) : C2"
```

1.5 ... and More

For completeness we mention that of course OSCAR can also directly produce a dihedral group if asked for; by default this produces a pc-group but this can be changed.

```
julia> dihedral_group(10)
Pc group of order 10

julia> dihedral_group(PermGroup, 10)
Permutation group of degree 5
```

There are further types of groups available in OSCAR but for this book chapter we will consider only the types discussed so far.

While a lot of OSCAR's group theory capabilities are inherited from GAP [GAP] (and indeed anything available in GAP can be accessed in OSCAR, see Section 10), one of the compelling advantages of working in OSCAR is that it is very convenient to combine them with top-class algorithms for other areas. We will shortly describe various applications that leverage OSCAR's number theory capabilities. Other applications can be seen elsewhere in this book, e.g. in the chapter on Invariant Theory [DRS24] (cf. page 297), where invariants of permutation and matrix groups are computed by combining methods from commutative algebra with group theory tools.

2 Group Actions

An essential feature of groups is that they *act* on themselves and on many other kinds of sets and objects. We already saw this in the form of symmetry groups which act via linear transformations or vertex permutations on n-gons or other geometric objects. A (right) *action* of a group G on a set Ω is a function $f : \Omega \times G \to \Omega$, $(x, g) \mapsto x.g$ such that $x.1_G = x$ and $(x.g).h = x.(gh)$ for all $x \in \Omega$, $g, h \in G$, where 1_G is the identity element of G.

In OSCAR, actions can be implemented via any `Julia` function matching this signature. However, a pure Julia function does not capture G and Ω. To deal with this OSCAR uses *G-sets* which are `Julia` objects that can be thought of as *being* the domain Ω, but stored together with an acting group G and an action function f. Focusing on the domain instead of the action function better matches how one usually works with and thinks about group actions. On the technical side, combining everything into a single object allows us to cache properties of the action and more.

Some examples of G-sets in OSCAR include:

- the natural G-sets of permutation and matrix groups,
- the set of right cosets of a subgroup U of G,
- conjugacy classes of elements or of subgroups of G,
- vertex or face orbits of the action of a symmetry group on a polyhedron.

Here is the G-set of the natural action of a permutation group:

```
julia> gset(alternating_group(4))
G-set of
  Alt(4)
  with seeds 1:4
```

A slightly more advanced example is the G-set of (right) cosets Ug, $g \in G$, of some subgroup U of G.

```
julia> G = dihedral_group(6);

julia> U = sub(G, [g for g in gens(G) if order(g) == 2])[1]
Pc group of order 2

julia> r = right_cosets(G, U)
Right cosets of
  pc group of order 2 in
  pc group of order 6
```

Note that r is a G-set. We can query it for G, its elements (i.e. the underlying set Ω) and the action function f.

```
julia> acting_group(r)
Pc group of order 6

julia> collect(r)
3-element Vector{GroupCoset{PcGroup, PcGroupElem}}:
 Right coset of U with representative <identity> of ...
```

```
 Right coset of U with representative f2
 Right coset of U with representative f2^2

julia> action_function(r)
* (generic function with 1736 methods)
```

As we can see from the last output, G acts on r via multiplication from the right, using the normal ($*$) to denote the operation, hence the large number of methods. We can now take any element of G and compute how it permutes the cosets (identified by the indices for which they appear in the list of right cosets) as follows.

```
julia> permutation(r, G[1])
(2,3)
```

This defines an *action homomorphism* $G \to S_3$. If that is all we need then we can also directly ask for it, without first creating a G-set:

```
julia> phi = right_coset_action(G,U)
Group homomorphism
  from pc group of order 6
  to permutation group of degree 3 and order 6

julia> phi(G[1]), phi(G[2])
((2,3), (1,2,3))
```

Equipped with this knowledge we can now write a function for finding a transitive permutation representation of a given group G of minimal degree: every transitive action of a group on a set Ω is equivalent to the action of the group on the cosets of a stabilizer G_a for some $a \in \Omega$. We can thus recover all transitive actions by iterating over all subgroups U of G (up to conjugacy suffices) and checking whether the action of G on the cosets G/U is faithful. Among these pick one with minimal index since $[G : U]$ is the degree of the resulting permutation action. Finally, as an optimization, if G is the symmetric or alternating group of Ω then we really can't do better and don't bother searching for a permutation representation of smaller degree.

Combining all that culminates in the following function, which takes a finite group G as input, computes all faithful transitive actions of G and from among those returns one of minimal degree:

```
julia> function optimal_transitive_perm_rep(G)
         is_natural_symmetric_group(G) && return hom(G,G,gens(G))
         is_natural_alternating_group(G) && return hom(G,G,gens(G))
         cand = []  # pairs (U,h) with U ≤ G and h a map G -> Sym(G/U)
         for C in subgroup_classes(G)
           U = representative(C)
           h = right_coset_action(G, U)
           is_injective(h) && push!(cand, (U, h))
         end
         return argmax(a -> order(a[1]), cand)[2]
       end;
```

We can test it with a small example:

```
julia> U = dihedral_group(8)
Pc group of order 8

julia> optimal_transitive_perm_rep(U)
Group homomorphism
  from pc group of order 8
  to permutation group of degree 4 and order 8
```

Note that `permutation_group` and `isomorphism` also allow conversion of an arbitrary finite group into an isomorphic permutation group, but they make no promise to produce something of minimal degree.

```
julia> isomorphism(PermGroup, U)
Group homomorphism
  from pc group of order 8
  to permutation group of degree 8 and order 8

julia> permutation_group(U)
Permutation group of degree 8 and order 8
```

3 Databases

For many research tasks it can be beneficial to have examples at hand. For this, libraries of (mainly finite) groups with specific properties are invaluable. Access to several of these via the cornerstone system `GAP` is directly available in `OSCAR`:

- perfect groups up to order two million, [Hul22; HP89];
- primitive groups up to degree 8191, via the `GAP` package `PrimGrp` [HRR23];
- "small" groups of order up to 2000 (except those of order 1024), and many infinite families beyond that, via the `GAP` packages `SmallGrp` [BEO23], `SOTGrps` [Pan23], `SglPPow` [VE22];
- transitive groups up to degree 48 (with the caveat that degree 32 and 48 data is huge and has to be downloaded separately; see [HRT22] to get an idea of the tremendous effort it took to compute it), via the `GAP` package `TransGrp` [Hul23];
- the Atlas of Group Representations [Wil+], a collection of permutation and matrix groups related to finite simple groups, via the `GAP` package `AtlasRep` [WPNBB24];
- the library of character tables from the `CTblLib` package [Bre24], which contains among others the character tables from the books [CCNPW85] and [JLPW95] (see for example Section 4).

Other `GAP` data libraries are in principle available in `OSCAR` via the `Julia-GAP` interface (see Section 10), and may get more direct support in the future. Note that also data libraries in `OSCAR` from other areas than group theory can be interesting in the context of groups, for example by considering the automorphism groups of objects, see page 144 in the chapter on Polyhedral Geometry [BJ24] (cf. page 139) for an example

All of these databases are the result of hard work by many researchers over many years. Regretfully, we cannot give a comprehensive bibliography here. However, the manuals of the referenced packages include more comprehensive bibliographies on their origins.

Some of these databases are rather large, e.g. the transitive groups of degree 48 take up 30 gigabytes of disk space. By leveraging the power of `Julia` we can transparently download additional data files on demand from the internet. `Julia` also provides general facilities to manage such "artifacts" which we already use for some of these database. In addition, `Julia` provides convenient interfaces to "traditional" database systems (such as PostgreSQL or MongoDB) which we are actively working on integrating into future OSCAR releases to provide more fine grained and flexible access to large datasets. For example this should eventually make it possible to execute queries such as "How many transitive groups of degree 48 have point stabilizer isomorphic to a transitive group" on a server, then wait a bit and finally get a simple integer value as result.

As a very small demonstration how one might use these libraries, here we iterate over all transitive groups of degree $3 \leq d \leq 8$ which are not primitive, and print something for each one that admits a transitive permutation representation of smaller degree. For this we use the function `optimal_perm_rep` introduced in the previous section. Here, the notation (d, n) refers to the n-transitive group in degree d available via `transitive_group(d, n)`.

```
julia> for g in all_transitive_groups(degree => 3:8, !is_primitive)
         h = image(optimal_transitive_perm_rep(g))[1]
         if degree(h) < degree(g)
           id = transitive_group_identification(g)
           id_new = transitive_group_identification(h)
           println(id => id_new)
         end
       end
(6, 2) => (3, 2)
(6, 4) => (4, 4)
(6, 7) => (4, 5)
(6, 8) => (4, 5)
(8, 4) => (4, 3)
(8, 13) => (6, 6)
(8, 14) => (4, 5)
(8, 24) => (6, 11)
```

4 Character Theory

For many questions about groups and their representations, characters are an important tool. This section shows some applications of complex and Brauer characters.

4.1 Introduction

Let $\rho\colon G \to \mathrm{GL}(d, K)$ be a group homomorphism, where G is finite, and assume that K is a field of characteristic zero. Then ρ is called an ordinary (matrix) representation of G, and the *character* of ρ is defined as the map

$$\chi_\rho\colon G \to K, g \mapsto \mathrm{Tr}(\rho(g)).$$

The character χ_ρ is a much simpler object than ρ; it can be represented by a vector of values in K that is indexed by the conjugacy classes of elements in G. Still it contains a lot of information about ρ: The degree of χ_ρ (the value at the identity element of G) equals d; one can compute the characteristic polynomials of matrices $\rho(g)$ from χ_ρ plus information in which conjugacy classes the powers of given group elements lie; the kernel of ρ is the normal subgroup of G that is given by the union of conjugacy classes of elements g for which $\chi_\rho(g)$ is equal to d; ρ is absolutely irreducible if and only if the norm of χ_ρ is 1; two ordinary representations are equivalent if and only if their characters are equal.

A group with n conjugacy classes has exactly n absolutely irreducible representations over a suitably large extension of K, their characters form the *(ordinary) character table* of G, which can be represented by a square matrix of field elements; by convention, the rows are the characters, and the columns correspond to the conjugacy classes of elements in G.

To illustrate this we compute the character table of $G = \mathrm{SL}_2(\mathbb{F}_5)$ (also known as $\mathrm{SL}(2,5) = \mathrm{SL}_2(5)$ to group theorists).

```
julia> G = SL(2, 5)
SL(2,5)

julia> T = character_table(G)
```

We chose to omit the regular output of this command, and instead show in Figure 3 a nicely rendered version by using LaTeX code produced using `show(stdout, "text/latex", T)` and some minimal editing.

The rows above the irreducible characters form the header of the table. It starts with the centralizer orders of the conjugacy classes in factorized form, each row is labeled by the prime in question. Then comes a row of conjugacy class names $1a$, $10a$, $10b$, etc. labeling the columns of the table; each class name consists of the element order in the class and distinguishing

2	3	1	1	3	1	1	1	1	2
3	1	.	.	1	.	.	1	1	.
5	1	1	1	1	1	1	.	.	.
	$1a$	$10a$	$10b$	$2a$	$5a$	$5b$	$3a$	$6a$	$4a$
$2P$	$1a$	$5b$	$5a$	$1a$	$5b$	$5a$	$3a$	$3a$	$2a$
χ_1	1	1	1	1	1	1	1	1	1
χ_2	2	$\zeta_5^3+\zeta_5^2+1$	$-\zeta_5^3-\zeta_5^2$	-2	$-\zeta_5^3-\zeta_5^2-1$	$\zeta_5^3+\zeta_5^2$	-1	1	.
χ_3	2	$-\zeta_5^3-\zeta_5^2$	$\zeta_5^3+\zeta_5^2+1$	-2	$\zeta_5^3+\zeta_5^2$	$-\zeta_5^3-\zeta_5^2-1$	-1	1	.
χ_4	3	$-\zeta_5^3-\zeta_5^2$	$\zeta_5^3+\zeta_5^2+1$	3	$-\zeta_5^3-\zeta_5^2$	$\zeta_5^3+\zeta_5^2+1$	.	.	-1
χ_5	3	$\zeta_5^3+\zeta_5^2+1$	$-\zeta_5^3-\zeta_5^2$	3	$\zeta_5^3+\zeta_5^2+1$	$-\zeta_5^3-\zeta_5^2$	.	.	-1
χ_6	4	-1	-1	4	-1	-1	1	1	.
χ_7	4	1	1	-4	-1	-1	1	-1	.
χ_8	5	.	.	5	.	.	-1	-1	1
χ_9	6	-1	-1	-6	1	1	.	.	.

Fig. 3: Character table of $\mathrm{SL}_2(\mathbb{F}_5)$.

letters. Finally, the row labeled by $2P$ shows the 2nd power map, that is, it contains for each column the name of the class containing the 2nd powers of the elements from that class; more power maps may occur once they get computed. The character values themselves are shown as sums of roots of unity, where ζ_n represents the value $\exp(2\pi i/n)$; zeros are abbreviated by dots. Note that since each character value is a sum of roots of unity, the entries of the character table lie in a common abelian number field, in fact $\mathbb{Q}(\zeta_{|G|})$.

The character table of G contains a lot of information about G, some of which are obvious, others are highly non-trivial:

- The number of rows and columns is n;
- if G is abelian, nilpotent, supersolvable, solvable, simple, almost simple, perfect;
- the lattice of normal subgroups of G;
- whether the Sylow p-subgroups of G are abelian, see [MN21].

Each character χ of G is a sum of absolutely irreducible characters, and once we know the character table of G, we can compute their multiplicities in χ.

The point is that computing the character table of G is usually much easier than computing the irreducible representations of G (cf. Section 7). Character-theoretic arguments can be used to avoid dealing with representations at all, or at least to reduce the work with explicit representations.

If the characteristic of K is a prime p that divides $|G|$ then the situation is more involved. Traces of representing matrices are not sufficient to define characters in this case. Instead, one introduces *Brauer characters* that are defined on the p-regular elements of G, and take values in a field of characteristic zero.

4.2 Character Tables Related to the Atlas of Finite Groups

Since a character table can be used to answer many questions about its group, it makes sense to deal with character tables which do not actually store a group object – it is sufficient to know that a given character table belongs to a particular group, and we can study the group just via its table.

The famous Atlas of Finite Groups [CCNPW85] contains several hundred character tables of simple groups and their automorphic and central extensions, among them are all such tables for the 26 sporadic simple groups. The Atlas defines conventions for the names of the simple groups and their maximal subgroups, for the ordering of rows and columns in the character tables, etc.

All character tables from this book are electronically available in a character table library that can be accessed from `OSCAR`, with names that fit to the Atlas names, and with the same ordering of rows and columns as in the book. This approach has its limitations. For example, the function `gmodule` for constructing a representation affording a given character needs a group object associated to the character table, cf. the example in Section 8.

In fact, the character table library contains many more character tables than are shown in the printed Atlas. For example, many character tables of maximal subgroups of simple Atlas groups are available, together with the information how the classes of the subgroup fuse into the classes of the group; this allows one to restrict characters to subgroups and to induce characters from subgroups.

For those library character tables whose groups are contained in the group libraries described in Section 3, this information is precomputed and can be used to create a group object which has the same character table as the given one.

4.3 Application: Generation Questions

It is known that each finite simple group can be generated by two elements. A typical question in this context is from which conjugacy classes one can choose a pair of generators, and a character theoretic approach is as follows: The given group G is generated by elements in its conjugacy classes C_1, C_2 if and only if not all elements of $C_1 \times C_2$ lie in maximal subgroups of G. For any subgroup M of G the permutation character 1_M^G is nonzero at the class C if and only if $C \cap M$ is empty. If all maximal subgroups M of G have the property that 1_M^G is zero on C_1 or C_2 then we see that in fact *each pair* in $C_1 \times C_2$ generates G.

As an example, let us consider the sporadic simple Janko group $G = J_2$. The character tables of G and its maximal subgroups are available in the character table library, and we can compute the permutation characters 1_M^G.

```
julia> t = character_table("J2");

julia> mx = character_table.(maxes(t));

julia> pis = [trivial_character(s)^t for s in mx];
```

The subgroups in only two classes of maximal subgroups (of the indices 100 and 1800 in G) contain elements of order 7, and we see that these maximal subgroups do *not* contain elements of order 5.

```
julia> ords = orders_class_representatives(t);  println(ords)
[1, 2, 2, 3, 3, 4, 5, 5, 5, 5, 6, 6, 7, 8, 10, 10, 10, 10, 12, 15, 15]

julia> ord7 = findall(is_equal(7), ords);

julia> filt = filter(pi -> pi[ord7[1]] != 0, pis);

julia> degree.(filt)
2-element Vector{QQFieldElem}:
 100
 1800

julia> ord5 = findall(is_equal(5), ords);

julia> s = sum(filt);

julia> all(i -> s[i] == 0, ord5)
true
```

Thus we have shown that any pair of elements of the orders 5 and 7 in G is a generating set for G.

Usually the situation is not so straightforward. For example, let us consider the question whether G is generated by a pair of elements of the orders 2 and 3 such that their product has order 7; groups with this property are called *Hurwitz groups*.

We see that only the maximal subgroups of index 1800 contain elements in the conjugacy classes 2b (the second class of elements of order 2 in G) and 7a.

```
julia> (x -> [x["1a"], x["2b"], x["3a"], x["3b"]]).(filt)
2-element Vector{Vector{QQAbElem{AbsSimpleNumFieldElem}}}:
 [100, 0, 10, 4]
 [1800, 20, 0, 6]
```

Let M be such a maximal subgroup of index 1800. It does not contain elements in the class 3a, but this does not help us because no product of a 2b and a 3a element can have order 7.

```
julia> ord3 = findall(is_equal(3), ords);

julia> [class_multiplication_coefficient(t, 3, i, ord7[1]) for i in ord3]
2-element Vector{ZZRingElem}:
 0
 70
```

This leaves the class 3b. We see that G contains pairs of 2b and 3b elements whose product has order 7, and we know that the only maximal subgroups of G that may contain them are those of index 1800; these subgroups have the structure $L_3(2).2$, in the notation defined by the Atlas [CCNPW85], as the `identifier` value of their character table shows.[2]

```
julia> s = mx[findfirst(is_equal(filt[2]), pis)];

julia> identifier(s)
"L3(2).2"
```

From these subgroups, only class 6 lies in class 2b of G, and only the classes 3 and 5 lie in the classes 3b and 7a, respectively, of G. It turns out that the subgroups do not contain pairs of 2b and 3b elements whose order is 7.

```
julia> possible_class_fusions(s, t)
1-element Vector{Vector{Int64}}:
 [1, 2, 5, 6, 13, 3, 12, 14, 14]

julia> class_multiplication_coefficient(s, 6, 3, 5)
0
```

This proves that G is a Hurwitz group.

It should be noted that character-theoretic computations like the above ones are not guaranteed to answer the given questions. Here we were lucky, but otherwise we would have to think about other strategies.

5 Applying Finitely Presented Groups Towards Group Extension

Despite the undecidability of many problems in fp-groups in general, they are an essential tool in many algorithms. As an example, given a (permutation) group G, via some generators, a standard problem is to write an arbitrary element of G as a word in the given generators. This can easily be achieved via fp-groups:

```
julia> G = dihedral_group(PermGroup, 10);

julia> h = epimorphism_from_free_group(G)
Group homomorphism
  from free group of rank 2
  to permutation group of degree 5
```

This constructs, as before, the dihedral group with 10 elements, this time as a permutation group. It is generated by

[2] For non-group theorists: $L_3(2).2$ is a group which has a subgroup of index 2 isomorphic to $L_3(2) \cong \mathrm{GL}_3(\mathbb{F}_2) \cong \mathrm{SL}_3(\mathbb{F}_2) \cong \mathrm{PGL}_3(\mathbb{F}_2) \cong \mathrm{PSL}_3(\mathbb{F}_2)$, in fact $L_3(2).2 \cong \mathrm{PGL}_2(\mathbb{F}_7)$.

```
julia> gens(G)
2-element Vector{PermGroupElem}:
 (1,2,3,4,5)
 (2,5)(3,4)
```

To decompose an element in G into a word in those generators, we can compute a preimage under the epimorphism h (this is in general not unique):

```
julia> g = rand(G)
(1,4)(2,3)

julia> w = preimage(h, g)
x1^-1*x2^-1*x1^-3

julia> map_word(w, gens(G))  # evaluate w at generators to get back g
(1,4)(2,3)
```

Furthermore, the same map h can also be used to find all relations between the generators of G:

```
julia> k, mk = kernel(h);  # mk is a map from k to domain(h)

julia> ngens(k)
11

julia> mk(k[2])            # pick any kernel generator ...
x1*x2^-2*x1^-1

julia> G[1]*G[-2]^2*G[-1]  # ... and it evaluates to the identity
()
```

A second application is building group extensions: given groups U and V any group extension X satisfying a short exact sequence

$$1 \to U \to X \to V \to 1$$

can naturally be constructed as an fp-group, as a quotient of a free group of rank `ngens(U) + ngens(V)`. For split extensions, he relations are then derived from those in U, in V and the operation of V on U, in the general case one needs more information.

```
julia> U = cyclic_group(5); V = cyclic_group(2); F = free_group(2);

julia> q = quo(F, [F[1]^5, F[2]^2*F[1]])[1];

julia> describe(q)
"C10"

julia> q, mq = quo(F, [F[1]^5, F[2]^2, F[-1]*F[-2]*F[1]*F[2]*F[1]^2])
(Finitely presented group, Hom: F -> q)

julia> describe(q)
"D10"
```

For special classes of fp-groups, typically classes where each group element has an efficiently computable unique representative (a *normal form*), the decidability problems vanish. In those classes computations can be extremely efficient.

Notable examples are groups defined by a confluent rewriting system. A particularly nice example for these are *pc-groups*, which are fp-groups given by a special kind of presentation that allows for fast algorithms. On the other hand, pc-groups are always solvable.[3]

To make this concrete, a *finite pc-group* is an fp-group with generators $g_1, \ldots, g_n$ and defining relations of the form

- for all $1 \leq i \leq n$: *power relations* $g_i^{e_i}$ = some word in generators with indices greater than i,
- for every pair $1 \leq i < j \leq n$: *commutator relations* $[g_i, g_j]$ = some word in generators with indices greater than i.

The particular shape leads to normal forms for each group element of the form $g_1^{f_1} \cdots g_n^{f_n}$ with $0 \leq f_i < e_i$. In addition to this obvious advantage, there are also special algorithms for many fundamental problems in group theory available, see e.g. [HEO05, Chapter 8]

In order to use this, the obvious question is to decide if a given group can be converted into a pc-group (or admits a pc-presentation). This is the domain of solvable quotient, p-quotient and nilpotent quotient algorithms, and also related to the matrix group recognition problem: given a group, with limited knowledge, transform it into a more useful presentation.

Next we sketch how to exploit groups with "nice" presentations (such as pc-groups) to compute their cohomology.

6 Computing Cohomology

Let G be a group. Later we will assume it to be finite. In this section we discuss how to compute the second cohomology groups $H^2(G, M)$ for any G-module M.

Note that most implementations heavily restrict which kinds of modules are supported. In contrast, for the methods presented here we do not assume too much about the module: we need homomorphisms, kernels, images and direct sums and products. As such, our methods work easily for

- finitely presented abelian groups,
- (fin. dim.) vector spaces in general,
- but also e.g. modules over group rings.

[3] In fact every pc-group is even polycyclic, a stronger property than solvable. Conversely every polycyclic group is isomorphic to a pc-group.

The basic data type here is `GModule`. Such modules are created by giving a group, a module, and for each generator an automorphism of the module. For some natural G-modules special constructors are provided, for example for matrix representations. E.g. for a finite group we can ask for all irreducible representations in characteristic 0:

```
julia> irreducible_modules(symmetric_group(3))
3-element Vector{GModule}:
 G-module for Sym(3) acting on vector space of dimension 1 over abelian
 ↪  closure of Q
 G-module for Sym(3) acting on vector space of dimension 1 over abelian
 ↪  closure of Q
 G-module for Sym(3) acting on vector space of dimension 2 over abelian
 ↪  closure of Q
```

Or we can construct a module by prescribing an explicit action of the group on some abelian group. E.g. we can construct the permutation module of the symmetric group S_3 over C_2^3

```
julia> G = symmetric_group(3);

julia> A = abelian_group([2,2,2])
(Z/2)^3

julia> M = gmodule(G, [hom(A, A, permuted(gens(A), g)) for g in gens(G)])
G-module for G acting on A
```

Note that a similar construction would work for any abelian group A (with a suitable operation by G). Or, in this specific case, we could just as well have used `vector_space(GF(2), 3)`. Which we use may have implications for the runtime of subsequent computations with that module, but other than that either works fine.

Now that we know some ways to construct actual modules, we turn to cohomology.

6.1 H^2

The computation of $H^2(G, M)$ here is based, similar to [DH21] and [Hol06], on the interpretation as group extensions: each element in $H^2(G, M)$ corresponds to an extension

$$1 \to M \to X \to G \to 1$$

Elements in X can be represented as pairs (g, m) with the group law given by a 2-cochain $\sigma : G^2 \to M$ as

$$(g, m)(h, n) := (gh, m^h + n + \sigma(g, h))$$

The associativity of the group implies that σ has to satisfy the cochain condition:

$$\sigma(g, hk)\sigma(h, k) = \sigma(g, h)^k \sigma(gh, k), \quad \forall g, h, k \in G$$

This condition can also be used to find cochains directly. However, naively specifying a cochain would require choosing and storing images for $|G|^2$ many elements. How can we overcome this?

Suppose G is an fp-group, $G = \langle g_1, \ldots, g_n \mid r_1, \ldots, r_s \rangle$, and M is generated by $m_1, \ldots, m_\ell$. Then

$$X = \langle (g_1, 0), \ldots, (g_n, 0), (1, m_1), \ldots, (1, m_\ell) \rangle$$

so we have a generating set of X, consisting of (lifts of) generators for G and (images of) generators of M. The exact sequence implies that any relation in M also holds in X, and any relation in G, evaluated in X has to yield an element of M. We can thus lift the defining relations of M and G to defining relations for X. The resulting presentation is parametrised by elements χ of M^ℓ while each r_i yields a linear condition that has to be satisfied by χ. Conversely every $\chi \in M^\ell$ satisfying the relations induces a presentation for a group X satisfying an exact sequence

$$M \to X \to G \to 1$$

but in general $M \to X$ will not be injective. In order to achieve this, further restrictions on χ are needed. For a general presentation it is not known how to obtain those efficiently, but in some special cases this can be done:

- if the representation is a confluent rewriting system (RWS), then relations from "overlaps" are sufficient;
- in case of a pc-presentation, the "consistency" relations are well known.

It should be noted that pc-group relations are very close to a confluent rewriting system, in fact translating the pc-relations into a (monoid) RWS shows that the consistency relations are exactly those from the overlaps in the RWS.

Let A, B, C be words in the fixed generators of G such that in G we have two relations resp. rewriting rules $AB \to U$ and $BC \to V$. Then these translate to relations in X by adding suitable *tails* $\chi_1, \chi_2 \in M$ to produce

$$(AB, 0) \to (U, \chi_1), \quad \text{and} \quad (BC, 0) \to (V, \chi_2).$$

Applying these to $w = (ABC, 0)$ we get two possible reductions:

$$w \to (U, \chi_1)(C, 0) \to (UC, \chi_1^C), \quad \text{or} \quad w \to (A, 0)(V, \chi_2) \to (AV, \chi_2).$$

Since $AV = UC$ holds in G, we deduce that $\chi_1^C = \chi_2$. Repeating this process for suitable additional relations gives us the missing restrictions on the presentation parameter $\chi \in M^\ell$, and thus ultimately allows for an effective computation of $H^2(G, M)$.

7 Finding Irreducible Representations of Solvable Groups

A central question in group theory is to find linear representations of (finite) groups. In theory, this is an easy task: all absolutely irreducible representations occur as composition factors of the regular representation. This has been used extremely successfully in positive characteristic, via the family of so-called MeatAxe algorithms [Par84; Hol98] and their many refinements. The MeatAxe has been implemented several times and some version is readily available via the `GModule` interface:

```
julia> G = dihedral_group(10);

julia> M = regular_gmodule(GF(7), G);

julia> C = composition_factors_with_multiplicity(M)
3-element Vector{Any}:
 (G-module for G acting on vector space of dimension 1 over prime field
 ↪  of characteristic 7, 1)
 (G-module for G acting on vector space of dimension 1 over prime field
 ↪  of characteristic 7, 1)
 (G-module for G acting on vector space of dimension 4 over prime field
 ↪  of characteristic 7, 2)

julia> [is_absolutely_irreducible(x[1]) for x in C]
3-element Vector{Bool}:
 1
 1
 0
```

We can easily find the splitting field of the third factor, extend scalars, and then split it:

```
julia> phi = embed(GF(7), splitting_field(C[3][1]))
Morphism of finite fields
  from prime field of characteristic 7
  to finite field of degree 2 and characteristic 7

julia> M = extension_of_scalars(C[3][1], phi)
G-module for G acting on vector space of dimension 4 over finite field of
↪  degree 2 and characteristic 7

julia> composition_factors_with_multiplicity(M)
2-element Vector{Any}:
 (G-module for G acting on vector space of dimension 2 over finite field
 ↪  of degree 2 and characteristic 7, 1)
 (G-module for G acting on vector space of dimension 2 over finite field
 ↪  of degree 2 and characteristic 7, 1)
```

And we have all absolutely irreducible representations of G in characteristic 7.

The situation in characteristic 0 is different: while deciding (absolute) irreducibility is possible, splitting is hard, see [Ste12]. However, in the special case of pc-groups, this can be done in `OSCAR`, following [Brü98]: Let $G =$

$\langle g_1, \ldots, g_r \rangle$ be a pc-group, and set $U_i = \langle g_i, \ldots, g_r \rangle$. Then the pc-group definition tells us this is a descending series $G = U_1 \trianglerighteq U_2 \trianglerighteq \cdots \trianglerighteq U_{r+1} = 1$. Furthermore, it is straightforward to arrange for $[U_i : U_{i+1}] = e_i$ to be prime. The core idea is then that a representation: $\phi : U_{i+1} \to \mathrm{GL}(n, K)$ can easily be extended to a representation of U_i: Suppose there is $X \in \mathrm{GL}(n, K)$ such that setting $\phi(g_i) = X$ yields a valid extension of the representation. Then we must have for all $y \in U_{i+1}$ that

$$\phi(y^{g_i}) = \phi(y)^{\phi(g_i)} = \phi(y)^X \quad \Longleftrightarrow \quad \phi(y)X = X\phi(y^{g_i}).$$

Setting $p := e_i$, by definition $g_i^p \in U_{i+1}$, thus $\phi(g_i^p) = X^p \in \phi(U_{i+1})$. One can now verify that if d is any zero of the polynomial $t^p - 1$, then $\phi(g_{i+1}) := dX$ also extends ϕ. For distinct roots d the resulting extended representations are inequivalent.[4]

If on the other hand no such X exists, then inducing ϕ to U_i yields an irreducible representation of degree np over K.

Iterating this procedure yields a complete set of absolutely irreducible representations for G. In those cases where induced representations arise, we can compute which representations from the previous step are conjugate, and then by discarding duplicates ensure that the computed representations are pairwise inequivalent.

Note that in general the representations obtained this way are not defined over fields of minimal degree: Let $G = S_4$ of order 24. We can describe it as a pc-group with generators a, b, c and d, power relations

$$a^2, b^3, c^2, d^2$$

and commutator relations

$$[b, a] = b, \; [c, a] = cd, \; [d, a] = cd, \; [c, b] = cd, \; [d, b] = c, \; [d, c] = 1.$$

We start with the trivial 1-dimensional representation of $\langle c, d \rangle$.

```
julia> G = pc_group(symmetric_group(4));

julia> C = abelian_closure(QQ)[1];

julia> F = free_module(C, 1);

julia> s, ms = sub(G, [G[3], G[4]]);    # subgroup generated by c, d

julia> T = trivial_gmodule(s, F);
```

Next we try to extend this to the second generator b of (relative) order $e_2 = 3$. As ϕ at this point is trivial, X can be chosen as $[1]$. The distinct roots

[4] In order to get all roots, we may need to replace K by a larger field. But as we see, cyclotomic extensions suffice.

d of $t^3 - 1$ are precisely 1, ζ_3 and ζ_3^2. For this example we choose to extend ϕ via $b \mapsto [\zeta_3]$.

```
julia> z = root_of_unity(C, 3);

julia> ss, mss = sub(G, [G[2], G[3], G[4]]);

julia> M = gmodule(ss, [hom(F, F, [z*F[1]]), action(T, s[1]), action(T,
↪  s[2])])
G-module for ss acting on F
```

Now trying to extend further to a, we see that no $X \in \mathrm{GL}(1, K)$ satisfying the conditions from above exists: The commutator relations imply $b^a = b^2$, hence such an X would have to satisfy

$$\phi(b)X = X\phi(b^a) = X\phi(b^2) \iff [\zeta_3]X = X[\zeta_3^2]$$

which is impossible. So instead we apply the induction procedure to obtain a 2-dimensional representation:

```
julia> FF = free_module(C, 2);

julia> zm = 0*matrix(action(M, one(ss)));

julia> im = [hom(FF, FF, [matrix(action(M, x)) zm; zm matrix(action(M,
↪  preimage(mss, mss(x)^G[1])))]) for x in gens(ss)];
```

We obtain the new generator:

```
julia> pushfirst!(im, hom(FF, FF, matrix(C, [0 1; 1 0])));

julia> phi = gmodule(G, im);
```

However, we also have

```
julia> character(phi)
class_function(character table of G, QQAbElem{AbsSimpleNumFieldElem}[2,
↪  0, -1, 2, 0])
```

so the character is rational, and

```
julia> schur_index(ans)
1
```

So, it should be possible to realize this representation over $\mathbb{Q}$ already. Indeed, conjugating by

```
julia> T = matrix(C, [C(1) -z; z+1 z]);
```

The result is indeed rational, even integral.

```
julia> [matrix(x)^T for x in action(phi)]
4-element Vector{AbstractAlgebra.Generic.MatSpaceElem{QQAbElem{AbsSimple⌋
↪  NumFieldElem}}}:
 [1 0; -1 -1]
 [0 1; -1 -1]
 [1 0; 0 1]
 [1 0; 0 1]
```

It is well known that in general minimal degree realisation fields need not be abelian, let alone a full cyclotomic field, so finding the matrix T will in general involve enlarging the field first, see the $\mathrm{SL}_2(\mathbb{F}_5)$ example in the next section.

From a theoretical point of view, the procedure is straight forward, assuming the representation is absolutely irreducible. In this case, the representation defines a central simple algebra, where the center is the character field. Any such algebra, as an element in the Brauer group, can be represented by a 2-cocycle with values in a suitable number field. The representation can be realized over a subfield if and only if the restriction of the 2-cocycle to the fixed group of the subfield is trivial. In this case, explicitly writing the 2-cocycle as a coboundary as the image of some 1-cochain allows one to make this explicit. Algorithmically, this is done using the cohomology described earlier. To give slightly more details: Let $\psi : G \to \mathrm{GL}_n(K)$ be the representation and $\sigma : K \to K$ an automorphism fixing the character field. Then ψ^σ is equivalent to ψ, hence we can find a matrix $X_\sigma \in \mathrm{GL}_n(K)$ such that

$$X_\sigma \psi = \psi^\sigma X_\sigma.$$

Let τ be a second such automorphism, then so is $\sigma\tau$ and there are matrices $X_\tau, X_{\sigma\tau} \in \mathrm{GL}_n(K)$ such that

$$X_\tau \psi = \psi^\tau X_\tau \quad \text{and} \quad X_{\sigma\tau} \psi = \psi^{\sigma\tau} X_{\sigma\tau}.$$

But we also have

$$X_\sigma^\tau X_\tau \psi = X_\sigma^\tau \psi^\tau X_\tau = (X_\sigma \psi)^\tau X_\tau = (\psi^\sigma X_\sigma)^\tau X_\tau = \psi^{\sigma\tau} X_\sigma^\tau X_\tau.$$

Hence $X_\sigma^\tau X_\tau X_{\sigma\tau}^{-1}$ centralizes ψ and so $X_\sigma^\tau X_\tau = \mu_{\sigma,\tau} X_{\sigma\tau}$ for some scalar $\mu_{\sigma,\tau} \in K$. As it turns out, computing this for all (pairs of) automorphisms defines the 2-cycle describing the algebra in the Brauer group.

The techniques from the previous chapter (plus some number theory, see [Fie09]) can be used to find a 1-cochain λ such that $\mu_{\sigma,\tau} = \lambda_\sigma^\tau \lambda_\tau$, thus λ is then used to obtain a 1-cochain of matrices via $X_\sigma \lambda_\sigma$. By Hilbert-90, the 1-cochain of matrices is actually generated by a single element and this element can be used to obtain the representation over the smaller field.

8 Example: Constructing a Representation for $\mathrm{SL}_2(\mathbb{F}_5)$

In this section we return to the perfect group $G = \mathrm{SL}_2(\mathbb{F}_5)$ whose character table we already saw in Section 4, Figure 3. By inspecting this character table we see that there is a unique absolutely irreducible degree 6 representation of G. We now would like to construct it.

The degree 6 character is last in the character table. We compute the corresponding representation over the abelian closure of the rationals.

```
julia> R = gmodule(T[end])
G-module for G acting on vector space of dimension 6 over abelian closure
↪  of Q
```

To make this more manageable (the abelian closure of $\mathbb{Q}$ is a complicated object, much more so than a plain number field), rewrite it over some cyclotomic field (which one is used depends on how the representation was constructed).

```
julia> S = gmodule(CyclotomicField, R)
G-module for G acting on vector space of dimension 6 over cyclotomic
↪  field of order 5
```

The character in question is rational, it has Schur index 2, thus the representation can be written over a quadratic field. But which one? First we try to write it over subfields of the given field $\mathbb{Q}(\zeta_5)$, in particular over a smallest possible subfield:

```
julia> schur_index(T[end])
2

julia> gmodule_minimal_field(S)
G-module for G acting on vector space of dimension 6 over number field of
↪  degree 4 over QQ
```

We see that the representation cannot be written over any subfield – even though there has to be a quadratic field that "works". In order to find out which quadratic field works, we use more theory: the representation also defines an algebra, namely the subalgebra of $\mathbb{Q}(\zeta_5)^{6\times 6}$ generated by the image of the representation. Since the character (and hence the representation) is absolutely irreducible, the algebra is central simple with the center being the character field. Classically, those algebras are elements of the Brauer group of the center ($\mathbb{Q}$). Algebras defined by matrices in a given extension $K/\mathbb{Q}$ are said to be split by K, they are represented by an element of the *relative* Brauer group, $\mathrm{Br}(K/\mathbb{Q})$. So let's try that:

```
julia> B, mB = relative_brauer_group(base_ring(S), character_field(S));

julia> B
Relative Brauer group for cyclotomic field of order 5 over number field
↪  of degree 1 over QQ
```

Classically, the (relative) Brauer group has a well known structure: Let A be central simple over k (split by K), where k and K are number fields, then for each place p of K, we can consider the completion K_p at this place and the (local) algebra $A \otimes K_p$. This is now a central simple algebra over K_p and thus an element in the local Brauer group $\mathrm{Br}(K_p) \cong \mathbb{Q}/\mathbb{Z}$. Any such algebra is uniquely defined by the local invariant. Finally, by Hasse-Brauer-Noether

$$\mathrm{Br}(K/k) \cong \sum_p \mathbb{Q}/\mathbb{Z}$$

where the sum is running over all places p of K; for each place p, the local invariant is used. Since an element can have only finitely many non-zero local invariants, we can represent any such element by the places with non-trivial invariants, and the invariants.

Furthermore, this representation also helps us with the actual problem: the non-trivial invariants describe all possible splitting fields!

Here, `B(S)` performs those steps: it turns the representation `S` into an element of $\mathrm{Br}(\mathbb{Q}(\zeta_5)/\mathbb{Q})$.

```
julia> b = B(S)
Element of relative Brauer group of number field of degree 1 over QQ
  <2, 2> -> 1//2 + Z
  <5, 5> -> 0 + Z
  Complex embedding of number field -> 1//2 + Z
```

The output indicates that we have non-trivial invariants at 2 and the "infinite" place, with both values $1/2+\mathbb{Z}$ of order 2. So any splitting field must be totally complex and all completions over primes above 2 must also have even degree (the local Schur-indices at 2 and ∞ are 2).

By the theorem of Grunwald and Wang, there are (many) cyclic extensions of $\mathbb{Q}$ matching this, so we feed in the Brauer group element to obtain a splitting field:

```
julia> K = grunwald_wang(b)
Class field defined mod (<40, 360>, InfPlc{AbsSimpleNumField,
↪   AbsSimpleNumFieldEmbedding}[Infinite place corresponding to (Complex
↪   embedding corresponding to 1.00 of number field)]) of structure Z/2
```

As this field is a ray class field, that is a certain parametrization of an abelian extension of $\mathbb{Q}$, we have to convert this to an explicit extension of $\mathbb{Q}$:

```
julia> F, _ = absolute_simple_field(number_field(K))
(Number field of degree 2 over QQ, Map: F -> non-simple number field)
```

But now we have the next problem, we have a splitting field F, but the representation is over a completely different field! So we first construct the field L generated by F and the 5-th cyclotomic field.

```
julia> L, _, F_to_L = compositum(base_ring(S), F)
(Number field of degree 8 over QQ, Map: cyclotomic field of order 5 -> L,
↪   Map: F -> L)
```

We write our representation first over L, then over its quadratic subfield F.

```
julia> gmodule_over(F_to_L, gmodule(L, S))
G-module for G acting on vector space of dimension 6 over F
```

And thus we have achieved our goal: an absolutely irreducible degree 6 representation of $\mathrm{SL}_2(\mathbb{F}_5)$ over a number field of minimal degree 2 over $\mathbb{Q}$.

9 Generic Character Tables

We have already seen what a character table of a single finite group is. Some finite groups come in families. For example, the matrix groups $\mathrm{GL}_n(\mathbb{F}_q)$ or $\mathrm{SL}_n(\mathbb{F}_q)$, for $n > 1$ and q a prime power. These groups have many properties in common. It turns out that for a fixed rank (say $n = 2$) it is possible to parametrize the conjugacy classes and irreducible characters of these group in terms of q, and to write this down into a so-called *generic character table.* This was first done by Schur for $\mathrm{SL}_2(\mathbb{F}_q)$.

Generic character tables were so far exclusive to the CHEVIE system (see e.g. [GHLMP96] for details), which involves use of an old version of Maple which heavily restricts the access to this data. Attempts to port this to e.g. GAP in the past were never published.

This changes now with our implementation in Julia for which we imported the data from CHEVIE. Right now this code is developed in a separate repository. To use it, enter within a Julia interactive command-line (REPL):

```
julia> import Pkg; Pkg.add(name="GenericCharacterTables",
↪  version="0.4"); using GenericCharacterTables
```

In the following we replicate parts of [GHLMP96, Section 5.3], and strongly recommend to read that concurrently with the present text.

To start we load the generic character table for $\mathrm{SL}_3(q)$ with $q \not\equiv 1 \pmod 3$. To learn about the origin of the table, we could enter `printinfotab(T)`.

```
julia> T = generic_character_table("SL3.n1")
Generic character table SL3.n1
  of order q^8 - q^6 - q^5 + q^3
  with 8 irreducible character types
  with 8 class types
  with parameters (a, b, m, n)
```

As we can see this table has four parameters in addition to q. The entries of the table are "generalized cyclotomics", that is, linear combinations over $\mathbb{Q}(q)$ of symbolic "roots of unity" depending on the parameters listed above.

```
julia> T[4,4]
(q + 1)*exp(2πi((a*n)//(q - 1))) + exp(2πi((-2*a*n)//(q - 1)))
```

Denoting row 4 of the table by χ_4, we note that is not a single character, but a *character type*, describing a family of characters. We can compute their norms, scalar products, and more. For this demonstration we tensor the second character type χ_2 with itself.

```
julia> h = T[2] * T[2];
```

We may now attempt to decompose this character type h by computing its scalar product with the irreducible character types. This returns a "generic" scalar product, plus a (possibly empty) list of parameter exceptions for which the general result may not hold. For example:

```
julia> scalar_product(T[4], h)
0
With exceptions:
  2*n1 ∈ (q - 1)ℤ
```

This scalar product is 0 except when $q-1$ divides $2n_1$, where n_1 indicates the value of the parameter n for the first factor in the scalar product (i.e. the character type χ_4). For a "generic decomposition" we need to compute all the scalar products and exceptions.

```
julia> for i in 1:8 println("<$i, h> = ", scalar_product(T[i], h)) end
<1, h> = 1
<2, h> = 2
<3, h> = 2
<4, h> = 0
With exceptions:
  2*n1 ∈ (q - 1)ℤ
<5, h> = 0
With exceptions:
  2*n1 ∈ (q - 1)ℤ
<6, h> = 0
With exceptions:
  2*m1 - n1 ∈ (q - 1)ℤ
  m1 - 2*n1 ∈ (q - 1)ℤ
  m1 + n1 ∈ (q - 1)ℤ
  m1 ∈ (q - 1)ℤ
  m1 - n1 ∈ (q - 1)ℤ
  n1 ∈ (q - 1)ℤ
<7, h> = 0
With exceptions:
  n1 ∈ (q - 1)ℤ
<8, h> = 0
With exceptions:
  q*n1 ∈ (q^2 + q + 1)ℤ
  n1 ∈ (q^2 + q + 1)ℤ
  q*n1 + n1 ∈ (q^2 + q + 1)ℤ
```

This suggest a decomposition of $\chi_2 \otimes \chi_2$ into $\chi_1 + 2\chi_2 + 2\chi_3$ "in general". But comparing the respective degrees, we notice a discrepancy:

```
julia> degree(linear_combination([1,2,2],[T[1],T[2],T[3]]))
2*q^3 + 2*q^2 + 2*q + 1

julia> degree(h)
q^4 + 2*q^3 + q^2
```

To resolve this, we need to work through the exceptions. Recall that earlier we saw that $\langle \chi_4, \chi_2^2 \rangle = 0$ except when $q-1$ divides $2n_1$, where n_1 was the value of the parameter n used in χ_4. Further restrictions apply to n:

```
julia> parameters(T[4])
n ∈ {1,..., q - 1} except n ∈ (q - 1)ℤ
```

So n may take on any value between 1 and $q-1$ not divisible by $q-1$. Hence the only possible exception where $q-1$ divides n is $n = (q-1)/2$ which can only occur if q is odd. It thus makes sense to consider q odd and q even separately.

We demonstrate this for q even. Then $\langle \chi_4, \chi_2^2 \rangle = 0$ and with a similar argument $\langle \chi_5, \chi_2^2 \rangle = 0$. We now construct a copy of the table but with the congruence equation $q \equiv 0 \pmod 2$ applied:

```
julia> T2 = set_congruence(T; remainder=0, modulus=2);
```

Inspecting the list of exceptions for $\langle \chi_6, \chi_2^2 \rangle$, the first occurs when $q-1$ divides $m+n$. To study this case we specialize m for character type 6:

```
julia> (q, (a, b, m, n)) = parameters(T2);

julia> x = parameter(T2, "x");  # create an additional "free" variable

julia> s = specialize(T2[6], m, -n + (q-1)*x);  # force m = -n (mod q-1)
```

Recomputing the scalar product now gives a new result. Note that we have to map the character type h into the new table for this work.

```
julia> scalar_product(s, T2(h))
1
With exceptions:
  3*n1 ∈ (q - 1)ℤ
  2*n1 ∈ (q - 1)ℤ
```

The exceptions both cannot occur as q is even and the table we are considering is only defined for $q \not\equiv 1 \pmod 3$. By working through the other possible exceptions and irreducible character types, and handling duplicates, one finally obtains

$$\chi_2^2 = \chi_1 + 2\chi_2 + 2\chi_3 + \frac{1}{2}\sum_{n=1}^{q-2} \chi_6(n, q-1-n) + \frac{1}{2}\sum_{n=1}^{q} \chi_7(n(q-1)).$$

Where $\chi_6(n, q-1-n)$ indicates that the 6th character in the table T2 is a family on two parameters: n and $q-1-n$, while χ_7 depends on only one, namely $n(q-1)$. A similar result can be obtained for odd q albeit with a few more cases that need to be dealt with, but all in essentially the same manner.

10 Outlook

OSCAR can already now deal with many kinds of groups, and more group related objects and algorithms are being actively worked on, such as Coxeter groups, complex reflection groups, Lie algebras, and more.

The abilities of GAP are vast and extended by many packages. We currently do not wrap all of it nicely in OSCAR functions. However more will be made available over time, prioritized by our needs and the needs of OSCAR users

– but for that, we need to know what they are. So please don't hesitate to contact us to let us know what you are missing, or even contribute your own additions.

In the meantime, you don't need to wait: our interface between `Julia` and `GAP` allows executing arbitrary `GAP` code in a `Julia` session. The interface is bidirectional, making it very easy to mix and match GAP and `Julia` code. Global `GAP` variables can be accessed from `Julia` via `GAP.Globals`, for example

```
julia> GAP.Globals.SmallGroupsInformation(32)
```

calls a `GAP` function that prints an overview of the groups of order 32, and entering

```
julia> ?GAP.Globals.SmallGroupsInformation
```

shows the documentation of this function.

One can also load additional (perhaps not yet distributed) `GAP` packages into the `Julia` session, via `GAP.Packages.load`, and then use their functionality. Furthermore files containing `GAP` code can be loaded via `GAP.Globals.Read`.

Using `GAP.prompt()`, you can even get a regular `GAP` prompt inside the `Julia` REPL. For details about this and other features, please have a look at the documentation of the `GAP.jl` package.

Acknowledgements The authors acknowledge support by the German Research Foundation (DFG) – Project-ID 286237555 – within the SFB-TRR 195 "Symbolic Tools in Mathematics and their Applications", by MaRDI (Mathematical Research Data Initiative), funded by the Deutsche Forschungsgemeinschaft (DFG), project number 460135501, NFDI 29/1 "MaRDI – Mathematische Forschungsdateninitiative" and "SymbTools" funded by the state of Rheinland-Pfalz via the Forschungsinitiative.

References

[Ban56] P. A. Bandler. *A method for enumerating the cosets of an abstract group on a digital computer.* M.A. thesis, University of Manchester, 1956.

[BEO23] H. U. Besche, B. Eick, and E. O'Brien. *SmallGrp, The GAP Small Groups Library, Version 1.5.3.* GAP package. May 2023.

[Boo58] W. W. Boone. "The word problem". *Proc. Natl. Acad. Sci. USA* 44 (1958), pp. 1061–1065. DOI: 10.1073/pnas.44.10.1061.

[Bre24] T. Breuer. *CTblLib, The GAP Character Table Library, Version 1.3.7.* GAP package. Jan. 2024.

[Brü98] H. Brückner. "Algorithmen für endliche auflösbare Gruppen und Anwendungen". PhD thesis. RWTH Aachen, 1998.

[BJ24] T. Brysiewicz and M. Joswig. "Polyhedral Geometry". In: *The Computer Algebra System OSCAR. Algorithms and Examples.* Ed. by W. Decker, C. Eder, C. Fieker, M. Horn, and M. Joswig. Springer, 2024, pp. 139–164.

[Can69] J. J. Cannon. "Computers in group theory: A survey". *Comm. ACM* 12 (1969), pp. 3–12. DOI: 10.1145/362835.362837.

[CCNPW85] J. H. Conway, R. T. Curtis, S. P. Norton, R. A. Parker, and R. A. Wilson. *ATLAS of finite groups.* Maximal subgroups and ordinary characters for simple groups, With computational assistance from J. G. Thackray. Oxford University Press, Eynsham, 1985, pp. xxxiv+252.

[DRS24] W. Decker, L. Ramesh, and J. Schmitt. "Invariant Theory". In: *The Computer Algebra System OSCAR. Algorithms and Examples.* Ed. by W. Decker, C. Eder, C. Fieker, M. Horn, and M. Joswig. Springer, 2024, pp. 297–331.

[DH21] H. Dietrich and A. Hulpke. "Universal covers of finite groups". *J. Algebra* 569 (2021), pp. 681–712. DOI: 10.1016/j.jalgebra.2020.10.032.

[Fel61] H. Felsch. "Programmierung der Restklassenabzählung einer Gruppe nach Untergruppen". *Numer. Math.* 3 (1961), pp. 250–256. DOI: 10.1007/BF01386025.

[Fie09] C. Fieker. "Minimizing representations over number fields. II. Computations in the Brauer group". *J. Algebra* 322.3 (2009), pp. 752–765. DOI: 10.1016/j.jalgebra.2009.05.009.

[GAP] *GAP – Groups, Algorithms, and Programming, Version 4.13.0.* https://www.gap-system.org. The GAP Group, 2024.

[GHLMP96] M. Geck, G. Hiss, F. Lübeck, G. Malle, and G. Pfeiffer. "CHEVIE—a system for computing and processing generic character tables". In: vol. 7. 3. Computational methods in Lie theory (Essen, 1994). 1996, pp. 175–210. DOI: 10.1007/BF01190329.

[HRT22] D. Holt, G. Royle, and G. Tracey. "The transitive groups of degree 48 and some applications". *J. Algebra* 607 (2022), pp. 372–386. DOI: 10.1016/j.jalgebra.2021.06.018.

[HEO05] D. F. Holt, B. Eick, and E. A. O'Brien. *Handbook of Computational Group Theory.* Discrete Mathematics and Its Applications (Boca Raton). Chapman and Hall/CRC, 2005, p. 514. DOI: 10.1201/9781420035216.

[Hol06] D. F. Holt. "Cohomology and group extensions in Magma". In: *Discovering mathematics with Magma.* Vol. 19. Algorithms Comput. Math. Springer, Berlin, 2006, pp. 221–241. DOI: 10.1007/978-3-540-37634-7_10.

[Hol98] D. F. Holt. "The Meataxe as a tool in computational group theory". In: *The atlas of finite groups: ten years on (Birmingham, 1995).* Vol. 249. London Math. Soc. Lecture Note Ser. Cambridge University Press, 1998, pp. 74–81. DOI: 10.1017/CBO9780511565830.011.

[HP89] D. F. Holt and W. Plesken. *Perfect groups.* Oxford Mathematical Monographs. New York: The Clarendon Press Oxford University Press, 1989, pp. xii+364.

[Hul22] A. Hulpke. "The perfect groups of order up to two million". *Math. Comp.* 91 (2022), pp. 1007–1017. DOI: 10.1090/mcom/3684.

[Hul23] A. Hulpke. *TransGrp, Transitive Groups Library, Version 3.6.5.* GAP package. Dec. 2023.

[HRR23] A. Hulpke, C. Roney-Dougal, and C. Russell. *PrimGrp, GAP Primitive Permutation Groups Library, Version 3.4.4.* GAP package. Feb. 2023.

[JLPW95] C. Jansen, K. Lux, R. Parker, and R. Wilson. *An atlas of Brauer characters.* Vol. 11. London Mathematical Society Monographs. New Series. Appendix 2 by T. Breuer and S. Norton, Oxford Science Publications. New York: The Clarendon Press Oxford University Press, 1995, pp. xviii+327.

[MN21] G. Malle and G. Navarro. "Brauer's Height Zero Conjecture for principal blocks". *J. Reine Angew. Math.* 778 (2021), pp. 119–125. DOI: 10.1515/crelle-2021-0024.

[Nov55] P. S. Novikov. *Über die algorithmische Unentscheidbarkeit des Wortproblems in der Gruppentheorie.* Tr. Mat. Inst. Steklova 44, 140 S. (1955). 1955.

[Pan23] E. Pan. *SOTGrps, Constructing and identifying groups of small order type, Version 1.2.* GAP package. June 2023.

[Par84] R. A. Parker. "The computer calculation of modular characters (the meat-axe)". In: *Computational group theory (Durham, 1982).* Academic Press, London, 1984, pp. 267–274.

[Ste12] A. Steel. "Construction of Ordinary Irreducible Representations of Finite Groups". PhD thesis. University of Sydney, Jan. 2012.

[TC36] J. A. Todd and H. S. M. Coxeter. "A practical method for enumerating cosets of a finite abstract group". *Proceedings of the Edinburgh Mathematical Society* 5.1 (1936), pp. 26–34. DOI: 10.1017/S0013091500008221.

[VE22] M. Vaughan-Lee and B. Eick. *SglPPow, Database of groups of prime-power order for some prime-powers, Version 2.3.* GAP package. Nov. 2022.

[Wil+] R. A. Wilson et al. *ATLAS of Finite Group Representations.* https://atlas.math.rwth-aachen.de/Atlas/v3.

[WPNBB24] R. A. Wilson, R. A. Parker, S. Nickerson, J. N. Bray, and T. Breuer. *AtlasRep, A GAP Interface to the Atlas of Group Representations, Version 2.1.8.* GAP package. Jan. 2024.

Polyhedral Geometry

Taylor Brysiewicz and Michael Joswig

Abstract Assuming little familiarity with the subject, we give an introduction to computations in polyhedral geometry using OSCAR. In particular, we define polytopes, polyhedra, and polyhedral fans, and we give a brief overview about computing convex hulls and solving linear programs. Three detailed case studies are left for experts in polyhedral geometry. These are concerned with face numbers of random polytopes, constructions and properties of Gelfand–Tsetlin polytopes, and secondary polytopes.

Key words: Combinatorial properties, polytopes, polyhedra

1 Introduction

Polyhedral geometry is linear algebra over ordered fields. The interest in the topic comes from several applications. Polyhedra arise as feasible regions of linear programs, whence they occur naturally in linear and integer programming [Sch86]. Newton polytopes of multivariate polynomials arise, implicitly or explicitly, in algorithms for solving systems of polynomial equations [Stu96]; this has further consequences for toric [CLS11] and tropical geometry [MS15; Jos21]. There are further connections to commutative algebra, combinatorial topology and more; e.g., see [Sta96]. This wide range of applications lead to polyhedral geometry becoming a field of its own [Grü03; Zie95; DRS10]. The

Taylor Brysiewicz
Department of Mathematics, University of Western Ontario, London, Canada, e-mail: `tbrysiew@uwo.ca`

Michael Joswig
TU Berlin, Chair of Discrete Mathematics/Geometry, Berlin, Germany; Max-Planck-Institut für Mathematik in den Naturwissenschaften, Leipzig, Germany, e-mail: `joswig@math.tu-berlin.de`

W. Decker et al. (eds.), *The Computer Algebra System OSCAR*,
Algorithms and Computation in Mathematics 32,
https://doi.org/10.1007/978-3-031-62127-7_5

latter textbooks are recommended for further reading. Our conventions follow [JT13] and [Zie95].

From an algorithmic point of view, polyhedra are rather benign. Since linear algebra suffices for most computations, we immediately get exact algorithms over any (ordered) field for which we have an exact implementation of the arithmetic. Of course, the rational numbers form the prime examples, but real number fields [Grü03, p.94f] and fields of rational functions [JLLS16] occur in practice, too.

2 Polytopes and Polyhedra

A *polyhedron* in $\mathbb{R}^d$ is the intersection of finitely many affine closed half-spaces. Equivalently, polyhedra can be described by finitely many linear inequalities, whence they are precisely the feasible regions of linear programs. A *polytope* is a bounded polyhedron.

An (affine) hyperplane H *supports* a polyhedron P if H intersects P nontrivially, but H does not separate P, i.e., one of the two half-spaces defined by H contains all of P. The intersection of P with a supporting hyperplane is a *(proper) face* of P. Each polyhedron has finitely many faces. They are partially ordered by inclusion and by construction, are polyhedra themselves. Each polyhedron has a *dimension* r, defined as the dimension of its affine hull, and may be described as an r-polyhedron or r-polytope accordingly. We allow for polyhedra which are not full dimensional; in that case we have $r < d$.

Furthermore, we also allow the empty set as a polytope. In that case, there are no supporting hyperplanes nor proper faces. We use the convention $\dim \emptyset = -1$.

2.1 Construction and Basic Properties

The convex hull of finitely many points is a polytope and conversely; see [Zie95, §1.1] or [JT13, §3.1.3]. Let us see this in a first example. We create a polytope P which is the convex hull of five points in the plane.

```
julia> points = [1 0; 1 1; 0 1; -1 0; -1 -1];

julia> P = convex_hull(points)
Polyhedron in ambient dimension 2
```

Mathematically a lone pentagon is not so exciting. Here it serves to demonstrate how to call the most basic polyhedral functions in OSCAR, and we can see the full output. The *facets* of a polyhedron are its maximal proper faces. They yield a representation of the polyhedron in terms of a minimal number of linear inequalities. If a polyhedron is full-dimensional, then its minimal inequality description is unique.

```
julia> facets(P)
5-element SubObjectIterator{AffineHalfspace{QQFieldElem}} over the
↪  Halfspaces of R^2 described by:
-x_1 <= 1
-x_1 + x_2 <= 1
x_1 - 2*x_2 <= 1
x_1 <= 1
x_2 <= 1
```

The *f-vector* counts the faces per dimension. For the pentagon, we have five 0-dimensional faces, the *vertices* (which were our input), and the five 1-dimensional facets that we saw just above.

```
julia> f_vector(P)
2-element Vector{ZZRingElem}:
 5
 5

julia> volume(P)
5//2
```

As a compact subset of $\mathbb{R}^d$ any polytope has a finite Euclidean volume. The (Euclidean) volume in this planar example is the area. The points with integer coordinates in a polyhedron are its *lattice points*. In our example the lattice points of P are the five vertices along with the origin.

```
julia> lattice_points(P)
6-element SubObjectIterator{PointVector{ZZRingElem}}:
 [-1, -1]
 [-1, 0]
 [0, 0]
 [0, 1]
 [1, 0]
 [1, 1]
```

Note that all coefficients and metric values we computed so far are exact rational numbers. More generally, whenever a polytope is given by inequalities whose coefficients lie in some ordered field $\mathbb{F}$, then the vertex coordinates also lie in $\mathbb{F}$. Via triangulation the volume is a sum of determinant expressions in the vertex coordinates, hence this lies in $\mathbb{F}$ too. OSCAR allows coordinates in an arbitrary real number field; the corresponding coordinate type is `EmbeddedAbsSimpleNumFieldElem`.

2.2 Algorithmic Ingredients

Going back to our initial pentagon example, we saw that a polytope which is defined as the convex hull of finitely points, admits an inequality description in terms of its facets. The algorithmic problem of passing from the points to the inequalities is known as the *convex hull problem*. Several different methods are known and implemented, but the precise complexity status of the general convex hull problem is open [Sei18, Open Problem 26.3.4]. In polyhedral geometry practice, computing convex hulls is usually a crucial step and often the bottleneck. The algorithms and their implementations compared in the survey [Ass+17] are available in `polymake` and thus also in `OSCAR`.

Computing face lattices and f-vectors of polytopes starts with the incidences between the vertices and the facets, which are usually determined after (or simultaneously with) a convex hull computation. State of the art is an algorithm of Kaibel and Pfetsch [KP02], which allows for numerous variations [HJS19]. Kliem and Stump describe a fast algorithm for computing the f-vector from the vertex-facet incidences without constructing the covering relations of the entire face lattice [KS22]. For volume computation see Büeler, Enge and Fukuda [BEF00], for further details on several algorithms for polytopes see Kaibel and Pfetsch [KP03].

2.3 Linear Programs

Polytopes and polyhedra occur most naturally in linear optimization. The task of a *linear program* is to optimize an affine linear function $x \mapsto c^T x + k$, called the *objective function*, over the points x within some polyhedron $P \subseteq \mathbb{R}^d$ called the *feasible set*. For simplicity, we show examples just in the plane.

```
julia> P = polyhedron([1 5; 2 -1; -1 0; 0 -1], [20,10,0,0])
Polyhedron in ambient dimension 2

julia> LP = linear_program(P, [1,1], convention=:max)
Linear program
   max{c*x + k | x in P}
where P is a Polyhedron{QQFieldElem} and
   c=Polymake.LibPolymake.Rational[1 1]
   k=0

julia> optimal_value(LP)
100/11

julia> optimal_vertex(LP)
2-element PointVector{QQFieldElem}:
 70//11
 30//11
```

The data of a linear program consists of a polyhedron, an objective function, and a choice of `max` or `min`. Since the value of k in the objective function does not influence the optimal solutions, it is zero by default. In our example, the maximum is attained at a unique point, namely the vertex $(70/11, 30/11)$.

If the objective function is bounded, with optimum, say, μ, then the set $\{x \in P \mid c^T x + k = \mu\}$ of optimal solutions forms a face of the feasible region P. Conversely, each face of P arises in this way.

2.4 The Wonderful World of Convex Polytopes

Convex polytopes occur throughout mathematics.

The Platonic Solids and Their Friends

A d-polytope P is *regular* if its group of symmetries acts transitively on the flags of faces of P. Specializing to $d = 3$ this leads us to the *Platonic solids*, the most classical polytopes of all. Up to scaling (and rigid motions) there are exactly five Platonic solids. The simplest one is the regular tetrahedron.

```
julia> T = tetrahedron()
Polytope in ambient dimension 3

julia> vertices(T)
4-element SubObjectIterator{PointVector{QQFieldElem}}:
 [1, -1, -1]
 [-1, 1, -1]
 [-1, -1, 1]
 [1, 1, 1]
```

The remaining four Platonic solids are the regular 3-cube, the regular 3-dimensional cross polytope (also known as the octahedron), the dodecahedron, and the icosahedron.

```
julia> [ f_vector(P) for P in [T, cube(3), cross_polytope(3),
                               dodecahedron(), icosahedron()] ]
5-element Vector{Vector{ZZRingElem}}:
 [4, 6, 4]
 [8, 12, 6]
 [6, 12, 8]
 [20, 30, 12]
 [12, 30, 20]
```

The dodecahedron and icosahedron do not admit regular realizations with rational coordinates. `OSCAR` supports arbitrary real number fields with fixed embeddings. The coordinates are exact algebraic numbers; the floating-point values inside the brackets indicate approximations.

```
julia> D = dodecahedron()
Polytope in ambient dimension 3 with EmbeddedAbsSimpleNumFieldElem type
↪  coefficients

julia> vertices(D)[1]
3-element PointVector{EmbeddedAbsSimpleNumFieldElem}:
```

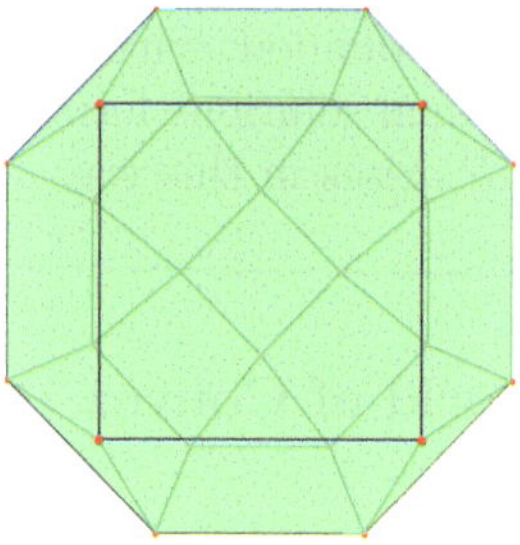

Fig. 1: Elongated square gyrobicupola. Picture generated via `visualize(johnson_solid(37))`.

```
1//2 (0.50)
1//4*sqrt(5) + 3//4 (1.31)
0 (0.00)
```

Natural generalizations of the Platonic solids vary, as does the terminology in the literature; here we follow [Grü03, Chapter 19]. For instance, a polytope is *semiregular* if each facet is regular and the group of symmetries acts transitively on the vertices. It turns out that, in dimension three, the semiregular polytopes consist of the Platonic solids, the prisms and antiprisms, plus 13 exceptional types known as the *Archimedean solids*.

This can be generalized further, by omitting the vertex-transitivity condition, and we arrive at *regular-faced* (or *uniform*) polytopes. In dimension three this gives us the 92 types of *Johnson solids*, in addition to the semiregular ones; see [Joh66; Zal69]. Here the polytopes of each *type* are unique, again up to scaling and rigid motions.

The *elongated square gyrobicupola* is the Johnson solid J_{37}; it is shown in Figure 1. We follow the naming conventions from [Zal69, Table 3]. That particular polytope plays a special role, as it is sometimes counted as a 14th Archimedean solid, because any two of its vertex figures are combinatorially equivalent. Yet it does not have a vertex transitive group of symmetries, so it is not an Archimedean solid according to our definition. The Johnson solid J_{37} has 26 facets, eight of which are regular triangles, and the remaining 16 facets are squares.

Like the dodecahedron, the polytope J_{37} does not admit a rational realization which is regular(-faced). However, each 3-dimensional polytope can be perturbed slightly into a rational polytope, without changing the combinatorial type. In contrast, in higher dimensions there are polytopes which do not admit any rational realization. A prominent example has been found by Perles (see [Grü03, p.94f]); in OSCAR this is available as `perles_nonrational_8_polytope()`.

We now turn to an example which shows how to browse and classify a set of polytopes. The following code iterates through the 92 Johnson solids, determines the vertex orbits, and collects in a multiset how many of them have a given number of orbits.

```
orbit_counts = MSet{Int}();

for k in 1:92
  J = johnson_solid(k)
  Aut = automorphism_group(J)[:on_vertices]
  push!(orbit_counts, length(orbits(Aut)))
end
```

We can sort the multiset by referring to the underlying dictionary.

```
julia> sort(collect(orbit_counts.dict))
14-element Vector{Pair{Int64, Int64}}:
  2 => 26
  3 => 24
  4 => 17
  5 => 8
  7 => 5
  8 => 2
  9 => 2
 12 => 1
 14 => 1
 15 => 1
 17 => 1
 20 => 1
 27 => 1
 29 => 2
```

The symmetry groups of the Johnson solids were discussed in [Joh66, §5].

Random Polytopes

Other interesting, yet totally different, classes of polytopes come from random constructions. To produce, e.g., the convex hull $R_{3,100}$ of 100 random points in $\mathbb{R}^3$ whose coordinates are normally distributed with mean 0 and standard deviation 1 is as easy as calling `convex_hull([randn(Float64, 3) for i in 1:100])` in OSCAR. Note that this would take the (random) coordinates of type `Float64` and convert them to exact rational numbers. Consequently, all subsequent computations are exact.

In practice, another class of random polytopes is quite valuable; see Section 4.1 below for a case study. These polytopes arise from taking the convex hull of points uniformly sampled from the unit sphere:

```
rand_spherical_polytope(3, 100)
```

Mathematically the situation is quite subtle. A random point on the unit sphere in $\mathbb{R}^d$ can be obtained by sampling from the d-variate standard normal distribution and projecting to the sphere; see [Knu98, §3.4.1.E.6] for the details. Converting to exact rational coordinates almost surely gives a point of norm different from one. Hence such a random point does not truly lie on $\mathbb{S}^{d-1}$. However, producing rational random points of norm one is not difficult in itself. We can achieve this by randomly sampling rational points in Euclidean space and employing stereographic projection, which is rational. This is also supported in OSCAR. Calling

```
rand_spherical_polytope(3, 100; distribution=:exact)
```

yields a polytope in $\mathbb{R}^3$ with exact rational coordinates of norm one. The resulting distribution is not precisely uniform on the sphere, but it can be approximated arbitrarily close. Moreover, this is more costly to compute. So there is a tradeoff here.

More Examples

As the random constructions above show, there is no shortage of polytopes. On the contrary, there are many more polytopes than meaningful names for them. Nonetheless, we want to show some constructions which are often useful. Polytopes can be obtained as convex hulls of point orbits of linear groups. The *regular permutahedron* is a standard example in $\mathbb{R}^4$.

```
julia> P3 = orbit_polytope([1,2,3,4], symmetric_group(4))
Polyhedron in ambient dimension 4

julia> dim(P3)
3
```

The input points lie in the affine hyperplane $\sum x_i = 10$, so the permutahedron has codimension one. It follows from the construction that the symmetric group acts naturally on this polytope. Yet, the full combinatorial automorphism group, i.e., the automorphism group of the face lattice, is even larger, by a factor of two.

```
julia> describe(combinatorial_symmetries(P3))
"C2 x S4"
```

The additional symmetry comes from the reflection at the linear hyperplane $x_1 - x_2 - x_3 + x_4 = 0$.

Polyhedral geometry in OSCAR is not restricted to bounded polyhedra. Given a collection $v_1, \ldots, v_k \in \mathbb{R}^d$, we may define their *positive hull*:

$$\operatorname{pos}(v_1, \ldots, v_k) = \{\lambda_1 v_1 + \lambda_2 v_2 + \cdots + \lambda_k v_k \mid \lambda_1, \ldots, \lambda_k \in \mathbb{R}_{\geq 0}\} \subseteq \mathbb{R}^d.$$

Subsets of $\mathbb{R}^d$ arising in this way are called *(polyhedral) cones*. Indeed, these sets are both cones (each is closed under positive linear combinations) and polyhedra (each is an intersection of finitely many half-spaces). As polyhedra, they have faces of various dimensions. Their one-dimensional faces are called *rays*, and a cone is *pointed* if it has the origin as a face. A cone is pointed if and only if it does not contain any line through the origin.

For instance, the following command defines the union of the $(+,+,+)$ and $(-,+,+)$ octants in $\mathbb{R}^3$.

```
julia> Q = convex_hull([0 0 0], [0 1 0; 0 0 1], [1 0 0])
Polyhedron in ambient dimension 3
```

The constructor has three arguments, corresponding to the decomposition $Q = P + C + L$, where P is a polytope, C is a cone, and L is a linear subspace. Recall that any polyhedron can be written in this way, and we may even assume that the cone C is pointed; e.g., see [Zie95, Theorem 1.2]. The polyhedron Q in our example does not have any vertices, but it has a unique minimal face, which is the vertical axis $x_1 = x_2 = 0$.

```
julia> vertices(Q)
0-element SubObjectIterator{PointVector{QQFieldElem}}

julia> minimal_faces(Q)
(base_points = PointVector{QQFieldElem}[[0, 0, 0]],
 lineality_basis = RayVector{QQFieldElem}[[1, 0, 0]])
```

Arbitrary affine subspaces can occur as the minimal faces of polyhedra. Here, an affine subspace is written as the sum of a point and a linear subspace.

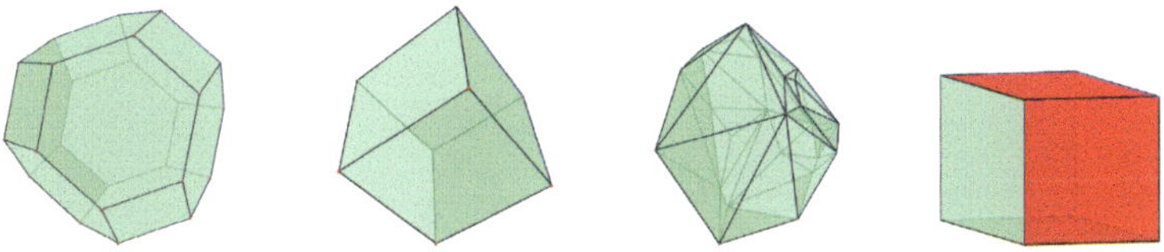

Fig. 2: Visual representations of $\mathcal{P}_3$, $\mathcal{C}_3$, $R_{3,100}$, and Q. Note that $\mathcal{P}_3$ is visualized within the hyperplane $x_1 + x_2 + x_3 + x_4 = 10$. The polyhedron Q is viewed within a box, and the unbounded portion of Q is coloured red. These figures were obtained via the command `visualize()`.

3 Polyhedral Fans

Here we move on to more involved objects from polyhedral geometry. A collection $\mathcal{F} = \{C_1, \ldots, C_n\}$ of cones in $\mathbb{R}^d$ is called a *fan* if

- Each face of each cone in $\mathcal{F}$ is, itself, an element of $\mathcal{F}$.
- For any two cones C_1, C_2 in $\mathcal{F}$, their intersection $C_1 \cap C_2$ is also an element of $\mathcal{F}$.

A fan in $\mathbb{R}^d$ is *complete* if the union of its cones is $\mathbb{R}^d$. The fans considered in this section are usually complete, however, incomplete fans appear frequently in other important areas of mathematics; for instance, in *tropical geometry* [MS15; Jos21] and *toric geometry* and [CLS11].

One easy way to produce a polyhedral fan in OSCAR is by passing a collection of rays, and an incidence of rays in cones: the (i, j)-th entry of the incidence matrix is 1 if cone i contains ray j and zero otherwise.

```
julia> Rays = [1 0; 1 1; 0 1; -1 0; 0 -1];

julia> Cones = IncidenceMatrix([[1,2], [2,3], [3,4], [4,5], [5,1]]);

julia> PF = polyhedral_fan(Cones, Rays)
Polyhedral fan in ambient dimension 2
```

Figure 3 shows the output of the command `visualize(PF)`. Visualizing fans is not entirely trivial because fans are unbounded. Mathematically, it would be satisfying to pass to the intersection with the unit sphere, yielding a spherical polytopal complex, which is always compact. Yet, since spherical polytopal complexes are non-linear, this poses other challenges to the visualization. In OSCAR, when visualizing, we truncate each ray to unit length, and replace each cone by the convex hull of the origin and its truncated rays. In this way, we always see an ordinary polytopal complex.

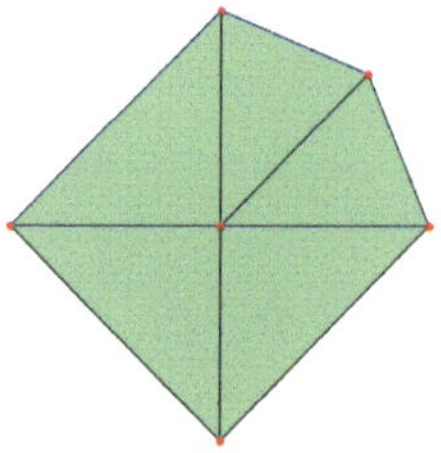

Fig. 3: The output of `visualize` applied to the polyhedral fan consisting of the cones $\mathrm{pos}(e_1, e_1 + e_2)$, $\mathrm{pos}(e_1 + e_2, e_2)$, $\mathrm{pos}(e_2, -e_1)$, $\mathrm{pos}(-e_1, -e_2)$, $\mathrm{pos}(-e_2, e_1)$.

Associated to any nonempty polyhedron $P \subseteq \mathbb{R}^d$ is its *normal fan* $\mathcal{N}(P)$ consisting of cones $N_F = \{\ell \in (\mathbb{R}^d)^* \mid \ell \text{ is maximized on } F\}$ for every nonempty face F of P. For example, the normal fan of the polytope

$$P = \mathrm{conv}\left(\begin{pmatrix}0\\0\end{pmatrix} \begin{pmatrix}2\\0\end{pmatrix} \begin{pmatrix}2\\1\end{pmatrix} \begin{pmatrix}1\\2\end{pmatrix} \begin{pmatrix}0\\2\end{pmatrix}\right)$$

obtained as the convex hull of five points, is exactly the fan displayed in Figure 3. The normal cone N_F comprises all those linear functions ℓ such that ℓ attains its maximum on P at some point in the face F. In this way, the normal fan of a polyhedron P encodes all of the information pertaining to linear programs over P. We note that the normal fan of a polytope is complete and that many different polytopes have the same normal fan.

4 Case Studies

The purpose of this section is to sketch some OSCAR use-cases which occur in research on polyhedral geometry. Some portions of the text require more substantial background in polyhedral geometry (and adjacent fields) than the previous sections.

4.1 Face Numbers of Random Polytopes

It is a major theme in geometric combinatorics to study the f-vectors of convex polytopes, and also of subclasses and generalizations. A first result in this direction is Euler's equation, which says that $f_0 - f_1 + f_2 = 2$ if $f = f(P)$ is the f-vector of a 3-polytope P. To simplify the discussion, let us assume that P is *simplicial*, i.e., each proper face is a simplex. Then a double-counting argument yields $2f_1 = 3f_2$, because each edge is contained in two facets, and each facet contains three edges (due to simpliciality). Taking into account that P must have at least four vertices, we see that the f-vectors of simplicial 3-polytopes with $f_0 = n$ vertices lie in the set

$$\left\{(n, 3n-6, 2n-4) \in \mathbb{N}^3 \;\middle|\; n \geq 4\right\} , \tag{1}$$

and it is an exercise to show that for each such f-vector, a suitable 3-polytope exists. The situation in higher dimensions is much more involved. A complete classification of the f-vectors of simplicial polytopes of arbitrary dimension was conjectured by McMullen [McM71] and established by Stanley [Sta80], Billera and Lee [BL81]. It is worth noting that Stanley's proof exploits the cohomological properties of the toric variety associated with a simplicial polytope in an essential way.

Now we want to test, empirically, to what extent we can explore f-vectors of simplicial polytopes by certain random constructions. We start with one example: 30 vertices in six dimensions.

```
julia> P = rand_spherical_polytope(6,30)
Polytope in ambient dimension 6

julia> show(f_vector(P))
ZZRingElem[30, 336, 1468, 2874, 2568, 856]
```

The construction picks 30 points uniformly at random on the unit sphere $\mathbb{S}^5$ in $\mathbb{R}^6$ and takes their convex hull. We note that, almost surely, this construction produces a simplicial polytope. We stress that since the output above is random, reproducing it with the corresponding commands would be incredibly unlikely.

Our goal is to see how this f-vector fits in the set of all possible f-vectors of simplicial 6-polytopes with 30 vertices. To this end we first convert that vector into a different coordinate system. The *h-vector* of the simplicial d-polytope P is given by

$$h_k = \sum_{i=0}^{k} (-1)^{k-i} \binom{d-i}{d-k} f_{i-1} \quad \text{for } 0 \le k \le d ,$$

where we use the convention $f_{-1} = 1$. In particular, we have $h_0 = 1$, $h_1 = f_0 - d$ and

$$h_d = f_{d-1} - f_{d-2} + f_{d-3} - \cdots + (-1)^{d-1} f_0 + (-1)^d = 1 ,$$

which, up to sign, is the reduced Euler characteristic of the boundary sphere $\partial P \approx \mathbb{S}^{d-1}$. Since the h-vector is obtained from the f-vector by an invertible linear transformation, it carries exactly the same information. Yet the h-vector satisfies $h_k = h_{d-k}$, which exhibits a redundancy known as the *Dehn–Sommerville equations*. The analogous symmetry in terms of the f-vector is more cumbersome; see [Zie95, §8.3].

```
julia> show(h_vector(P))
ZZRingElem[1, 24, 201, 404, 201, 24, 1]
```

The redundancy in the h-vector is eliminated by applying a final transformation

$$g_0 = h_0 = 1 \quad \text{and} \quad g_k = h_k - h_{k-1} \text{ for } 1 \le k \le \lfloor \tfrac{d}{2} \rfloor ,$$

to obtain the *g-vector* of P. For our example above it reads as follows.

```
julia> show(g_vector(P))
ZZRingElem[1, 23, 177, 203]
```

Now we run a small experiment, producing 100 random 6-polytopes with 30 vertices each, and keeping track of the g-vectors produced.

```
n_vertices = 30;
n_samples = 100;
g_vectors = Array{Int32}(undef,n_samples,2);

for i=1:n_samples
    RS = rand_spherical_polytope(6,n_vertices)
    g = g_vector(RS)
    g_vectors[i,1] = g[3] # notice index shift as Julia counts from 1
    g_vectors[i,2] = g[4]
end
```

It is instructive to plot a scatter diagram of the resulting (g_2, g_3)-pairs, using the standard `Julia` package `Plots`; the result is Figure 4.

```
using Plots
scatter(g_vectors[:,1], g_vectors[:,2],
        xlabel="g_2", ylabel="g_3", legend=false);
```

In our sample the g_2-values range between 159 and 192. This is well below the maximal possible value for g_2, which equals 276, by McMullen's Upper Bound Theorem.

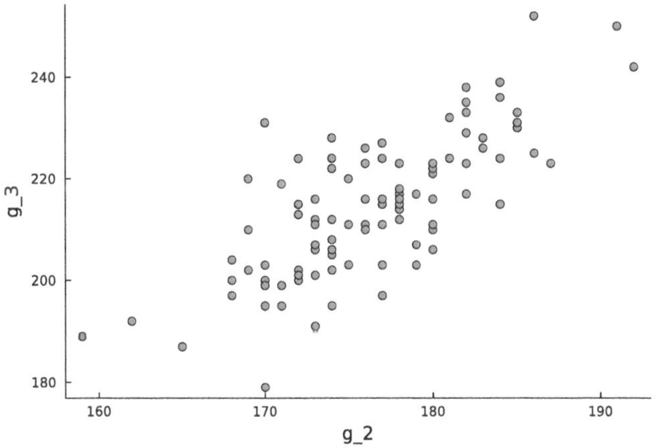

Fig. 4: g-vectors of simplicial random 6-polytopes with 30 vertices.

```
julia> min_g2 = minimum(g_vectors[:,1])
159

julia> max_g2 = maximum(g_vectors[:,1])
192

julia> show(upper_bound_g_vector(6,30))
[1, 23, 276, 2300]
```

4.2 Gelfand–Tsetlin Polytopes

Gelfand–Tsetlin polytopes arise in the representation theory of Lie groups. For background on the subject, see [FH91]. Given an irreducible representation V_λ of $\mathrm{GL}(n)$ with highest weight $\lambda = (\lambda_1 \geq \ldots \geq \lambda_n)$, its restriction to a subgroup $\mathrm{GL}(n-1) \hookrightarrow \mathrm{GL}(n)$ is no longer irreducible, but decomposes as $\bigoplus_\mu V_\mu$. Here, each V_μ is an irreducible $\mathrm{GL}(n-1)$ representation indexed by a partition μ of $(n-1)$, and each V_μ appears with multiplicity one. The indexing set of the direct sum is given by partitions μ which *interlace* λ, that is,

$$\lambda_1 \geq \mu_1 \geq \lambda_2 \geq \cdots \geq \mu_{n-1} \geq \lambda_n .$$

Fixing a chain of subgroups $\mathrm{GL}(1) \subset \mathrm{GL}(2) \subset \cdots \subset \mathrm{GL}(n)$, one may iterate this restriction process, terminating at $\dim(V_\lambda)$-many representations of $\mathrm{GL}(1)$, each of dimension one. A *Gelfand–Tsetlin* diagram encodes the successive partitions indexing the representations appearing in such a chain of restrictions.

For the combinatorics of Gelfand–Tsetlin polytopes, we follow the exposition of [Pos09, §15]. A *Gelfand–Tsetlin diagram* is a triangular array (see Figure 5) of $\binom{n+1}{2}$ real numbers $(p_{ij} \mid i, j \geq 1, i + j \leq n + 1)$ satisfying:

- The first row is weakly decreasing: $p_{11} \geq p_{12} \geq \cdots \geq p_{1n}$.
- Each row weakly *interlaces* the row above it: $p_{ij} \geq p_{(i+1)j} \geq p_{i(j+1)}$.

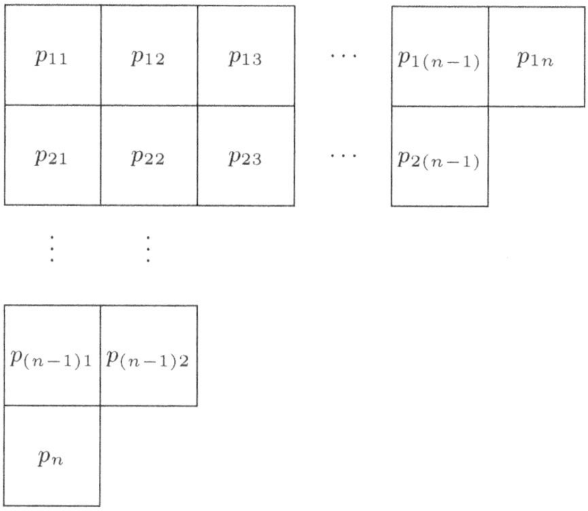

Fig. 5: A Gelfand–Tsetlin diagram.

Given $\lambda = (\lambda_1 \geq \lambda_2 \geq \cdots \geq \lambda_n) \in \mathbb{R}^n$, the *Gelfand–Tsetlin polytope* $\mathrm{GT}(\lambda)$ is the set of all Gelfand–Tsetlin diagrams in $\mathbb{R}^{\binom{n+1}{2}}$ with λ as its first row. These are indeed polytopes, defined by inequalities of the form $p_{ij} \geq p_{(i+1)j} \geq p_{i(j+1)}$. When λ is an integer partition, the lattice points of $\mathrm{GT}(\lambda)$ are in bijection with chains of weights of irreducible representations appearing in successive restrictions as described above. Hence, the number of lattice points in $\mathrm{GT}(\lambda)$ is the dimension of V_λ. In fact, these lattice points may be taken as a basis of V_λ, called a *Gelfand–Tsetlin basis.*

4.2.1 An Explicit Gelfand–Tsetlin Polytope and the Weyl Dimension Formula

The following code computes $\mathrm{GT}(3, 1, 1)$ and its six lattice points. Each lattice point corresponds to a diagram in Figure 6. Additionally, we compute the *Ehrhart polynomial* of the polytope $\mathrm{GT}(3, 1, 1)$: the univariate polynomial $\mathrm{ehr}(\mathrm{GT}(3, 1, 1))$ whose evaluation at $x = k$ counts the number of lattice points in the k-th dilate of $\mathrm{GT}(3, 1, 1)$.

```
julia> lambda = Partition([3,1,1])
[3, 1, 1]

julia> GT = gelfand_tsetlin_polytope(lambda)
Polyhedron in ambient dimension 6

julia> lattice_points(GT)
6-element SubObjectIterator{PointVector{ZZRingElem}}:
 [3, 1, 1, 1, 1, 1]
 [3, 1, 1, 2, 1, 1]
 [3, 1, 1, 2, 1, 2]
 [3, 1, 1, 3, 1, 1]
 [3, 1, 1, 3, 1, 2]
 [3, 1, 1, 3, 1, 3]

julia> ehrhart_polynomial(GT)
2*x^2 + 3*x + 1

julia> volume(project_full(GT))
2
```

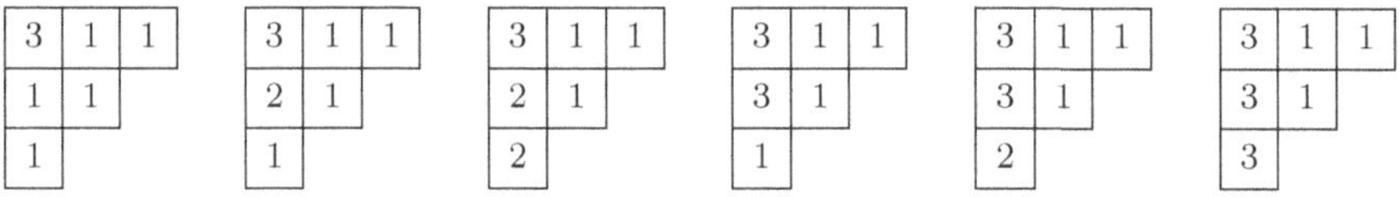

Fig. 6: The six Gelfand–Tsetlin diagrams with integer content and $(3, 1, 1)$ as their top row.

The example above shows that $\dim(V_{(3,1,1)}) = |(\mathrm{GT}(3,1,1) \cap \mathbb{Z}^6)| = 6$ and more generally, that $\dim(V_{(3k,k,k)}) = \mathrm{ehr}(\mathrm{GT}(3,1,1))(k) = 2k^2 + 3k + 1$ as shown by the Ehrhart polynomial. Alternatively, this Ehrhart computation may be verified directly from the Weyl-dimension formula

$$\mathrm{ehr}(\mathrm{GT}(\lambda))(k) \;=\; \dim(V_{k\lambda}) \;=\; \prod_{1 \le i,j \le n} \frac{k\lambda_i - k\lambda_j + i - j}{j - i} \;.$$

From this point-of-view, the volume of $\mathrm{GT}(\lambda)$ is the top coefficient of $\mathrm{ehr}(\mathrm{GT}(\lambda))$:

$$\mathrm{vol}(\mathrm{GT}(\lambda)) \;=\; \prod_{1 \le i,j \le n} \frac{\lambda_i - \lambda_j}{j - i} \;. \tag{2}$$

Calculating the volume of $\mathrm{GT}(\lambda)$ directly involves computing a triangulation of $\mathrm{GT}(\lambda)$, however, evaluating (2) is essentially instantaneous.

4.2.2 Generalized Gelfand–Tsetlin Polytopes and Demazure Characters

In [PS09], Postnikov and Stanley introduce *generalized Gelfand–Tsetlin polytopes*, which are constructed from a partition λ as before, and a permutation σ in the symmetric group $\mathfrak{S}_n$. Given a permutation $\sigma \in \mathfrak{S}_n$, its *code* is the vector whose i-th coordinate enumerates inversions of σ greater than i:

$$\operatorname{code}(\sigma) = (c_1, \ldots, c_n) \quad \text{where} \quad c_i = \left|\{j \mid i < j \leq n\ \sigma_i > \sigma_j\}\right|$$

Let $w_0 = (1\, n\, n-1 \ldots, 1) \in \mathfrak{S}_n$ be the long word. The generalized Gelfand–Tsetlin polytope $\mathrm{GT}(\lambda, \sigma)$ is the intersection of $\mathrm{GT}(\lambda)$ with the linear constraints that the first $c_i(w_0 \cdot \sigma)+1$ boxes in the i-th column of a Gelfand–Tsetlin diagram are equal.

For example, the permutation $w_0^2 = (1\,2\,3) \in \mathfrak{S}_3$ has $\operatorname{code}(w_0^2) = (1, 1, 0)$ as its code. Taking $\lambda = (3, 1, 1)$ as before, the polytope $\mathrm{GT}(\lambda, w_0)$ is the intersection of $\mathrm{GT}(\lambda)$ with the equalities $p_{11} = p_{21}$ and $p_{12} = p_{22}$. One can see explicitly that this polytope contains three lattice points, namely, the last three Gelfand–Tsetlin diagrams in Figure 6: each of these diagrams has the property that the first two numbers in the first two columns are equal.

```
julia> w0 = @perm (1,3,2)
(1,3,2)

julia> genGT = gelfand_tsetlin_polytope(lambda,w0)
Polyhedron in ambient dimension 6

julia> lattice_points(genGT)
3-element SubObjectIterator{PointVector{ZZRingElem}}:
 [3, 1, 1, 3, 1, 1]
 [3, 1, 1, 3, 1, 2]
 [3, 1, 1, 3, 1, 3]
```

Associated to λ and σ is the so called *Demazure module* $V_{\lambda,\sigma}$ of a generalized flag manifold. Corollary 15.2 of [PS09] states that when σ is 312-avoiding, the character $\mathrm{ch}_{\lambda,\sigma}(z)$ of $V_{\lambda,\sigma}$ is given by a sum over the lattice points in $\mathrm{GT}(\lambda, \sigma)$:

$$\mathrm{ch}_{\lambda,\sigma}(z_1, \ldots, z_n) = \sum_{P \in \mathrm{GT}(\lambda,\sigma) \cap \mathbb{Z}^{\binom{n+1}{2}}} z^{\mathrm{weight}(P)} \tag{3}$$

where $\mathrm{weight}(P)$ is the vector which encodes the increase in the sum of the rows of P (from the bottom to the top). For example, the first Gelfand–Tsetlin diagram in Figure 6 has weight $(1-0, 2-1, 5-2) = (1, 1, 3)$. The corresponding monomial, if this diagram were to index a summand of (3), is $z_1^1 z_2^1 z_3^1$. The following code computes $\mathrm{ch}_{\lambda,\sigma}(z_1, z_2, z_3)$ for $\lambda = (3, 1, 1)$ and $\sigma = (1, 3, 2)$:

```
julia> dc = demazure_character(lambda,w0)
x1^3*x2*x3 + x1^2*x2^2*x3 + x1*x2^3*x3
```

```
julia> dc(1,1,1)
3
```

This final evaluation shows that

$$\dim(V_{(3,1,1),(1,3,2)}) = |\,\mathrm{GT}((3,1,1),(1,3,2)) \cap \mathbb{Z}^6| = \mathrm{ch}_{(3,1,1),(1,3,2)}(1,1,1) \ .$$

Corollary 14.6 of [PS09] states that this is also equal to the determinant of the matrix of binomial coefficients

$$\dim V_{\lambda,\sigma} \ = \ \det\left(\binom{\lambda_i + n - c_i(w_0\cdot\sigma) - i}{n - c_i(w_0\cdot\sigma) - j}\right)_{i,j=1}^{n}$$

which evaluates, in the case of $\lambda = (3,1,1)$ and $\sigma = w_0 = (1,3,2)$ to

$$\dim(V_{(3,1,1),(1,3,2)}) \ = \ \det\begin{pmatrix} 4 & 1 & 0 \\ 1 & 1 & 1 \\ 0 & 0 & 1 \end{pmatrix} \ = \ 3 \ .$$

4.3 Hyperdeterminants and Secondary Polytopes

Amongst all polynomials of the form

$$F = \sum_{v\in\{0,1\}^n} c_v x_1^{v_1} x_2^{v_2} \cdots x_n^{v_n}, \quad c_v \in \mathbb{C},$$

those which have singular zero-sets form a hypersurface in the 2^n-dimensional vector space of (complex) coefficients. This hypersurface $\mathcal{D}_{2,2,\ldots,2}$ is called the $(2\times 2\times\cdots\times 2)$-hyperdeterminant. The polynomial $F \in \mathbb{C}[c][x]$ is supported on the monomials whose exponent vectors $\mathcal{A}$ form the vertices of the n-cube $\mathcal{C}_n = [0,1]^3$. In other words, for nonzero choices of the coefficients c, the resulting polynomial has $\mathcal{C}_n$ as its *Newton polytope*. From this point-of-view, $\mathcal{D}_{2,2,\ldots,2}$ is realized as the $\mathcal{A}$-*discriminant* of the n-cube.

A seemingly unrelated construction associated to $\mathcal{C}_n$ is that of its *secondary polytope*. The secondary polytope of a point configuration $\mathcal{A}$ is a combinatorial object derived from information about the triangulations of $\mathrm{conv}(\mathcal{A})$. If $\mathcal{T}$ is a triangulation of the convex hull of $\mathcal{A}$ (involving only the points in $\mathcal{A}$) then its *GKZ-vector* is

$$\phi_{\mathcal{A}}(\mathcal{T}) \ = \ \sum_{a\in\mathcal{A}} \sum_{T\in\mathcal{T}|a\in T} \mathrm{vol}(T) e_a \ .$$

The *secondary polytope* of $\mathcal{A}$ is the convex hull of all such GKZ-vectors [DRS10, Theorem 5.1.9].

The goal of this section is to illustrate the relationship between the secondary polytope of the vertices of $\mathcal{C}_n$ and the $(2\times2\times\cdots\times2)$-hyperdeterminant: as shown in [GKZ08, Section 11], the secondary polytope of $\mathcal{C}_n$ is the Minkowski sum of the Newton polytopes of the hyperdeterminants of the faces of $\mathcal{C}_n$. Our process closely follows the exposition in [HSYY08].

4.3.1 The $2 \times 2 \times 2$ Hyperdeterminant

The polynomial $\mathcal{D}_{2,2,2}$ can be easily calculated using the `eliminate` command:

```
julia> C = cube(3, 0, 1);

julia> R, x, c = polynomial_ring(QQ, :x=>1:3, :c=>(0:1,0:1,0:1));

julia> f = sum(prod(x[i]^Int(p[i]) for i=1:3)
                * c[(Vector{Int}(p)+[1,1,1])...] for p=lattice_points(C))
x[1]*x[2]*x[3]*c[1, 1, 1] + x[1]*x[2]*c[1, 1, 0] +
x[1]*x[3]*c[1, 0, 1] + x[1]*c[1, 0, 0] +
x[2]*x[3]*c[0, 1, 1] + x[2]*c[0, 1, 0] +
x[3]*c[0, 0, 1] + c[0, 0, 0]

julia> I = ideal(R, vcat([derivative(f,t) for t = x], [f]));

julia> D222 = eliminate(I,x)[1]
c[0, 0, 0]^2*c[1, 1, 1]^2 - 2*c[0, 0, 0]*c[1, 0, 0]*c[0, 1, 1]*c[1, 1, 1]
↪ - [...] + c[1, 1, 0]^2*c[0, 0, 1]^2
```

Multiplying D_{222} by the determinants corresponding to all proper faces of $\mathcal{C}_3$ produces a polynomial E_{222} in 8 variables called the *principal* $2 \times 2 \times 2$ *determinant*. Observe that we need to shift the indices by the vector $(1, 1, 1)$ because indices in `Julia` are one-based.

```
julia> CubeFacets = [lattice_points(F) for F in faces(C,2)];

julia> VariableIndices = [[c+[1,1,1] for c=Vector{Vector{Int}}(F)]
                            for F=CubeFacets];

julia> FacialDeterminants = [c[v[1]...]*c[v[4]...]-c[v[2]...]*c[v[3]...]
                               for v=VariableIndices]
6-element Vector{QQMPolyRingElem}:
 c[0, 0, 0]*c[0, 1, 1] - c[0, 1, 0]*c[0, 0, 1]
 c[1, 0, 0]*c[1, 1, 1] - c[1, 1, 0]*c[1, 0, 1]
 c[0, 0, 0]*c[1, 0, 1] - c[1, 0, 0]*c[0, 0, 1]
 c[0, 1, 0]*c[1, 1, 1] - c[1, 1, 0]*c[0, 1, 1]
 c[0, 0, 0]*c[1, 1, 0] - c[1, 0, 0]*c[0, 1, 0]
 c[0, 0, 1]*c[1, 1, 1] - c[1, 0, 1]*c[0, 1, 1]

julia> E222 = D222*prod(FacialDeterminants);
```

The Newton polytope of E_{222} is a 4-polytope P_{222} in $\mathbb{R}^8$ with f-vector $(74, 152, 100, 22)$.

```
julia> NP = newton_polytope(E222);

julia> dim(NP)
4

julia> show(f_vector(NP))
ZZRingElem[74, 152, 100, 22]
```

Often one chooses to illustrate a 4-polytope like P_{222} in 3-dimensional space via a *Schlegel diagram*: an image in $\mathbb{R}^3$ which shows the combinatorial configuration of the 22 facets of P_{222} via a polyhedral subdivision of a 3-polytope into 21 polyhedra. The 22-nd facet of P_{222} is understood to correspond to the complement of this polytope. We provide two snapshots of this Schlegel diagram in Figure 7.

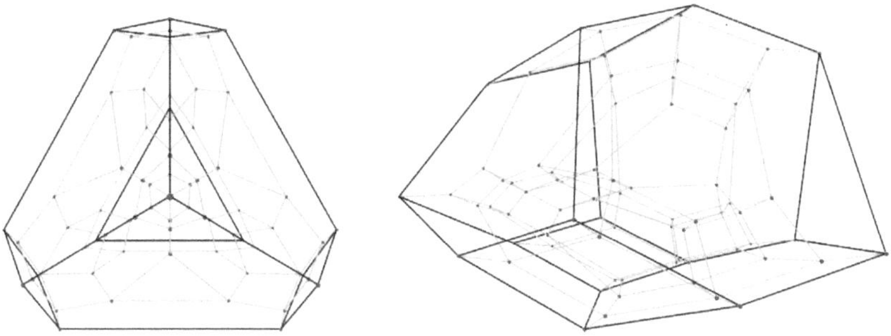

Fig. 7: Two views of a Schlegel diagram for the 4-polytope P_{222} in $\mathbb{R}^8$.

4.3.2 The Secondary Polytope and Triangulations of $\mathcal{C}_3$

The secondary polytope of a given polytope P is interpreted in OSCAR and Polymake to be the secondary polytope associated to the vertices of P. Constructing such an object in OSCAR is straightforward.

```
julia> C = cube(3,0,1)
Polytope in ambient dimension 3

julia> SP = secondary_polytope(C)
Polytope in ambient dimension 8
```

The main connection with the previous section on hyperdeterminants is that the secondary polytope of $\mathcal{C}_3$ *is* the Newton polytope of E_{222}, up to some minor coordinate manipulations.

```
julia> V = point_matrix(vertices(NP))[:, 4:11].+1;

julia> SP == convex_hull(V)
true
```

The elimination which gave us the polynomial D_{222} remains in the original ring, with eleven variables. So the code above projects out the first three coordinates. Adding the constant 1 to each coefficient of each vertex is related to the index shift.

Alternatively, one may take the definitional path towards computing a secondary polytope by computing all triangulations of the vertices of P, finding their GKZ-vectors, and taking the convex hull of those.

```
julia> AllTriangulations = all_triangulations(C);

julia> Tri_as_SOP = [subdivision_of_points(C,T)
                     for T in AllTriangulations];

julia> GKZ_Vectors = [gkz_vector(T) for T in Tri_as_SOP];

julia> SP_from_GKZ = convex_hull(GKZ_Vectors)
Polyhedron in ambient dimension 8

julia> SP_from_GKZ == SP
```

The raw output for a triangulation in OSCAR is an array of arrays: each array at the first level corresponds to a simplex and hence contains the indices of the vertices of that simplex. For many computations which require great efficiency, such a representation is preferred. However, to obtain contextual properties of a triangulation T, like the GKZ-vector, one must provide the relevant polytope P to OSCAR. Warning: the ordering of the vertices of P matters when constructing a subdivision of points, since the raw triangulation only refers to indices!

```
julia> T = AllTriangulations[1]
6-element Vector{Vector{Int64}}:
 [1, 2, 3, 5]
 [2, 3, 4, 5]
 [2, 4, 5, 6]
 [3, 4, 5, 7]
 [4, 5, 6, 7]
 [4, 6, 7, 8]

julia> S = subdivision_of_points(C,T)
Subdivision of points in ambient dimension 3

julia> is_regular(S)
true

julia> show(min_weights(S))
[3, 1, 1, 0, 0, 0, 0, 1]
```

The above evaluation checks that the triangulation T is regular, that is, it is induced as the lower hull of a lifting of the vertices of C by some weight function. Since it is regular, we also compute such a weight function. Finally, we showcase how to visualize a triangulation, making full-use of the visualization features `transparency` and the famous `explode` feature (see Figure 8).

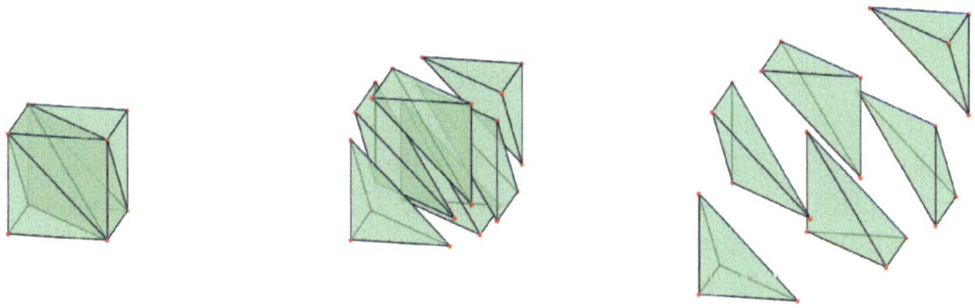

Fig. 8: Three views of a triangulation of the 3-cube with varying parameter choices for `explode`.

```
julia> IM = IncidenceMatrix(T);

julia> V = matrix(QQ, vertices(C));

julia> PC = polyhedral_complex(IM,V);

julia> visualize(PC)
```

The action of the symmetries of the cube induces an action on the GKZ-vectors by permutation of their coordinates. We compute the minimal representatives for the orbits these GKZ-vectors along with the number of elements in each orbit, as done in [HSYY08].

```
julia> G = automorphism_group(C)[:on_vertices]
Permutation group of degree 8

julia> OrbitRepresentatives=
        unique([minimum(gset(G,permuted,[v]))
          for v in GKZ_Vectors])
6-element Vector{Vector{QQFieldElem}}:
 [1, 3, 3, 5, 5, 3, 3, 1]
 [1, 3, 3, 5, 6, 2, 2, 2]
 [1, 3, 4, 4, 6, 2, 1, 3]
 [1, 4, 4, 3, 6, 1, 1, 4]
 [1, 5, 5, 1, 5, 1, 1, 5]
 [2, 2, 2, 6, 6, 2, 2, 2]

julia> OrbitSizes =
        [length(
          filter(x->minimum(gset(G,permuted,[x]))==u,
          GKZ_Vectors)) for u in OrbitRepresentatives];

julia> show(OrbitSizes)
[12, 24, 24, 8, 2, 4]
```

We close this section with some general remarks about computing triangulations and secondary polytopes; see also [DRS10]. This is a truly demanding task and computationally infeasible for most point sets. For instance, the 4-dimensional cube is known to have 87 959 448 regular triangulations, which come in 235 277 orbits. That is to say, the secondary polytope of C_4, which is 11-dimensional, has nearly 90 million vertices. The precise number of (regular) triangulations of C_n is unknown for $n \geq 5$. State of the art software for finding all triangulations is `TOPCOM` [Ram02]. For finding all *regular* triangulations `mptopcom` [JJK18] holds the current records. These two packages share parts of their code bases, both run in parallel, but there are significant technical differences. `mptopcom` has been developed with large computer clusters in mind, and this makes the software not very suitable for integration into `OSCAR`. Nonetheless, finding all orbits of regular triangulations of C_4 with `mptopcom` takes less than 10 minutes on a standard desktop computer. On the other hand, the native implementation of secondary polytopes in `OSCAR` is little more than a proof of concept. Pfeifle gave an explicit construction of the secondary polytope of C_3 [Pfe00].

5 Computing Convex Hulls

The following two computational tasks are fundamental to polyhedral geometry. The *convex hull problem* asks to convert a finite set $V \subset \mathbb{R}^d$ into finitely many affine half spaces whose intersection is the polytope $\operatorname{conv}(V)$. The *dual convex hull problem* is the reverse operation, which is algorithmically equivalent to its primal incarnation thanks to cone polarity. These problems enjoy a close relationship to linear programming: computing all of the vertices of a polytope, given inequalities, is the same as solving all (generic) linear programs with these constraints. For the geometric background of this discussion, see the textbooks [Grü03; Zie95; JT13]. In order to be computationally accessible, the coordinates are usually restricted to suitable subfields of $\mathbb{R}$. The most important example is $\mathbb{Q}$, but as we saw earlier, real algebraic number fields commonly occur too. Generally speaking, any ordered field works, e.g., ordered fields of rational functions occur at the intersection of optimization and tropical geometry [Jos21, §8.5].

Here we collect some general wisdom about convex hull computations, and how this is reflected in `OSCAR`. Many different convex hull algorithms are known and implemented; see [Sei18] for a survey. It is important to understand, that different algorithms may exhibit very different performance, depending on the kind of input. We illustrate this on one small example, which is not even the tip of the iceberg. Since computations are cached in the polyhedron object, organizing a fair race involves defining the *same* random polytope before each command. One achieves this by using the same seed value in the random construction.

```
julia> P = rand_spherical_polytope(6, 500; seed=11)
Polytope in ambient dimension 6

julia> Polymake.prefer("ppl") do
         @time n_facets(P)
       end
205.427732 seconds (7.13 G allocations: 321.966 GiB, 0.68% gc time)
58163

julia> P = rand_spherical_polytope(6, 500; seed=11);

julia> Polymake.prefer("beneath_beyond") do
         @time n_facets(P)
       end
 61.090983 seconds (335.63 M allocations: 5.309 GiB, 0.08% gc time)
58163

julia> P = rand_spherical_polytope(6, 500; seed=11);

julia> Polymake.prefer("libnormaliz") do
         @time n_facets(P)
       end
 11.167154 seconds (19.12 M allocations: 811.301 MiB, 0.07% gc time)
58163
```

OSCAR's implementations and interfaces are inherited from polymake. Here, we use the Parma Polyhedral Library (default) [BHZ08], polymake's own implementation beneath_beyond, and libnormaliz [BIRSS]. Observe that, in addition to the running time, the memory consumption varies considerably. For a more comprehensive analysis, with more drastic examples and different winners see [ABS97; Ass+17]. The timings are taken on a laptop with an Intel i7-1165G7 processor (4700MHz maximum clock) running on Manjaro Linux, with kernel 6.4.6-1, OSCAR v0.12.2.

Acknowledgements Michael Joswig is supported by MaRDI (Mathematical Research Data Initiative), funded by the Deutsche Forschungsgemeinschaft (DFG), project number 460135501, NFDI 29/1 "MaRDI – Mathematische Forschungsdateninitiative", and by the SFB-TRR 195–286237555 "Symbolic Tools in Mathematics and their Application". Taylor Brysiewicz is supported by an NSERC Discovery grant (RGPIN-2023-03551).

References

[Ass+17] B. Assarf et al. "Computing convex hulls and counting integer points with polymake ". *Math. Program. Comput.* 9.1 (2017), pp. 1–38. DOI: 10.1007/s12532-016-0104-z.

[ABS97] D. Avis, D. Bremner, and R. Seidel. "How good are convex hull algorithms?" In: vol. 7. 5-6. 11th ACM Symposium on Computational Geometry (Vancouver, BC, 1995). 1997, pp. 265–301. DOI: 10.1016/S0925-7721(96)00023-5.

[BHZ08] R. Bagnara, P. M. Hill, and E. Zaffanella. "The Parma Polyhedra Library: Toward a Complete Set of Numerical Abstractions for the Analysis and Verification of Hardware and Software Systems". *Science of Computer Programming* 72.1–2 (2008), 3–21. DOI: 10.1016/j.scico.2007.08.001.

[BL81] L. J. Billera and C. W. Lee. "A proof of the sufficiency of McMullen's conditions for f-vectors of simplicial convex polytopes". *J. Combin. Theory Ser. A* 31.3 (1981), pp. 237–255. DOI: 10.1016/0097-3165(81)90058-3.

[BIRSS] W. Bruns, B. Ichim, T. Römer, R. Sieg, and C. Söger. *Normaliz. Algorithms for rational cones and affine monoids.* Available at https://www.normaliz.uni-osnabrueck.de.

[BEF00] B. Büeler, A. Enge, and K. Fukuda. "Exact volume computation for polytopes: a practical study". In: *Polytopes—combinatorics and computation.* Vol. 29. DMV Sem. Birkhäuser, Basel, 2000, pp. 131–154.

[CLS11] D. A. Cox, J. B. Little, and H. K. Schenck. *Toric varieties.* Vol. 124. Graduate Studies in Mathematics. American Mathematical Society, Providence, RI, 2011. DOI: 10.1090/gsm/124.

[DRS10] J. A. De Loera, J. Rambau, and F. Santos. *Triangulations.* Vol. 25. Algorithms and Computation in Mathematics. Structures for algorithms and applications. Berlin: Springer-Verlag, 2010, pp. xiv+535.

[FH91] W. Fulton and J. Harris. *Representation theory.* Vol. 129. Graduate Texts in Mathematics. A first course, Readings in Mathematics. Springer-Verlag, New York, 1991, pp. xvi+551. DOI: 10.1007/978-1-4612-0979-9.

[GKZ08] I. M. Gelfand, M. M. Kapranov, and A. V. Zelevinsky. *Discriminants, resultants and multidimensional determinants.* Modern Birkhäuser Classics. Reprint of the 1994 edition. Birkhäuser Boston, Inc., Boston, MA, 2008, pp. x+523.

[Grü03] B. Grünbaum. *Convex polytopes.* Second. Vol. 221. Graduate Texts in Mathematics. Prepared and with a preface by Volker Kaibel, Victor Klee and Günter M. Ziegler. New York: Springer-Verlag, 2003, pp. xvi+468.

[HJS19] S. Hampe, M. Joswig, and B. Schröter. "Algorithms for tight spans and tropical linear spaces". *J. Symbolic Comput.* 91 (2019), pp. 116–128. DOI: 10.1016/j.jsc.2018.06.016.

[HSYY08] P. Huggins, B. Sturmfels, J. Yu, and D. S. Yuster. "The hyperdeterminant and triangulations of the 4-cube". *Math. Comp.* 77.263 (2008), pp. 1653–1679. DOI: 10.1090/S0025-5718-08-02073-5.

[Joh66] N. W. Johnson. "Convex polyhedra with regular faces". *Canadian J. Math.* 18 (1966), pp. 169–200. DOI: 10.4153/CJM-1966-021-8.

[JJK18] C. Jordan, M. Joswig, and L. Kastner. "Parallel enumeration of triangulations". *Electron. J. Combin.* 25.3 (2018), Paper 3.6, 27. DOI: 10.37236/7318.

[Jos21] M. Joswig. *Essentials of tropical combinatorics.* Vol. 219. Graduate Studies in Mathematics. Providence, RI: American Mathematical Society, 2021.

[JLLS16] M. Joswig, G. Loho, B. Lorenz, and B. Schröter. "Linear programs and convex hulls over fields of Puiseux fractions". In: *Proceedings of MACIS 2015, Berlin, November 11–13, 2015. LNCS 9582.* Springer, 2016, pp. 429–445. DOI: 10.1007/978-3-319-32859-1_37.

[JT13] M. Joswig and T. Theobald. *Polyhedral and Algebraic Methods in Computational Geometry.* Springer, 2013.

[KP02] V. Kaibel and M. E. Pfetsch. "Computing the face lattice of a polytope from its vertex-facet incidences". *Comput. Geom.* 23.3 (2002), pp. 281–290. DOI: 10.1016/S0925-7721(02)00103-7.

[KP03] V. Kaibel and M. E. Pfetsch. "Some algorithmic problems in polytope theory". In: *Algebra, geometry, and software systems.* Springer, Berlin, 2003, pp. 23–47.

[KS22] J. Kliem and C. Stump. "A new face iterator for polyhedra and for more general finite locally branched lattices". *Discrete Comput. Geom.* 67.4 (2022), pp. 1147–1173. DOI: 10.1007/s00454-021-00344-x.

[Knu98] D. E. Knuth. *The art of computer programming. Vol. 2.* Seminumerical algorithms, Third edition. Addison-Wesley, Reading, MA, 1998, pp. xiv+762.

[MS15] D. Maclagan and B. Sturmfels. *Introduction to tropical geometry.* Vol. 161. Graduate Studies in Mathematics. American Mathematical Society, Providence, RI, 2015, pp. xii+363. DOI: 10.1090/gsm/161.

[McM71] P. McMullen. "The numbers of faces of simplicial polytopes". *Israel J. Math.* 9 (1971), pp. 559–570. DOI: 10.1007/BF02771471.

[Pfe00] J. Pfeifle. "Secondary polytope of the 3-cube". *Electronic Geometry Models* (2000). No. 2000.09.031, http://www.eg-models.de/models/Polytopes/Secondary_Polytopes/2000.09.031/_direct_link.html.

[Pos09] A. Postnikov. "Permutohedra, associahedra, and beyond". *Int. Math. Res. Not. IMRN* 6 (2009), pp. 1026–1106. DOI: 10.1093/imrn/rnn153.

[PS09] A. Postnikov and R. Stanley. "Chains in the Bruhat order". *J Algebr Comb* 29 (2009), pp. 133–174.

[Ram02] J. Rambau. "TOPCOM: Triangulations of point configurations and oriented matroids." In: *Mathematical software. Proceedings of the 1st international congress, Beijing, China, August 17–19, 2002.* Singapore: World Scientific, 2002, pp. 330–340.

[Sch86] A. Schrijver. *Theory of linear and integer programming.* Wiley-Interscience Series in Discrete Mathematics. John Wiley & Sons, Ltd., Chichester, 1986, pp. xii+471.

[Sei18] R. Seidel. "Convex hull computations". In: *Handbook of Discrete and Computational Geometry.* Ed. by C. D. Tóth, J. E. Goodmann, and J. O'Rourke. 3rd edition. CRC Press, 2018. Chap. 26.

[Sta96] R. P. Stanley. *Combinatorics and commutative algebra.* Second. Vol. 41. Progress in Mathematics. Birkhäuser Boston, Inc., Boston, MA, 1996, pp. x+164.

[Sta80] R. P. Stanley. "The number of faces of a simplicial convex polytope". *Adv. in Math.* 35.3 (1980), pp. 236–238. DOI: `10.1016/0001-8708(80)90050-X`.

[Stu96] B. Sturmfels. *Gröbner bases and convex polytopes.* Vol. 8. University Lecture Series. American Mathematical Society, Providence, RI, 1996, pp. xii+162. DOI: `10.1090/ulect/008`.

[Zal69] V. A. Zalgaller. *Convex polyhedra with regular faces.* Vol. Vol. 2. Seminars in Mathematics, V. A. Steklov Mathematical Institute, Leningrad. Translated from Russian. Consultants Bureau, New York, 1969, pp. vi+95.

[Zie95] G. M. Ziegler. *Lectures on polytopes.* Vol. 152. Graduate Texts in Mathematics. New York: Springer-Verlag, 1995, pp. x+370.

Part III
Applications

Algorithms for Gromov–Witten Invariants of Elliptic Curves

Janko Böhm, Firoozeh Dastur, Alain Hoffmann, Hannah Markwig, and Ali Traore

Abstract This chapter presents an enhanced algorithm for exploring mirror symmetry in elliptic curves through the correspondence of algebraic and tropical geometry, focusing on Gromov–Witten invariants of elliptic curves and, in particular, Hurwitz numbers. We present a new highly efficient algorithm for computing generating series for these numbers. We have implemented the algorithm both using `Singular` and `OSCAR`. The implementations significantly outperform the current method provided in `Singular`. The `OSCAR` implementation, benefiting in particular from just-in-time compilation, again outperforms the implementation of the new algorithm in `Singular` by far. This advancement in computing the Gromov–Witten invariants facilitates a study of number theoretic and geometric properties of the generating series, including quasi-modularity and homogeneity.

Key words: Mirror symmetry, elliptic curves, Feynman integrals, tropical geometry, Hurwitz numbers, quasimodular forms

Janko Böhm
Fachbereich Mathematik, RPTU Kaiserslautern-Landau, e-mail: `boehm@mathematik.uni-kl.de`

Firoozeh Dastur
Fraunhofer ITWM and Fachbereich Mathematik, RPTU Kaiserslautern-Landau, e-mail: `firoozeh.aga@itwm.fraunhofer.de`

Alain Hoffmann
Fachbereich Mathematik, RPTU Kaiserslautern-Landau, e-mail: `hoffmann@mathematik.uni-kl.de`

Hannah Markwig
Fachbereich Mathematik, Eberhard Karls Universität Tübingen, e-mail: `hannah@math.uni-tuebingen.de`

Ali Traore
Fachbereich Mathematik, RPTU Kaiserslautern-Landau, e-mail: `atraore@rptu.de`

W. Decker et al. (eds.), *The Computer Algebra System OSCAR*,
Algorithms and Computation in Mathematics 32,
https://doi.org/10.1007/978-3-031-62127-7_6

1 Introduction

Mirror symmetry is a deep duality relation motivated by physics. It relates invariants of a manifold and its "mirror manifold" in a way allowing an exchange of methods. Tropical methods find a way into mirror symmetry most prominently in the famous Gross-Siebert program [GS06].

Here, we present the special case of elliptic curves, for which mirror symmetry is best understood [Dij95]. The tropical side of mirror symmetry for elliptic curves was discovered within the Collaborative Research Center SFB-TRR 195 [BBBM17; BGM22a; BGM22b], which is also responsible for the development of `OSCAR`.

One consequence of mirror symmetry for elliptic curves is the equality of the generating function of Hurwitz numbers (resp. more generally, of descendant Gromov–Witten invariants) to certain Feynman integrals which are complex analytic path integrals whose construction is governed by combinatorics.

Hurwitz numbers are traditional enumerative invariants that count covers of Riemann surfaces satisfying fixed ramification data. Their definition goes back to Hurwitz himself and was used to study the irreducibility of the famous moduli space of smooth genus g curves, a fundamental object in algebraic geometry. In modern mathematics, Hurwitz numbers provide fruitful interactions between several mathematical areas such as geometry, topology, representation theory, combinatorics and mathematical physics [CM16].

Tropical geometry can be viewed as a degenerate version of algebraic geometry. It has been used successfully to solve enumerative problems in algebraic geometry [Mik05]. In particular, by counting *tropical covers* (see Definition 5), we can determine Hurwitz numbers via a so-called *correspondence theorem* [CJM10; BBBM17]. It allows to translate any computations we perform involving tropical covers to traditional Hurwitz theory.

In this chapter, we describe how one can use tropical mirror symmetry for elliptic curves in order to compute generating series for tropical Hurwitz numbers of an elliptic curve and their generalizations (up to fixed order) by means of computing Feynman integrals. We discuss explicit algorithms and implementations to handle this task. An algorithm computing Hurwitz numbers via Feynman integrals has been introduced in [BBBM17]. It works by computing coefficients in the Laurent series expansion of the propagator of the Feynman integral. Based on this algorithm, we develop in this chapter a significantly improved version which shows a better performance when applied to current research problems. This algorithm works by assigning to each cover a so-called *flip signature* to each cover.Covers with the same signature lead to the same integral. Further improvements are achieved by direct computation of the coefficients of the Laurent series, avoiding explicit expansion altogether. We have created two structurally identical implementations of the algorithm, one using the computer algebra system `Singular` and one using `OSCAR`. Both implementations outperform the previous algorithm of [BBBM17] in its implementation in `Singular` significantly, which demonstrates the structural improvement in the algorithm. The OSCAR implementation again by far outperforms the implementation of the new algorithm in `Singular` (which partially relies on the interpreted `Singular` language). We attribute this performance improvement to the just-in-time (JIT) compilation features, which turn at run-time easy-to-comprehend `Julia` code into code which executes almost as fast machine code, but also to superior arithmetic in OSCAR. This demonstrates the potential of the OSCAR platform.

2 Mirror Symmetry for Elliptic Curves

2.1 Hurwitz Numbers

Hurwitz numbers count branched covers of non-singular curves with a given ramification profile over fixed points. Here, we consider covers of elliptic curves. Hurwitz numbers are topological invariants, in particular they do not depend on the position of the branch points. Moreover, since all complex elliptic curves are homeomorphic to the real torus, numbers of covers of an elliptic curve do not depend on the choice of the base curve. We thus fix an arbitrary complex elliptic curve $\mathcal{E}$.

Let $\mathcal{C}$ be a non-singular curve of genus g and let $\phi : \mathcal{C} \to \mathcal{E}$ be a cover. We denote by d be the degree of ϕ, i.e., the number of preimages of a generic point in $\mathcal{E}$. For our purpose, it is sufficient to consider covers which are *simply ramified*, that is, over any branch point exactly two sheets of the map come together and all others stay separate. In other words, the *ramification profile* (that is, the partition of the degree indicating the multiplicities of the inverse images of a branch point) of a simple branch point is $(2, 1, \ldots, 1)$. It follows from the Riemann–Hurwitz formula (see e.g. [Har77], Corollary IV.2.4) that a simply ramified cover of $\mathcal{E}$ has exactly $2g - 2$ branch points. Two covers $\varphi : \mathcal{C} \to \mathcal{E}$ and $\varphi' : \mathcal{C}' \to \mathcal{E}$ are isomorphic if there exists an isomorphism of curves $\phi : \mathcal{C} \to \mathcal{C}'$ such that $\varphi' \circ \phi = \varphi$.

Definition 1 (Hurwitz Numbers) Fix $2g-2$ points $p_1, \ldots, p_{2g-2}$ in $\mathcal{E}$. We define the *Hurwitz number* $N_{d,g}$ to be the weighted number of (isomorphism classes of) simply ramified covers $\phi : \mathcal{C} \to \mathcal{E}$ of degree d, where $\mathcal{C}$ is a connected curve of genus g, and the branch points of ϕ are the points p_i for $i = 1, \ldots, 2g - 2$. We count each such cover ϕ with weight $|\operatorname{Aut}(\phi)|$.

Remark 2 In this definition, we choose the convention to fix a marking of the branch points p_i. In the literature, this is not always the case, leading to a factor of $(2g - 2)!$ when compared to our definition.

Definition 3 We package the Hurwitz numbers of Definition 1 into a generating series as follows:

$$F_g(q) := \sum_{d=1}^{\infty} N_{d,g} q^{2d}.$$

2.2 Tropical Hurwitz Numbers

Definition 4 (Tropical Curves) A (generic) *tropical curve* C (without ends) is a connected, finite, trivalent, metric graph. Its *genus* is given by its first Betti number. The *combinatorial type* of a curve is its homeomorphism class, that is, the underlying graph without the lengths of edges.

A tropical elliptic curve consists of one edge forming a circle of a certain length. We will fix a tropical elliptic curve E (for example, one having length 1). Moreover, to fix notation, in the following we denote by C a tropical curve of genus g and combinatorial type Γ.

Definition 5 (Tropical Covers) A map $\pi : C \to E$ is a *tropical cover* of E if it is continuous, non-constant, integer affine on each edge and respects the *balancing condition* at every vertex $P \in C$:

For an edge e of C denote by w_e the *weight of* e, that is, the absolute value of the slope of $\pi_{|e}$. Consider a (small) open neighbourhood U of $p = \pi(P)$. Then U combinatorially consists of p together with two rays r_1 and r_2, left and right of p. Let V be the connected component of $\pi^{-1}(U)$ which contains P. Then V consists of P together with three adjacent rays. We say that π is balanced at P if

$$\sum_{R \text{ maps to } r_1} w_R = \sum_{R \text{ maps to } r_2} w_R,$$

where R runs over the three rays adjacent to P and rays inherit their weights from the corresponding edges.

The *degree* of a cover is the weighted number of preimages of a generic point: For all $p \in E$ not having any vertex of C as preimage we have

$$d = \sum_{P \in C : \pi(P) = p} w_{e_P},$$

where e_P is the edge of C containing P. The images of the vertices of C are called the *branch points* of π.

Example 6 A tropical cover of degree 4 with a genus 2 source curve is depicted in Figure 1 (taken from [BBBM17]). The red numbers close to the vertex P are the weights of the corresponding edges, the black numbers denote the lengths. The cover is balanced at P since there is an edge of weight 3 leaving in one direction and an edge of weight 2 plus an edge of weight 1 leaving in the opposite direction.

We can see that the length of an edge of C is determined by its weight and the length of its image. We will therefore not specify edge lengths in the following.

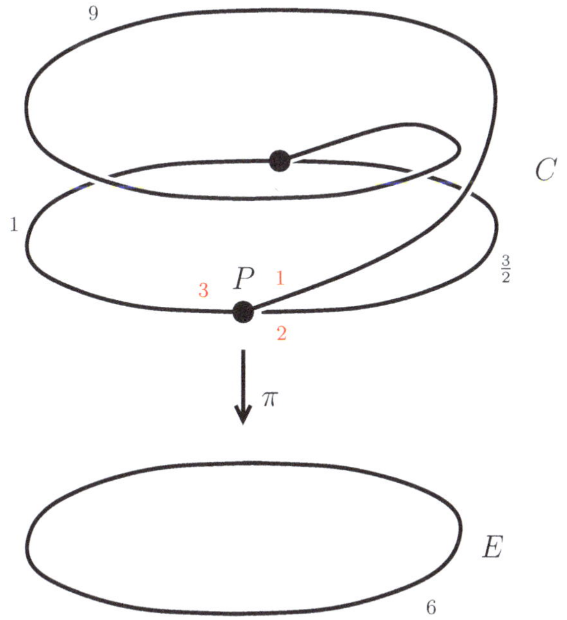

Fig. 1: A tropical cover of degree 4 with genus 2 source curve.

Definition 7 (Isomorphisms of Curves and Covers) An *isomorphism of tropical curves* is an isometry of metric graphs. Two covers $\pi : C \to E$ and $\pi' : C' \to E$ are isomorphic if there is an isomorphism of curves $\phi : C \to C'$ such that $\pi' \circ \phi = \pi$.

As usual when counting tropical objects we have to weight them with a certain multiplicity.

Definition 8 (Multiplicities) The multiplicity of a cover $\pi : C \to E$ is defined to be

$$\text{mult}(\pi) := \frac{1}{|\operatorname{Aut}(\pi)|} \prod_e w_e,$$

where the product goes over all edges e of C and $\operatorname{Aut}(\pi)$ is the automorphism group of π.

Definition 9 (Tropical Hurwitz Numbers) Fix $2g - 2$ branch points $p_1, \ldots, p_{2g-2}$ in the tropical, elliptic curve E. The *tropical Hurwitz number*

$N_{d,g}^{\text{trop}}$ is the weighted number of isomorphism classes of degree d covers $\pi : C \to E$ having their branch points at the p_i, where C is a curve of genus g:

$$N_{d,g}^{\text{trop}} := \sum_{\pi:C\to E} \text{mult}(\pi).$$

Theorem 10 (Correspondence Theorem) *The algebraic and tropical Hurwitz numbers of simply ramified covers of an elliptic curve coincide (see Definition 1 and 9), i.e., we have*

$$N_{d,g}^{\text{trop}} = N_{d,g}.$$

For a proof, see the extension of the result of [BBM11] in [BBBM17].

2.3 Feynman Integrals

Definition 11 (The Propagator) We define the *propagator*

$$P(z,q) := \frac{1}{4\pi^2}\wp(z,q) + \frac{1}{12}E_2(q^2)$$

in terms of the Weierstraß-$\wp$-function and the Eisenstein series

$$E_2(q) := 1 - 24\sum_{d=1}^{\infty}\sigma(d)q^d.$$

Here, $\sigma = \sigma_1$ denotes the sum-of-divisors function $\sigma(d) = \sum_{m|d} m$.

The variable q above should be considered as a coordinate of the moduli space of elliptic curves and the variable z is the complex coordinate of a fixed elliptic curve. (More precisely, $q = e^{i\pi\tau}$, where $\tau \in \mathbb{C}$ is the parameter in the upper half plane in the well-known definition of the Weierstraß-$\wp$-function.)

Definition 12 (Feynman Graphs and Integrals) A *Feynman graph Γ* of genus g is a trivalent connected graph of genus g. For a Feynman graph, we fix throughout a reference labeling $x_1, \ldots, x_{2g-2}$ of the $2g-2$ trivalent vertices and $q_1, \ldots, q_{3g-3}$ of the edges of Γ, respectively.

For an edge q_k of Γ connecting the vertices x_i and x_j, we define a function

$$P_k := P(z_i - z_j, q),$$

where P denotes the propagator of Definition 11. (The function P_k is invariant under re-ordering of z_i and z_j.) Pick a total ordering Ω of the vertices and starting points of the form $iy_1, \ldots, iy_{2g-2}$ in the complex plane, where the y_j are pairwise distinct small real numbers. We define integration paths $\gamma_1, \ldots, \gamma_{2g-2}$ by

$$\gamma_j : [0,1] \to \mathbb{C} : t \mapsto iy_j + t,$$

such that the order of the real coordinates y_j of the starting points of the paths equals Ω. We then define the integral

$$I_{\Gamma,\Omega}(q) := \int_{z_j \in \gamma_j} \prod_{k=1}^{3g-3} (-P_k). \tag{1}$$

Finally, we define

$$I_\Gamma(q) = \sum_{\Omega} I_{\Gamma,\Omega}(q),$$

where the sum runs over all $(2g-2)!$ orders of the vertices.

The following relates Hurwitz numbers and Feynman integrals and can be viewed as a consequence of mirror symmetry (see Theorem 3 of [Dij95]):

Theorem 13 (Mirror Symmetry for Elliptic Curves) *Let $g > 1$. For the definition of the invariants, see Definitions 1 and 12. We have*

$$F_g(q) = \sum_{d=1}^{\infty} N_{d,g} q^{2d} = \sum_{\Gamma} I_\Gamma(q) \cdot \frac{1}{|\operatorname{Aut}(\Gamma)|},$$

where $\operatorname{Aut}(\Gamma)$ denotes the automorphism group of Γ and the sum goes over all trivalent graphs Γ of genus g.

Theorem 14 *After coordinate change, the propagator $P(x,q)$ of Definition 11 with $x = e^{i\pi z}$ equals*

$$P(x,q) = -\frac{x^2}{(x^2-1)^2} - \sum_{n=1}^{\infty} \left(\sum_{d|n} d \left(x^{2d} + x^{-2d} \right) \right) q^{2n}.$$

For a proof, see [BBBM17, Theorem 2.22].

To better relate Feynman integrals with counts of tropical covers, we label the sources of our tropical covers just like a Feynman graph and consider labeled tropical covers. This way, we can also refine the notion of degree by considering the degree of each edge of the source curve. Also, we can add information about the edge in the definition of the Feynman integral in Equation 1 by making the propagator associated to edge k dependent on a formal variable q_k, see Definition 25 below.

Definition 15 We fix a base point p_0 in E. For a tuple $\underline{a} = (a_1, \ldots, a_{3g-3})$ of non-negative integers, we define $N^{\text{trop}}_{\underline{a},g}$ to be the weighted number of labeled tropical covers $\hat{\pi} : C \to E$ of degree $\sum_{i=1}^{3g-3} a_i$ where C has genus g, such that $\hat{\pi}$ has its branch points at the prescribed positions and satisfying the condition

$$\#(\hat{\pi}^{-1}(p_0) \cap q_i) \cdot w_i = a_i$$

for all $i = 1, \ldots, 3g-3$.

Theorem 16 (Refined Version of Tropical Mirror Symmetry) *Let $g > 1$. We have*

$$F_g(q_1, \ldots, q_{3g-3}) = \sum_{\underline{a}} N^{\text{trop}}_{\underline{a},g} q^{2d} = \sum_{\Gamma} I_\Gamma(q_1, \ldots, q_{3g-3}),$$

that is, the coefficient of the monomial $q^{2 \cdot a}$ in $I_\Gamma(q_1, \ldots, q_{3g-3})$ equals $N^{\text{trop}}_{\underline{a},\Gamma}$.

Lemma 17 *Fix a Feynman graph Γ, an order Ω , and a tuple $(q_1, \ldots, q_{3g-3})$ as in Definition 14. We express the coefficient of $q^{2 \cdot \underline{a}}$ in $I_{\Gamma,\Omega}(q_1, \ldots, q_{3g-3})$. Assume that $a_k = 0$, and that the edge q_k connects the two vertices x_{k_1} and x_{k_2}. Choose the order of labeling of x_{k_1} and x_{k_2} such that Ω implies $\left|\frac{x_{k_1}}{x_{k_2}}\right| < 1$ for the starting points on the integration paths. Then the coefficient of $q^{2 \cdot \underline{a}}$ equals the constant term of the series*

$$\prod_{k|a_k=0} \left(\sum_{w_k=1}^{\infty} w_k \left(\frac{x_{k_1}}{x_{k_2}} \right)^{2w_k} \right) \cdot \prod_{k|a_k \neq 0} \left(\sum_{w_k|a_k} w_k \left(\left(\frac{x_{k_1}}{x_{k_2}} \right)^{2w_k} + \left(\frac{x_{k_2}}{x_{k_1}} \right)^{2w_k} \right) \right). \tag{2}$$

For a Taylor series

$$F(x_1, \ldots, x_n) = \sum_{\substack{1 \leq j \leq n \\ a_j \geq 0}} \alpha(a_1, \ldots, a_n) x_1^{a_1}, \ldots, x_n^{a_n}$$

we write

$$\text{coeff}_{[x_1^{a_1}, \ldots, x_n^{a_n}]}(F) = \alpha(a_1, \ldots, a_n).$$

Lemma 18 *Fix a Feynman graph Γ and an order Ω as in Definition 12 and write*

$$P_{\Gamma,\Omega} = \prod_{k=1}^{3g-3} \left(-P \left(\frac{x_{k_1}}{x_{k_2}}, q \right) \right).$$

Then we have

$$I_{\Gamma,\Omega}(q) = \text{coeff}_{[x_1^0, \ldots, x_{2g-2}^0]}(P_{\Gamma,\Omega}).$$

2.4 Descendant Gromov–Witten Invariants

Theorem 13 can be generalized to descendant Gromov–Witten invariants. Following Okounkov–Pandharipande [OP06], these are related to covers with more complicated ramification profiles.

A *stable map* of degree d from a curve of genus g to $\mathcal{E}$ with n markings is a map $f : \mathcal{C} \to \mathcal{E}$, where $\mathcal{C}$ is a connected projective curve with at worst nodal singularities, and with n distinct nonsingular marked points $x_1, \dots, x_n \in \mathcal{C}$, such that $f_*([\mathcal{C}]) = d[\mathcal{E}]$ and f has a finite group of automorphisms. The moduli space of stable maps, denoted $\overline{\mathcal{M}}_{g,n}(\mathcal{E}, d)$, is a proper Deligne-Mumford stack of virtual dimension $2g - 2 + n$ [Beh97; BF97]. The i-th evaluation morphism is the map $\mathrm{ev}_i : \overline{\mathcal{M}}_{g,n}(\mathcal{E}, d) \to \mathcal{E}$ that sends a point $[\mathcal{C}, x_1, \dots, x_n, f]$ to $f(x_i) \in \mathcal{E}$. The i-th cotangent line bundle $\mathbb{L}_i \to \overline{\mathcal{M}}_{g,n}(\mathcal{E}, d)$ is defined by a canonical identification of its fiber over a moduli point $(\mathcal{C}, x_1, \dots, x_n, f)$ with the cotangent space $T^*_{x_i}(\mathcal{C})$. The first Chern class of the cotangent line bundle is called a *psi class* ($\psi_i = c_1(\mathbb{L}_i)$).

Definition 19 Fix g, n, d and let $k_1, \dots, k_n$ be non-negative integers with

$$k_1 + \ldots + k_n = 2g - 2.$$

The *stationary descendant Gromov–Witten invariant* $\langle \tau_{k_1}(pt) \ldots \tau_{k_n}(pt) \rangle_{g,n}^{\mathcal{E},d}$ is defined by

$$\langle \tau_{k_1}(pt) \ldots \tau_{k_n}(pt) \rangle_{g,n}^{\mathcal{E},d} = \int_{[\overline{\mathcal{M}}_{g,n}(\mathcal{E},d)]^{\mathrm{vir}}} \prod_{i=1}^{n} ev_i^*(pt) \psi_i^{k_i}, \tag{3}$$

where pt denotes the class of a point in $\mathcal{E}$.

In order to define tropical multiplicities, we also need to discuss relative descendant Gromov–Witten invariants of $\mathbb{P}^1$. They are constructed using moduli spaces of *relative stable maps* $\overline{\mathcal{M}}_{g,n}(\mathbb{P}^1, \mu, \nu, d)$, where part of the data specified are the ramification profiles μ and ν which we fix over 0 resp. $\infty \in \mathbb{P}^1$. The preimages of 0 and ∞ are marked. A detailed discussion of spaces of relative stable maps to $\mathbb{P}^1$ and their boundary is not necessary for our purpose, we refer to [Vak08]. We write

$$\langle \mu | \tau_{k_1}(pt) \ldots \tau_{k_n}(pt) | \nu \rangle_{g,n}^{\mathbb{P}^1,d} = \int_{[\overline{\mathcal{M}}_{g,n}(\mathbb{P}^1,\mu,\nu,d)]^{\mathrm{vir}}} \prod_{i=1}^{n} \mathrm{ev}_i^*(pt) \psi_i^{k_i}. \tag{4}$$

Tropically, adding descendants amounts to allowing vertices of higher genus and higher valency. For this purpose, we first generalize the definition of abstract tropical curve:

Definition 20 An *abstract tropical curve* is a connected metric graph Γ, such that edges leading to leaves (called *ends*) have infinite length, together with a genus function $g : \Gamma \to \mathbb{Z}_{\geq 0}$ with finite support. Locally around a point p, Γ is homeomorphic to a star with r half rays. The number r is called the *valence* of the point p and is denoted by $\mathrm{val}(p)$. The vertex set of Γ consists of the nonzero points of the genus function and the points of valence different from 2. The vertices of valence greater than 1 are called *inner vertices*. Besides *edges*, we introduce the notion of *flags* of Γ. A flag is a pair (V, e) of a vertex V and an edge e incident to it. ($V \in \partial e$). Edges that are not ends are required to have finite length and are referred to as *bounded* or *internal* edges.

A *marked tropical curve* is a tropical curve whose leaves are labeled. An isomorphism of a tropical curve is an isometry respecting the leaf markings and the genus function. The *genus* of a tropical curve Γ is given by

$$g(\Gamma) = h_1(\Gamma) + \sum_{p \in \Gamma} g(p).$$

The *combinatorial type* is the equivalence class of tropical curves obtained by identifying any two tropical curves which differ only by edge lengths.

Definition 5 introducing tropical covers easily extends to allow such more general source curves.

Definition 21 (Psi- and Point Conditions) We say that a tropical cover $\pi : \Gamma_1 \to \Gamma_2$ with a marked end i *satisfies a psi-condition* with power k at i, if the vertex V adjacent to the marked end i has valency $k + 3 - 2g(V)$. We say $\pi : \Gamma_1 \to \Gamma_2$ *satisfies the point conditions* $p_1, \ldots, p_n \in \Gamma_2$ if

$$\{\pi(1), \ldots, \pi(n)\} = \{p_1, \ldots, p_n\}.$$

Fix g, n, d and let $k_1, \ldots, k_n$ be non-negative integers with

$$k_1 + \ldots + k_n = 2g - 2.$$

Let $\pi : \Gamma \to E$ be a tropical cover such that Γ is of genus g and has n marked ends. Fix n distinct points $p_1, \ldots, p_n \in E$. Assume that at the marked end i, a psi-condition with power k_i is satisfied, and that the point conditions are satisfied. The marked ends must be adjacent to different vertices, since they satisfy different point conditions. It follows from an Euler characteristic argument incorporating the valencies imposed by the psi-conditions that Γ has exactly n vertices, each adjacent to one marked end.

Locally at the marked end i, the cover sends the vertex to an interval consisting of two flags f and f'. We define the *local vertex multiplicity* $\mathrm{mult}_i(\pi)$ to be a one-point relative descendant Gromov–Witten invariant:

$$\mathrm{mult}_i(\pi) = \langle \mu_f | \tau_{k_i}(pt) | \mu_{f'} \rangle_{g_i,1}^{\mathbb{P}^1, d_i}, \tag{5}$$

where g_i denotes the genus of the vertex adjacent to the marked end i, d_i its local degree, and μ_f resp. $\mu_{f'}$ the ramification profiles above the two flags of the image interval.

We define the multiplicity of π to be

$$\frac{1}{|\operatorname{Aut}(\pi)|} \cdot \prod_i \mathrm{mult}_i(\pi) \cdot \prod_e \omega(e). \tag{6}$$

Note that all ends of a tropical cover of E are contracted ends, with image points the points p_i which we fix as conditions in E.

Definition 22 (Tropical Stationary Descendant Gromov–Witten Invariant of E) For g, n, d, $k_1, \ldots, k_n$ as above, define the *tropical stationary descendant Gromov–Witten invariant*

$$\langle \tau_{k_1}(pt) \ldots \tau_{k_n}(pt) \rangle_{g,n}^{\mathcal{E},d,\mathrm{trop}}$$

to be the weighted count of tropical genus g degree d covers of E with n marked points satisfying point and psi-conditions as above, each counted with its multiplicity as defined in (6).

For descendant invariants, a correspondence theorem holds as well:

Theorem 23 (Correspondence Theorem for Descendants) *A stationary descendant Gromov–Witten invariant of $\mathcal{E}$ coincides with its tropical counterpart:*

$$\langle \tau_{k_1}(pt) \ldots \tau_{k_n}(pt) \rangle_{g,n}^{\mathcal{E},d} = \langle \tau_{k_1}(pt) \ldots \tau_{k_n}(pt) \rangle_{g,n}^{E,d,\mathrm{trop}}.$$

For a proof, see [CJMR18, Theorem 3.2.1].

Generating series of descendant invariants can also be expressed in terms of Feynman integrals. However to accommodate this setting, we as well need to extend our notion of Feynman integrals.

Definition 24 (The Propagator and the $\mathcal{S}$-Function) We define the *propagator* as a (formal) series in x and q,

$$P(x,q) = \sum_{w=1}^{\infty} w \cdot x^w + \sum_{a=1}^{\infty} \left(\sum_{w|a} w \left(x^w + x^{-w}\right) \right) q^a$$

and the $\mathcal{S}$-function as series in z,

$$\mathcal{S}(z) = \frac{\sinh(z/2)}{z/2}.$$

We also introduce a further formal power series in q, which should be viewed as the propagator for loop edges:

$$P^{\mathrm{loop}}(q) = \sum_{a=1}^{\infty} \left(\sum_{w|a} w \right) q^a.$$

Analogous to the way we generalized our notion of abstract tropical curves which can serve as sources of our tropical covers, we have to generalize the notion of Feynman graphs. We allow any graph Γ without ends with n vertices which are labeled $x_1, \ldots, x_n$ and with labeled edges $q_1, \ldots, q_r$. By convention, we assume that $q_1, \ldots, q_s$ are loop edges and $q_{s+1}, \ldots, q_r$ are non-loop edges.

Definition 25 (Feynman Integrals) Let Γ be a Feynman graph. Let Ω be an order of the n vertices of Γ. For $k > s$, denote the vertices adjacent to the (non-loop) edge q_k by x_{k^1} and x_{k^2}, where we assume $x_{k^1} < x_{k^2}$ in Ω. We define the *Feynman integral* for Γ and Ω to be

$$I_{\Gamma,\Omega}(q) = \mathrm{Coef}_{[x_1^0 \dots x_n^0]} \prod_{k=1}^{s} P^{\mathrm{loop}}(q) \cdot \prod_{k=s+1}^{r} P\Big(\frac{x_{k^1}}{x_{k^2}}, q\Big).$$

$$I_{\Gamma,\Omega}(q_1,\dots,q_r) = \mathrm{Coef}_{[x_1^0 \dots x_n^0]} \prod_{k=1}^{s} P^{\mathrm{loop}}(q_k) \prod_{k=s+1}^{r} P\Big(\frac{x_{k^1}}{x_{k^2}}, q_k\Big).$$

Finally, we set

$$I_\Gamma(q) = \sum_{\Omega} I_{\Gamma,\Omega}(q),$$

where the sum goes over all $n!$ orders of the vertices of Γ.

If we assume $|x| < 1$ to express the q-constant coefficient of the (non-loop) propagator (that is, the first sum appearing in the propagator series in Definition 11) as the rational function $\frac{x^2}{(x^2-1)^2}$ (using geometric series expansion), we can view the series from which we take the $x_1^0 \dots x_n^0$-coefficient in the Feynman integral above as a function on a Cartesian product of elliptic curves. The Feynman integral then becomes a path integral in complex analysis as above. Note that using the change of coordinates $x = e^{i\pi u}$ the (non-loop) propagator has the form of Definition 11.

Definition 26 (Feynman Integrals With Vertex Contributions) Let Γ be a Feynman graph, and equip it with an additional genus function $\underline{g}$ associating a non-negative integer g_i to every vertex x_i. Let Ω be an order of the n vertices of Γ. We adapt our notion of propagators from Definitions 24 and 25 to include vertex contributions: for non-loop edges, we set

$$\begin{aligned}\tilde{P}(\frac{x_{k^1}}{x_{k^2}}, q) &= \sum_{w=1}^{\infty} \mathcal{S}(wz_{k^1})\mathcal{S}(wz_{k^2}) \cdot w \cdot \left(\frac{x_{k^1}}{x_{k^2}}\right)^w \\ &+ \sum_{a=1}^{\infty} \left(\sum_{w|a} \mathcal{S}(wz_{k^1})\mathcal{S}(wz_{k^2}) \cdot w \cdot \left(\left(\frac{x_{k^1}}{x_{k^2}}\right)^w + \left(\frac{x_{k^2}}{x_{k^1}}\right)^w \right) \right) \cdot q^a.\end{aligned}$$

For loop-edges connecting the vertex x_{k^1} to itself, we set

$$\tilde{P}^{\mathrm{loop}}(q) = \sum_{a=1}^{\infty} \left(\sum_{w|a} \mathcal{S}(wz_{k^1})^2 \cdot w \right) q^a.$$

The new variables z_{k_i} are introduced for each vertex in order to take care of the genus contribution.

We define the *Feynman integral with vertex contributions* for Γ, $\underline{g}$ and Ω to be

$$I_{\Gamma,\underline{g},\Omega}(q) = \mathrm{Coef}_{[z_1^{2g_1}\ldots z_n^{2g_n}]}\,\mathrm{Coef}_{[x_1^0\ldots x_n^0]}\prod_{i=1}^{n}\frac{1}{\mathcal{S}(z_i)}\prod_{k=1}^{s}\tilde{P}^{\mathrm{loop}}(q)\prod_{k=s+1}^{r}\tilde{P}(\frac{x_{k^1}}{x_{k^2}},q).$$

Again, we set

$$I_{\Gamma,\underline{g}}(q) = \sum_{\Omega} I_{\Gamma,\underline{g},\Omega}(q)$$

where the sum goes over all $n!$ orders of the vertices.

With this, we can generalize Theorem 13 to the version involving descendants:

Theorem 27 (Mirror Symmetry for E, Version With Descendants) *Fix $g \geq 2$, $n \geq 1$ and $k_1, \ldots, k_n \geq 1$ satisfying $k_1 + \ldots + k_n = 2g - 2$.*

We can express the series of descendant Gromov–Witten invariants of E in terms of Feynman integrals,

$$\sum_{d\geq 1}\langle\tau_{k_1}(pt)\ldots\tau_{k_n}(pt)\rangle_{g,n}^{\mathcal{E},d}q^d = \sum_{(\mathrm{ft}(\Gamma),\underline{g})}\frac{1}{|\,\mathrm{Aut}(\mathrm{ft}(\Gamma),\underline{g})|}I_{\Gamma,\underline{g}}(q),$$

*where Γ is a Feynman graph with a genus function $\underline{g}$, such that the vertex x_i has genus g_i and valency $k_i + 2 - 2g_i$, and such that $h^1(\Gamma) + \sum g_i = g$, and where we consider automorphisms of unlabeled graphs (*ft *is the forgetful map that forgets all labels of a Feynman graph Γ) respecting the genus function.*

For a proof, see [Li11; BGM22b].

3 Computation of Generating Series

In this section, we discuss our improved algorithm for computing tropical Hurwitz numbers and Gromov–Witten invariants via evaluation of Feynman integrals. The previous algorithm is implemented in `Singular` in the library `ellipticcovers.lib` [BBBM18]. It is based on evaluating the propagator product as a Laurent series in the variables x_i in the order given by Ω. For $k > s$, denoting the vertices adjacent to the (non-loop) edges q_k by x_{k^1} and x_{k^2} in any order, and fixed degree partition $\underline{a}$ over the base point, we evaluate

$$I_{\Gamma,\Omega,\underline{a}} = \mathrm{Coef}_{[x^0_{\Omega(1)}\cdots x^0_{\Omega(n)}]}\,\mathrm{Coef}_{\underline{a}}\prod_{k=1}^{s}P^{\mathrm{loop}}(q_k)\prod_{k=s+1}^{r}P\Big(\frac{x_{k^1}}{x_{k^2}},q_k\Big)$$

by extracting the constant coefficient in the x_i and sum over all Ω. Vertex contributions are introduced as in Definition 26. For more details, see also [BBBM17].

3.1 Basic Improved Algorithm

Before turning to the flip signature, we first describe a structural improvement to the above mentioned method. Let us consider an edge labeled q_k in the graph Γ, connecting vertices x_i and x_j. The propagator associated with this edge will be a polynomial in the variable q_k. However, we are specifically interested in extracting the coefficient of the term $q_k^{2a_k}$, where a_k is the k-th entry of $\underline{a}$. Write $N = \sum_{i=1}^{3g-3} a_i$. By Lemma 18 and Equation (2), the coefficient of $q^{2\cdot\underline{a}}$ in $I_{\Gamma,\Omega}(q_1, \dots, q_{3g-3})$ can explicitly be described as follows. For $a_k > 0$, we have

$$\sum_{w_k | a_k} w_k \left(\left(\frac{x_{k_1}}{x_{k_2}} \right)^{w_k} + \left(\frac{x_{k_2}}{x_{k_1}} \right)^{w_k} \right) = \frac{\sum_{w_k | a_k} w_k (x_{k_1}^{N+w_k} x_{k_2}^{N-w_k} + x_{k_1}^{N-w_k} x_{k_2}^{N+w_k})}{x_{k_1}^N x_{k_2}^N}$$

and for $a_k = 0$, we obtain

$$\sum_{w_k=1}^{N} w_k \left(\frac{x_{k_1}}{x_{k_2}} \right)^{w_k} = \frac{\sum_{w_k=1}^{N} w_k \cdot x_{k_1}^{N+w_k} x_{k_2}^{N-w_k}}{x_{k_1}^N x_{k_2}^N}.$$

In our computations, the focus lies on computing the numerator of this representation of the propagator. The denominator shift $x_i^N x_j^N$ can be taken into account separately within the function `coefficient_of_term` which determines the coefficient of a specific monomial $q^{2\cdot\underline{a}}$ by extracting the coefficient of the $x_i^N x_j^N$ term of the numerator.

3.2 Generating Series for Hurwitz Numbers

3.2.1 Flip Signature

In our approach, we introduce a flip signature with respect to the vertex ordering Ω to group covers having the same Feynman integral $I_{\Gamma,\Omega}$ together into a single computation. This grouping provides a further significant speedup of the computation of generating series of Gromov–Witten invariants via Feynman integrals $I_\Gamma = \sum_\Omega I_{\Gamma,\Omega}$ by identifying how many permutations Ω lead to the same signature, and determining the integral for one representative for each signature.

The permutation Ω determines which vertices of the edges will occur in the numerator and which will occur in the denominator of the corresponding propagator. While there are $(2g-2)!$ orders of branch points, there are at most 2^{2g-2} possible flip signatures, two possibilities for each edge of the graph. Moreover, the number of flip signatures is typically even smaller due to symmetries in the propagator product.

We define the flip signature as follows: We first find the *sign* of an edge q_k for a given vertex ordering Ω. We set the sign of an edge q_k to be 1, if $\Omega(i) < \Omega(j)$, otherwise it is set to be -1.

For a given graph Γ, order Ω, and branch type $\underline{a} = (a_1, \ldots, a_{3g-3})$, Algorithm 1 `flip_signature` determines the flip signature as a vector of length $3g-3$ assigning each non-loop edges the value -1 in case the degree of the edge is zero and its sign is -1, and the value 0 otherwise. To detect loop edges we assign in this case the value -2.

Algorithm 1: `flip_signature`

Input: Graph Γ, permutation Ω, branch type $\underline{a}$.
Output: Flip signature b.

```
begin
    E = edges of Γ (x_{k_1} - e_k - x_{k_2})
    foreach e_k in E do
        foreach a_k in a do
            if a_k = 0 then
                if preimg(Ω, x_{k_1}) < preimg(Ω, x_{k_2}) then
                    b_k = -1
                else
                    b_k = 0
            if x_{k_1} = x_{k_2} then
                b_k = -2
    return b
```

3.2.2 Multiplicities

Algorithm 2 `signature_and_multiplicities` determines how often each flip signature occurs over all Ω. This is then called the *multiplicity* of the flip signature. Note that we only permute the vertices of edges, where a flip can occur. For the other vertices, the order will not change the flip signature. Hence, we count over a subgroup of the permutation group of all vertices, and therefore need to adjust the multiplicities of the flip signatures by the corresponding index.

Algorithm 2: `signature_and_multiplicities`

Input: Graph Γ, branch type $\underline{a}$.
Output: Vector of flip signatures and their multiplicity

begin
  E = edges of Γ $(x_{k_1} - -e_k - -x_{k_2})$
  V = set of all vertices of edges k of Γ with $a_k = 0$
  $P = S(V)$ all permutations of V
  D = empty dictionary
  foreach $\Omega \in P$ **do**
    $y = \operatorname{sgn}(\Gamma, \Omega, a)$
    if *y key of D* **then**
      $D[y] = D[y] + 1$
    else
      $D[y] = 1$
  multiply each value of D by $\frac{(\#\text{ vertices of } \Gamma)!}{|P|}$
  return D

3.2.3 Feynman Integral

Building on signatures and multiplicities, we now proceed to the computation of the Feynman integral for a fixed branch type $\underline{a}$, which is one coefficient in the multivariate generating series and is denoted by $I_{\Gamma,\underline{a}}$. We also give an algorithm to compute the multivariate generating series up to a given degree d, which is denoted by $I_{\Gamma,d}$. The respective algorithms are Algorithm 3 `feynman_integral_branchtype` and Algorithm 4 `feynman_integral_degree`.

The functions `const_term` and `non_const_term` compute the q-constant term of the propagator multiplied by $x_i^n x_j^n$ and the coefficient of q_k^{2d} of the propagator multiplied by $x_i^n x_j^n$, respectively. For given n they are given as

$$\text{const\_term}(x_i, x_j, n) = \sum_{w=1}^{n} w \cdot x_i^{n+w} x_j^{n-w}$$

and

$$\text{non\_const\_term}(x_i, x_j, q_k, d, n) = \sum_{w|d} w \cdot (x_i^{n+w} x_j^{n-w} + x_i^{n-w} x_j^{n+w}) \cdot q_k^{2d}.$$

Here we can choose $n = \sum_{k=1}^{3g-3} a_k$, since any variable x_i^d with degree $d < -n$ or $n < d$ can not contribute to the constant term $x_1^0 \cdot \ldots \cdot x_{2g-2}^0$.

Algorithm 3: `feynman_integral_branchtype`

Input: Graph Γ, branch type $\underline{a}$.
Output: $I_{\Gamma,\underline{a}}$

begin
    E = edges of Γ
    $N = \sum_{i=1}^{3g-3} a_i$
    D = signatures_and_multiplicities($\Gamma, \underline{a}$)
    $p = 0$
    foreach $f \in \mathrm{keys}(D)$ **do**
        $tmp = 1$
        for j **from** 1 **to** size(f) **do**
            if $k_j = -1$ **then**
                $tmp = tmp \cdot \mathrm{const\_term}(x_{E_{j,1}}, x_{E_{j,2}}, N)$
            else if $k_j = 0$ **then**
                $tmp = tmp \cdot \mathrm{const\_term}(x_{E_{j,2}}, x_{E_{j,1}}, N)$
            else
                $tmp = tmp \cdot \mathrm{non\_const\_term}(x_{E_{j,1}}, x_{E_{j,2}}, q_j, f_j, N)$
        $p = p + D(f) \cdot \mathrm{Coef}_{[x_1^N \cdots x_{2g-2}^N]}(tmp)$
    return p

Algorithm 4: `feynman_integral_degree`

Input: Graph Γ, maximum degree d.
Output: $I_{\Gamma,d}$

begin
    $sum = 0$ **for** j **from** 1 **to** d **do**
        A = partitions of d into #(edges of Γ) summands.
        foreach $a \in A$ **do**
            sum=sum+ feynman_integral_branchtype(Γ, a)
    return sum

3.3 Generating Series for Descendant Gromov–Witten Invariants

The improved algorithms from Section 3.1 have a straight-forward generalization to the case of Gromov–Witten invariants with Psi-classes, following the formula given at the beginning of Section 3. We omit the respective algorithms here, but give example and timings at [BDHMT23].

4 Timings

In this section, we evaluate the performance of our implementation for various graphs and degrees d in the Tables 1 and 2. We compare

- the implementation of the original algorithm in the `Singular` library `ellipticcovers.lib` in the function `gromovWitten` (denoted in the tables by Singular-1)
- the implementation of our new algorithm (as described in the current paper) in the function `feynman_integral` included in an updated version of `ellipticcovers.lib` (denoted by Singular-2); and
- its `OSCAR`-based implementation in the function `feynman_integral` of the package `GromovWitten.jl` [BDHMT23] (denoted by OSCAR).

The package `GromovWitten.jl` is available at

`https://github.com/singular-gpispace/GromovWitten`.

All computations have been done sequentially on an M-core with 3.2 MHz. We have verified that all answers are consistent. The tables show all data collected in the respective configuration. All numbers are in seconds rounded to 3 digits.

We consider the source curves depicted in Figures 2–5. To generate the graphs considered in the timings, the following code can be used:

```
julia> using GromovWitten

# caterpillar genus 2
julia> G =  feynman_graph([(1,2),(1,2),(1,2)]);

# caterpillar genus 3
julia> G = feynman_graph([(1,2),(1,2),(2,4),(1,3),(3,4),(3,4)]);

# caterpillar genus 4
julia> G = feynman_graph([(1,2),(1,2),(1,3),(2,4),(3,4),(3,5),(4,6),(5,6
↪ ),(5,6)]);

# star graph K_{1,3}
julia> G = feynman_graph([(1,2),(1,3),(1,4),(2,4),(2,3),(3,4)]);
```

For the timings, see Tables 1 and 2, and for a visualization of the timings, see Figures 4 and 5. We observe that the new algorithm presented in this paper significantly outperforms the previous one, and that the `OSCAR` implementation of that algorithm significantly outperforms the `Singular` implementation again, although it is structurally identical. The demonstrates the potential of the new `OSCAR` platform. The `OSCAR` implementation allows one to compute, in comparable time, at least double the degree in the expansion of the formal generating series than the `Singular` implementation, which in turn can achieve about three times the degree compared to the old algorithm. The at least six-fold or more increase in degree allows for example to study homogeneity properties of quasi modular representations of the generating series for many graphs of larger genus which were previously not tractable. There is ongoing work on parallelization of the computation.

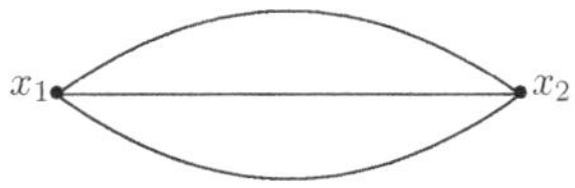

FIGURE 2. Caterpillar genus 2.

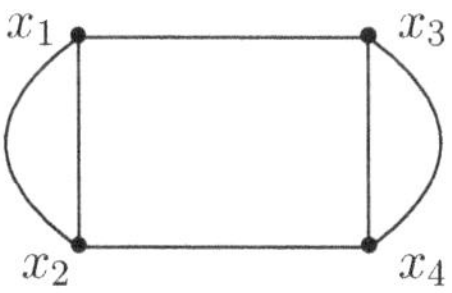

FIGURE 3. Caterpillar genus 3.

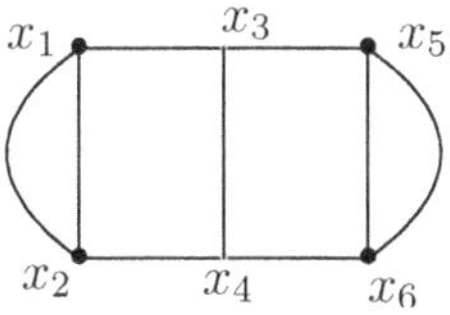

FIGURE 4. Caterpillar genus 4.

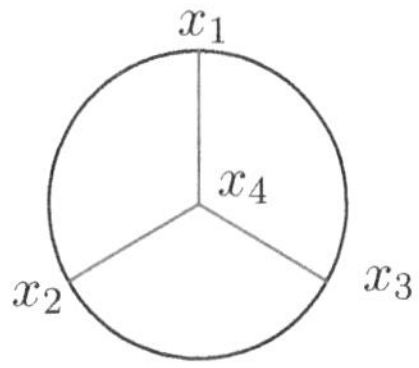

FIGURE 5. Star graph $K_{1,3}$.

degree d	Caterpillar genus 2			Caterpillar genus 3		
	Singular-1	Singular-2	OSCAR	Singular-1	Singular-2	OSCAR
1	0.01	0.01	0.00001	0.10	0.10	0.0003
2	0.70	0.02	0.00006	0.74	0.35	0.001
3	1.23	0.021	0.0001	10.20	0.78	0.004
4	10.21	0.030	0.00016	72.41	1.54	0.012
5	51.84	0.032	0.0002	343.03	2.76	0.03
6	173.94	0.041	0.0003	1300	4.72	0.68
7	345.16	0.044	0.00035	4030	7.94	0.14
8	570.06	0.051	0.0004	–	13.6	0.26
9	819.31	0.053	0.0005	–	23.9	0.47
10	1060	0.06	0.0007	–	41.9	0.82
11	1287	0.07	0.0007	–	72.3	1.35
12	1386	0.08	0.0010	–	122	2.24
13	1714	0.09	0.0011	–	211	3.48
14	2433	0.10	0.0014	–	351	5.62
15	3376	0.11	0.0015	–	559	8.49
16	4390	0.12	0.0018	–	940	12.9
17	–	0.14	0.0019	–	1 450	20.7
18	–	0.15	0.0023	–	2 300	27.6
19	–	0.17	0.0024	–	3 440	40.5
20	–	0.19	0.0029	–	5 130	58.7

Table 1: Timings for caterpillar source curves of genus 2 and genus 3.

degree d	Caterpillar genus 4			Star graph $K_{1,3}$		
	Singular-1	Singular-2	OSCAR	Singular-1	Singular-2	OSCAR
1	32.16	8.24	0.021	0.08	0.17	0.0007
2	4016	35.0	0.12	1.20	0.42	0.003
3	–	136	0.83	18.1	0.92	0.012
4	–	2240	5.53	124	1.77	0.039
5	–	32 890	29.0	549	3.19	0.112
6	–	–	121	1 990	5.80	0.27
7	–	–	419	6 080	10.3	0.56
8	–	–	2 420	18 200	19.7	1.10
9	–	–	3 490	–	38.0	1.94
10	–	–	38 700	–	72.0	3.21
11	–	–	–	–	130	5.28
12	–	–	–	–	238	8.50
13	–	–	–	–	424	13.8
14	–	–	–	–	716	20.2
15	–	–	–	–	1 190	28.6
16	–	–	–	–	1 880	40.8
17	–	–	–	–	2 970	57.3
18	–	–	–	–	4 777	79.2
19	–	–	–	–	7 103	107
20	–	–	–	–	10 620	144

Table 2: Timings for caterpillar genus 4 and $K_{1,3}$ source curves.

4.1 Example Computation

In this section, we provide example code for computing the Hurwitz numbers for a caterpillar genus 3 source curve.

```
julia> using GromovWitten

julia> G = feynman_graph([(1, 3),(1, 2),(1, 2),(2, 4),(3, 4),(3, 4)])
FeynmanGraph([(1, 3), (1, 2), (1, 2), (2, 4), (3, 4), (3, 4)])

julia> F = feynman_integral(G);

julia> a = [0, 2, 1, 0, 0, 1];

julia> feynman_integral_branch_type(F, a)
256*q[2]^4*q[3]^2*q[6]^2

julia> feynman_integral_degree(F, 3)
```

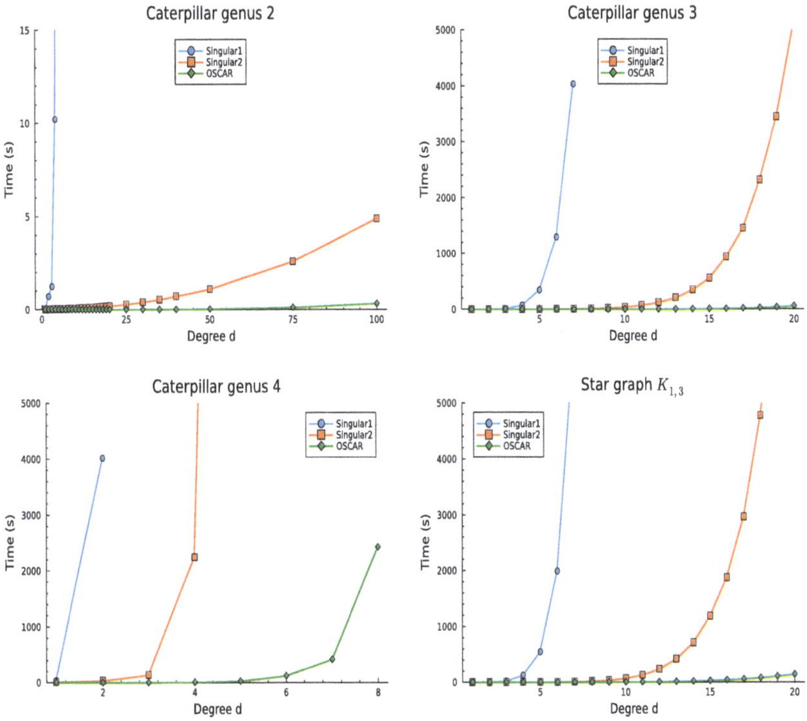

Fig. 4: Visualization of the timings for Singular-1, Singular-2, and OSCAR.

```
288*q[1]^6 + 32*q[1]^4*q[2]^2 + 32*q[1]^4*q[3]^2 + 32*q[1]^4*q[5]^2 +
↪  32*q[1]^4*q[6]^2 + 8*q[1]^2*q[2]^2*q[5]^2 + 8*q[1]^2*q[2]^2*q[6]^2 +
↪  8*q[1]^2*q[3]^2*q[5]^2 + 8*q[1]^2*q[3]^2*q[6]^2 + 24*q[2]^6 +
↪  152*q[2]^4*q[3]^2 + 8*q[2]^4*q[5]^2 + 8*q[2]^4*q[6]^2 +
↪  152*q[2]^2*q[3]^4 + 32*q[2]^2*q[3]^2*q[5]^2 + 32*q[2]^2*q[3]^2*q[6]^2
↪  + 32*q[2]^2*q[4]^4 + 8*q[2]^2*q[4]^2*q[5]^2 + 8*q[2]^2*q[4]^2*q[6]^2 +
↪  8*q[2]^2*q[5]^4 + 32*q[2]^2*q[5]^2*q[6]^2 + 8*q[2]^2*q[6]^4 +
↪  24*q[3]^6 + 8*q[3]^4*q[5]^2 + 8*q[3]^4*q[6]^2 + 32*q[3]^2*q[4]^4 +
↪  8*q[3]^2*q[4]^2*q[5]^2 + 8*q[3]^2*q[4]^2*q[6]^2 + 8*q[3]^2*q[5]^4 +
↪  32*q[3]^2*q[5]^2*q[6]^2 + 8*q[3]^2*q[6]^4 + 288*q[4]^6 +
↪  32*q[4]^4*q[5]^2 + 32*q[4]^4*q[6]^2 + 24*q[5]^6 + 152*q[5]^4*q[6]^2 +
↪  152*q[5]^2*q[6]^4 + 24*q[6]^6

julia> f = feynman_integral_degree_sum(F, 3)
```

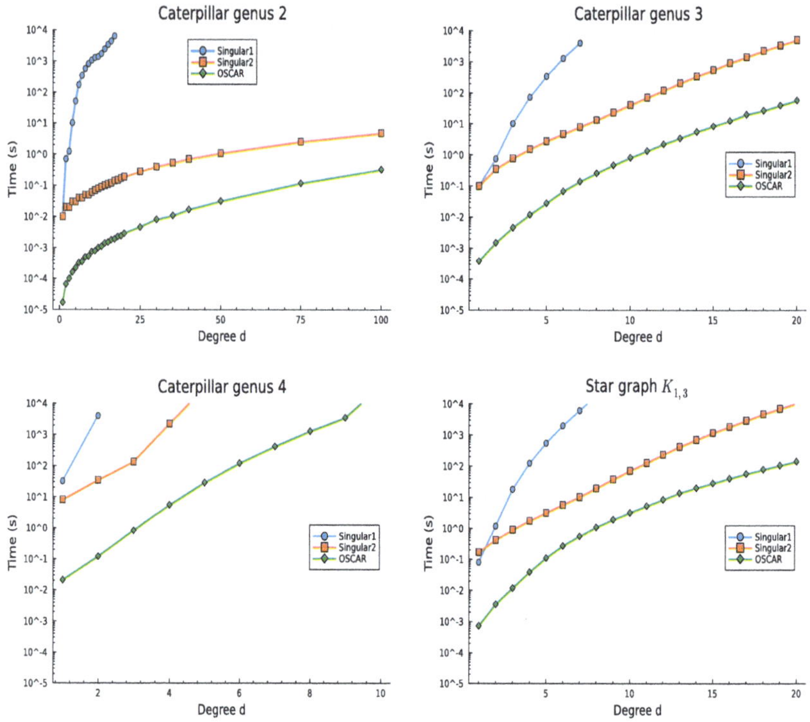

Fig. 5: Visualization of the timings for Singular-1, Singular-2, and OSCAR (logarithmic time scale).

```
288*q[1]^6 + 32*q[1]^4*q[2]^2 + 32*q[1]^4*q[3]^2 + 32*q[1]^4*q[5]^2 +
↪  32*q[1]^4*q[6]^2 + 8*q[1]^4 + 8*q[1]^2*q[2]^2*q[5]^2 +
↪  8*q[1]^2*q[2]^2*q[6]^2 + 8*q[1]^2*q[3]^2*q[5]^2 +
↪  8*q[1]^2*q[3]^2*q[6]^2 + 24*q[2]^6 + 152*q[2]^4*q[3]^2 +
↪  8*q[2]^4*q[5]^2 + 8*q[2]^4*q[6]^2 + 152*q[2]^2*q[3]^4 +
↪  32*q[2]^2*q[3]^2*q[5]^2 + 32*q[2]^2*q[3]^2*q[6]^2 + 8*q[2]^2*q[3]^2 +
↪  32*q[2]^2*q[4]^4 + 8*q[2]^2*q[4]^2*q[5]^2 + 8*q[2]^2*q[4]^2*q[6]^2 +
↪  8*q[2]^2*q[5]^4 + 32*q[2]^2*q[5]^2*q[6]^2 + 8*q[2]^2*q[6]^4 +
↪  24*q[3]^6 + 8*q[3]^4*q[5]^2 + 8*q[3]^4*q[6]^2 + 32*q[3]^2*q[4]^4 +
↪  8*q[3]^2*q[4]^2*q[5]^2 + 8*q[3]^2*q[4]^2*q[6]^2 + 8*q[3]^2*q[5]^4 +
↪  32*q[3]^2*q[5]^2*q[6]^2 + 8*q[3]^2*q[6]^4 + 288*q[4]^6 +
↪  32*q[4]^4*q[5]^2 + 32*q[4]^4*q[6]^2 + 8*q[4]^4 + 24*q[5]^6 +
↪  152*q[5]^4*q[6]^2 + 152*q[5]^2*q[6]^4 + 8*q[5]^2*q[6]^2 + 24*q[6]^6

julia> substitute(f)
1792*q[1]^6 + 32*q[1]^4
```

Acknowledgements Funded by the Deutsche Forschungsgemeinschaft (DFG, German Research Foundation) – Project-ID 286237555 – TRR 195. Janko Böhm also acknowledges support of Potentialbereich *SymbTools – Symbolic Tools in Mathematics and their Application* of the Forschungsinitiative Rheinland-Pfalz. The work of Firoozeh Dastur was supported by Fraunhofer ITWM, Kaiserslautern. The work of Ali Traore was supported by the German Academic Exchange Service (DAAD) through the Mathematics in Industry and Commerce (MIC) program.

References

[Beh97] K. Behrend. "Gromov-Witten invariants in algebraic geometry". *Invent. Math.* 127.3 (1997), pp. 601–617.

[BF97] K. Behrend and B. Fantechi. "The intrinsic normal cone". *Invent. Math.* 128 (1997), pp. 45–88.

[BBM11] B. Bertrand, E. Brugallé, and G. Mikhalkin. "Tropical open Hurwitz numbers". *Rend. Semin. Mat. Univ. Padova* 125 (2011), pp. 157–171. DOI: 10.4171/RSMUP/125-10.

[BBBM18] J. Böhm, K. Bringmann, A. Buchholz, and H. Markwig. *ellipticcovers.lib. A* SINGULAR *4 library for Gromov–Witten invariants of elliptic curves.* 2018.

[BBBM17] J. Böhm, K. Bringmann, A. Buchholz, and H. Markwig. "Tropical mirror symmetry for elliptic curves". *J. Reine Angew. Math.* 732 (2017), pp. 211–246.

[BDHMT23] J. Böhm, F. Dastur, A. Hoffmann, H. Markwig, and A. Traore. *GromovWitten.jl. An* OSCAR*-based package computing Gromov–Witten invariants of elliptic curves.* https://github.com/singular-gpispace/GromovWitten. 2023.

[BGM22a] J. Böhm, C. Goldner, and H. Markwig. "Counts of (tropical) curves in $E \times \mathbb{P}^1$ and Feynman integrals". *Ann. Inst. Henri Poincaré D* 9.1 (2022). Preprint, arXiv:1812.04936, pp. 121–158.

[BGM22b] J. Böhm, C. Goldner, and H. Markwig. "Tropical mirror symmetry in dimension one". *SIGMA* 18 (2022). Preprint, arXiv:1809.10659, p. 046.

[CJM10] R. Cavalieri, P. Johnson, and H. Markwig. "Tropical Hurwitz numbers". *J. Algebr. Comb.* 32.2 (2010), pp. 241–265. DOI: 10.1007/s10801-009-0213-0.

[CJMR18] R. Cavalieri, P. Johnson, H. Markwig, and D. Ranganathan. "A graphical interface for the Gromov-Witten theory of curves". In: *Algebraic geometry: Salt Lake City 2015.* Vol. 97. Proc. Sympos. Pure Math. Amer. Math. Soc., Providence, RI, 2018, pp. 139–167.

[CM16] R. Cavalieri and E. Miles. *Riemann surfaces and algebraic curves.* Vol. 87. London Mathematical Society Student Texts. A first course in Hurwitz theory. Cambridge University Press, 2016, pp. xii+183. DOI: 10.1017/CBO9781316569252.

[Dij95] R. Dijkgraaf. "The moduli space of curves". In: vol. 129. Progr. Math. Birkhäuser Boston, 1995. Chap. Mirror symmetry and elliptic curves, pp. 149–163.

[GS06] M. Gross and B. Siebert. "Mirror Symmetry via logarithmic degeneration data I". *J. Differential Geom.* 72 (2006). arXiv:math.AG/0309070, pp. 169–338.

[Har77] R. Hartshorne. *Algebraic Geometry.* Springer, 1977.

[Li11] S. Li. "BCOV theory on the elliptic curve and higher genus mirror symmetry". Preprint, arXiv:1112.4063. 2011.

[Mik05] G. Mikhalkin. "Enumerative tropical geometry in $\mathbb{R}^2$". *J. Amer. Math. Soc.* 18 (2005), pp. 313–377.

[OP06] A. Okounkov and R. Pandharipande. "Gromov–Witten theory, Hurwitz theory, and completed cycles". *Ann. of Math.* 163.2 (2006), pp. 517–560.

[Vak08] R. Vakil. "The moduli space of curves and Gromov–Witten theory". In: *Enumerative invariants in algebraic geometry and string theory.* Springer, 2008, pp. 143–198.

Toric Geometry

Martin Bies and Lars Kastner

Abstract This is an introduction to computations in toric geometry using OSCAR. We focus on computing vanishing sets for line bundle cohomology using cohomCalg, constructions of toric varieties from triangulations as well as resolution of singularities via star subdivisions.

Key words: Toric varieties, Newton polyhedra

1 Introduction

Toric varieties play an important role at the intersection of algebraic geometry and combinatorics. Their combinatorial aspects give rise to combinatorial algorithms for questions from algebraic geometry. Often these algorithms perform much better than their purely algebraic counterparts. At the same time, toric varieties are general enough such that they can provide counterexamples in many cases. Thus, toric varieties are an important testing ground for conjectures from algebraic geometry.

We begin with a review of the basics of toric geometry in Section 2, including the relation between polyhedral fans and normal toric varieties. This review aims to make the following sections accessible to non-experts. Readers may also consult [Tel22; CLS11] for an introduction to toric geometry. In the subsequent Section 3, we discuss implementation details. Our implementation starts out as a wrapper around polymake's application fulton for toric varieties. Based

Martin Bies
RPTU Kaiserslautern-Landau, Kaiserslautern, Germany, e-mail: martin.bies@rptu.de

Lars Kastner
TU Berlin, Chair of Discrete Mathematics/Geometry, Berlin, Germany, e-mail: kastner@math.tu-berlin.de

W. Decker et al. (eds.), *The Computer Algebra System OSCAR*,
Algorithms and Computation in Mathematics 32,
https://doi.org/10.1007/978-3-031-62127-7_7

on this, other functionality from the other OSCAR cornerstones is tied in and added via the `@attributes` mechanism. By now, all of the OSCAR cornerstones provide some functionality for toric varieties.

Toric geometry is commonly employed to construct solutions to string theory. One particular offspring of this avenue is the `cohomCalg` algorithm[1] [BJRR10], which can be used to compute the dimension of line bundle cohomology. This method, validated in [Jow11], has seen wide applications [RR10; Rah12]. Section 4 details the use of `cohomCalg` for computing line bundle cohomology vanishing sets, as initially described in [Bie18]. This result connects to studies of immaculate line bundles on toric varieties [ABKW20].

Similarly, toric varieties are also applied in constructing Calabi–Yau manifolds. Famously, the toric setups for Calabi–Yau 2- and 3-folds are classified by studying reflexive polytopes in dimensions 3 and 4 [KS98; KS00]. One connects a reflexive polytope to a toric variety by triangulating the lattice points of the polytope. We turn to such constructions of toric varieties in Section 5. In particular, we discuss the construction of toric varieties based on their *GLSM-charges* [Wit88] – a strategy well-known in the string theory community but far less well-known to the mathematics community.

Section 6 discusses the resolution of singularities in toric varieties via star subdivisions. Yet again, let us mention an application in the context of string theory. F-theory, a special domain of string theory, is devoted to relating the physics of string theory to the geometry and resolution of singular elliptic fibrations. Many insights in F-theory are rooted in toric studies, because the generally difficult resolution tasks are typically a lot more manageable in the toric context. In this chapter, we will explain the relation between star subdivisions of a polyhedral fan and the resolution of singularities in a toric variety. For more details on the applications thereof in the context of F-theory, the interested reader may wish to consult the F-theory chapter [BT24] (cf. page 451).

This chapter concludes with Section 7, where we give an outlook on the future development of toric geometry in OSCAR.

2 Preliminaries

This section introduces the most important definitions and notation from [CLS11]. For polyhedral objects, like cones, fans, and polyhedra, we use the conventions from the polyhedral chapter [BJ24] (cf. page 139).

As is common in most computer algebra systems, we will only deal with the lattice $M = \mathbb{Z}^n$ and its dual lattice $N = \mathbb{Z}^n$, which are linked by the standard scalar product. Conversion from other lattices is left to the user, but

[1] `http://wwwth.mppmu.mpg.de/members/blumenha/cohomcalg/`

can easily be achieved with standard normal forms provided by other parts of OSCAR.

2.1 Polytopes, Cones, Fans and Toric Varieties

Definition 1 ([CLS11, Def. 1.2.3]) Given a polyhedral cone $\sigma \subseteq \mathbb{Q}^n$, its *dual cone* is defined by

$$\sigma^\vee = \{m \in \mathbb{Q}^n \mid \langle m, u \rangle \geq 0 \text{ for all } u \in \sigma\},$$

where $\langle \cdot, \cdot \rangle \colon \mathbb{Q}^n \times \mathbb{Q}^n \to \mathbb{Q}$ is the standard inner product.

As a short side remark let us mention, that the command for dualizing a cone is called `polarize` for compatibility reasons with the polyhedral part of OSCAR.

Now we can obtain a semigroup from $\sigma^\vee$ by intersecting it with the lattice $\mathbb{Z}^n$. This semigroup is finitely generated, and a minimal generating set is the *Hilbert basis* of $\sigma^\vee$. Details can be found in the chapter on polyhedral geometry [BJ24] (cf. page 139).

Definition 2 ([CLS11, Thm. 1.2.18]) The *affine normal toric variety* U_σ associated to σ is the spectrum of the semigroup algebra associated to the semigroup $\sigma^\vee \cap \mathbb{Z}^n$, i.e.

$$U_\sigma := \mathrm{Spec}(\mathbb{C}[\sigma^\vee \cap \mathbb{Z}^n]).$$

The term *toric* in toric varieties stems from these varieties having the algebraic torus $\mathbb{C}^*$ (or a suitable power thereof) as a dense open subset. Therefore, we use the field of complex numbers in the above expression. Further details can be found in [CLS11].

The reason for first dualizing, then taking the semigroup and finally applying Spec to the semigroup algebra is that we want the functor taking cones to affine normal toric varieties to be covariant. Only then we can extend this functor to polyhedral fans and use the face structure for gluing the affine patches together.

Example 3 As an example pick $\sigma = \{0\}$. The dual $\sigma^\vee$ is the entire space $\mathbb{Q}^n$. Thus the semigroup $\sigma^\vee \cap \mathbb{Z}^n$ is $\mathbb{Z}^n$ itself. Now if we construct the affine normal toric variety U_σ we get $\mathrm{Spec}(\mathbb{C}[\mathbb{Z}^n])$. This is exactly the *algebraic torus* $T_N := (\mathbb{C}^*)^n$ itself. Since every pointed cone contains the origin as a face, this shows the inclusion of the algebraic torus T_N in every affine normal toric variety.

Furthermore the map $\{0\} \hookrightarrow \sigma$ gives rise to the inclusion $\sigma^\vee \cap \mathbb{Z}^n \hookrightarrow \mathbb{Z}^n$. If we take the map

$$\begin{array}{ccc} \sigma^\vee \cap \mathbb{Z}^n & \to & (\sigma^\vee \cap \mathbb{Z}^n) \times \mathbb{Z}^n \\ u & \mapsto & (u, u) \end{array}$$

and apply the semigroup algebra and Spec functors to this morphism, then the resulting map $U_\sigma \times (\mathbb{C}^*)^n \to U_\sigma$ encodes the action of the torus on U_σ.

Example 4 Let σ be the cone spanned by the two vectors $(1,0)$ and $(-1,2)$. Then the dual cone $\sigma^\vee$ is spanned by the vectors $(0,1)$ and $(2,1)$. The Hilbert basis of $\sigma^\vee$ consists of the two primitive ray generators and the additional point $(1,1)$. Thus we have

$$U_\sigma = \operatorname{Spec}\left(\mathbb{C}[x,y,z]/(xz-y^2)\right) = \mathbb{V}(xz-y^2) \subset \mathbb{C}^3,$$

where x has the multidegree $(0,1)$, y has the multidegree $(1,1)$ and the multidegree of z is $(2,1)$. As per usual, $\mathbb{V}$ denotes the vanishing set of the provided ideal.

Note that the multidegrees of x, y and z also determine the torus action:

$$\begin{array}{rcl} \mathbb{V}(xz-y^2) \times (\mathbb{C}^*)^2 & \to & \mathbb{V}(xz-y^2) \\ ((x,y,z)\quad , \quad (t_1,t_2)) & \mapsto & (t_2 \cdot x,\ t_1 \cdot t_2 \cdot y,\ t_1^2 \cdot t_2 \cdot z). \end{array}$$

In OSCAR this construction can be realized as follows:

```
julia> sigma = positive_hull([1 0; -1 2])
Polyhedral cone in ambient dimension 2

julia> Usigma = affine_normal_toric_variety(sigma)
Normal toric variety

julia> toric_ideal(Usigma)
Ideal generated by
  -x1*x2 + x3^2

julia> hilbert_basis(weight_cone(Usigma))
3-element SubObjectIterator{PointVector{ZZRingElem}}:
 [2, 1]
 [0, 1]
 [1, 1]
```

Definition 5 ([CLS11, Thm. 3.1.5]) Let Σ be a polyhedral fan, then the *normal toric variety* X_Σ associated to Σ is

$$X_\Sigma := \left(\coprod_{\sigma \in \Sigma} U_\sigma\right) / \sim,$$

where $\sim$ refers to gluing along faces, i.e. if $\tau \in \Sigma$ is a face of the cones $\sigma^1, \sigma^2 \in \Sigma$, then we get maps $U_\tau \to U_{\sigma^1}$ and $U_\tau \to U_{\sigma^2}$. We use these maps as a gluing datum.

Example 6 For convenience the constructors for normal toric varieties in OSCAR have been modeled such that they take the same arguments as the constructor for polyhedral fans. Hence to obtain the toric variety associated to the normal fan of the square one can proceed as follows:

```
julia> R = [1 0; 0 -1; -1 0; 0 1];

julia> IM = IncidenceMatrix([[1,2],[2,3],[3,4],[1,4]]);

julia> ntv = normal_toric_variety(IM, R)
Normal toric variety

julia> is_complete(ntv)
true

julia> is_smooth(ntv)
true
```

The rays are the rows of the matrix `R` and the rows of the incidence matrix `IM` indicate which collections of rays generate (maximal) cones.

Given a polytope P, one can take the normal fan Σ_P of P and then the associated toric variety. This is the *normal toric variety* X_P associated to P. Note that there is also a different construction [CLS11, Def. 2.1.1] associating a toric variety to a polytope via its lattice points. For very ample P, one can prove that the two definitions coincide. Indeed, the definition of X_P in [CLS11, Def. 2.3.14] immediately employs a very ample multiple of P.

Example 7 Two general classes of examples are given by $P = \Delta^n$ the n-dimensional simplex giving rise to $X_{\Delta^n} = \mathbb{P}^n$, and $P = C^n = [0,1]^n$ the n-dimensional cube, giving rise to $X_{C^n} = (\mathbb{P}^1)^n$. To the reader unfamiliar with toric varieties we recommend working out the details of these examples for $n = 2$ in the language of polyhedral fans.

Example 8 Via the polyhedral part of `OSCAR`, detailed in the polyhedral chapter [BJ24] (cf. page 139), many standard polytopes are directly accessible. For constructing the toric variety from the three-dimensional cube, one can proceed as follows:

```
julia> c = cube(3)
Polytope in ambient dimension 3

julia> tv = normal_toric_variety(c)
Normal toric variety
```

The *Orbit-Cone-Correspondence* [CLS11, Thm. 3.2.6] is key for many insights in toric geometry. It states that the face lattice of Σ encodes the orbit structure of the toric variety X_Σ. In the same spirit, see for example [CLS11, Thm. 3.1.19], many properties of the polyhedral fan Σ correspond directly to properties of the toric variety X_Σ. For example

- X_Σ is smooth if and only if Σ smooth, that is every cone $\sigma \in \Sigma$ is smooth.
- X_Σ is an orbifold if and only if Σ is simplicial.
- X_Σ is compact if and only if Σ is complete.

This showcases that polyhedral geometry is a powerful tool in analyzing toric varieties, making `polymake` the backbone of our implementation in `OSCAR`.

2.2 Divisor Class Group and Picard Group

Consider a toric variety X_Σ associated to the fan $\Sigma \subseteq \mathbb{Q}^n$. The Orbit-Cone Correspondence [CLS11, Thm. 3.2.6] relates k-dimensional cones $\sigma \in \Sigma$ to $(n-k)$-dimensional torus orbits in X_Σ. Let $\Sigma(1)$ denote the set of one-dimensional cones of Σ, i.e. the rays. Then every $\rho \in \Sigma(1)$ yields the codimension one orbit $\mathrm{O}(\rho)$, whose closure $\overline{\mathrm{O}(\rho)}$ is a torus-invariant prime divisor on X_Σ. We denote this divisor by D_ρ. Consequently, the group of torus-invariant Weil divisors on X_Σ is given by

$$\mathrm{Div}_{T_N}(X_\Sigma) = \bigoplus_{\rho \in \Sigma(1)} \mathbb{Z}D_\rho \subseteq \mathrm{Div}(X_\Sigma)\,. \tag{1}$$

By [CLS11, Prop. 4.1.2], we associate to $m \in M$ the divisor,

$$\mathrm{div}(\chi^m) = \sum_{\rho \in \Sigma(1)} \langle m, u_\rho \rangle \cdot D_\rho\,, \tag{2}$$

where u_ρ is the primitive ray generator of the ray $\rho \in \Sigma(1)$. Often one will consider divisors up to linear equivalence, and here the key takeaway is that every divisor class has a torus-invariant representative. Recall that X_Σ is always a normal variety, owing to its origin in a polyhedral fan Σ. Consequently, the Cartier divisors are special Weil divisors, $\mathrm{CDiv}_{T_N}(X_\Sigma) \subseteq \mathrm{Div}(X_\Sigma)$. Furthermore, $\mathrm{div}(\chi^m) \in \mathrm{CDiv}_{T_N}(X_\Sigma)$. This leads to the following result:

Theorem 9 ([CLS11, Thm. 4.1.3 and Thm. 4.2.1]) *Let Σ be a polyhedral fan such that the rays of Σ span the entire ambient space $\mathbb{Q}^n$. Then we have the following exact sequences:*

$$\begin{array}{ccccccc} 0 \to M \to & \mathrm{Div}_{T_N}(X_\Sigma) & \to & \mathrm{Cl}(X_\Sigma) & \to 0 \\ \| & \uparrow & & \uparrow & \\ 0 \to M \to & \mathrm{CDiv}_{T_N}(X_\Sigma) & \to & \mathrm{Pic}(X_\Sigma) & \to 0 \end{array},$$

where the first map is the map div *described above.*

In Section 4, our focus will be on two categories of toric varieties: those that are smooth and complete, and those that are simplicial and projective. Being compact, these varieties have no torus factor, leading to $\mathrm{Pic}(X_\Sigma) \cong \mathbb{Z}^r$, where r is the rank of $\mathrm{Pic}(X_\Sigma)$. Thus, in this context, a line bundle can be canonically represented as a tuple of integers.

2.3 Homogeneous Coordinate Ring and Chow Ring

While much more general, toric varieties behave in certain ways similarly to projective spaces. In particular, there is a homogeneous coordinate ring and ideals in this ring provide valuable insights into the toric variety.

Definition 10 Introduce one coordinate x_ρ for every $\rho \in \Sigma(1)$. This enables us to define the homogeneous coordinate ring S – often dubbed the *Cox ring* – via

$$S = \mathbb{C}\left[x_\rho \mid \rho \in \Sigma(1)\right] . \tag{3}$$

The Cox ring is graded by $\mathrm{Cl}(\Sigma)$, by assigning to x_ρ the divisor class $[D_\rho] \in \mathrm{Cl}(X_\Sigma)$. The irrelevant ideal $B(\Sigma)$ in the Cox ring contains important information about the toric variety X_Σ.

Definition 11 The irrelevant ideal is defined as

$$B(\Sigma) = \left\langle x^{\hat{\sigma}} \,\middle|\, \sigma \in \Sigma \right\rangle , \qquad x^{\hat{\sigma}} = \prod_{\rho \notin \sigma(1)} x_\rho . \tag{4}$$

Yet another ideal of ample importance is the Stanley-Reisner ideal $\mathcal{I}(\Sigma)$. For instance, this ideal is key for the `cohomCalg` algorithm, which we will discuss in detail in Section 4. The Stanley-Reisner ideal $\mathcal{I}(\Sigma)$ and the irrelevant ideal $B(\Sigma)$ of a toric variety X_Σ should not be confused. While the irrelevant ideal is a homogeneous ideal in the Cox ring, the Stanley-Reisner ideal need in general not be homogeneous with respect to the grading of the Cox ring by $\mathrm{Cl}(X_\Sigma)$.

Definition 12 Introduce one coordinate x_i for every $\rho_i \in \Sigma(1)$ and define the polynomial ring $\mathbb{C}\left[x_i \,|\rho_i \in \Sigma(1)\right]$. The Stanley-Reisner ideal is defined as

$$\mathcal{I}(\Sigma) = \langle x_{i_1} \cdots x_{i_s} \mid i_j \text{ distinct and } \rho_{i_1} + \cdots + \rho_{i_s} \text{ not a cone of } \Sigma \rangle . \tag{5}$$

It is illustrative to compute both ideals for the projective space. In `OSCAR`, we achieve this as follows:

```
julia> PP2 = projective_space(NormalToricVariety, 2)
Normal toric variety

julia> irrelevant_ideal(PP2)
Ideal generated by
  x3
  x1
  x2

julia> stanley_reisner_ideal(PP2)
Ideal generated by
  x1*x2*x3
```

In projective space $\mathbb{P}^2$, it is well-known that the three homogeneous coordinates cannot vanish simultaneously. This important fact is captured in the Stanley-Reisner ideal. Specifically, the expression $x_1 \cdot x_2 \cdot x_3 \in \mathcal{I}(\Sigma)$ serves as the algebraic representation of the statement $\mathbb{V}(x_1, x_2, x_3) = \emptyset$. Indeed, it turns out that the Stanley-Reisner ideal contains valuable information about the intersection theory on X_Σ. The latter is based on the Chow ring $A(X_\Sigma)$, and we have the following result [CLS11, Thm. 12.5.3]:

Theorem 13 *Let X_Σ be a complete and simplicial toric variety, then the Chow ring of X_Σ satisfies*

$$A(X_\Sigma) \cong \mathbb{Q}\left[x_i \,|\rho_i \in \Sigma(1)\right] / \left(\mathcal{I}(\Sigma) + \mathcal{J}(\Sigma)\right) , \tag{6}$$

where $\mathcal{I}(\Sigma)$ is the Stanley-Reisner ideal, and $\mathcal{J}(\Sigma)$ the ideal of linear relations. The latter is generated by the linear forms:

$$\sum_{i=1}^{r} \langle m, u_i \rangle \cdot x_i . \tag{7}$$

Even if X_Σ is not complete, this result holds. We then refer to $A(X_\Sigma)$ as the combinatorial Chow ring of X_Σ [Peg14]. Here is an example.

```
julia> v = normal_toric_variety(IncidenceMatrix([[1], [2], [3]]), [[1,
↪  0], [0, 1], [-1, -1]])
Normal toric variety

julia> is_complete(v)
false

julia> chow_ring(v)
Quotient
  of multivariate polynomial ring in 3 variables x1, x2, x3
    over rational field
  by ideal (x1 - x3, x2 - x3, x1*x2, x1*x3, x2*x3)
```

The combinatorial Chow ring of toric varieties is linked to the Chow ring of certain matroids by the results in [FY04]. The interested reader may wish to consult the matroid chapter [CKS24] (cf. page 351).

3 Data Types and Tools From Combinatorics

As explained in the previous section, toric varieties have strong ties to the combinatorial world. Most importantly, recall that in the case that a toric variety is normal, it can be completely described as a polyhedral fan. Consequently, this combinatorial object defines an algebro-geometric object, a scheme, and thereby creates a bridge to the scheme-theoretic world of the algebraic geometry chapter [BDS24] (cf. page 21). Note in particular, that this includes non-normal toric varieties, which are not discussed in this chapter, nor covered by the toric geometry implementations in OSCAR.

Similar correspondences exist for other algebro-geometric objects if they are assumed to be torus-invariant. We provide a rich toolbox of such objects with the combinatorial objects utilized for computation in the background. Among others, the following are supported:

1. `NormalToricVariety` encodes a normal toric variety.
2. `AffineNormalToricVariety` encodes an affine normal toric variety.
3. `ToricDivisor` encodes a torus-invariant divisor on a normal toric variety.
4. `ToricDivisorClass` encodes the divisor class of a torus-invariant divisor in the divisor class group.
5. `ToricLineBundle` encodes a line bundle on a toric variety associated to a torus-invariant divisor. Equivalently, this structure encodes the linear equivalence class of the torus-invariant divisor in the Picard group.

While we will not discuss them in this chapter for sake of brevity, there are a number of yet more specialized structures, including the following:

6. `ToricMorphism` encodes a morphism among two normal toric varieties, which is induced from a grid morphism that is compatible with the fans of the two varieties in question.
7. `RationalEquivalenceClass` encodes the equivalence class of an algebraic cycle, where equivalence is given by rational equivalence.
8. `ClosedSubvarietyOfToricVariety` encodes a closed subvariety of a simplicial toric variety. Such subvarieties are given by homogeneous ideals in the Cox ring – the coordinate ring – of the variety in question.

Example 14 Every normal toric variety has an underlying polyhedral fan Σ. Accessing Σ can be done by calling `polyhedral_fan` on the normal toric variety. For convenience, most polyhedral functions of Σ may also be called directly on a normal toric variety:

```
julia> pp1 = projective_space(NormalToricVariety, 1)
Normal toric variety

julia> tv = pp1 * pp1
Normal toric variety

julia> rays(tv)
4-element SubObjectIterator{RayVector{QQFieldElem}}:
 [1, 0]
 [-1, 0]
 [0, 1]
 [0, -1]

julia> maximal_cones(IncidenceMatrix, tv)
4×4 IncidenceMatrix
[1, 3]
[1, 4]
```

```
[2, 3]
[2, 4]

julia> Sigma = polyhedral_fan(tv)
Polyhedral fan in ambient dimension 2
```

Many properties of toric varieties are expensive to compute, e.g. the toric ideal of an affine toric variety makes it necessary to compute the Hilbert basis of a cone, before arriving at the algorithm for a toric ideal which in itself is expensive too. Hence it is useful to cache these things with the object they are tied to. Some properties are already being cached by `polymake`, but `OSCAR` vastly extends the capabilities via the other cornerstones. Hence, we define our structs with the macro `@attributes`, to introduce such a cache for things which are not already cached by `polymake`.

4 Vanishing Sets From `cohomCalg`

Here we focus on toric varieties that are either smooth and complete or simplicial and projective. These are the toric varieties, for which the `cohomCalg` algorithm calculates line bundle cohomology dimensions [BJRR10; Jow11; RR10; Rah12]. We review the essential concepts behind `cohomCalg`, which culminates in its application to computing vanishing sets [Bie18]. This computational method for vanishing sets, valuable but to our knowledge not yet widely employed, relates to findings on immaculate line bundles in [ABKW20].

4.1 The `cohomCalg` Algorithm

4.1.1 Combinatorial Data of the Stanley-Reisner Ideal $\mathcal{I}(\Sigma)$

The `cohomCalg` algorithm places significant emphasis on the Stanley-Reisner ideal $\mathcal{I}(\Sigma)$. Let us write $\mathcal{I}(\Sigma) = \langle S_1, \ldots, S_N \rangle$. For each subset

$$A = \{\alpha_1, \ldots, \alpha_k\} \subseteq \{1, \ldots, N\}, \qquad 1 \leq k \leq N, \tag{1}$$

we define Q_A^k as the set consisting of all coordinates present in $\{S_{\alpha_1}, \ldots, S_{\alpha_k}\}$.

The structure of these sets Q_A^k is evaluated by use of the *c-degree*, defined as $c(Q_A^k) := |Q_A^k| - k$. It is noteworthy that different sets $B = \{\beta_1, \ldots, \beta_k\}$, which are also subsets of $\{1, \ldots, N\}$ and distinct from A, can result in identical sets $Q_A^k = Q_B^k$. The quantity $c^i(Q_A^k)$ is defined as the number of all subsets B that satisfy the conditions:

$$|B| = k, \quad Q_A^k = Q_B^k, \quad i = c(Q_B^k)\,. \tag{2}$$

It turns out that the integers $c^i(Q^k_A)$ are the dimensions of $\mathbb{Q}$-vector spaces $\mathcal{C}^i\left(Q^k_A\right)$, which collectively form a complex that was explored in large detail in the references [RR10; Jow11]:

$$\cdots \to 0 \to \mathcal{C}^0(Q^k_A) \to \mathcal{C}^1(Q^k_A) \to \cdots \to \mathcal{C}^d(Q^k_A) \to \ldots . \quad (3)$$

The `cohomCalg` algorithm is instrumental in determining the homology dimension $h^i(Q^k_A)$ at each position i within this complex.

4.1.2 From Combinatorial Data to Line Bundle Cohomology

Let us select a line bundle $\mathcal{L} \in \mathrm{Pic}(X_\Sigma)$. Recall that we focus on smooth and complete toric varieties, as well as simplicial and projective toric varieties. As we revised in Section 2.2, the Picard group of any such toric variety is free, implying that $\mathcal{L}$ can simply be thought of as an element in $\mathbb{Z}^r$, with r being the rank of $\mathrm{Pic}(X_\Sigma)$.

In the context of the `cohomCalg` algorithm, an important discovery is that only the sets $Q^k_A = \{y_1, y_2, \ldots, y_k\}$, for which $h^i(Q^k_A)$ is non-zero, contribute to the dimension of the i-th line bundle cohomology $H^i(X_\Sigma, \mathcal{L})$ [BJRR10; RR10; Jow11]. The aggregate of these contributions yields the overall dimension of $H^i(X_\Sigma, \mathcal{L})$, which we will denote as $h^i(X_\Sigma, \mathcal{L})$. To compute the specific contribution from a set Q^k_A to $h^i(X_\Sigma, \mathcal{L})$, we first express $\mathcal{L}$ as $\mathcal{O}_{X_\Sigma}(\alpha)$. In this expression, α is a tuple of integers, which – as we recalled above – uniquely fixes the line bundle $\mathcal{L}$. Next, we represent Q^k_A as $\{y_1, \ldots, y_k\}$ and identify all homogeneous coordinates outside Q^k_A as $\{x_1, \ldots, x_m\}$. The contribution from Q^k_A is calculated by multiplying $h^i(Q^k_A)$ with the number of rationoms of degree α, expressible in the form:

$$\frac{T\left(x_1 \ldots, x_n\right)}{y_1 \cdot y_2 \cdot \ldots \cdot y_k \cdot W\left(y_1, \ldots, y_k\right)}, \quad (4)$$

where T and W are appropriately chosen monomials.

4.1.3 Example – Line Bundle Cohomology on $\mathbb{P}^1 \times \mathbb{P}^1$

Let us focus on $\mathbb{P}^1 \times \mathbb{P}^1$ and compute $h^1\left(\mathbb{P}^1 \times \mathbb{P}^1, \mathcal{L}\right)$ for $\mathcal{L} = \mathcal{O}_{\mathbb{P}^1 \times \mathbb{P}^1}(-3, 1)$. To this end, let us assume that the Cox ring $S(\mathbb{P}^1 \times \mathbb{P}^1) = \mathbb{C}[x_1, x_2, x_3, x_4]$ is graded with respect to $\mathrm{Cl}(\mathbb{P}^1 \times \mathbb{P}^1) = \mathbb{Z}^2$ by

$$\deg(x_1) = \deg(x_2) = (1, 0)\,, \qquad \deg(x_3) = \deg(x_4) = (0, 1)\,. \quad (5)$$

The Stanley-Reisner ideal satisfies $\mathcal{I}(\mathbb{P}^1 \times \mathbb{P}^1) = \langle x_1 x_2, x_3 x_4 \rangle = \langle S_1, S_2 \rangle$, leading to the combinatorial data – including the c-degrees $c^i(Q^k_A)$ and $h^i(Q^k_A)$ – as listed in table 1. Note that only $Q^1_{\{1\}}$ and $Q^1_{\{2\}}$ have $h^1(Q^k_A) \neq 0$. The

sets	c-degree	$c^i\left(Q^k_A\right)$	$h^i\left(Q^k_A\right)$
$Q^0_\emptyset = \emptyset$	0	$(1,0,0)$	$(1,0,0)$
$Q^1_{\{1\}} = \{x_1, x_2\}$	1	$(0,1,0)$	$(0,1,0)$
$Q^1_{\{2\}} = \{x_3, x_4\}$	1	$(0,1,0)$	$(0,1,0)$
$Q^2_{\{1,2\}} = \{x_1, x_2, x_3, x_4\}$	2	$(0,0,1)$	$(0,0,1)$

Table 1: `cohomCalg`'s combinatorial data of $\mathbb{P}^1 \times \mathbb{P}^1$.

contribution of $Q^1_{(2)}$ to $h^1\left(\mathbb{P}^1 \times \mathbb{P}^1, \mathcal{L}\right)$ is

$$\underbrace{h^1\left(Q^1_{\{2\}}\right)}_{=1} \cdot \left| \left\{ R(x_i) = \frac{W(x_3, x_4)}{x_1 x_2 T(x_1, x_2)} \;\middle|\; \deg\left(R(x_i)\right) = (-3, 1) \right\} \right| , \tag{6}$$

where T, W must be suitable monomials. It may be readily verified that $Q^1_{\{2\}}$ contributes a 4 to $h^1\left(\mathbb{P}^1 \times \mathbb{P}^1, \mathcal{L}\right)$. Similarly, one readily verifies that $Q^1_{\{1\}}$ contributes a 0 to $h^1\left(\mathbb{P}^1 \times \mathbb{P}^1, \mathcal{L}\right)$. So in summary we have:

$$h^1\left(\mathbb{P}^1 \times \mathbb{P}^1, \mathcal{L}\right) = h^1(Q^1_1) \cdot 0 + h^1(Q^1_2) \cdot 4 = 4\,. \tag{7}$$

We reproduce this result with `OSCAR` as follows:

```
julia> P1 = projective_space(NormalToricVariety, 1)
Normal toric variety

julia> l = toric_line_bundle(P1*P1, [-3,1])
Toric line bundle on a normal toric variety

julia> cohomology(l, 1)
4
```

4.2 Vanishing Sets

In this section, we outline the algorithm detailed in [Bie18] for the computation of the vanishing set $V^i(X_\Sigma)$:

$$V^i(X_\Sigma) = \left\{ \mathcal{L} \in \mathrm{Pic}(X_\Sigma) \mid h^i(X_\Sigma, \mathcal{L}) = 0 \right\} , \quad 0 \leq i \leq \dim(X_\Sigma)\,. \tag{8}$$

The algorithm applies the following steps:

1. Calculate the Cox ring $S(X_\Sigma)$ and the Stanley-Reisner ideal $\mathcal{I}(\Sigma)$.
2. Utilize `cohomCalg` to identify the set $\{S_1, \ldots, S_l\} := \left\{ Q^k_A \middle| h^i(Q^k_A) \neq 0 \right\}$.
3. For each $m \in \{1, 2, \ldots, l\}$, denote the variables in S_m by $y_1, \ldots, y_k$ and the remaining variables of $S(X_\Sigma)$ by $x_{k+1}, \ldots, x_n$. Observe that S_m does not contribute to $h^i(X_\Sigma, \mathcal{O}_{X_\Sigma}(\alpha))$ iff $\alpha \notin \mathbf{o}_m + \mathrm{Span}_{\mathbb{Z}_{\geq 0}}(\mathcal{B}_m)$, where

$$\mathbf{o}_m := -\sum_{i=1}^{k} \deg(y_i)\,, \tag{9}$$

$$\mathcal{B}_m := \{-\deg(y_1), \ldots, -\deg(y_k), \deg(x_{k+1}), \ldots, \deg(x_n)\}\,. \tag{10}$$

Determine $\{(\mathcal{B}_1, \mathbf{o}_1), (\mathcal{B}_2, \mathbf{o}_2), \ldots, (\mathcal{B}_l, \mathbf{o}_l)\}$ and return the vanishing set $V^i(X_\Sigma)$ as follows:

$$V^i(X_\Sigma) = \mathrm{Pic}(X_\Sigma) - \bigcup_{m=1}^{l} \left(\mathbf{o}_m + \mathrm{Span}_{\mathbb{Z}_{\geq 0}}(\mathcal{B}_m)\right). \tag{11}$$

OSCAR utilizes this algorithm to compute the vanishing sets $V^i\left(\mathbb{P}^1 \times \mathbb{P}^1\right)$:

```
julia> P1 = projective_space(NormalToricVariety, 1)
Normal toric variety

julia> v0, v1, v2 = vanishing_sets(P1*P1)
3-element Vector{ToricVanishingSet}:
 Toric vanishing set for cohomology indices [0]
 Toric vanishing set for cohomology indices [1]
 Toric vanishing set for cohomology indices [2]
```

For $V^0(\mathbb{P}^1 \times \mathbb{P}^1)$ there is a single polyhedron with defining constraints:

```
julia> print_constraints(polyhedra(v0)[1])
-x_1 <= 0
-x_2 <= 0
```

Hence, we have $V^0(\mathbb{P}^1 \times \mathbb{P}^1) = \mathbb{Z}^2 - P^0$ where P^0 are the lattice points in the first quadrant. We illustrate this vanishing set in Figure 1. Equivalently, we conclude that $\mathcal{O}_{\mathbb{P}^1 \times \mathbb{P}^1}(m_1, m_2)$ has global sections iff $m_1, m_2 \geq 0$, extending the well-known fact $h^0\left(\mathbb{P}^1, \mathcal{O}_{\mathbb{P}^1}(m)\right) \neq 0$ iff $m \geq 0$. For $V^1(\mathbb{P}^1 \times \mathbb{P}^1)$ we find:

```
julia> print_constraints(polyhedra(v1)[1])
-x_2 <= 0
x_1 <= -2

julia> print_constraints(polyhedra(v1)[2])
-x_1 <= 0
x_2 <= -2
```

Hence, $V^1(\mathbb{P}^1 \times \mathbb{P}^1) = \mathbb{Z}^2 - P^1_{(1)} - P^{(1)}_{(2)}$, which we also illustrate in Figure 1. It remains to repeat these steps for $V^2(\mathbb{P}^1 \times \mathbb{P}^1)$:

```
julia> print_constraints(polyhedra(v2)[1])
x_2 <= -2
x_1 <= -2
```

We display the ensuing vanishing set in Figure 1. It is illustrative to recall $K_{\mathbb{P}^1 \times \mathbb{P}^1} = \mathcal{O}(-2, -2)$, so that one can read-off Serre duality from this picture. It is also worth pointing out the analogy to $h^2(\mathbb{P}^1, \mathcal{O}_{\mathbb{P}^1}(m)) \neq 0$ iff $m \leq -2$.

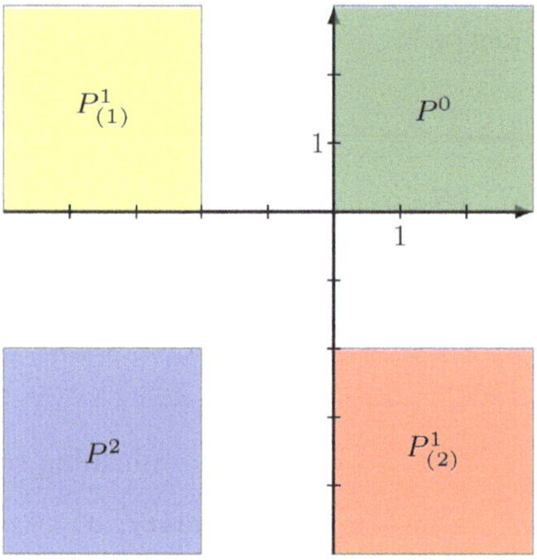

Fig. 1: The vanishing sets $V^i(\mathbb{P}^1 \times \mathbb{P}^1) = \mathrm{Pic}(X_\Sigma) - \bigcup_{j \in J} \left(P^i_{(j)} \cap \mathbb{Z}^2 \right)$.

5 Triangulations

The seminal works [KS98; KS00] have classified toric setups for Calabi–Yau manifolds in dimensions two and three by examining reflexive polytopes in dimensions three and four. The connection between a polytope and its associated toric varieties is established through the triangulation of the polytope's lattice points. The significance of such triangulation tasks has recently spurred the development of the software *CYTools* [DRM22]. Given the burgeoning interest in this area, we aim to delineate the current state of triangulations within `OSCAR`.

Motivated by applications in physics, we initiate our discussion somewhat unconventionally by selecting an integer-valued matrix g. In string theory, the matrix g is commonly referred to as GLSM-charges [Wit88], a terminology also adopted by `cohomCalg`. Subsequently, we define a (possibly empty) family $\mathcal{F}$ of toric varieties, subject to the following constraints:

1. Each $X_\Sigma \in \mathcal{F}$ must have no torus factor.
2. The Cox rings of every $X_\Sigma \in \mathcal{F}$ are graded by g.

As discussed in Section 2, the first condition ensures the existence of a short exact sequence:

$$0 \to N \xrightarrow{r} \mathrm{Div}_T(X_\Sigma) \xrightarrow{\mathrm{g}} \mathrm{Cl}(X_\Sigma) \to 0\,, \tag{1}$$

where the second map is defined by the matrix g. Consequently, the map $N \xrightarrow{r} \mathrm{Div}_T(X_\Sigma)$ is computed as the kernel, and the rows of the mapping matrix are interpreted as the ray generators of the toric varieties $X_\Sigma \in \mathcal{F}$.

The challenge then becomes identifying all possible fans, which may be subject to additional conditions, constructed from these ray generators. This endeavor aligns perfectly with the process of triangulation. To this end, one views the ray generators as vertices of a polytope, seeking fine triangulations of this polytope centered at the origin. Recall that fine means that every vertex of said polytope is included in at least one cell of the triangulation. In addition, by enforcing the origin as center, we know that all cells of the triangulation include the origin. If required, one can further refine the triangulations, e.g. by opting for regular triangulations, thereby ensuring that all $X_\Sigma \in \mathcal{F}$ are projective varieties. Here is a minimal example that yields solely $\mathbb{P}^2$:

```
julia> normal_toric_varieties_from_glsm([[1,1,1]])
1-element Vector{NormalToricVariety}:
 Normal toric variety
```

Here is a slightly more complex example:

```
julia> normal_toric_varieties_from_glsm([[1,1,-1,-1]])
2-element Vector{NormalToricVariety}:
 Normal toric variety
 Normal toric variety
```

These two varieties can be interpreted as representing the two phases of a conifold transition.

Triangulations can rapidly result in a vast number of combinatorial possibilities. For instance, the reflexive polytope Δ_8° in the three-dimensional Kreuzer-Skarke list [KS98] is known to have on the order of $\mathcal{O}(10^{15})$ triangulations [HT17]. Related computationally intensive triangulation tasks were successfully undertaken using OSCAR, as demonstrated in [BCDO22].

6 Blowup Resolution Meets Star Subdivision

In algebraic geometry, the resolution of singularities is of great importance in understanding and simplifying complex geometric structures. Singularities are, in the simplest case, points at which a geometric object fails to be well-behaved in some manner, such as having a well-defined tangent space. The resolution of these singularities means to transform a space with singularities into one that is smooth, yet preserving the essence of the original space. The pioneering work by Hironaka in the 1960s established the possibility of resolution of singularities in characteristic zero [Hir64]. This breakthrough provided a foundation upon which much of modern algebraic geometry is built.

One of the main obstacles in the resolution of singularities is the inherent complexity of singular geometric structures. Singularities can be wildly intricate and unpredictable, making their resolution a non-trivial task. Techniques for the resolution of singularities include blow-ups. In the simplest case, a blow-up replaces a point by a higher-dimensional space, thereby 'smoothing out' the singularity. As illustrated before, in the context of toric geometry, many of the challenges involved are more manageable, owing to the combinatorial and polyhedral background of toric geometry. This is also true regarding singularities. The resolution of singular toric varieties "merely" requires blow-ups, which are mirrored on the polyhedral/combinatorial side by star subdivisions. The central result is as follows [CLS11, Thm. 11.1.9]:

Theorem 15 *Every fan Σ has a refinement Σ' with the following properties:*

1. *Σ' is smooth.*
2. *Σ' contains every smooth cone of Σ.*
3. *Σ' is obtained from Σ by a sequence of star subdivisions.*

A *star subdivision* is a geometric operation that introduces a new ray in a fan, effectively refining and smoothing the original fan. Specifically, a star subdivision at a ray ρ replaces each cone containing ρ with cones that are smaller but more numerous, enhancing the smoothness of the fan. This process is crucial in resolving singularities as it leads to a more refined and smoother fan, which corresponds to a less singular, or even smooth, toric variety. A *refinement of a fan Σ* to a fan Σ', is a process where each cone in Σ is contained in some cone of Σ'. This refinement process is essential for ensuring that the new fan Σ' maintains the overall structure of Σ while improving its properties, like smoothness, as ensured by star subdivisions.

While the theorem significantly aids in resolving singularities in toric varieties, it is crucial to note that the process is not entirely systematic. The act of selecting which ray to add during a star subdivision introduces complexity. This selection is often challenging and has substantial implications on the outcome. Additionally, the resolution process usually involves multiple star subdivisions, and the order in which these are performed can affect the result. While these subtleties are less pronounced than in more general algebraic geometry scenarios, they highlight that even in toric geometry, resolving singularities involves intricate decision-making and can yield varying results based on the chosen approach. The interested reader may wish to consult the discussion in the F-theory chapter [BT24] (cf. page 451) for involved examples. In this chapter, we restrict ourselves to a simple and instructive example:

```
julia> v = normal_toric_variety(IncidenceMatrix([[1,2]]), [1 1; -1 1]);

julia> set_coordinate_names(v, ["u1", "u2"])

julia> is_smooth(v)
false

julia> toric_ideal(v)
Ideal generated by
  -x1*x2 + x3^2
```

These lines of code create an affine and non-smooth toric variety, that we may think of as $\mathbb{V}(x_1 x_2 - x_3^2) \subset \mathbb{A}^3$, which immediately showcases the singularity. Alternatively, the educated eye may also "spot" this singularity from the polyhedral cone C, to which the variety is associated. Specifically, the ray generators of this cone are as follows:

```
julia> rays(maximal_cones(v)[1])
2-element SubObjectIterator{RayVector{QQFieldElem}}:
 [1, 1]
 [-1, 1]
```

Hence, $(0, 1) \in C$ but cannot be expressed as $\mathbb{Z}$-linear combination of the ray generators of C. This is another way to see the singularity of v.

To resolve this singularity, we perform a star subdivision of the fan along $(0, 1)$, which means that we add $(0, 1)$ as ray generator and split the cone C in half along this ray. This is achieved with the command

```
julia> bu = blow_up(v, [0,1])
Toric blowdown morphism
```

We can of course use `OSCAR` to inspect the resulting variety:

```
julia> v2 = domain(bu)
Normal toric variety

julia> is_smooth(v2)
true

julia> is_affine(v2)
false
```

Note that the resulting toric variety is no longer affine. But we can still get a good impression of what happened from looking at its Cox ring:

```
julia> cox_ring(v2)
Multivariate polynomial ring in 3 variables over QQ graded by
  u1 -> [1]
  u2 -> [1]
  e -> [-2]
```

The subdivision introduced a new toric divisor, often termed an exceptional divisor. This divisor corresponds to the homogeneous coordinate e in the Cox ring of the blown-up toric variety v_2. If we consider the vanishing locus $V(e)$, then, based of the grading of the Cox ring, we can expect that this locus is isomorphic to $\mathbb{P}^1_{[u_1 \,:\, u_2]}$. So in comparison to the original variety v, we have added a projective space $\mathbb{P}^1_{[u_1 \,:\, u_2]}$. Indeed, we can see that this projective space avoids the singularity that appears in v at $u_1 = u_2 = 0$. To this end, we merely have to inspect the Stanley-Reisner ideal:

```
julia> stanley_reisner_ideal(v2)
Ideal generated by
  u1*u2
```

So, as $u_1 \cdot u_2 \in \mathcal{I}(v_2)$, we have $\mathbb{V}(u_1, u_2) = \emptyset$ in v_2 – as expected.

7 Summary and Outlook

OSCAR is an actively developed, evolving software, inviting contributions, especially from young developers. Its toric varieties interface has become stable recently, offering users a reliable experience without disrupting workflows. We are planning to enhance its toric functionality, including coherent sheaves, leveraging the equivalence between sheaves and finitely presented graded modules in smooth, complete toric varieties [CLS11]. Algorithms for computing sheaf cohomologies are also proposed [Bie18], with future developments to encompass specialized algorithms for vector bundle cohomologies [Pay08; ABKW20; AP20; AW21].

In F-theory applications, OSCAR faces challenges, particularly in toric implementations. Efforts are underway to enhance these implementations for various computational tasks. The significance of toric geometry in F-theory, particularly in constructing solutions that align with modern accelerator experiments, is notable, as seen in the *Quadrillion F-theory Standard Models* using toric K3 surfaces [KS98]. For a detailed discussion on F-theory in OSCAR, refer to the F-theory chapter [BT24] (cf. page 451).

The CYTools software [DRM22], designed specifically for triangulating 4-dimensional reflexive polytopes – an essential task in finding Calabi–Yau manifold realizations from toric geometry – could benefit from enhancements provided by either mptopcom [JJK18] or TOPCOM [Ram02]. In heterotic string theory, the cohomCalg application for line bundle cohomology on CICYs is notable, with recent developments in topological formulæ[BCDL20; BCDL21]. The potential for extending these formulæ and utilizing vanishing sets computed with cohomCalg remains an open question.

Recently, based on OSCAR's toric geometry capabilities, a Julia-package dedicated to computing Gromov-Witten invariants has been released [Mur23]. Future studies will explore its expansion and alignment with other efforts, as detailed in chapter [BDHMT24] (cf. page 168).

Acknowledgements Martin Bies and Lars Kastner express their gratitude and appreciation for the support provided by the OSCAR team, led by Claus Fieker, Max Horn, Michael Joswig, and Wolfram Decker. Martin Bies acknowledges financial support from the Forschungsinitiative des Landes Rheinland-Pfalz through the project *SymbTools – Symbolic Tools in Mathematics and their Application*. Lars Kastner is thankful for the funding received from *MaRDI – Mathematical research initiative* of the German Research Foundation (DFG), project number 460135501, NFDI 29/1. This work was supported by the SFB-TRR 195 *Symbolic Tools in Mathematics and their Application* of the German Research Foundation (DFG).

References

[ABKW20] K. Altmann, J. Buczyński, L. Kastner, and A.-L. Winz. *Immaculate line bundles on toric varieties.* 2020. DOI: 10.4310/PAMQ.2020.v16.n4.a12.

[AP20] K. Altmann and D. Ploog. *Displaying the cohomology of toric line bundles.* 2020. DOI: 10.1070/IM8948.

[AW21] K. Altmann and F. Witt. *The structure of exceptional sequences on toric varieties of Picard rank two.* 2021. DOI: 10.48550/ARXIV.2112.14637.

[Bie18] M. Bies. "Cohomologies of coherent sheaves and massless spectra in F-theory". PhD thesis. Heidelberg U., Feb. 2018. DOI: 10.11588/heidok.00024045.

[BCDO22] M. Bies, M. Cvetič, R. Donagi, and M. Ong. "Brill-Noether-general limit root bundles: absence of vector-like exotics in F-theory Standard Models". *JHEP* 11 (2022), p. 004. DOI: 10.1007/JHEP11(2022)004.

[BT24] M. Bies and A. P. Turner. "F-Theory Applications". In: *The Computer Algebra System OSCAR. Algorithms and Examples.* Ed. by W. Decker, C. Eder, C. Fieker, M. Horn, and M. Joswig. Springer, 2024, pp. 451–472.

[BJRR10] R. Blumenhagen, B. Jurke, T. Rahn, and H. Roschy. "Cohomology of line bundles: A computational algorithm". *Journal of Mathematical Physics* 51.10 (Oct. 2010), p. 103525. DOI: 10.1063/1.3501132.

[BDHMT24] J. Böhm, F. Dastur, A. Hoffmann, H. Markwig, and A. Traore. "Algorithms for Gromov–Witten Invariants of Elliptic Curves". In: *The Computer Algebra System OSCAR. Algorithms and Examples.* Ed. by W. Decker, C. Eder, C. Fieker, M. Horn, and M. Joswig. Springer, 2024, pp. 167–190.

[BDS24] J. Böhm, W. Decker, and F.-O. Schreyer. "Commutative Algebra and Algebraic Geometry". In: *The Computer Algebra System OSCAR. Algorithms and Examples.* Ed. by W. Decker, C. Eder, C. Fieker, M. Horn, and M. Joswig. Springer, 2024, pp. 21–79.

[BCDL20] C. R. Brodie, A. Constantin, R. Deen, and A. Lukas. "Index Formulae for Line Bundle Cohomology on Complex Surfaces". *Fortsch. Phys.* 68.2 (2020), p. 1900086. DOI: 10.1002/prop.201900086. arXiv: 1906.08769 [hep-th].

[BCDL21] C. R. Brodie, A. Constantin, R. Deen, and A. Lukas. "Topological Formulae for the Zeroth Cohomology of Line Bundles on del Pezzo and Hirzebruch Surfaces". *Complex Manifolds* 8.1 (2021), p. 223. DOI: 10.1515/coma-2020-0115. arXiv: 1906.08363 [math.AG].

[BJ24] T. Brysiewicz and M. Joswig. "Polyhedral Geometry". In: *The Computer Algebra System OSCAR. Algorithms and Examples.* Ed. by W. Decker, C. Eder, C. Fieker, M. Horn, and M. Joswig. Springer, 2024, pp. 139–164.

[CKS24] D. Corey, L. Kühne, and B. Schröter. "Matroids". In: *The Computer Algebra System OSCAR. Algorithms and Examples.* Ed. by W. Decker, C. Eder, C. Fieker, M. Horn, and M. Joswig. Springer, 2024, pp. 351–368.

[CLS11] D. A. Cox, J. B. Little, and H. K. Schenck. *Toric varieties.* Vol. 124. Graduate Studies in Mathematics. American Mathematical Society, Providence, RI, 2011. DOI: 10.1090/gsm/124.

[DRM22] M. Demirtas, A. Rios-Tascon, and L. McAllister. *CYTools: A Software Package for Analyzing Calabi–Yau Manifolds.* 2022. DOI: 10.48550/ARXIV.2211.03823.

[FY04] E. M. Feichtner and S. Yuzvinsky. "Chow rings of toric varieties defined by atomic lattices". *Invent. Math.* 155.3 (Mar. 2004), pp. 515–536. DOI: 10.1007/s00222-003-0327-2.

[HT17] J. Halverson and J. Tian. "Cost of seven-brane gauge symmetry in a quadrillion F-theory compactifications". *Phys. Rev. D* 95.2 (2017), p. 026005. DOI: 10.1103/PhysRevD.95.026005. arXiv: 1610.08864 [hep-th].

[Hir64] H. Hironaka. "Resolution of Singularities of an Algebraic Variety Over a Field of Characteristic Zero". *Ann. of Math.* 79 (1964), pp. 109–326. DOI: 10.2307/1970547.

[JJK18] C. Jordan, M. Joswig, and L. Kastner. "Parallel Enumeration of Triangulations". *The Electronic Journal of Combinatorics* (2018). https://polymake.org/doku.php/mptopcom. DOI: 10.37236/7318.

[Jow11] S.-Y. Jow. "Cohomology of toric line bundles via simplicial Alexander duality". *Journal of Mathematical Physics* 52.3 (Mar. 2011), p. 033506. DOI: 10.1063/1.3562523.

[KS98] M. Kreuzer and H. Skarke. "Classification of reflexive polyhedra in three-dimensions". *Adv. Theor. Math. Phys.* 2 (1998), pp. 853–871. DOI: 10.4310/ATMP.1998.v2.n4.a5.

[KS00] M. Kreuzer and H. Skarke. "Complete classification of reflexive polyhedra in four-dimensions". *Adv. Theor. Math. Phys.* 4 (2000), pp. 1209–1230. DOI: 10.4310/ATMP.2000.v4.n6.a2. arXiv: hep-th/0002240.

[Mur23] G. Muratore. *Computations of Gromov–Witten invariants of toric varieties.* 2023. arXiv: 2309.03741 [math.AG].

[Pay08] S. Payne. "Moduli of toric vector bundles". *Compositio Mathematica* 144.5 (Sept. 2008), pp. 1199–1213. DOI: 10.1112/s0010437x08003461.

[Peg14] C. Pegel. "Chow Rings of Toric Varieties". Refereed by Prof. Dr. Eva Maria Feichtner and Dr. Emanuele Delucchi. MA thesis. Faculty of Mathematics: University of Bremen, Sept. 2014.

[Rah12] T. Rahn. "Heterotic target space dualities with line bundle cohomology". May 2012.

[Ram02] J. Rambau. "`TOPCOM`: Triangulations of point configurations and oriented matroids." In: *Mathematical software. Proceedings of the 1st international congress, Beijing, China, August 17–19, 2002.* Singapore: World Scientific, 2002, pp. 330–340.

[RR10] H. Roschy and T. Rahn. "Cohomology of line bundles: Proof of the algorithm". *Journal of Mathematical Physics* 51.10 (Oct. 2010), p. 103520. DOI: `10.1063/1.3501135`.

[Tel22] S. Telen. *Introduction to Toric Geometry.* 2022. arXiv: `2203.01690 [math.AG]`.

[Wit88] E. Witten. "Topological Sigma Models". *Commun. Math. Phys.* 118 (1988), p. 411. DOI: `10.1007/BF01466725`.

Algorithms for GIT-Fans of Affine Torus Actions

Janko Böhm and Thomas Breuer

Abstract The GIT-fan assigned to an algebraic variety with the action of an algebraic group is a polyhedral fan enumerating all GIT-quotients in the sense of Mumford. We discuss an algorithm to compute the GIT-fan for torus actions on affine varieties with finite symmetries, and its implementation in OSCAR. Relying on Gröbner basis techniques to decide monomial containment, computations with rational polyhedral cones, and expanding orbits with respect to actions of permutation groups, the algorithm combines computational methods from commutative algebra, convex geometry and group theory. Consequently, OSCAR, for the first time, provides an open-source framework that seamlessly manages every stage of the algorithm. We illustrate the algorithm through examples.

Key words: Geometric invariant theory, group actions, Gröbner bases, polyhedral geometry, Mori chamber decomposition, GIT-fan

1 Introduction

In this section, we discuss an algorithm which combines elements of Gröbner bases, polyhedral calculations, and algorithmic group theory to solve a problem from geometric invariant theory. Due to the interdisciplinary nature of the algorithm, OSCAR is the perfect environment for an implementation which is convenient both for the user and the programmer.

Janko Böhm
RPTU Kaiserslautern-Landau, e-mail: `boehm@mathematik.uni-kl.de`

Thomas Breuer
Lehrstuhl für Algebra und Zahlentheorie, RWTH Aachen University, e-mail: `thomas.breuer@math.rwth-aachen.de`

W. Decker et al. (eds.), *The Computer Algebra System OSCAR*,
Algorithms and Computation in Mathematics 32,
https://doi.org/10.1007/978-3-031-62127-7_8

Recall that the key goal of Mumford's Geometric Invariant Theory (GIT) is to assign to a given algebraic variety X, together with the action of an algebraic group G, a quotient space $X/\!\!/G$ with reasonable properties [MFK94]. This situation is commonly encountered when dealing with a variety X that parametrizes a specific class of algebraic varieties. The action of G on X could then describe isomorphisms between varieties.

This approach comes with some issues: The first problem is that the orbit space X/G in general is not a good candidate for $X/\!\!/G$ since we would like the quotient to carry the structure of an algebraic variety. The orbit space might fail to inherit a natural structure of an algebraic variety, for example, due to non-trivial stabilizers. See Figure 1, where the orbit space consists of two distinct points.

$$\mathbb{C} \times \mathbb{C}^* \to \mathbb{C}, \qquad (x,t) \mapsto tx$$

Fig. 1: Orbit space.

A potential solution for this is to define the quotient $X/\!\!/G$ for affine X as the spectrum of the invariant ring,

$$X/\!\!/G = \operatorname{Spec} K[X]^G,$$

which is indeed finitely generated. For more general X, we can glue the quotients of an affine covering.

Next, we encounter the problem that the quotient $X/\!\!/G$ might not carry much information:

Example 1 For instance, consider $X = \mathbb{C}^2$ and the component-wise multiplication action of

$$\mathbb{C}^2 \times \mathbb{C}^* \to \mathbb{C}^2, \qquad ((x,y),t) \mapsto (tx,ty)$$

Then the quotient $X/\!\!/\mathbb{C}^*$ is isomorphic to a solitary point. Considering the open subset $U = X \setminus \{(0,0)\}$ the respective quotient is the projective line $U/\!\!/\mathbb{C}^* = \mathbb{P}^1$, see Figure 2.

In general, there are many choices for the open subsets $U \subseteq X$, which in turn lead to different quotients $U/\!\!/G$. To describe this behaviour, Dolgachev and Hu [DH98] introduced the **GIT-fan**, which is a polyhedral fan[1] that describes the variation of GIT-quotients.

[1] Recall that a **polyhedral fan** is a finite collection of strongly convex rational polyhedral cones such that faces of cones are again elements of the fan, and the intersection of any two cones is a common face.

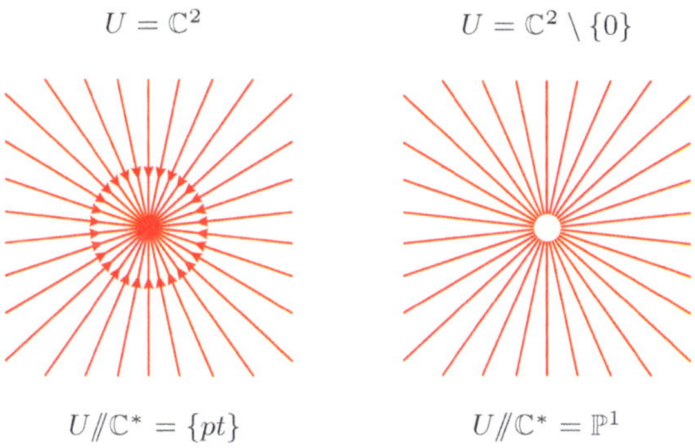

Fig. 2: Different quotients for different choices of open subsets.

2 The GIT-Fan

We focus on the action of an algebraic torus $G = (\mathbb{C}^*)^k$ on a integral affine variety $X \subset \mathbb{C}^r$. Our setup consist of

- a prime ideal $\mathfrak{a} \subset \mathbb{C}[T_1, \ldots, T_r]$ defining X, and
- a matrix $Q = (q_{ij})_{i,j} \in \mathbb{Z}^{r \times k}$ of full rank such that $\mathfrak{a}$ is homogeneous with respect to the grading $\deg(T_i) = (q_{ij})_j \in \mathbb{Z}^k$.

Example 2 The action of $\mathbb{C}^*$ on $\mathbb{C}^2$, $\mathbb{C}^2 \times \mathbb{C}^* \to \mathbb{C}^2, \quad ((x, y), t) \mapsto (tx, ty)$ is encoded by the matrix

$$Q = \begin{pmatrix} 1 \\ 1 \end{pmatrix} \in \mathbb{Z}^{2\times 1}.$$

Write $A = \mathbb{C}[T_1, \ldots, T_r]/\mathfrak{a}$ and A_w for the degree w graded part of A.

Definition 3 The **orbit monoid** for a point $x \in X$ is defined as

$$S(x) := \{w \in \operatorname{im}_{\mathbf{Z}} Q \mid \exists f \in A_w \text{ with } f(x) \neq 0\}.$$

and the **orbit cone**

$$\omega(x) := \operatorname{cone}(S(x)) \subset \operatorname{im}_{\mathbf{Q}} Q$$

is the convex polyhedral cone generated by the orbit monoid.[2]

Definition 4 To every $w \in \operatorname{im}_{\mathbf{Q}} Q$ we associate the torus invariant open **set of semistable points**

$$X^{ss}(w) = \{x \in X \mid \exists n > 0,\, f \in A_{nw} \text{ with } f(x) \neq 0\} \subset X$$

[2] In the image definition, we regard Q as a map between the row modules $\mathbb{Q}^r \to \mathbb{Q}^k$, respectively row vector spaces $\mathbb{Q}^r \to \mathbb{Q}^k$, via multiplication with Q from the right. So the image is the row span.

and the **GIT-cone**

$$\lambda(w) = \bigcap_{w \in \omega(x)} \omega(x)$$

Theorem 5 *[BH06, Prop. 2.9] With the notation from above:*

1. $\Omega = \{\omega(x) \mid x \in X\}$ *is finite.*
2. *There exists a good quotient* $X^{ss}(w)/\!\!/G$.
3. $\lambda(w) = \lambda(w') \Leftrightarrow X^{ss}(w) = X^{ss}(w')$, *and*
4. $\Lambda(\mathfrak{a}, Q) = \{\lambda(w) \mid w \in \operatorname{im} Q\}$ *is a rational polyhedral fan, the* ***GIT-fan***.

Example 6 For $\mathbb{C}^2 \times \mathbb{C}^* \to \mathbb{C}^2, \quad ((x,y),t) \mapsto (tx, ty), \quad Q = \begin{pmatrix} 1 \\ 1 \end{pmatrix}, \quad \mathfrak{a} = 0.$

$U_1 = \mathbb{C}^2$
$U_2 = \mathbb{C}^2 \setminus \{0\}$ $\quad \Lambda(\mathfrak{a}, Q) =$ 

3 Computing GIT-Fans

In case of a torus acting on an affine variety, Berchthold and Hausen [BH06] and Keicher [Kei12] have developed a method for computing the orbit cones. With the notation from above:

Definition 7 A face γ of the orthant $\mathbb{Q}^r_{\geq 0}$ is called an **$\mathfrak{a}$-face** if

$$\mathfrak{a}_\gamma = \mathfrak{a}|_{T_i = 0 \text{ for } e_i \notin \gamma}$$

does not contain a monomial, that is,

$$\mathfrak{a}_\gamma : \langle T_1 \cdot \ldots \cdot T_r \rangle^\infty \neq \langle 1 \rangle$$

Proposition 8 *[BH06] The* ***orbit cones*** *are the* $Q(\gamma) = \operatorname{cone}\{q_i \mid e_i \in \gamma\}$ *with* γ *an* $\mathfrak{a}$*-face.*

4 Fast Monomial Containment Test

Since determining whether an ideal contains a monomial or not via a saturation computation is rather expensive, a specialized algorithm for this problem is needed, which is based on a modification of Buchberger's algorithm. In the following, let K be a field. We denote by $\mathrm{L}_>(f)$ the leading monomial of $f \in K[x_1, \ldots, x_n]$ with respect to the monomial ordering $>$. As a first step towards a specialized algorithm, we extend a result of [Stu96, Lemma 12.1] from the case of the degree reverse lexicographic ordering to a more general setting, which is also suitable to handle multigradings as considered in our setting.

Proposition 9 *[BKR20, Prop. 3.1] Let $>$ be a monomial ordering on $R = K[Y_1, \ldots, Y_n]$ and $\mathcal{G}$ a Gröbner basis of the ideal I in R. Suppose that for all $f \in \mathcal{G}$*

$$Y_n \mid f \iff Y_n \mid \mathrm{L}_>(f).$$

Then

$$\left\{ \frac{f}{Y_n^i} \,\middle|\, f \in \mathcal{G} \text{ and } i \geq 0 \text{ maximal such that } Y_n^i \mid f \right\}$$

is a Gröbner basis for $I : Y_n^\infty$.

Based on this result, while executing Buchberger's algorithm, we can already divide by variables whenever a new element of the Gröbner basis is found as a non-zero remainder of an S-polynomial (see also [BKR20, Sec. 3]. This leads to Algorithm 1.

Algorithm 1 *To compute $I : \langle Y_1 \cdot \ldots \cdot Y_n \rangle^\infty$, replace any remainder $r \neq 0$ in Buchberger's algorithm by*

$$\frac{r}{Y_1^{a_1} \cdot \ldots \cdot Y_n^{a_n}} \quad \text{where } a_j \text{ is maximal such that } Y_j^{a_j} \mid r.$$

5 Symmetry Groups of Torus Actions

Definition 10 A **symmetry group** of the action of $(\mathbb{C}^*)^k$ on X is a subgroup $G \subset S_r$ of the symmetric group such that there are group actions

$$\begin{aligned} \mathbb{C}[T_1, \ldots, T_r] \times G &\to \mathbb{C}[T_1, \ldots, T_r], & (T_j, \sigma) &\mapsto c_{\sigma,j} \cdot T_{j^\sigma} \\ \mathbb{Q}^r \times G &\to \mathbb{Q}^r, & (e_j, \sigma) &\mapsto e_{j^\sigma} \\ \mathbb{Q}^k \times G &\to \mathbb{Q}^k, & (v, \sigma) &\mapsto v \cdot A_\sigma \end{aligned}$$

with $A_\sigma \in \mathrm{GL}(k, \mathbb{Q})$ and $c_\sigma \in (\mathbb{C}^*)^r$ such that $\mathfrak{a} \cdot G = \mathfrak{a}$ and that for each $\sigma \in G$ the following diagram is commutative:

$$\begin{array}{ccc} e_j & \longmapsto & e_{j^\sigma} \\ \mathbb{Q}^r & \longrightarrow & \mathbb{Q}^r \\ Q \downarrow & & \downarrow Q \\ \mathbb{Q}^k & \underset{A_\sigma}{\longrightarrow} & \mathbb{Q}^k \end{array}$$

The permutation action of a symmetry group G on the row indices of Q induces permutation actions of G on the $\mathfrak{a}$-faces, the orbit cones, and the GIT-cones, and matrix actions on the orbit cones and GIT-cones via multiplication with the matrices A_σ, for $\sigma \in G$.

6 An Example With D_4-Symmetry

We first illustrate the concept of a symmetry of an action on a simple example, a quadric with the symmetry group of a square.

Example 11

$$\mathfrak{a} = \langle T_1T_3 - T_2T_4 \rangle \subset \mathbb{C}[T_1, \ldots, T_4], \quad Q = \begin{pmatrix} 1 & 1 \\ -1 & 1 \\ -1 & -1 \\ 1 & -1 \end{pmatrix}, \quad \deg(T_i) = (q_{i,1}, q_{i,2})$$

$$D_4 \cong G = \langle (1,2)(3,4), (1,2,3,4) \rangle \subset S_4$$

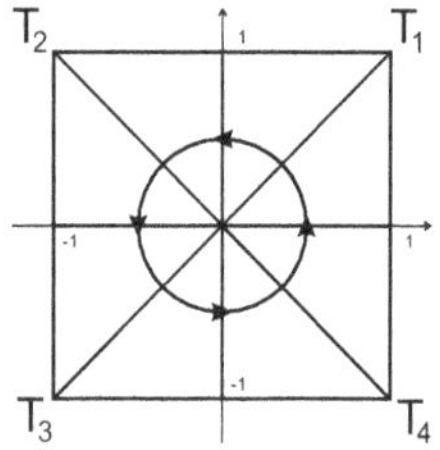

$$A_{(1,2)(3,4)} = \begin{pmatrix} -1 & 0 \\ 0 & 1 \end{pmatrix} \qquad A_{(1,2,3,4)} = \begin{pmatrix} 0 & 1 \\ -1 & 0 \end{pmatrix}$$

7 Representation of GIT-Cones

We can encode each GIT-cone λ by the set of orbit cones that contain λ. In other words,

$$h_\Omega \colon \Lambda(\mathfrak{a}, Q) \to \{0,1\}^\Omega, \qquad \lambda \mapsto \begin{bmatrix} \Omega \to & \{0,1\} \\ \vartheta \mapsto & \begin{cases} 1 & \lambda \subset \vartheta \\ 0 & \lambda \not\subset \vartheta \end{cases} \end{bmatrix}$$

defines a perfect hash function for cones. This is compatible with the actions of a given symmetry group G in the sense that the map

$$\{0,1\}^\Omega \times G \to \{0,1\}^\Omega, \qquad (b, g) \mapsto \begin{bmatrix} \Omega \to \{0,1\} \\ \vartheta \mapsto b(\vartheta^g) \end{bmatrix}$$

defines an action of G on $\{0,1\}^\Omega$ such that

$$h_\Omega(\lambda)^g = h_\Omega(\lambda \cdot A_g)$$

holds.

8 Symmetric GIT-Algorithm

Algorithm 2 computes a system of representatives of the G-orbits on the set $\Lambda(\mathfrak{a}, Q)(k)$ of maximal dimensional GIT-cones, which is an efficient data structure from which all cones of the GIT-fan are then derived.

To find the system of representatives, the algorithm commences by finding a maximal dimensional starting cone as the intersection of all orbit cones containing an interior vector in the support of the fan (chosen generic enough to not lie in a lower dimensional GIT cone). We then use the fact that the fan is connected in codimension one, passing from a given cone to all of its neighbors by traversing its facets. When finding a neighboring cone as an intersection of orbit cones containing an interior vector, we check if the cone is contained in an orbit of a previously computed cone. Once all facets of cones have been transversed or are on the boundary of the support of the GIT-fan, the algorithm stops.

Algorithm 2 (GIT-Fan)

Input: *An ideal* $\mathfrak{a} \subset \mathbb{C}[T_1, \ldots, T_r]$ *and a matrix* $Q = (q_{ij}) \in \mathbb{Z}^{r \times k}$ *such that* $\mathfrak{a}$ *is homogeneous with respect to the grading* $\deg(T_i) = (q_{ij})_j$*, moreover a finite symmetry group* G *of the torus action defined by* Q *and* $\mathfrak{a}$ *in the sense of Section 5.*

Output: $\Lambda(\mathfrak{a}, Q)$

$\mathcal{S}$:= system of representatives of G-orbits of $\mathrm{faces}(\mathbb{Q}^r_{\geq 0})$
$\mathcal{A} := \{\gamma \in \mathcal{S} \mid \gamma$ *is* $\mathfrak{a}$*-face*$\}$
$\Omega := \bigcup_{\gamma \in \mathcal{A}} Q(\gamma) \cdot G$
$\Omega :=$ *set of minimal elements of* $\Omega(k)$
Choose $w_0 \in Q(\mathbb{Q}^r_{\geq 0})$ *such that* $\dim(\lambda(w_0)) = k$.
$\mathcal{C} := \{\}$, $\mathcal{H} := \{\}$, $\mathcal{H}_{todo} := \{h_\Omega(\lambda(w_0))\}$
while *there is* $h \in \mathcal{H}_{todo}$ ***do***
  $\lambda := h_\Omega^{-1}(h)$
  $\mathcal{C} := \mathcal{C} \cup \{\lambda\}$, $\mathcal{H} := \mathcal{H} \cup \{h\}$, $\mathcal{H}_{todo} := \mathcal{H}_{todo} - \{h\}$
  for all η *face of* λ *interior facet with inner normal* v ***do***
    Find $w \in Q(\mathbb{Q}^r_{\geq 0})$ *s.t.* η *face of* $\lambda(w)$ *is facet and* $-v \in \lambda(w)^\vee$.
    if $h_\Omega(\lambda(w)) \cdot G \cap \mathcal{H} = \emptyset$ ***then***
      $\mathcal{H}_{todo} := \mathcal{H}_{todo} \cup \{h_\Omega(\lambda(w))\}$
return *Fan* $\Lambda(\mathfrak{a}, Q)$ *such that* $\Lambda(\mathfrak{a}, Q)(k) = \mathcal{C} \cdot G$.

9 Example: GIT-Fan for $\mathbb{G}(2, 5)$

Example 12 As an example, we consider the Deligne-Mumford compactification $\overline{M}_{0,5}$ of the moduli space of 5-pointed stable curves of genus zero, which has a natural action of the symmetric group S_5. The space $\overline{M}_{0,5}$ is an instance where the moduli space of pointed stable curves of genus zero is known to be a *Mori dream space*, that is, its Cox ring is finitely generated. The corresponding GIT-fan characterizes the birational geometry of $\overline{M}_{0,5}$.

Consider an algebraically closed field K of characteristic zero. The Cox ring of $\overline{M}_{0,5}$ is in this case indeed isomorphic to the coordinate ring $R = K[T_1, \ldots, T_{10}]/\mathfrak{a}$ of the affine cone over the Grassmannian $\mathbb{G}(2,5)$ where the ideal $\mathfrak{a}$ is generated by the Plücker relations and the i-th row of the matrix Q is the degree $\deg(T_i) \in \mathbb{Z}^5$; this determines the natural $\mathbb{Z}^5$-grading of R.

```
julia> using Oscar.GITFans

julia> Q = [
            1  1  0  0  0
            1  0  1  1  0
            1  0  1  0  1
            1  0  0  1  1
            0  1  0  0 -1
            0  1  0 -1  0
            0  1 -1  0  0
            0  0  1  0  0
            0  0  0  1  0
            0  0  0  0  1
            ];

julia> n = nrows(Q);

julia> Qt, T = polynomial_ring(QQ, :T => 1:n);

julia> D = free_abelian_group(ncols(Q));

julia> w = [D(Q[i, :]) for i = 1:n];

julia> R, T = grade(Qt, w);

julia> a = ideal([
                T[5]*T[10] - T[6]*T[9] + T[7]*T[8],
                T[1]*T[9]  - T[2]*T[7] + T[4]*T[5],
                T[1]*T[8]  - T[2]*T[6] + T[3]*T[5],
                T[1]*T[10] - T[3]*T[7] + T[4]*T[6],
                T[2]*T[10] - T[3]*T[9] + T[4]*T[8],
            ]);
```

We observe that there is a natural S_5-symmetry for the $H \cong (K^*)^5$-action on $V(a)$, which is inherited from the Grassmannian and its interpretation as the moduli space of 5-pointed stable curves. The symmetry group $S_5 \cong G \subseteq S_{10}$ is generated by (2,3)(5,6)(9,10), (1,5,9,10,3)(2,7,8,4,6).

```
julia> perms_list = [ [1,3,2,4,6,5,7,8,10,9], [5,7,1,6,9,2,8,4,10,3] ];

julia> sym = symmetric_group(n);

julia> G, emb = sub([sym(x) for x in perms_list]...);

julia> G
```

```
Permutation group of degree 10

julia> describe(G)
"S5"

julia> iso = isomorphism(G, symmetric_group(5));

julia> [iso(x) for x in gens(G)]
2-element Vector{PermGroupElem}:
 (1,3)
 (1,5,4,3,2)
```

We compute the matrix action $G \to \mathrm{GL}(5, \mathbb{Q}), \sigma \mapsto A_\sigma$.

```
julia> matrix_action = GITFans.action_on_target(Q, G)
Group homomorphism
  from permutation group of degree 10 and order 120
  to matrix group of degree 5 over QQ

julia> x = gen(G, 2)
(1,5,9,10,3)(2,7,8,4,6)

julia> Ax = matrix_action(x)
[-1   1   -1   -1   -2]
[ 1   0    1    1    1]
[ 1   0    0    1    1]
[ 0   0    0    0    1]
[ 1   0    1    0    1]

julia> matrix(QQ, Q[Vector{Int}(x), 1:5]) == matrix(QQ, Q) * Ax
true
```

We compute the orbit cones as projections of $\mathfrak{a}$-faces, and partition the set of orbit cones into orbits under the G-action.

```
julia> collector_cones = GITFans.orbit_cones(a, Q, G);

julia> length(collector_cones)
4

julia> orbit_list = GITFans.orbit_cone_orbits(collector_cones,
↪  matrix_action);

julia> println(map(length, orbit_list))
[10, 15, 10, 1]
```

We also compute the permutation action on the orbit cones in terms of group homomorphisms.

```
julia> perm_actions = GITFans.action_on_orbit_cone_orbits(orbit_list,
↪  matrix_action);

julia> perm_actions[1]
Group homomorphism
```

```
  from permutation group of degree 10 and order 120
  to permutation group of degree 10
```

Now we compute the GIT-fan in terms of a set of orbit representatives of maximal dimensional GIT-cones under the action of G, where the GIT-cones are described via hashes encoding the cones as intersections of orbit cones. The function also returns the incidence relation of the orbits of GIT-cones. There are six maximal cones, up to G-symmetry.

```
julia> q_cone = positive_hull(Q)
Polyhedral cone in ambient dimension 5

julia> (hash_list, edges) = GITFans.fan_traversal(orbit_list, q_cone,
↪  perm_actions);

julia> length(hash_list)
6

julia> println(edges)
Set([[4, 6], [2, 3], [3, 5], [1, 2], [3, 4]])
```

The fan object itself is created as follows. We can compute its F-vector.

```
julia> fanobj = GITFans.hashes_to_polyhedral_fan(orbit_list, hash_list,
↪  matrix_action)
Polyhedral fan in ambient dimension 5

julia> println(f_vector(fanobj))
ZZRingElem[20, 110, 240, 225, 76]
```

We show how to obtain cones of the fan in a given dimension, and how to compute further information about the fan.

```
julia> c = cones(fanobj, 5)[1]
Polyhedral cone in ambient dimension 5

julia> rays(c)
5-element SubObjectIterator{RayVector{QQFieldElem}}:
 [0, 1, 0, 0, -1]
 [1, 0, 1, 1, 0]
 [1, 1, 0, 0, 0]
 [1, 1, 0, 1, 0]
 [1, 1, 1, 0, 0]

julia> dim(fanobj)
5

julia> n_rays(fanobj)
20

julia> n_maximal_cones(fanobj)
76

julia> n_cones(fanobj)
```

```
671

julia> is_pointed(fanobj)
true

julia> is_complete(fanobj)
false
```

If we are interested only in the GIT-fan then we can combine the above steps in one function call.

```
julia> fanobj = GITFans.git_fan(a, Q, G)
Polyhedral fan in ambient dimension 5
```

If we are interested in the adjacency graph of maximal-dimensional GIT-cones then we can expand the orbits and compute the adjacency relation. We get a graph with 76 vertices and 180 edges.

```
julia> expanded = GITFans.orbits_of_maximal_GIT_cones(orbit_list,
↪  hash_list, matrix_action);

julia> orbit_lengths = map(length, expanded); println(orbit_lengths)
[1, 10, 30, 20, 10, 5]

julia> sum(orbit_lengths)
76

julia> maxcones = vcat( expanded... );

julia> full_edges = GITFans.edges_intersection_graph(maxcones, size(Q, 2)
↪  - 1);

julia> length(full_edges)
180
```

We plot the adjacency graphs of the maximal-dimensional GIT-cones and their orbits.

```
julia> full_graph = Graph{Undirected}(length(maxcones));

julia> [add_edge!(full_graph, e...) for e in full_edges];

julia> visualize(full_graph)

julia> orbit_graph = Graph{Undirected}(length(hash_list));

julia> [add_edge!(orbit_graph, e...) for e in edges];

julia> visualize(orbit_graph)
```

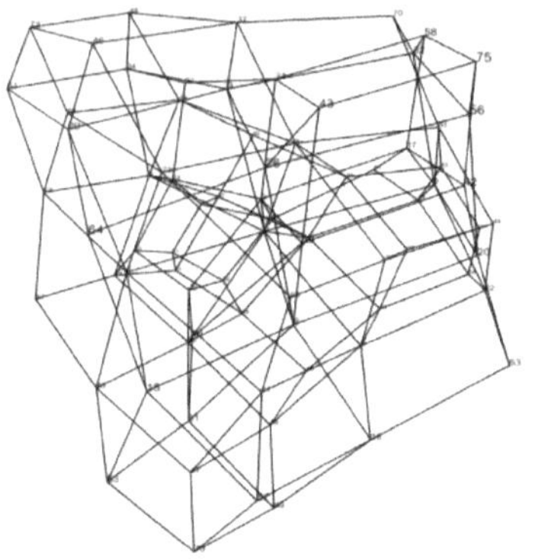

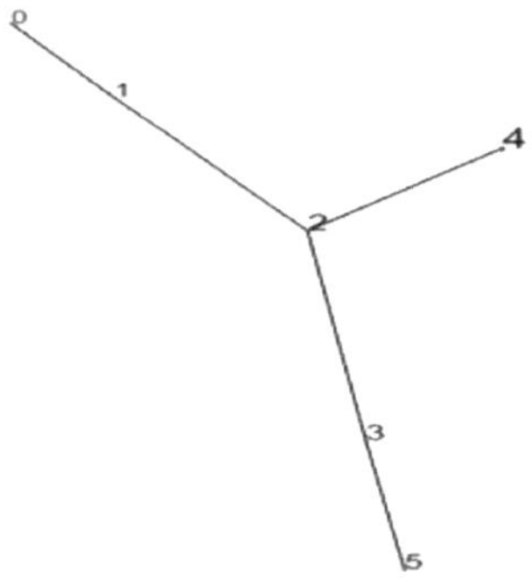

We could compute the GIT-fan without using the symmetry group G, that is, replace G by the trivial group in the computations. This would yield the same result but would take longer.

Acknowledgements Funded by the Deutsche Forschungsgemeinschaft (DFG, German Research Foundation) – Project-ID 286237555 – TRR 195. Janko Böhm also acknowledges support of Potentialbereich *SymbTools – Symbolic Tools in Mathematics and their Application* of the Forschungsinitiative Rheinland-Pfalz.

References

[BH06] F. Berchtold and J. Hausen. "GIT-Equivalence beyond the ample cone". *Michigan Mathematical Journal* 54.3 (2006), pp. 483–516. DOI: 10.1307/mmj/1163789912.

[BKR20] J. Böhm, S. Keicher, and Y. Ren. "Computing GIT-fans with symmetry and the Mori chamber decomposition of $\overline{M}_{0,6}$". *Math. Comp.* 89.326 (2020), pp. 3003–3021. DOI: 10.1090/mcom/3546.

[DH98] I. V. Dolgachev and Y. Hu. "Variation of geometric invariant theory quotients". *Inst. Hautes Études Sci. Publ. Math.* 87 (1998). With an appendix by Nicolas Ressayre, pp. 5–56.

[Kei12] S. Keicher. "Computing the GIT-fan". *Int. J. Algebra Comput.* 22.7 (2012), p. 1250064. DOI: 10.1142/S0218196712500646.

[MFK94] D. Mumford, J. Fogarty, and F. Kirwan. *Geometric invariant theory.* Third. Vol. 34. Ergebnisse der Mathematik und ihrer Grenzgebiete (2) [Results in Mathematics and Related Areas (2)]. Springer-Verlag, Berlin, 1994, pp. xiv+292.

[Stu96] B. Sturmfels. *Gröbner bases and convex polytopes.* Vol. 8. University Lecture Series. American Mathematical Society, Providence, RI, 1996, pp. xii+162. DOI: 10.1090/ulect/008.

Elliptic Fibrations on Vinberg's Most Algebraic K3 Surface

Simon Brandhorst and Matthias Zach

Abstract We explain how to use OSCAR to find all elliptic fibrations (up to automorphism) on a given surface and compute their Weierstrass models. This is illustrated for Vinberg's most algebraic K3 surface, the unique K3 surface of Picard rank 20 and discriminant 3.

Key words: Elliptic fibration, K3 surface, automorphisms

1 Introduction

A K3 surface over a field k is a complete non-singular surface [1] X/k with

$$h^1(X, \mathcal{O}_X) = 0, \quad \omega_X \cong \mathcal{O}_X.$$

Examples are double covers of the projective plane $\mathbb{P}^2$ branched over a smooth sextic, and smooth quartics in $\mathbb{P}^3$, smooth complete intersections of a quadric and a cubic in $\mathbb{P}^4$. If A is an abelian surface, e.g. the Jacobian of a genus 2 curve, then the minimal model of $A/\pm 1$ is a K3 surface, a so called Kummer K3 surface.

K3 surfaces have a central place in algebraic geometry. They can be seen as a generalization of elliptic curves and as the two-dimensional incarnation of Calabi–Yau and Hyperkähler manifolds. As such they have been studied extensively over the past decades and their geometry is by now reasonably well understood, see [Huy16]. For instance the Tate conjecture (see [Cha13;

Simon Brandhorst
Universität des Saarlandes, e-mail: `brandhorst@math.uni-sb.de`

Matthias Zach
RPTU Kaiserslautern-Landau, e-mail: `zach@mathematik.uni-kl.de`

[1] separated, geometrically integral scheme of finite type over k and dimension 2

W. Decker et al. (eds.), *The Computer Algebra System OSCAR*,
Algorithms and Computation in Mathematics 32,
https://doi.org/10.1007/978-3-031-62127-7_9

LMS14; Mad15; KM16] for k a finite field) and the cone conjecture [Ste85; LM18] are both known for K3 surfaces yet wide open in general. For *elliptically fibered*[2] K3 surfaces, the situation is under even better control; for example the Tate conjecture for elliptic K3 surfaces was known already in the '70s [AS73].

A distinctive feature of K3 surfaces is that a single K3 surface X may admit several elliptic fibrations. Such a fibration $\pi\colon X \to \mathbb{P}^1$ is determined by the class of a fiber of π in the Néron–Severi lattice $\mathrm{NS}(X)$. We call it an *elliptic class* [3]. Shioda and Kuwata [KS08] posed the following problems: Given an elliptic class $f \in \mathrm{NS}(X)$ in the Néron–Severi group of X determine

(1) the elliptic parameter t for f;
(2) the defining equation of the elliptic curve $E/k(t)$;
(3) the Mordell–Weil lattice of $E/k(t)$ and its generators;
(4) all the elliptic fibrations $\pi' : X \to \mathbb{P}^1$ up to isomorphism.

These problems have been solved to varying degree on several (families of) K3 surfaces. The automorphism group of a Kummer K3 surface X of the Jacobian of a genus two curve C (whose Jacobian $J(C)$ has no extra endomorphisms) was computed by Kondo in [Kon98]. This was used by Kumar in [Kum14] to compute representatives for the set of elliptic (and genus one) fibrations on X modulo $\mathrm{Aut}(X)$ and their elliptic parameters. There are 25 elliptic fibrations on a very general Jacobian Kummer K3 X up to automorphisms. Similar classifications on other (families of) K3 surfaces have been worked out by Elkies and Schütt [ES15], Nishiyama [Nis96], Kuwata and Shioda [KS08] largely by hand. In [FV22], Festi and Veniani give a formula for the number of elliptic fibrations up to automorphisms. The first named author and Elkies used the techniques to find equations for Lehmer's automorphism of minimal topological entropy on a K3 surface.

We demonstrate how to solve Shioda's and Kuwata's problems using OSCAR by considering Vinberg's "most algebraic" K3 surface $Y_1/\mathbb{C}$. The proofs of the algorithms involved and the details of the implementation shall appear elsewhere.

2 The Most Algebraic K3 Surface

Vinberg's most algebraic K3 surface Y_1 is the unique complex K3 surface of Picard number 20 and discriminant 3 [Vin83]. In what follows we will illustrate (some part of) the theory of K3 surfaces and their elliptic fibrations by considering Y_1 as a running example.

[2] See Definition 1 below

[3] See Section 2.2 for a definition

2.1 Elliptic Fibrations

In this section we fix our notation on elliptic surfaces and recall some definitions. For a detailed account of the theory of elliptic surfaces we refer to [Mir89; SS19]. For simplicity, we assume throughout that the characteristic of the base field k is not 2 or 3.

Definition 1 A genus-one fibration on a surface X is a surjective map $\pi\colon X \to C$ to a smooth curve C such that the generic fiber is a smooth curve of genus one. If π has a section O, we call the triple (X, π, O) an elliptic fibration and O the zero section.

Any choice of a zero section turns the generic fiber of π into an elliptic curve E over $K = k(C)$, the function field of C. If π and O are clear from the context, they are omitted from the notation and we call X simply an elliptic surface. For a K3 surface X, only the base $C = \mathbb{P}^1$ is possible because any regular 1-form on the base pulls back to a regular 1-form on X and K3 surfaces have no regular 1-forms.

We say that (X, π) is *relatively minimal*, if X is smooth and the fibers of π do not contain any (-1)-curves. Since such curves can be contracted to a smooth point, an arbitrary genus-one fibration (X', π') with smooth total space X' can be blown down to a relatively minimal one. Conversely, given an elliptic curve E/K one can construct a relatively minimal elliptic fibration whose generic fiber is E. This elliptic fibration is called the Kodaira-Néron model of E and is unique up to isomorphisms [SS19, Thm. 5.19]. For $C = \mathbb{P}^1$ we will see later how to construct it in OSCAR.

Let

$$E_1/\mathbb{C}(t) : y^2 = x^3 + t^5(t-1)^2 \tag{1}$$

and $\pi\colon Y_1 \to \mathbb{P}^1$ the relatively minimal elliptic fibration defined by E_1. Then we will see that Y_1 is Vinberg's most algebraic K3 surface.

In OSCAR we define E_1 as an elliptic curve over $\mathbb{Q}(t)$.

```
julia> Qt, t = polynomial_ring(QQ, :t)
(Univariate polynomial ring in t over QQ, t)

julia> Qtf = fraction_field(Qt)
Fraction field
  of univariate polynomial ring in t over QQ

julia> E = elliptic_curve(Qtf, [0,0,0,0,t^5*(t-1)^2])
Elliptic curve with equation
y^2 = x^3 + t^7 - 2*t^6 + t^5
```

The Kodaira-Néron Model Y_1 of E_1 has a contraction to a Weierstrass model S_1 given by the homogeneous equation $Y^2 = X^3 + t^5(t-s)^2s^5$ of degree 12 in weighted projective space $\mathbb{P}(1,1,2,3)$ with coordinates (s, t, X, Y). Later we construct S in a certain projective bundle of $\mathbb{P}^1$. At this point we remark that S_1 has an E_8 singularity in the fibers over $[0 : 1]$ and $[1 : 0]$ each and an A_2 singularity in the fiber over $[1:1]$. The minimal resolution of these singularities is the contraction $Y_1 \to S_1$. The exceptional divisors are contained in the respective fibers. This is how reducible fibers may appear. They will play an important role in the next section.

2.1.1 Fibers and the Trivial Lattice

Since any complex K3 surface X is simply connected, the notions of linear, algebraic, and numerical equivalence agree and yield $\mathrm{Pic}(X) \cong \mathrm{NS}(X) \cong \mathrm{Num}(X)$ [Huy16]. This is also true in positive characteristic. Equipped with the intersection pairing $\mathrm{NS}(X)$ is a *quadratic integer lattice*. Its rank is called the Picard number of X. Computing generators of $\mathrm{NS}(X)$ or even its rank is in general a hard problem — just like computing rational points on elliptic curves.

On an elliptic fibration some part of $\mathrm{NS}(X)$ is readily computable, the so called *trivial lattice* $\mathrm{Triv}(X, \pi, O)$ spanned by the zero section and components of the reducible fibers. The classes $o, f \in \mathrm{NS}(X)$ of the zero section O and any fiber F of π satisfy $f^2 = 0$, $o^2 = -2$ and $f.o = 1$, i,e. they span a so called *hyperbolic plane* in $\mathrm{NS}(X)$. An elliptic fibration may have reducible fibers. The reducible fibers of π which are disjoint from the zero section O decompose as a direct sum of (negative definite) ADE-root lattices — as predicted by Kodaira classification of singular fibers [Kod63, Thm. 6.2] and Tate's algorithm [Tat75] to compute them. Tate's algorithm implies that we can compute the Kodaira types of the singular fibers by factoring the j-invariant and the discriminant of E and comparing this with the table [Sil94, p.365] (outside of characteristic 2 and 3). In our case $j(E) = 0$.

```
julia> factor(Qt, discriminant(E))
-432 * (t - 1)^4 * t^10
```

Since the base of the fibration is $\mathbb{P}^1$, we should view the discriminant as a homogeneous polynomial of degree $12l$ in $[t:s]$ where $l > 0$ comes from the construction of the Kodaira-Néron model. For X a K3 surface, $l = 2$ and the discriminant has degree 24 which is – and this is not a coincidence – the topological Euler number of any K3 surface.

We are in the affine chart $s = 1$ and therefore the homogeneous polynomial for the discriminant is

$$\Delta(E) = -432(t-s)^4 t^{10} s^{10}.$$

We see that the fibration π has two singular fibers of type II^* (an extended E_8 configuration) at $[0:1]$ and $[1:0]$, as well as a fiber of type IV (forming an extended A_2 configuration) at $[1:1]$.

The components not meeting the zero section form an E_8, E_8, A_2 configuration. Therefore the trivial lattice is $\mathrm{Triv}(\pi) \cong U \oplus E_8 \oplus E_8 \oplus A_2$ which is of rank 20 and determinant -3. Since the maximum Picard number of a K3 surface in characteristic 0 is 20 and -3 is the minimal possible determinant, we conclude that $\mathrm{NS}(X) = \mathrm{Triv}(\pi)$ is the full Néron–Severi lattice. Therefore Y_1 is indeed Vinberg's most algebraic K3 surface [Vin83].

2.2 Elliptic Fibrations and the Automorphism Group

A distinctive feature of K3 surfaces is that they may admit several elliptic fibrations. They are in bijection with the extremal rays of the Nef cone.

Theorem 2 *[PŠ71, §3 Theorem 1][SS19, Lemma 11.25] Let X be a K3 surface and $f \in \mathrm{NS}(X)$ a primitive and Nef divisor class with $f^2 = 0$. Then the map $\pi_{|f|}$ induced by the linear system $|f|$ is a genus 1 fibration. The ADE-type of its singular fibers is given by the ADE-type of the root sublattice of $(f^\perp/\mathbb{Z}f)$. If there is a class $e \in \mathrm{NS}(X)$ with $e.f = 1$, then the fibration has a section.*

We call a primitive and Nef divisor class $f \in \mathrm{NS}(X)$ an *elliptic class* and $\pi^*(t/s) \in k(X)$ an elliptic parameter where t/s is a generator of the function field of the base $\mathbb{P}^1$. Let us point out that if $f \neq 0$ is isotropic but not Nef, then it can be mapped to an elliptic class f' by an element of the Weyl group. Therefore, by Mayer's theorem, any K3 surface of Picard number at least 5 has a genus one fibration.

To classify all elliptic fibrations on X up to automorphism, we need a description of its automorphism group. Over $\mathbb{C}$, the strong Torelli type theorem implies that $\mathrm{Aut}(X)$ can be identified with the group of effective, integral Hodge isometries of $H^2(X,\mathbb{Z})$, i.e. isometries that preserve the intersection product, the Nef cone $\mathrm{nef}(X)$, and the integral Hodge structure on $H^2(X,\mathbb{Z})$. We refer to [Huy16, Section 7] for details. More generally, for an arbitrary field k and any K3 surface X/k the map

$$\Psi\colon \mathrm{Aut}(X) \to O(\mathrm{NS}(X), \mathrm{nef}(X)) \cong O(\mathrm{NS}(X))/W(\mathrm{NS}(X)) \tag{2}$$

$$\phi \mapsto (\phi^*)^{-1} \tag{3}$$

has finite kernel and cokernel. Here $W(\mathrm{NS}(X))$ denotes the Weyl group, that is the group generated by all reflections in (-2)-vectors of $\mathrm{NS}(X)$. Denote by $\mathrm{aut}(X)$ the image of Ψ. For $Y_1/\mathbb{C}$ we know (from the K3-Torelli theorem) that Ψ is an isomorphism. To solve Shioda and Kuwata's problems, we have to compute $\mathrm{aut}(X)$, $\mathrm{nef}(X)$, and

$$\{f \in \mathrm{NS}(X) \mid f \text{ is elliptic }\}/\,\mathrm{aut}(X).$$

Using an implementation in OSCAR of Shimada's algorithm [Shi15], which builds on ideas of Borcherds' and Vinberg, we compute $\mathrm{aut}(Y_1)$ as well as a fundamental domain of its action on the Nef cone.

First we set up the Néron–Severi lattice.

```
julia> R = rescale(root_lattice([(:E, 8), (:E, 8), (:A, 2)]), -1);

julia> U = integer_lattice(gram=ZZ[0 1; 1 -2]);

julia> NS, _ = direct_sum([U, R]);
```

Next we need to find an ample class, to identify the ample cone as a Weyl-chamber. The ample class must intersect each curve (in particular the ones corresponding to the basis vectors) positively.

```
julia> e = matrix(ZZ,1,20,ones(Int,20));e[1,1]=51;

julia> ample = e*inv(gram_matrix(NS));
```

We confirm that the class is indeed in the interior of a Weyl-chamber, by checking that it is not contained in any hyperplane orthogonal to a (-2)-class.

```
julia> minimum(rescale(orthogonal_submodule(NS, ample),-1))
4
```

We let OSCAR compute generators for the automorphism group $\text{aut}(Y_1)$.

```
julia> Xaut, Xchambers, Xrational_curves =
↪  K3_surface_automorphism_group(NS, ample);
```

The return value `Xchambers` is a list of polyhedral cones $C_1, \dots C_n$ which are contained in the Nef cone. Let $\text{Aut}(C_i) := \text{Aut}(X)_{C_i}$ be the stabilizer of C_i. This stabilizer acts on the points of C_i. Let $P_i \subseteq C_i$ be its fundamental domain. A fundamental domain of the action of $\text{Aut}(X)$ on the Nef cone is the union of the P_i. In our case $n = 1$.

```
julia> C = Xchambers[1]
Chamber  in dimension 20 with 36 walls
```

The walls $W = r^\perp$ of C with $r^2 = -2$ and $r \in \text{NS}(X)$ are classes of smooth rational curves. The rays f of C with $f^2 = 0$ give elliptic fibrations. We now compute the rays of C modulo $\text{Aut}(C)$.

```
julia> Xelliptic_classes = [f for f in rays(C) if 0 ==
↪  f*gram_matrix(NS)*transpose(f)];

julia> Xelliptic_classes_orb =
↪  orbits(gset(matrix_group(aut(C)),(x,g)->x*matrix(g),
↪  Xelliptic_classes));

julia> length(Xelliptic_classes_orb)
6
```

Thus modulo $\text{Aut}(Y_1/\mathbb{C})$ there are exactly 6 elliptic fibrations on $Y_1/\mathbb{C}$, this recovers a result of Nishiayama, Watari and Braun (see the introduction [FV22]) We will see later they are all defined over $\mathbb{Q}$ – because so is the full Néron–Severi lattice of Y_1.

In what follows we aim to compute different Weierstrass-models of Vinberg's K3 surface which realize the different elliptic fibrations. Moreover, we wish to identify their minimal desingularizations by explicit rational maps. It turns out that such a rational map can be arbitrarily complicated, meaning that only writing down the required rational functions to describe them might fill several pages. Therefore, it is preferable to work with a class of simple, better controlled isomorphisms and factor the more complicated ones in terms of these. A first bound on the complexity of a rational map between two elliptic fibrations is the intersection number of their associated elliptic classes:

Definition 3 Let $p \in \mathbb{N}$ be prime. Two elliptic classes $f_1, f_2 \in \mathrm{NS}(X)$ are called p-neighbors if $f_1.f_2 = p$.

In fact one can show that the corresponding frame lattices of f_1 and f_2 can be seen as p-neighbors in the sense of Kneser if $\det(\mathrm{NS}(X))$ is coprime to p [BE23].

The simplest non-trivial case is when two elliptic classes are 2-neighbors. Therefore, to compute all the sought for Weierstrass models and the corresponding elliptic parameters, we first single out 6 representatives $f_1, \ldots f_6$, which we can reach from our initial fibration by 2-neighbor steps.

```
julia> f1 = identity_matrix(ZZ, 20)[1:1,:];

julia> f1_2neighbors = [f for f in Xelliptic_classes if 2 ==
↪  (f1*gram_matrix(NS)*transpose(f))[1,1]];

julia> f2, f3, _ = f1_2neighbors;

julia> f2_2neighbors = [f for f in Xelliptic_classes if 2 ==
↪  (f2*gram_matrix(NS)*transpose(f))[1,1]];

julia> fibers = [f1,f2,f3];

julia> todo = [o for o in Xelliptic_classes_orb if !any(f in o for f in
↪  fibers)];

julia> while length(todo) > 0
         o = popfirst!(todo)
         f = findfirst(x-> x in o, f2_2neighbors)
         push!(fibers, f2_2neighbors[f])
       end

julia> fibers = [vec(collect(i*basis_matrix(NS))) for i in fibers]
6-element Vector{Vector{QQFieldElem}}:
 [1, 0, 0, 0, 0, 0, 0, 0, 0, 0, 0, 0, 0, 0, 0, 0, 0, 0, 0, 0]
 [4, 2, -4, -7, -10, -8, -6, -4, -2, -5, -2, -4, -6, -5, -4, -3, -2, -3,
 ↪ -1, -1]
 [4, 2, -4, -7, -10, -8, -6, -4, -2, -5, -4, -7, -10, -8, -6, -4, -2, -5,
 ↪ 0, 0]
 [6, 3, -5, -10, -15, -12, -9, -6, -3, -8, -5, -10, -15, -12, -9, -6, -3,
 ↪ -8, -1, -1]
```

```
[10, 5, -8, -16, -24, -20, -15, -10, -5, -12, -7, -14, -21, -17, -13,
↪ -9, -5, -11, -3, -2]
[6, 3, -4, -8, -12, -10, -8, -6, -3, -6, -4, -8, -12, -10, -8, -6, -3,
↪ -6, -2, -1]
```

Now we compute their Mordell–Weil rank, the torsion subgroup of the Mordell–Weil group, and the ADE-type of their singular fibers. This recovers a result of Nishiyama [Nis96, Table 1.1].

```
julia> [fibration_type(NS, f) for f in fibers]
6-element Vector{Tuple{Int64, FinGenAbGroup, Vector{Tuple{Symbol,
↪  Int64}}}}:
 (0, Z/1, [(:A, 2), (:E, 8), (:E, 8)])
 (1, Z/2, [(:D, 10), (:E, 7)])
 (0, Z/2, [(:A, 2), (:D, 16)])
 (1, Z/3, [(:A, 17)])
 (0, Z/4, [(:A, 11), (:D, 7)])
 (0, Z/3, [(:E, 6), (:E, 6), (:E, 6)])
```

We compute their mutual intersections $(f_i.f_j)_{1\leq i,j\leq 6}$

```
julia> F = transpose(matrix((reduce(hcat, fibers)))); F * gram_matrix(NS)
↪  * transpose(F)
[0  2  2  3  5  3]
[2  0  2  2  2  2]
[2  2  0  2  5  4]
[3  2  2  0  2  3]
[5  2  5  2  0  2]
[3  2  4  3  2  0]
```

and see that $f_1.f_2 = f_1.f_3 = 2$ and $f_2.f_i = 2$ for $i \neq 2$. Therefore, we can reach f_2, f_3 by a single 2-neighbor step from f_1 and f_4, f_5, f_6 by another one from f_2.

2.3 The (Relatively) Minimal Model

To solve the next problem on Shimada's list, we return from lattices back to algebraic geometry.

In OSCAR we can construct $\pi : Y_1 \to \mathbb{P}^1$ as a covered scheme which is represented internally by patches and gluing data, just like in Hartshorne's book Algebraic Geometry [Har77]. Recall that $E_1/\mathbb{Q}(t)$ is the generic fiber of π_1. The (singular) Weierstrass model S_1 of (Y_1, π_1) is the zero locus of a section of the projective bundle $\mathbb{P}(\mathcal{O}_{\mathbb{P}^1}(-2l) \oplus \mathcal{O}_{\mathbb{P}^1}(-3l) \oplus \mathcal{O}_{\mathbb{P}^1})$ where $l = 2$ since Y_1 is a K3 surface [SS19, §5.13].

```
julia> Y1 = elliptic_surface(E, 2)
Elliptic surface
  over rational field
with generic fiber
  -x^3 + y^2 - t^7 + 2*t^6 - t^5
```

```
julia> S = weierstrass_model(Y1)[1]
Scheme
  over rational field
with default covering
  described by patches
    1: scheme(-(x//z)^3 + (y//z)^2 - t^7 + 2*t^6 - t^5)
    2: scheme((z//x)^3*t^7 - 2*(z//x)^3*t^6 + (z//x)^3*t^5 -
    ↪  (z//x)*(y//x)^2 + 1)
    3: scheme((z//y)^3*t^7 - 2*(z//y)^3*t^6 + (z//y)^3*t^5 - (z//y) +
    ↪  (x//y)^3)
    4: scheme((x//z)^3 - (y//z)^2 + s^7 - 2*s^6 + s^5)
    5: scheme((z//x)^3*s^7 - 2*(z//x)^3*s^6 + (z//x)^3*s^5 -
    ↪  (z//x)*(y//x)^2 + 1)
    6: scheme((z//y)^3*s^7 - 2*(z//y)^3*s^6 + (z//y)^3*s^5 - (z//y) +
    ↪  (x//y)^3)
  in the coordinate(s)
    1: [(x//z), (y//z), t]
    2: [(z//x), (y//x), t]
    3: [(z//y), (x//y), t]
    4: [(x//z), (y//z), s]
    5: [(z//x), (y//x), s]
    6: [(z//y), (x//y), s]
```

We compute a resolution of singularities $\psi\colon Y_1 \to S_1$ by performing a series of blowups in one singular point at a time. This is fully automated and leads to a total of 17 blowups.

```
julia> piS = weierstrass_contraction(Y1)
Composite morphism of[...]
```

Note that this is less elegant than Tate's algorithm but it works for any surface with isolated ADE singularities. We use the resolution to compute the exceptional divisors, i.e. the trivial lattice,

```
julia> basisNSY1, gramTriv = trivial_lattice(Y1);
```

as well as the singular fibers.

```
julia> [(i[1],i[2]) for i in reducible_fibers(Y1)]
3-element Vector{Tuple{Vector{QQFieldElem}, Tuple{Symbol, Int64}}}:
 ([1, 1], (:A, 2))
 ([0, 1], (:E, 8))
 ([1, 0], (:E, 8))
```

Since the computation of the full Néron–Severi lattice is a difficult problem, we work with a sublattice called the algebraic lattice in OSCAR. It is spanned by fiber components and the sections supplied to the constructor. Recall that for Vinberg's K3 surface Y_1 the trivial lattice agrees with $\mathrm{NS}(Y_1)$, hence so does the algebraic lattice.

```
julia> basisNSY1, _, NSY1 = algebraic_lattice(Y1);
```

The first return value consists of the chosen basis of the algebraic lattice. For brevity we only show the first 5 elements.[4]

```
julia> basisNSY1
20-element Vector{Any}:
 Fiber over (2, 1)
 section: (0 : 1 : 0)
 component A2_1 of fiber over (1, 1)
 component A2_2 of fiber over (1, 1)
 component E8_1 of fiber over (0, 1)
```

They are represented as effective Weil divisors in OSCAR.

```
julia> basisNSY1[1]
Effective weil divisor Fiber over (2, 1)
  on elliptic surface with generic fiber -x^3 + y^2 - t^7 + 2*t^6 - t^5
with coefficients in integer Ring
given as the formal sum of
  1 * sheaf of ideals
```

2.4 Divisors and Linear Systems on Elliptic K3 Surfaces

Our goal is to compute the linear systems $|f_i|$, $i \in \{2, 3, 4, 6\}$. To this end we need a better control on the Picard group. Here the elliptic fibration comes to help.

Recall that $K = k(C)$ is the function field of the base curve of an elliptic fibration $\pi\colon X \to C$. The K-rational points $E(K)$ of E form a finitely generated abelian group group, which is called the Mordell–Weil group of E. A section of the fibration π is a map $s\colon C \to X$ with $\pi \circ s = id_C$. We identify s with its image $s(C)$ and call it a section as well. The generic point of a section is a K-rational point of E. Conversely, the closure of a K-rational point of E in X is a section. This gives a bijection between the set of K-rational points $E(K)$ and the set of sections of π. We use it to give the set of sections the structure of an abelian group which we denote by $\mathrm{Mwl}(\pi)$ and its addition by $\oplus$.

We call a prime divisor *vertical* if its image under π is a point and otherwise *horizontal*. Any fibration structures the Néron–Severi lattice into a vertical part and a horizontal part. The advantage of an *elliptic* fibration is that a horizontal divisor is linearly equivalent to a sum of sections (modulo vertical divisors).

Theorem 4 *[Shi90, Theorem 1.3] Let $\pi\colon X \to C$ be an elliptic fibration. Then there is an isomorphism of groups*

$$\mathrm{Mwl}(\pi) \cong \mathrm{NS}(X)/\operatorname{Triv}(X, \pi)$$

[4] Due to implementation details, the order in which the divisors appear might differ from the output presented here.

and an isometry of lattices

$$\mathrm{Mwl}(\pi)_{free} \cong \mathrm{NS}(X)/\overline{\mathrm{Triv}(X,\pi)}.$$

Here $\overline{\mathrm{Triv}(X,\pi)}$ denotes the primitive closure of the trivial lattice in $\mathrm{NS}(X)$. This theorem results in the so called *Shioda–Tate formula* which relates the Picard number of X with the Mordell–Weil rank of the generic fiber of π.

$$\rho(X) = 2 + \sum_{\nu\in\mathbb{P}^1}(n_\nu - 1) + \mathrm{rk}\,\mathrm{MW}(\pi)$$

where n_ν is the number of irreducible components of the fiber $\pi^{-1}(\nu)$. In our case we find that $E/\mathbb{Q}(t)$ has zero Mordell–Weil rank.

On the side of divisors the theorem tells us the following (see [Shi90, Lemma 5.1]: Let $A \in \mathrm{Div}(X)$ be a divisor, then there is a linear equivalence

$$A \sim V + (n-1)O + P$$

where P is some section, V is a vertical divisor, i.e. a sum of fiber components, and $n = A.F$. Thus when we want to compute linear systems on an elliptic surface, it suffices to consider the linear systems of this form.

We wish to find $F_i \in \mathrm{Div}(X)$ in the above form with $[F_i] = f_i \in \mathrm{NS}(X)$. We start with F_2. So first we have to identify the abstract $\mathrm{NS}(X)$ that we computed with the actual geometric one on X spanned by divisors.

```
julia> b, I =
↪  Oscar._is_equal_up_to_permutation_with_permutation(gram_matrix(NS),
↪  gram_matrix(NSY1))
(true, [1, 2, 19, 20, 3, 4, 5, 6, 7, 8, 9, 10, 11, 12, 13, 14, 15, 16,
↪  17, 18])

julia> @assert gram_matrix(NSY1) == gram_matrix(NS)[I,I]
```

This identification is just a permutation of the basis.

Since the divisor V is vertical, it is dominated by some divisor of fibers $V \leq V' = n_1V_1 + \dots n_kV_k$ where $V_i = \pi^*(p_i)$ is a fiber over a point $p_i \in \mathbb{P}^1$ for $i \in \{1,\dots k\}$. Note that in our case $n = 2$ because we work with two neighbors. OSCAR can compute V' and V for us. We display V.

```
julia> Oscar.horizontal_decomposition(Y1, fibers[2][I])[2]
Effective weil divisor
  on elliptic surface with generic fiber -x^3 + y^2 - t^7 + 2*t^6 - t^5
with coefficients in integer Ring
given as the formal sum of
  2 * component E8_7 of fiber over (0, 1)
  2 * section: (0 : 1 : 0)
  1 * component A2_0 of fiber over (1, 1)
  1 * component E8_0 of fiber over (1, 0)
  2 * component E8_4 of fiber over (0, 1)
  2 * component E8_3 of fiber over (0, 1)
  1 * component E8_2 of fiber over (0, 1)
```

```
  2 * component E8_0 of fiber over (0, 1)
  2 * component E8_6 of fiber over (0, 1)
  2 * component E8_5 of fiber over (0, 1)
  1 * component E8_8 of fiber over (0, 1)
```

The larger linear system $|V' + O + P|$ is now of a simple form and computed in [BE23]. Then OSCAR can figure out the linear equations cutting out the subsystem $|V + O + P|$. Since F_2 is an elliptic class, this linear system is of (projective) dimension one generated by two sections $v, w \in k(Y_1)$. Their quotient $u = v/w \in k(Y_1)$ is an elliptic parameter. The fibration $\pi_2 \colon Y_1 \to \mathbb{P}^1$ defined by F_2 is thus given by $P \mapsto (1 : u(P))$

```
julia> representative(elliptic_parameter(Y1, fibers[2][I]))
(x//z)//(t^3 - t^2)
```

with $u = x/(t^3 - t^2)$ in Weierstrass coordinates.

2.5 A 2-Neighbor Step

To compute the fibrations of F_i for $i = 4, 5, 6$, we need to pass to a Weierstrass model of the fibration F_2. Using the elliptic parameter $u_i = u_i(x, y, t)$ OSCAR can find a suitable change of coordinates such that the generic fiber of F_i is given as double cover of $\mathbb{P}^1_{k(u_i)}$ ramified over four points. An affine chart of this double cover is given by $g_i = y_i^2 - q(x_i) \in k(u_i)[x_i, y_i]$ with $q_i(x_i)$ a separable polynomial of degree 3 or 4. It describes a curve of genus one over $k(u_i)$, but not yet an elliptic curve, because in general it may not have a rational point. The following command produces a pair `(g, phi1)` consisting of g_2, together with a morphism $\varphi_1 \colon k[x, y, t] \to (k(u_2))(x_2, y_2)$ taking the Weierstrass equation for the original Y_1 to an element with g_2 as the relevant factor of the numerator.

```
julia> g, phi1 = two_neighbor_step(Y1, fibers[2][I]); g
-t^3*x^3 + t^3*x^2 - x + y^2
```

To compute a Weierstrass form we choose the rational point P, given by $x_2 = y_2 = 0$, of the generic fiber and move it to infinity. This produces an elliptic surface Y_2 and another homomorphism $\varphi_2 : k(t_2)(x_2, y_2) \to k(Y_2)$ encoding the necessary change of coordinates:

```
julia> kt2 = base_ring(parent(g)); P = kt2.([0,0]);

julia> Y2, phi2 = elliptic_surface(g, P; minimize=false); Y2
Elliptic surface
  over rational field
with generic fiber
  -x^3 + t^3*x^2 - t^3*x + y^2
and relatively minimal model
  scheme over QQ covered with 47 patches
```

From our computations in Section 2.2 we already know that the corresponding elliptic fibration π_2 has a D_{10} and an E_7 fiber, a 2-torsion section and Mordell–Weil rank one. To find a section of infinite order, we note that f_2 intersects some fiber components of π_1 with multiplicity one. Therefore they are sections of π_2. We inform OSCAR about the section.

```
julia> E2 = generic_fiber(Y2); t2 = gen(kt2);

julia> P2 = E2([t2^3, t2^3]); set_mordell_weil_basis!(Y2, [P2]);
```

The upshot is that we have two elliptic surfaces $\pi_i\colon Y_i \to \mathbb{P}^1$, $i = 1, 2$, and a birational map $\phi\colon Y_2 \dashrightarrow Y_1$ which is given in Weierstrass coordinates of Y_1 and Y_2 by

$$(x, y, t) \mapsto \left(\frac{-tx + t}{x^3}, \frac{xy - y}{x^5}, \frac{1}{x}\right).$$

With OSCAR this goes as follows:

```
julia> U2 = weierstrass_chart_on_minimal_model(Y2); U1 =
↪  weierstrass_chart_on_minimal_model(Y1);

julia> imgs = phi2.(phi1.(ambient_coordinates(U1))) # k(Y1) -> k(Y2)
3-element Vector{AbstractAlgebra.Generic.Frac{QQMPolyRingElem}}:
 (-(x//z)*t + t)//(x//z)^3
 ((x//z)*(y//z) - (y//z))//(x//z)^5
 1//(x//z)

julia> phi_rat = morphism_from_rational_functions(Y2, Y1, U2, U1, imgs);
```

Since the surfaces in question have Nef canonical class, we know from theory that the birational map is an isomorphism. Let us tell OSCAR that ϕ is a morphism.

```
julia> set_attribute!(phi_rat, :is_isomorphism=>true);
```

However actually computing ϕ as a morphism on some given chart can be prohibitively expensive. This poses some serious computational challenges.

Next, we compute the matrix representing the pullback $\phi^*\colon \mathrm{NS}(Y_1) \to \mathrm{NS}(Y_2)$. This is done by pulling back the equations on carefully chosen charts. Then we use the intersection pairing.

```
julia> pullbackDivY1 = [pullback(phi_rat, D) for D in basisNSY1];

julia> B = [basis_representation(Y2, D) for D in pullbackDivY1];
```

At this point `B` is a list of rational vectors. We turn it into a matrix and perform some sanity checks.

```
julia> B = matrix(QQ, 20, 20, reduce(vcat, B)); NSY2 =
↪  algebraic_lattice(Y2)[3];

julia> NSY1inY2 = lattice(ambient_space(NSY2),B);

julia> @assert NSY1inY2 == NSY2 && gram_matrix(NSY1inY2) ==
↪  gram_matrix(NSY1)
```

Finally, we have the basis representation of the f_i in $\mathrm{NS}(Y_2)$.

```
julia> fibers_in_Y2 = [f[I]*B for f in fibers]
6-element Vector{Vector{QQFieldElem}}:
 [4, 2, 0, 0, 0, 0, 0, 0, 0, -4, -4, -8, -7, -6, -5, -4, -3, -2, -1, 0]
 [1, 0, 0, 0, 0, 0, 0, 0, 0, 0, 0, 0, 0, 0, 0, 0, 0, 0, 0, 0]
 [5, 2, -2, -3, -4, -3, -2, -1, -2, -5, -4, -8, -7, -6, -5, -4, -3, -2,
 ↪ -1, 0]
 [4, 2, -2, -4, -6, -9//2, -3, -3//2, -7//2, -3, -5//2, -5, -9//2, -4,
 ↪ -7//2, -3, -5//2, -2, -3//2, 0]
 [2, 1, -1, -2, -3, -2, -1, 0, -2, -1, -1, -2, -2, -2, -2, -2, -2, -1, 0,
 ↪ 1]
 [2, 1, 0, 0, 0, 0, 0, 0, 0, -2, -2, -4, -4, -4, -4, -3, -2, -1, 0, 1]
```

2.6 The Remaining Fibrations

With computations similar to the above we can now compute the elliptic parameters as well as the generic fibers for $\pi_i\colon Y_i \to \mathbb{P}^1$ for $i = 3, \ldots 6$.

First we compute π_3 starting from Y_1.

```
julia> f3 = fibers[3][I]; representative(elliptic_parameter(Y1, f3))
(x//z)//t^2

julia> g3a, phi3a = two_neighbor_step(Y1, f3); g3a
-x^3 + (-t^3 + 2)*x^2 - x + y^2
```

Since we have the pullback action ϕ^*, we are in a position to compute π_i for $i = 4, 5, 6$ from Y_2.

```
julia> [representative(elliptic_parameter(Y2, f)) for f in
↪  fibers_in_Y2[4:6]]
3-element Vector{AbstractAlgebra.Generic.FracFieldElem{QQMPolyRingElem}}:
 (y//z)//((x//z)*t)
 ((y//z) + t^3)//((x//z)*t - t^4)
 ((y//z) + t^3)//((x//z) - t^3)

julia> g,_ = two_neighbor_step(Y2, fibers_in_Y2[4]);g
-1//4*x^4 - 1//2*t^2*x^3 - 1//4*t^4*x^2 + x + y^2

julia> R = parent(g); K_t = base_ring(R);

julia> (x,y) = gens(R); P = K_t.([0,0]); # rational point

julia> g, _ = transform_to_weierstrass(g, x, y, P);

julia> E4 = elliptic_curve(g, x, y)
Elliptic curve with equation
y^2 = x^3 + 1//4*t^4*x^2 - 1//2*t^2*x + 1//4

julia> g,_ = two_neighbor_step(Y2, fibers_in_Y2[5]);g
t^2*x^3 + (-1//4*t^4 + 2*t)*x^2 + x + y^2
```

```
julia> R = parent(g); K_t = base_ring(R);

julia> (x,y) = gens(R); P = K_t.([0,0]); # rational point

julia> g, _ = transform_to_weierstrass(g, x, y, P);

julia> E5 = elliptic_curve(g, x, y)
Elliptic curve with equation
y^2 = x^3 + (1//4*t^4 - 2*t)*x^2 + t^2*x

julia> g,_ = two_neighbor_step(Y2, fibers_in_Y2[6]);g
(t^2 + 2*t + 1)*x^3 + y^2 - 1//4*t^4

julia> R = parent(g); K_t = base_ring(R); t = gen(K_t);

julia> (x,y) = gens(R); P = K_t.([0,1//2*t^2]); # rational point

julia> g, _ = transform_to_weierstrass(g, x, y, P);

julia> E6 = elliptic_curve(g, x, y)
Elliptic curve with equation
y^2 + (-t^2 - 2*t - 1)//t^4*y = x^3
```

Let us summarize the equations we found in the following theorem.

Theorem 5 *Let $Y/\mathbb{C}$ be the singular K3 surface of discriminant 3. By [Nis96; FV22] it admits exactly* 6 *elliptic fibrations $E_1, \dots E_6$ up to automorphisms. Weierstrass models of their generic fibers, the Kodaira types of their singular fibers and their Mordell–Weil groups are listed in the following table:*

i	E_i	*Kodaira type*	*Mordell–Weil group*
1	$y^2 = x^3 + t^5(t-1)^2$	$2 \times II^*, IV$	*0*
2	$y^2 = x^3 - t^3x^2 + t^3x$	I_6^*, III^*	$\mathbb{Z}/2\mathbb{Z} \times \mathbb{Z}$
3	$y^2 = x^3 + (t^3-2)x^2 + x$	I_{12}^*, I_3	$\mathbb{Z}/2\mathbb{Z}$
4	$y^2 = x^3 + t^4x^2 - 8t^2x + 16$	I_{18}	$\mathbb{Z}/3\mathbb{Z} \times \mathbb{Z}$
5	$y^2 = x^3 + (t^4 - 8t)x^2 + 16t^2x$	I_{12}, I_3^*	$\mathbb{Z}/4\mathbb{Z}$
6	$y^2 - (t^2 + 2t + 1)t^8y = x^3$	$3 \times IV^*$	$\mathbb{Z}/3\mathbb{Z}$

To sum up, we have computed all elliptic fibrations up to isomorphism on Vinberg's K3 surface as well as their elliptic parameters and generic fibers in Weierstrass form. Moreover, we can explicitly write down the isomorphisms between the different desingularizations of the Weierstrass models and put them to use to transform between the respective bases of the Néron–Severi lattice.

Acknowledgements This work is funded by the Deutsche Forschungsgemeinschaft (DFG, German Research Foundation) – Project-ID 286237555 – TRR 195.

References

[AS73] M. Artin and H. P. F. Swinnerton-Dyer. "The Shafarevich-Tate conjecture for pencils of elliptic curves on $K3$ surfaces". *Invent. Math.* 20 (1973), pp. 249–266. DOI: 10.1007/BF01394097.

[BE23] S. Brandhorst and N. D. Elkies. "Equations for a K3 Lehmer map". *J. Algebraic. Geom.* (2023).

[Cha13] F. Charles. "The Tate conjecture for $K3$ surfaces over finite fields". *Invent. Math.* 194.1 (2013), pp. 119–145. DOI: 10.1007/s00222-012-0443-y.

[ES15] N. D. Elkies and M. Schütt. "Genus 1 fibrations on the supersingular K3 surface in characteristic 2 with Artin invariant 1". *Asian J. Math.* 19.3 (2015), pp. 555–581. DOI: 10.4310/AJM.2015.v19.n3.a7.

[FV22] D. FESTI and D. C. VENIANI. "Counting elliptic fibrations on K3 surfaces". *Journal of the Mathematical Society of Japan* (2022), pp. 1–31. DOI: 10.2969/jmsj/88178817.

[Har77] R. Hartshorne. *Algebraic geometry.* Vol. No. 52. Graduate Texts in Mathematics. Springer-Verlag, New York-Heidelberg, 1977, pp. xvi+496.

[Huy16] D. Huybrechts. *Lectures on K3 surfaces.* Vol. 158. Cambridge Studies in Advanced Mathematics. Cambridge University Press, 2016, pp. xi+485. DOI: 10.1017/CBO9781316594193.

[KM16] W. Kim and K. Madapusi Pera. "2-adic integral canonical models". *Forum Math. Sigma* 4 (2016), Paper No. e28, 34. DOI: 10.1017/fms.2016.23.

[Kod63] K. Kodaira. "On Compact Analytic Surfaces: II". *Ann. of Math.* 77.3 (1963), pp. 563–626.

[Kon98] S. Kondo. "The automorphism group of a generic Jacobian Kummer surface". *J. Algebraic Geom.* 7.3 (1998), pp. 589–609.

[Kum14] A. Kumar. "Elliptic fibrations on a generic Jacobian Kummer surface". *J. Algebraic Geom.* 23.4 (2014), pp. 599–667. DOI: 10.1090/S1056-3911-2014-00620-2.

[KS08] M. Kuwata and T. Shioda. "Elliptic parameters and defining equations for elliptic fibrations on a Kummer surface". In: *Algebraic geometry in East Asia—Hanoi 2005.* Vol. 50. Adv. Stud. Pure Math. Math. Soc. Japan, Tokyo, 2008, pp. 177–215. DOI: 10.2969/aspm/05010177.

[LM18] M. Lieblich and D. Maulik. "A note on the cone conjecture for K3 surfaces in positive characteristic". *Math. Res. Lett.* 25.6 (2018), pp. 1879–1891. DOI: 10.4310/MRL.2018.v25.n6.a9.

[LMS14] M. Lieblich, D. Maulik, and A. Snowden. "Finiteness of K3 surfaces and the Tate conjecture". *Ann. Sci. Éc. Norm. Supér. (4)* 47.2 (2014), pp. 285–308. DOI: 10.24033/asens.2215.

[Mad15] K. Madapusi Pera. "The Tate conjecture for K3 surfaces in odd characteristic". *Invent. Math.* 201.2 (2015), pp. 625–668. DOI: 10.1007/s00222-014-0557-5.

[Mir89] R. Miranda. *The basic theory of elliptic surfaces.* Dottorato di Ricerca in Matematica. [Doctorate in Mathematical Research]. ETS Editrice, Pisa, 1989, pp. vi+108.

[Nis96] K.-i. Nishiyama. "The Jacobian fibrations on some $K3$ surfaces and their Mordell-Weil groups". *Japan. J. Math. (N.S.)* 22.2 (1996), pp. 293–347. DOI: 10.4099/math1924.22.293.

[PŠ71] I. I. Pjateckiı-Šapiro and I. R. Šafarevič. "Torelli's theorem for algebraic surfaces of type K3". *Izv. Akad. Nauk SSSR Ser. Mat.* 35 (1971), pp. 530–572.

[SS19] M. Schütt and T. Shioda. *Mordell–Weil lattices.* Vol. 70. Ergebnisse der Mathematik und ihrer Grenzgebiete. 3. Folge. A Series of Modern Surveys in Mathematics [Results in Mathematics and Related Areas. 3rd Series. A Series of Modern Surveys in Mathematics]. Springer, Singapore, 2019, pp. xvi+431. DOI: 10.1007/978-981-32-9301-4.

[Shi15] I. Shimada. "An algorithm to compute automorphism groups of $K3$ surfaces and an application to singular $K3$ surfaces". *Int. Math. Res. Not. IMRN* 22 (2015), pp. 11961–12014. DOI: 10.1093/imrn/rnv006.

[Shi90] T. Shioda. "On the Mordell–Weil lattices". *Comment. Math. Univ. St. Pauli* 39.2 (1990), pp. 211–240.

[Sil94] J. H. Silverman. *Advanced topics in the arithmetic of elliptic curves.* Vol. 151. Graduate Texts in Mathematics. Springer-Verlag, New York, 1994, pp. xiv+525. DOI: 10.1007/978-1-4612-0851-8.

[Ste85] H. Sterk. "Finiteness results for algebraic $K3$ surfaces". *Math. Z.* 189.4 (1985), pp. 507–513. DOI: 10.1007/BF01168156.

[Tat75] J. Tate. "Algorithm for determining the type of a singular fiber in an elliptic pencil". In: *Modular functions of one variable, IV (Proc. Internat. Summer School, Univ. Antwerp, Antwerp, 1972).* Vol. Vol. 476. Lecture Notes in Math. Springer, Berlin-New York, 1975, pp. 33–52.

[Vin83] E. B. Vinberg. "The two most algebraic $K3$ surfaces". *Math. Ann.* 265.1 (1983), pp. 1–21. DOI: 10.1007/BF01456933.

An Atlas of Orthogonal Representations

Thomas Breuer, Gabriele Nebe, and Richard Parker

Abstract Let G be a finite group and $\rho : G \to \mathrm{GL}(2n, F)$ be an absolutely irreducible orthogonal representation of even degree over a finite field F. Then $\rho(G)$ embeds into $\mathrm{GO}^+(2n, F)$ or $\mathrm{GO}^-(2n, F)$. We describe methods to decide which case holds for ρ, and use them to determine most of the orthogonal discriminants of the absolutely irreducible orthogonal representations of even degree that are listed in the ATLAS of Finite Groups [CCNPW85].

Key words: Orthogonal representations of finite groups

In memory to our friend and colleague Richard Parker, who sadly passed away after the preparation of this chapter

1 Introduction

The ATLAS of Finite Groups [CCNPW85] and the ATLAS of Brauer Characters [JLPW95] contain the ordinary and modular character tables of finite simple groups, their covering groups and automorphism groups. These characters classify the absolutely irreducible representations ρ of the group G, the building blocks of all group homomorphisms of G into a linear group. Often $\rho(G)$ lies in a smaller classical group, such as the symplectic or unitary group, or an orthogonal group. In even dimension n there are two possible orthogonal groups over a finite field F, $\mathrm{GO}^+(n, F)$ and $\mathrm{GO}^-(n, F)$.

Thomas Breuer
Lehrstuhl für Algebra und Zahlentheorie, RWTH Aachen University, e-mail: `thomas.breuer@math.rwth-aachen.de`

Gabriele Nebe
Lehrstuhl für Algebra und Zahlentheorie, RWTH Aachen University, e-mail: `nebe@math.rwth-aachen.de`

W. Decker et al. (eds.), *The Computer Algebra System OSCAR*,
Algorithms and Computation in Mathematics 32,
https://doi.org/10.1007/978-3-031-62127-7_10

During the past two years, the authors compiled a list of additional data, the *orthogonal discriminants* of the even degree indicator + characters. Over finite fields these are $O+$ resp. $O-$ according to whether $\rho(G)$ is a subgroup of GO^+ or GO^-. Note that these questions make sense only if one considers the representations over finite fields (and number fields), contrary to the situation in many representation theoretical results, where one considers only representations over algebraically closed fields.

The computational task is to determine the orthogonal discriminants (as far as possible) of absolutely irreducible representations of Atlas groups.

The results are collected in the text file [BBBNP23].

The data rely on the notation and the ordering of character tables in the ATLAS of Finite Groups [CCNPW85], in the ATLAS of Brauer Characters [JLPW95], and in the character table library that belongs to the `OSCAR` system, as a part of the `GAP` system. More generally, the names of groups and characters as well as the notation to describe irrational values from character fields in characteristic zero are compatible with the functions in `GAP` and `OSCAR` that deal with characters and character tables.

Section 2 introduces the notion of *orthogonally stable* characters and the necessary facts about characters, quadratic forms, and indicators. The methods for computing orthogonal discriminants are then described in Section 3, and Section 4 shows two examples. Finally, Section 5 lists further applications of our results.

2 Theoretical Background

2.1 Characters

Let G be a finite group. Any group homomorphism $\rho\colon G \to \mathrm{GL}(n, K)$, for some field K, is called a (matrix) *representation* of G.

Put $T_\rho\colon G \to K$, $g \mapsto \mathrm{Tr}(\rho(g))$. If the characteristic of K is zero then $\chi_\rho := T_\rho$ is called an *ordinary character*. In this case, two representations are equivalent if and only if they have the same character. The *character field* of the character χ is $F(\chi) = \mathbb{Q}(\{\chi(g); g \in G\})$. Since each matrix $\rho(g)$ is diagonalizable, where the diagonal entries are roots of unity, $F(\chi)$ is contained in some cyclotomic field $\mathbb{Q}(\zeta_N)$, where $\zeta_N = \exp(2\pi i/N)$ for some divisor N of $|G|$.

If the characteristic of K is a prime p then we consider only the situation that K is a finite extension of its prime field $\mathbb{F}_p$. The map T_ρ is then called a *Frobenius character*, and the character field $F(\chi) = \mathbb{F}_p(\{\chi(g); g \in G\})$ of a Frobenius character χ is a finite field. Frobenius characters do in general not determine their representations up to equivalence.

In order to relate representations in characteristic zero and in finite characteristic p, we define the *Brauer character* of a representation $\rho\colon G \to \mathrm{GL}(n, K)$, where K is a finite extension of $\mathbb{F}_p$, as a map on the set $G_{p'}$ of those elements in G that have order coprime to p, as follows.

For each element $g \in G_{p'}$, the matrix $\rho(g)$ is conjugate to a diagonal matrix $\text{diag}(\epsilon_1, \ldots, \epsilon_n)$. Let q be a power of p such that $\mathbb{F}_q$ contains all eigenvalues of all $\rho(g)$ for $g \in G_{p'}$. The multiplicative group $\mathbb{F}_q^\times$ is cyclic, we first choose a generator z and define the group isomorphism $\eta_0 \colon \langle \zeta_{q-1} \rangle \to \mathbb{F}_q^\times$ by $\eta_0(\zeta_{q-1}) = z$. Then we define $\eta_q \colon \mathbb{Z}[\zeta_{q-1}] \to \mathbb{F}_q$ as the unique ring homomorphism with the property $\eta_q(\zeta_{q-1}) = z$. The *Brauer character* of ρ at g is defined as $\varphi_\rho(g) = \eta_0^{-1}(\epsilon_1) + \cdots + \eta_0^{-1}(\epsilon_n)$. Note that $\eta_q(\varphi_\rho(g)) = \chi_\rho(g)$, that is, the Brauer character of ρ determines the Frobenius character of ρ.

Note that the Brauer character values depend on our choice of the generator z of $\mathbb{F}_q^\times$. We want to consider many different groups and their Brauer characters at the same time, thus we have to choose the maps η_q compatibly for various powers q of p (see Remark 1).

An ordinary or Brauer character is called *absolutely irreducible* if it is not the sum of two characters. We denote the set of absolutely irreducible ordinary characters of G by $\text{Irr}(G)$, and the set of absolutely irreducible Brauer characters of G in characteristic p by $\text{IBr}_p(G)$. The cardinalities of $\text{Irr}(G)$ and $\text{IBr}_p(G)$ are equal to the numbers of conjugacy classes of elements in G and in $G_{p'}$, respectively.

Each character can be written uniquely as a sum of absolutely irreducible characters, with nonnegative integer coefficients. Moreover, the restriction of each ordinary character to $G_{p'}$ yields a Brauer character; this is described by the p-modular *decomposition matrix* $D_p = [d_{\chi,\varphi}]$ of G, whose rows and columns are indexed by $\chi \in \text{Irr}(G)$ and $\varphi \in \text{IBr}_p(G)$, respectively, where $\chi_{G_{p'}} = \sum_{\varphi \in \text{IBr}_p(G)} d_{\chi,\varphi} \varphi$.

If p does not divide $|G|$ then $G_{p'} = G$ holds, in this case regarding ordinary characters as p-Brauer characters defines a bijection from $\text{Irr}(G)$ to $\text{IBr}_p(G)$; thus after reordering $\text{IBr}_p(G)$ we have $D_p = I$ is the unit matrix.

Remark 1 The choice of η_q can be interpreted as the choice of a series of prime ideals in the cyclotomic fields $\mathbb{Q}[\zeta_{q-1}]$, and hence of prime ideals in the character fields of the ordinary characters compatible with the action of the Galois group on $\text{Irr}(G)$ (for more details see [NP23, Section 6]). These prime ideals do play a crucial role when we use the decomposition matrix to deduce restrictions on the orthogonal discriminants as illustrated in [NP23, Section 7.1] and also Section 3.1.2 below.

If the characteristic p divides the group order, then representations are not necessarily (equivalent to) the direct sum of irreducible representations; the Brauer character χ of a representation ρ only determines the composition factors of ρ. Choosing a composition series the matrices in $\rho(G)$ are block triangular matrices where the diagonal blocks give the action of G on the composition factors. In particular we get the following remark.

Remark 2 For any $a \in KG$ the characteristic polynomial of $\rho(a)$ does not depend on the representation ρ of G but only on its character χ. In particular $\det_\chi := \det \circ \rho : KG \to K, a \mapsto \det(\rho(a))$ only depends on the character χ.

2.1.1 Some Notation

We briefly recall the most important abbreviations for character values as they are used in [CCNPW85]. For more details see [CCNPW85, Section 7.10]. Character values are expressed as sums of roots of unity, e.g. $z_N = \zeta_N$ and $y_N = \zeta_N + \zeta_N^{-1}$. The superscript ^{*k} means the same sum where each root of unity is replaced by its k-th power. The names b_N, c_N, ... usually denote irrationalities in the N-th cyclotomic number field that have degree 2, 3, ... over the rationals.

2.2 Quadratic Forms

Let K be a field and V a finite dimensional vector space over K. A *quadratic form* is a map $Q : V \to K$ such that $Q(av) = a^2 Q(v)$ for all $v \in V, a \in K$ and such that its associated *polarisation*

$$B_Q : V \times V \to K, B_Q(v, w) := Q(v + w) - Q(v) - Q(w)$$

is a K-bilinear form. The quadratic form is called *non-degenerate*, if its polarisation is a non-degenerate symmetric bilinear form. As $2Q(v) = B_Q(v, v)$, one recovers the quadratic form from the symmetric bilinear form B_Q if $\mathrm{char}(K) \neq 2$. This can be used to define the *discriminant* of the quadratic form as $(-1)^a \det(B_Q)(K^\times)^2$, where $a = \dim(V)(\dim(V) - 1)/2$ and $\det(B_Q)$ is the determinant of a Gram matrix of B_Q. For fields of characteristic 2 the discriminant is replaced by the Arf invariant (see [KMRT98, page xix], [Kne02, Section 10]).

2.2.1 Finite Fields

Over finite fields dimension and discriminant are separating invariants of the isometry classes of quadratic forms. A classification of quadratic forms over finite fields is well known (see [Kne02, Chapter IV]): So let K be a finite field and $Q : V \to K$ a non-degenerate quadratic form. If the characteristic of K is odd, then the space (V, B_Q) has an orthogonal basis and for each even dimension there are exactly two isometry classes of non-degenerate quadratic forms according to their two possible discriminants $\in K^\times/(K^\times)^2$. If the characteristic of K is 2, then B_Q is a non-degenerate symplectic form and hence the dimension of any non-degenerate quadratic space is even.

Over any finite field there are exactly two non-degenerate quadratic spaces of dimension 2, the *hyperbolic plane*

$$\mathbf{H} := (\langle e, f\rangle, Q) \text{ with } Q(ae + bf) = ab$$

and the *norm form* $\mathbf{N} := (F, N_{F/K})$ where F/K is the field extension of degree 2. Every quadratic space of dimension $2n$ is an orthogonal sum of copies of $\mathbf{H}$ and $\mathbf{N}$. As $\mathbf{N} \perp \mathbf{N} \cong \mathbf{H} \perp \mathbf{H}$ there are hence two isometry classes of such quadratic spaces of even dimension

$$Q_{2n}^+ :=\perp^n \mathbf{H} \text{ and } Q_{2n}^- :=\perp^{n-1} \mathbf{H} \perp \mathbf{N}.$$

In odd characteristic the discriminant of Q_{2n}^+ is a square and the discriminant of Q_{2n}^- is a non-square.

Definition 3 For all finite fields we denote the discriminant of Q_{2n}^+ by $O+$ and the discriminant of Q_{2n}^- by $O-$.

The *orthogonal groups* of non-degenerate quadratic spaces over a field K with q elements are denoted by

$$\mathrm{GO}_{2n}^+(q) = O(Q_{2n}^+),\ \mathrm{GO}_{2n}^-(q) := O(Q_{2n}^-),\ \text{and } \mathrm{GO}_{2n+1}(q)$$

where the latter only occurs for odd q, and is the orthogonal group of any odd dimensional quadratic space (V, Q). Note that if $\dim(V) = 2n + 1$ is odd, then

$$\operatorname{disc}(V, \epsilon Q) = \epsilon \operatorname{disc}(V, Q)$$

and $O(V, Q) = O(V, \epsilon Q)$ for any $\epsilon \in K^\times$.

2.2.2 Hermitian Forms

Given a Galois extension L/K of degree 2 and an L-vector space V of finite dimension n. Restriction of scalars turns V into a K-vector space V_K of dimension $2n$. Any Hermitian form $H : V \times V \to L$ defines a quadratic form $Q_H : V \to K, v \mapsto H(v, v)$. The discriminant of this quadratic form is determined directly by the extension L/K (see [Sch85, page 350], [NP23, Proposition 3.12]):

Proposition 4 *Let (V, H) be a non-degenerate Hermitian L-vector space of dimension n.*

(a) If char$(K) \neq 2$ then write $L = K[\sqrt{\delta}]$. Then $\operatorname{disc}(Q_H) = \delta^n (K^\times)^2$.
(b) If K is a finite field in any characteristic then $\operatorname{disc}(Q_H) = O+$ *if n is even and* $\operatorname{disc}(Q_H) = O-$ *if n is odd.*

2.3 The Indicator of an Irreducible Character

Let χ be an irreducible ordinary character or Brauer character and let $\rho : G \to \mathrm{GL}(V)$ be an absolutely irreducible representation with character χ. Then the character of the contragredient representation $\rho^\vee : G \to \mathrm{GL}(V^*)$ is the complex conjugate character $\overline{\chi}$. If $\chi = \overline{\chi}$ then any isomorphism $\varphi : V \to V^* = \mathrm{Hom}(V, K)$ gives rise to a G-invariant bilinear form on V defined by $B'(v, w) := \varphi(v)(w)$. As the radical of an invariant form is a submodule of V this form $B := B'$ is either skew-symmetric or $B(v, w) := B'(v, w) + B'(w, v)$ is a symmetric non-degenerate G-invariant bilinear form. In characteristic 2 we need to distinguish whether B is the polarisation of a G-invariant quadratic form (indicator $+$) or not (indicator $-$).

Definition 5 The *indicator* of χ is defined as

- $\circ$ if χ takes non real values.
- $+$ if $\chi = \mathbf{1}$ is the trivial character or χ is real and the form B comes from a G-invariant quadratic form on V.
- $-$ if χ is real and B is not the polarisation of a G-invariant quadratic form on V.

2.4 Orthogonally Stable Characters

Given a representation $\rho : G \to \mathrm{GL}(V)$ we use

$$\mathcal{Q}(\rho) := \{Q : V \to K \text{ quad. form } \mid Q(vg) = Q(v) \text{ for all } g \in G, v \in V\}$$

to denote the space of G-invariant quadratic forms in ρ. Then ρ is called *orthogonal*, if $\mathcal{Q}(\rho)$ contains a non-degenerate quadratic form. A character χ of G is called *orthogonal* if there is an orthogonal representation affording χ.

An orthogonal character χ is *orthogonally stable*, if there is a square class Δ of the character field of χ such that for all representations $\rho : G \to \mathrm{GL}_{\chi(1)}(K)$ of G affording the character χ all non-degenerate quadratic forms in $\mathcal{Q}(\rho)$ have discriminant $\Delta(K^\times)^2$. (Note that K may be larger than the character field of χ.) Then $\Delta =: \mathrm{disc}(\chi)$ is called the *orthogonal discriminant* of χ. Clearly orthogonally stable characters and their orthogonal constituents have even degree, but this is the only restriction for being orthogonally stable:

Theorem 6 *(see [NP23, Theorem 5.15]) An orthogonal character χ is orthogonally stable, if and only if all indicator $+$ constituents of χ have even degree.*

The main result of [Neb22b] shows that even though there might be no representation ρ over the character field with character χ, there is always such a square class of the character field that gives the orthogonal discriminant of an orthogonally stable character.

If $\chi = \chi_1 + \chi_2$ is the sum of two orthogonally stable characters then $\mathrm{disc}(\chi) = \mathrm{disc}(\chi_1)\,\mathrm{disc}(\chi_2)$ (see [NP23, Proposition 5.17] for a precise formulation taking into account the different character fields). So it suffices to determine the orthogonal discriminants of the *orthogonally simple* characters ([NP23, Section 5.3]).

Remark 7 The orthogonally simple characters χ are

+ Absolutely irreducible characters χ of even degree and indicator +.
∘ The sum $\chi = \psi + \overline{\psi}$ of a pair of complex conjugate characters of indicator ∘: Then $K(\psi) = K(\chi)[\sqrt{\delta}]$ and $\mathrm{disc}(\chi) = \delta^{\psi(1)}(K(\chi)^\times)^2$ by Proposition 4.
− $\chi = 2\psi$ for an indicator − self-dual character and $\mathrm{disc}(\chi) = 1$.

Starting from the character table of G with all indicators known it hence suffices to compute the orthogonal discriminants of the absolutely irreducible even degree characters of indicator +.

3 Methods

3.1 Theoretical Methods

3.1.1 p-Groups

The paper [Neb22a] gives a formula for the orthogonal discriminant of an orthogonally stable ordinary character χ of a p-group P. The idea is described easily for odd primes p. Given a non-trivial absolutely irreducible representation ρ of P, the image $\rho(P)$ is a non-trivial p-group and hence has a non-trivial center. As ρ is absolutely irreducible, the center acts as scalar matrices. Hence the character field of ρ contains the cyclotomic field $\mathbb{Q}[\zeta_p]$ and one may use Proposition 4 to obtain the orthogonal discriminant of $\rho + \overline{\rho}$:

The maximal real subfield of $\mathbb{Q}[\zeta_p]$ is generated by $y_p := \zeta_p + \zeta_p^{-1}$. Choose $\delta_p \in \mathbb{Q}[y_p] =: Z^+$ such that $\mathbb{Q}[\zeta_p] = Z^+[\sqrt{\delta_p}]$. For $p \equiv 3 \pmod 4$ one may choose $\delta_p = -p$, in general the totally negative generator $\delta_p = (\zeta_p - \zeta_p^{-1})^2 = y_p^{*2} - 2$ of the prime ideal over p is a possible choice.

The character χ is orthogonally stable, if and only if χ does not contain the trivial character as a constituent. Let K denote the character field of χ, put $K_1 := K \cap Z^+$, and $a := [Z^+ : K_1]$. Then $2a$ divides $\chi(1)$.

Theorem 8 *(see [Neb22a, Theorem 4.3, Theorem 4.7]) Let χ be an orthogonally stable character of a p-group P and let K_1, a be as above.*

- *If p is odd then* $\mathrm{disc}(\chi) = N_{Z^+/K_1}(\delta_p)^{\chi(1)/(2a)}(K^\times)^2$.

- *For* $p \equiv 3 \pmod 4$ *this reads as* $\mathrm{disc}(\chi) = (-p)^{\chi(1)/2}$.
- *If* $p = 2$ *then* $\mathrm{disc}(\chi) = (-1)^{\chi(1)/2}$.

3.1.2 Modular Reduction

The discriminant of an ordinary character χ is a square class $\mathrm{disc}(\chi) = \delta(K^\times)^2$ of the character field $K = F(\chi)$. It hence determines a unique field extension $\mathrm{Disc}(\chi) := K[\sqrt{\delta}]$ of degree 1 or 2 of the character field. This field extension is called the *discriminant field* of χ.

Theorem 9 *(see [NP23, Theorem 6.4]) Let* χ *be an orthogonally stable ordinary character. If the reduction of* χ *modulo the prime* $\wp$ *(cf. Remark 1) is orthogonally stable then* $\wp$ *is unramified in the discriminant field extension* $\mathrm{Disc}(\chi)/K$.

Mild extra conditions allow one to read off $\mathrm{disc}(\chi \pmod{\wp})$ from the decomposition behaviour (split or inert) of $\wp$ in the discriminant field extension $\mathrm{Disc}(\chi)/K$. These extra conditions are always satisfied if $\wp$ does not divide the group order and allow one to determine the modular orthogonal discriminants from the ordinary ones for those primes.

Corollary 10 *The only primes that might ramify in* $\mathrm{Disc}(\chi)/K$ *are the prime divisors of the group order. This yields a finite a priori list of possibilities for* $\mathrm{disc}(\chi)$.

For characters in blocks with cyclic defect group, even more is true. We only give the conclusion for defect 1:

Remark 11 (see [NP23, Theorem 6.10]) If χ is an irreducible character in a block of defect 1, then also the converse of Theorem 9 holds: $\wp$ is ramified in $\mathrm{Disc}(\chi)/K$ if and only if the reduction of χ modulo $\wp$ is not orthogonally stable.

[NP23, Section 7.1] exclusively uses the modular decomposition matrices and the methods described above to determine all orthogonal discriminants for the sporadic simple group J_1. Another example where this strategy works well is given in the next section.

3.1.3 The Orthogonal Discriminants of $R(27)$

The finite simple group $R(27)$ is a twisted group of Lie type, the centraliser of an outer automorphism in $G_2(27)$. The order of $R(27)$ is $2^3 \cdot 3^9 \cdot 7 \cdot 13 \cdot 19 \cdot 37$, and there are no even degree indicator + absolutely irreducible 3-Brauer characters. All modular and ordinary orthogonal discriminants of $R(27)$ are determined by the p-modular decomposition matrices for the primes $p = 2, 7, 13, 19$, and 37 as shown in the following table.

χ	$F(\chi)$	disc(χ)	mod 2	mod 7	mod 13	mod 19	mod 37
$13832abcdef$	f_{37}	1	$O+$	$O+$	$O+$	$O+$	$O+$
$18278a$	$\mathbb{Q}$	-3	$O-$	$O+, O+$	$O+$	$O+$	$O+$
$18278bcd$	y_7	-3	$O-$	$O+$	$O+$	$O+$	$O+$
$19684abcdef$	y_{13}	$3(2-y_{13})$	$O-$	$O-$	$1+19683$	$O-$	$O-$
$19684ghijkl$	y_{13}	$3(2-y_{13})$	$O-$	$O-$	$703+18981$	$O-$	$O-$
$26936abc$	c_{19}	1	$O+$	$O+$	$O+$	$O+, O+, O+$	$O+$

The first column gives the ordinary absolutely irreducible orthogonal character in the form $\chi(1)ab...$, the second one its character field (in ATLAS notation see Section 2.1.1) followed by a representative of the orthogonal discriminant disc(χ). We group the Galois conjugate characters into one row. The next columns, headed by mod p, indicate the p-modular reduction of χ, where we list the orthogonal discriminants of the orthogonally simple constituents.

By Theorem 9 the discriminant field extension is unramified at all primes but possibly at the ones dividing 3 for all absolutely irreducible characters of degree $\neq 19684$. For the 12 characters of degree 19684, Remark 11 implies that the discriminant field extension is ramified at the prime dividing 13 and possibly at the two primes dividing 3. In all cases this yields a unique discriminant field from which one obtains the orthogonal discriminants of the ordinary irreducible characters of indicator $+$. These allow one to read off the modular orthogonal discriminants of their modular reductions and hence all orthogonal discriminants for all irreducible p-Brauer characters χ of indicator $+$ that do lift. Only the following three exceptions do not lift:

(a) $p = 2$, $\chi(1) = 16796$. Here χ occurs with multiplicity 1 in a permutation character of degree 19684 which decomposes as

$$2 \cdot \mathbf{1} + 2 \cdot 702 + 741ab + 16796.$$

The following argument can also be found in [GW97, Section 1]: Let $V \cong \mathbb{F}_2^{19684}$ be the permutation module and $e := v_1 + \ldots + v_{19684}$ the canonical fixed vector in V. The subspace $e^{\perp}$ consists of even weight vectors and half of the weight mod 2 is an S_{19684}-invariant quadratic form on $e^{\perp}$ with radical $\langle e \rangle$. Hence it induces a non-degenerate quadratic form Q on $e^{\perp}/\langle e \rangle$, which is of orthogonal discriminant $O-$, as $19684 \equiv 4 \pmod 8$. Now $e^{\perp}/\langle e \rangle = 2 \cdot 702 + 741ab + 16796$ is an orthogonally stable module for $R(27)$. The irrationality of $741a$ is z_3, so $741ab$ contributes $O-$ to this sum leaving $O+$ for the orthogonal discriminant of 16796.

(b) $p = 7$, $\chi(1) = 16796$. Here χ occurs in the 7-modular reduction of $\mathcal{X}_{15} = 741ab + 16796$. As $z_3 \in \mathbb{F}_7$, the orthogonal discriminant of $741ab$ is $O+$ and hence the orthogonal discriminant of 16796 is also $O+$.
(c) $p = 19$, $\chi(1) = 19682$. Here χ occurs in the 19-modular reduction of $\mathcal{X}_{33} = 1443ab + 2184ab + 19682$ which is orthogonally stable. The character fields of $1443a$ and $2184a$ are both $\mathbb{F}_{19}[z_3] = \mathbb{F}_{19}$ so the orthogonal discriminant of χ is $O+$.

3.2 Reduction to Simple Groups

3.2.1 Groups With a Non-Trivial Center

By Schur's Lemma, central elements act as scalars on irreducible representations, in particular, it is enough to consider cyclic central subgroups. If the exponent of the center of G is strictly bigger than 2 then all faithful irreducible characters of G are non-real, i.e. of indicator $\circ$, and Proposition 4 can be used to determine orthogonal discriminants. For central elements of order 2 we use the spinor norm to deduce discriminants:

Given a non-degenerate quadratic form $Q : V \to K$, the *spinor norm* defines a group homomorphism from the orthogonal group of Q into $K^\times/(K^\times)^2$, a group of exponent 2, where the spinor norm of a reflection along vector v equals $Q(v)$ (see [Kne02]). Over a field K of characteristic not 2, the space V has an orthonormal basis $(v_1, \ldots, v_n)$. The orthogonal mapping $-\mathrm{id}_V$ is the product of the reflections along the v_i and hence its spinor norm is $\prod_{i=1}^n Q(v_i) = 2^{-n} \det(Q)$.

Theorem 12 *(see for instance [Neb99, Section 3.1.2]) Let χ be an orthogonally stable character of a finite group G in characteristic not 2 and let ρ be a faithful representation of G affording χ*

- *If there is $g \in G$ with $\rho(g)^2 = -\mathrm{id}$ then* $\mathrm{disc}(\chi) = (-1)^{\chi(1)/2}$.
- *If $[G : G']$ is odd and $-\mathrm{id} \in \rho(G)$ then* $\mathrm{disc}(\chi) = (-1)^{\chi(1)/2}$.

3.2.2 Split Extensions

Given a finite group G and an outer automorphism α of order 2 the split extension $H := G : 2$ has a pseudo presentation

$$G : \langle \alpha \rangle = \langle G, h \mid hgh^{-1} = \alpha(g), h^2 = 1 \rangle.$$

Given an orthogonal character χ of G such that $\chi \circ \alpha \neq \chi$, Clifford theory shows that there is a unique irreducible character $\mathcal{X}$ of H such that $\mathcal{X}_{|G} = \chi + \chi \circ \alpha$. As $\mathcal{X}(H \setminus G) = \{0\}$, the character field F of $\mathcal{X}$ is contained in the character field K of χ.

Theorem 13 *(see [Neb22b, Theorem 4.3]) Assume that the characteristic is not 2. If* $K = F$ *then* $\mathrm{disc}(\mathcal{X}) = (-1)^{\chi(1)}(F^\times)^2$. *Otherwise* $K = F[\sqrt{\delta}]$ *is a quadratic extension of* F *and* $\mathrm{disc}(\mathcal{X}) = (-\delta)^{\chi(1)}(F^\times)^2$.

Note that in the case that χ is already orthogonally stable, then $\mathrm{disc}(\chi) = \mathrm{disc}(\chi \circ \alpha)$ and $\mathrm{disc}(\mathcal{X}) = N_{K/F}(\mathrm{disc}(\chi)) \in (K^\times)^2 \cap F$.

3.2.3 Non-Split Extensions

The following table lists all those examples of characters of almost simple Atlas groups H of the structure $G.2$, such that our methods (Theorem 13 and restriction to the normal subgroup G) do not suffice to compute the orthogonal discriminant of χ from that of an irreducible constituent ψ of χ_G.

H	G	χ	i	$\mathbb{Q}(\chi)$	$\mathbb{Q}(\psi)$	$\mathrm{disc}(\chi)$
$L_2(16).4$	$L_2(16).2$	$34a$	15	$\mathbb{Q}$	$\mathbb{Q}(b_5)$	-1
$L_2(16).4$	$L_2(16).2$	$34b$	16	$\mathbb{Q}$	$\mathbb{Q}(b_5)$	-5
$U_3(4).4$	$U_3(4).2$	$78a$	10	$\mathbb{Q}$	$\mathbb{Q}(b_5)$	-5
$U_3(4).4$	$U_3(4).2$	$78b$	11	$\mathbb{Q}$	$\mathbb{Q}(b_5)$	-1

The orthogonal discriminants can be computed in these cases as follows.

The group $H = L_2(16).4$ is a subgroup of $S_4(4).2$, the irreducible characters of degree 50 of $S_4(4).2$ have orthogonal discriminant -17, and the restrictions of these characters to G are orthogonally stable, and decompose as $16a + 34a$ and $16c + 34a$, respectively. Both $16a$ and $16c$ have orthogonal discriminant 17, thus $34a$ has orthogonal discriminant -1. Analogously, the irreducible character $34c$ of $S_4(4).2$, which has orthogonal discriminant -5, restricts to $34b$ of H, which thus also has orthogonal discriminant -5.

The group $H = U_3(4).4$ is a subgroup of $G_2(4).2$, the irreducible character $350a$ of $G_2(4).2$ has orthogonal discriminant -13, its restriction to H is orthogonally stable and decomposes as $78a + 52abcd + 64a$, where $52abcd$ and $64a$ have orthogonal discriminants 1 and 65, respectively, thus $78a$ has orthogonal discriminant -5. Analogously, the irreducible character $78a$ of $G_2(4).2$, which has orthogonal discriminant -1, restricts to $78b$ of H, which thus also has orthogonal discriminant -1.

3.3 Direct Methods

Given an orthogonal representation ρ affording the character χ one can determine $\mathcal{Q}(\rho)$ either by solving a system of linear equations or by applying the Reynolds operator (see [PS96] for a more sophisticated approach). Then it is straightforward to compute the orthogonal discriminant $\mathrm{disc}(\chi)$.

If the characteristic of the underlying field K is not 2, there is no need to determine $\mathcal{Q}(\rho)$, as we can compute $\mathrm{disc}(\chi)$ as the discriminant of the adjoint involution:

3.3.1 The Natural Involution on the Group Algebra

Let K be a field of characteristic not 2. Inverting the group elements defines a natural involution $^\circ$ on KG, i.e. $(\sum_{g\in G} a_g g)^\circ = \sum_{g\in G} a_g g^{-1}$. Then $KG = KG^- \oplus KG^+$ where $KG^\epsilon = \{a \in KG \mid a^\circ = \epsilon a\}$. Now let ρ be an orthogonal representation of G and choose a non-degenerate $Q \in \mathcal{Q}(\rho)$. The condition $B_Q(\rho(g)v, \rho(g)w) = B_Q(v,w)$ for all $g \in G, v, w \in V$ shows that $\rho(a^\circ) = \rho(a)^{ad}$ for all $a \in KG$, where ^{ad} is the adjoint involution of B_Q. To see this fix a basis of V and work with matrices. Let B be the Gram matrix of B_Q. Then $\rho(g)B\rho(g)^{tr} = B$ and hence $B\rho(g)^{tr}B^{-1} = \rho(g^{-1})$ for all $g \in G$, thus

$$\rho(a^\circ) = B\rho(a)^{tr}B^{-1} \text{ for all } a \in KG.$$

In particular $XB = -BX^{tr}$ for all $X \in \rho(KG^-)$. As the determinant of a skew symmetric matrix is always a square, we conclude that $\det(X)(K^\times)^2 = \det(B)(K^\times)^2$. By Remark 2, this determinant only depends on the character of ρ, so we conclude the following lemma.

Lemma 14 *The orthogonal character χ is orthogonally stable if and only if there is $X \in KG^-$ with $\det_\chi(X) \neq 0$. Then, $\mathrm{disc}(\chi) = (-1)^{\chi(1)/2}\det_\chi(X)$.*

In practice, one finds a suitable X as the sum of at most three matrices $g - g^{-1}$, where g are randomly chosen elements of order at least 3 in $\rho(G)$.

3.3.2 Condensation Methods

Lemma 14 also allows one to compute the orthogonal discriminant of a character using well established condensation techniques (see [Ryb90]). To analyse the composition factors $S_1, \dots, S_t$ of a KG-module V one computes a suitable idempotent $e \in KG$. The *condensed module* Ve is then a module for $eKGe$ with composition factors $\{S_i e \mid 1 \leq i \leq t\} \setminus \{0\}$. The main problem here is that a K-algebra generating set $\{g_1, \dots, g_s\}$ of KG does not necessarily condense to a K-algebra generating set $\{eg_ie \mid 1 \leq i \leq s\}$, the map $a \mapsto eae$ is only a vector space homomorphism and even the condensed algebra is in general too big to compute a basis.

In practise we use fixed point condensation in permutation representations V with respect to a suitable subgroup H whose order is not divisible by the characteristic of K. In view of Section 3.1.1, we choose $H = P$ to be either a Sylow p-subgroup of G (for p odd), or $H = P'P^2$, where P is a Sylow 2-subgroup of G, and $e := \frac{1}{|H|}\sum_{h\in H} h$. Then for any orthogonal KG-module

V, the restriction of $V(1-e)$ to the Sylow p-subgroup P is orthogonally stable and its discriminant can be computed with the formula in Section 3.1.1.

We start with a big permutation representation $V := 1_U^G$. Then, a basis for Ve is given by the H-orbit sums $\sum o_1, \ldots, \sum o_m$, and for $g \in G$, the matrix of $ege = (a_{ij})_{i,j=1}^m$ satisfies

$$a_{ij} = \frac{1}{|o_i|} |\{x \in o_i \mid xg \in o_j\}|.$$

As $e^\circ = e$, the algebra $eKGe$ inherits the natural involution $^\circ : ege \mapsto eg^{-1}e = eg^{tr}e$. The dimensions of the composition factors of Ve and their multiplicities can be predicted by character theoretic methods.

In our applications we took 5-10 random group elements g_i, and computed the K-algebra $A := \langle eg_ie, eg_i^{-1}e = (eg_ie)^\circ \rangle$. The composition factors of the A-module Ve are obtained using meataxe methods. We check, whether these do have the predicted dimension and then compute an element $a = -a^\circ$ in A acting as a unit X on such a composition factor Se. Then Lemma 14 together with Section 3.1.1 allow us to deduce the orthogonal discriminant of S as

$$\operatorname{disc}(S) = (-1)^{\dim(Se)/2} \det(X) \operatorname{disc}(S(1-e)_{|P}).$$

To obtain the orthogonal discriminant for number fields K it is essential to use Corollary 10 to compile a finite list of possible orthogonal discriminants, as meataxe methods only perform well for finite fields. Given this list of possible discriminants we compute enough p-modular reductions (usually for small primes p not dividing the group order) of $\operatorname{disc}(S)$ to conclude the exact value in $K^\times/(K^\times)^2$.

The largest permutation module V handled so far is the one of degree 108, 345, 600 of the Harada Norton group. Using fixed point condensation with the Sylow 5-subgroup of HN, we obtain a module Ve of dimension 7008. As Ve is an $e\mathbb{Z}[\frac{1}{5}]HNe$-module, we are free to reduce this module modulo all primes $\neq 5$ to compute and analyse the composition factors.

A more sophisticated implementation of the meataxe should be able to handle even larger examples.

3.3.3 Summary

Direct methods in characteristic $\neq 2$ usually compute the discriminant of the natural involution to deduce the orthogonal discriminant of χ. In characteristic 2 these do not work and, in particular, we do not have a provable method to use condensation techniques for computing orthogonal discriminants. Here, we compute the Gram matrix of the invariant quadratic form in the original representation, and use it to compute the discriminant. (The implementation in `GAP` uses an algorithm due to Jon Thackray.)

- Many matrix representations are publicly available via the ATLAS of Group Representations [Wil+]. The data file marks these entries with `"AGR"`.
- We can reduce the permutation representations that are available via the ATLAS of Group Representations [Wil+] modulo primes dividing the group order, compute their absolutely irreducible constituents, and determine the orthogonal discriminants of those that are orthogonal and have even degree. The data file marks these entries with `"const(desc)"` where `desc` is the identifier of the permutation representation.
- Many representations have been constructed by Richard Parker in order to compute the orthogonal discriminant. The data file marks these entries with `"RP"`.
- The orthogonal discriminants that have been obtained by Gabriele Nebe using condensation methods as described in Section 3.3.2 are marked by `"GNcond"`.
- In certain cases decomposition matrices allow us to conclude orthogonal discriminants using Theorem 9. Entries obtained in such a way are marked by `"GN"`.

3.4 Character Theoretic Methods

Here the idea is to use only the character table of the given character χ plus information from the character table library, concerning (character tables of) subgroups and overgroups. This information, for example known orthogonal discriminants of related characters, may suffice to deduce the orthogonal discriminant of χ. The advantage of this approach is that checking these criteria is cheap, but the disadvantage is that they need not yield the answer.

The following criteria are used. (The string in brackets is used to mark those entries in the data file for which the criterion in question yields the value.)

Group order (`"order"`): In positive characteristic, if the orthogonal discriminant of χ with character field F is $O+$ ($O-$) then the order of G divides that of $\mathrm{GO}^+(\chi(1), F)$ ($\mathrm{GO}^-(\chi(1), F)$). This condition determines the orthogonal discriminant in some cases.

```
julia> ch = character_table("Co2", 2)[2];

julia> degree(ch)
22

julia> Oscar.OrthogonalDiscriminants.od_from_order(ch)
(true, "O+")
```

Group automorphisms (`"grpaut(n)"`): For a character χ of the group G and a group automorphism σ of G, the character χ^σ is defined by $\chi^\sigma(g) =$

$\chi(g^\sigma)$, for $g \in G$. If χ has an orthogonal discriminant then χ^σ has the same orthogonal discriminant.

Galois action ("galaut(n)"): For a character χ of the group G, and a field automorphism σ of the character field of χ, the character χ^σ is defined by $\chi^\sigma(g) = \chi(g)^\sigma$, for $g \in G$. In characteristic zero, if χ has orthogonal discriminant d then χ^σ has orthogonal discriminant d^σ. In positive characteristic, if χ has an orthogonal discriminant then χ^σ has the same orthogonal discriminant.

Transitive permutation characters ("permchar"): If π is a transitive permutation character of G, i. e., there is a subgroup H of G such that π is the induced character 1_H^G, then $\chi = \pi - 1_G$ is the character of a rational representation that fixes a symmetric bilinear form of determinant $\pi(1)$. If χ is orthogonally stable then its orthogonal discriminant is $(-1)^{\chi(1)/2}\pi(1)$ (modulo squares). If χ is absolutely irreducible then this yields the value, otherwise it yields a condition on the orthogonal discriminants of the constituents of χ.

Eigenvalues ("ev"): Assume that χ is either an ordinary character, or a p-modular Brauer character for an odd prime p. If χ is orthogonal, and if there is $g \in G$ such that a representation ρ affording χ map g to a matrix that does not have an eigenvalue ± 1, then the restriction of χ to the subgroup $\langle g \rangle$ is orthogonally stable, and has determinant $\det(\rho(g) - \rho(g^{-1}))$, modulo squares, see [Neb22b, Cor. 4.2]. (This is a special case of the criterion from Section 3.3.1.) Note that the eigenvalues of $\rho(g)$, and hence, the determinant can be computed from the power map information that belongs to the character table of G.

```
julia> ch = character_table("Co3", 3)[2];

julia> degree(ch)
22

julia> Oscar.OrthogonalDiscriminants.od_from_eigenvalues(ch)
(true, "O+")
```

Jantzen–Schaper formula ("specht"): The ordinary irreducible representations of the symmetric group on n points are parameterized by the partitions of n, and the determinant of the bilinear form that is fixed by the representing matrices for the partition λ can be expressed in terms of λ, via the Jantzen–Schaper formula [Mat99, p. 5.33]. This yields the orthogonal discriminants of those characters of the alternating group on n points that extend to the symmetric group. We are interested in the cases $5 \leq n \leq 13$.

```
julia> ch = character_table("A12")[26];

julia> degree(ch)
1728
```

```
julia> Oscar.OrthogonalDiscriminants.od_for_specht_module(ch)
(true, "1")
```

Restriction to p-subgroups (`"syl(p)"`): Let p be an odd prime, and let χ be a character in characteristic different from p. The restriction χ_P of χ to a p-subgroup P of G is orthogonally stable if and only if the trivial character of P is not a constituent of χ_P, and the orthogonal discriminant of χ_P can be computed in terms of $\chi(1)$ and the character field of χ_P (see [Neb22a, Section 4.1] and Section 3.1.1). Note that in order to check whether χ_P is orthogonally stable, it is sufficient to know the permutation character 1_P^G, we do not need the character table of P.

```
julia> ch = character_table("R(27)")[16];

julia> degree(ch)
18278

julia> Oscar.OrthogonalDiscriminants.od_from_p_subgroup(ch, 3)
(true, "-3")
```

Restriction to subgroups (`"rest(...)"`, `"ext(...)"`): If H is a subgroup of G whose character table is known, and if the restriction χ_H is orthogonally stable then we can argue as follows. If the orthogonal discriminants of the constituents of χ_H are known, then we can deduce that of χ; in this case, the data file contains the label `"ext(...)"`. If the orthogonal discriminant of χ is known, then we get a condition on the orthogonal discriminants of the constituents of χ_H; for example, if all of them except one are already known, then we can deduce the missing one; in this case, the data file contains the label `"rest(...)"`.

Regard ordinary characters as Brauer characters (`"lift(+...)"`): Let χ be a p-modular Brauer character. If χ is the restriction of an ordinary character whose orthogonal discriminant is known, then reducing this value modulo p often yields the orthogonal discriminant of χ. If χ is a constituent of the restriction of an ordinary character whose orthogonal discriminant is known, then reducing this value modulo p often yields the orthogonal discriminant of χ if the discriminants of the other constituents are known.

Tensor products (`"tensor(...)"`): [Neb99, Section 3.1.3] lists formulae for the determinants of the invariant bilinear forms of tensor products $\chi \cdot \psi$ and of symmetric squares $\chi^{2+} - 1_G$ and antisymmetric squares χ^{2-}. In those cases where these tensor products and symmetrizations are orthogonally stable, this yields conditions on the orthogonal discriminants of their constituents, as in the above criteria.

Consistency checks: Often an orthogonal discriminant can be computed with several criteria, and the results must be consistent. A posteriori, also those conditions about constituents of restrictions, tensor products, p-modular reductions that were not sufficient to deduce the orthogonal discriminants can be used for consistency checks.

4 Examples

In total, we compiled data for ATLAS groups containing almost 20 000 orthogonal discriminants, that can be displayed in more than 1 000 ordinary and Brauer character tables that are available in the OSCAR character table library. To illustrate the output and some of the methods, we give a few examples in this section.

4.1 The Orthogonal Discriminants of $G_2(3)$

The 2-modular Brauer character table of the simple group $G_2(3)$ together with the stored orthogonal discriminants can be displayed as follows.

```
julia> Oscar.OrthogonalDiscriminants.show_with_ODs(
         character_table("G2(3)", 2))
G2(3)mod2

              2   6   3   3    .  1  1  .  .  .  .   .   .
              3   6   6   6    6  4  4  .  3  3  3   .   .
              7   1   .   .    .  .  .  1  .  .  .   .   .
             13   1   .   .    .  .  .  .  .  .  .   1   1

                 1a  3a  3b   3c 3d 3e 7a 9a 9b 9c 13a 13b
             2P  1a  3a  3b   3c 3d 3e 7a 9a 9c 9b 13b 13a
             3P  1a  1a  1a   1a 1a 1a 7a 3c 3c 3c 13a 13b
             7P  1a  3a  3b   3c 3d 3e 1a 9a 9b 9c 13b 13a
            13P  1a  3a  3b   3c 3d 3e 7a 9a 9b 9c  1a  1a
      d OD    2
 X_1 1        +   1   1   1    1  1  1  1  1  1  1   1   1
 X_2 1 O-     +  14   5   5   -4  2 -1  .  2 -1 -1   1   1
 X_3 2        o  64  -8  -8    1  4 -2  1  1  A /A  -1  -1
 X_4 2        o  64  -8  -8    1  4 -2  1  1 /A  A  -1  -1
 X_5 1 O-     +  78  -3  -3   -3 -3  6  1  .  .  .   .   .
 X_6 1 O+     +  90   9   9    9  .  . -1  .  .  .  -1  -1
 X_7 1 O-     +  90  -9  18    .  3 -3 -1 -3  .  .  -1  -1
 X_8 1 O-     +  90  18  -9    .  3 -3 -1 -3  .  .  -1  -1
 X_9 1 O-     + 378  -9  -9    9 -3 -6  .  3  .  .   1   1
X_10 2 O+     + 448  16  16  -11 -2 -2  .  1  1  1   B  B*
X_11 2 O+     + 448  16  16  -11 -2 -2  .  1  1  1  B*   B
X_12 1 O+     + 832 -32 -32   -5  4  4 -1  1  1  1   .   .

A = 3z_3 + 1
/A = -3z_3 - 2
B = -z_13^11 - z_13^8 - z_13^7 - z_13^6 - z_13^5 - z_13^2 - 1
B* = z_13^11 + z_13^8 + z_13^7 + z_13^6 + z_13^5 + z_13^2
```

The new data are contained in the column headed by OD. Here we give the type of the invariant quadratic form as described in Definition 3. For instance this allows us to read off that the image of the 14-dimensional absolutely irreducible 2-modular representation of $G_2(3)$ is contained in $O_{14}^-(\mathbb{F}_2)$.

The group $G_2(3)$ is one of the interesting examples where all ordinary and modular orthogonal discriminants can be obtained directly from the known decomposition matrix.

```
julia> show_OD_info("G2(3)")
G2(3):  2^6*3^6*7*13
--------------------

 i| chi|     K|disc|           2|3|            7|      13
--+----+------+----+------------+-+-------------+--------
 2| 14a|     Q|  -3|         14a| |          14a|     14a
  |    |      |    |          0-| |           0+|      0+
--+----+------+----+------------+-+-------------+--------
 5| 78a|     Q|  -3|         78a| |          78a|     78a
  |    |      |    |          0-| |           0+|      0+
--+----+------+----+------------+-+-------------+--------
 9|104a|     Q|  21|     14a+90a| |     (def. 1)|    104a
  |    |      |    |      0-, 0+| |             |      0-
--+----+------+----+------------+-+-------------+--------
10|168a|     Q|  13|     78a+90a| |         168a|(def. 1)
  |    |      |    |      0-, 0+| |           0-|
--+----+------+----+------------+-+-------------+--------
11|182a|     Q|  -3| 14a+78a+90c| |         182a|    182a
  |    |      |    |  0-, 0-, 0-| |           0+|      0+
--+----+------+----+------------+-+-------------+--------
12|182b|     Q|  -3| 14a+78a+90b| |         182b|    182b
  |    |      |    |  0-, 0-, 0-| |           0+|      0+
--+----+------+----+------------+-+-------------+--------
15|448a|Q(b13)|   1|        448a| |         448a|14a+434a
  |    |      |    |          0+| |           0+|  0+, 0+
--+----+------+----+------------+-+-------------+--------
16|448b|Q(b13)|   1|        448b| |         448b|14a+434a
  |    |      |    |          0+| |           0+|  0+, 0+
--+----+------+----+------------+-+-------------+--------
17|546a|     Q|  -3|78a+90b+378a| |         546a|    546a
  |    |      |    |  0-, 0-, 0-| |           0+|      0+
--+----+------+----+------------+-+-------------+--------
18|546b|     Q|  -3|78a+90c+378a| |         546b|    546b
  |    |      |    |  0-, 0-, 0-| |           0+|      0+
--+----+------+----+------------+-+-------------+--------
19|728a|     Q|   1|    14a+378a| |         728a|    728a
  |    |      |    |      0-, 0-| |           0+|      0+
--+----+------+----+------------+-+-------------+--------
20|728b|     Q|   1|    14a+378a| |         728b|    728b
  |    |      |    |      0-, 0-| |           0+|      0+
--+----+------+----+------------+-+-------------+--------
23|832a|     Q|   1|        832a| |64ab+78a+626a|    832a
  |    |      |    |          0+| |   0+, 0+, 0+|      0+
```

The function `show_OD_info` collects the information about ordinary and modular orthogonal discriminants that are stored in our data. The rows of the table correspond to the ordinary indicator + characters χ of even degree. The first column lists the ATLAS number of χ followed by the degree. Then we give the character field $\mathbb{Q}(\chi)$, and column four displays a representative of $\mathrm{disc}(\chi)$. The following columns are headed by the prime divisors p of the group order. If $\chi \pmod p$ is orthogonally stable, then we give the corresponding character degrees of the p-modular constituents of χ and their corresponding orthogonal discriminants. The entry "(def. 1)" means that $\chi \pmod p$ is not orthogonally stable but has defect 1, from which we know that p is ramified in the discriminant field extension by Remark 11.

For the case of $G_2(3)$ all orthogonal discriminants can be obtained from the decomposition matrices: For the ordinary characters we know from Theorem 9 that the discriminant field extension is unramified at all primes but possibly those that divide 3, except for the characters number 9 and 10 where we know that 7 respectively 13 are ramified. In all cases, this yields a unique possibility for the respective quadratic extension $\mathrm{Disc}(\chi)$ of the character field. Let us illustrate the consideration for the two non-rational characters number 15 and 16. Here the character field is $\mathbb{Q}(\sqrt{13})$ and we know that the discriminant field is either the character field or a totally real quadratic extension of $\mathbb{Q}(\sqrt{13})$ that is unramified at all primes but possibly those dividing 3. There are no such quadratic extensions, as can be computed with the commands

```
julia> K, _ = quadratic_field(13)
(Real quadratic field defined by x^2 - 13, sqrt(13))

julia> ray_class_field(3*maximal_order(K))
Class field defined mod (<3, 3>, InfPlc{AbsSimpleNumField,
↪  AbsSimpleNumFieldEmbedding}[]) of structure Z/1
```

This way we get all the ordinary orthogonal discriminants. The p-modular reductions allow us to find all the modular discriminants from the ordinary ones as we illustrate for the prime $p = 2$:

As -3 is not a 2-adic square, the prime 2 is inert in the quadratic extension $\mathbb{Q}(\sqrt{-3})/\mathbb{Q}$ and hence the orthogonal discriminant of the 2-modular reductions of χ_i are $O-$ for $i \in \{2, 5, 11, 12, 17, 18\}$. A similar argument yields discriminant $O-$ for the 2-modular reduction of χ_9 and χ_{10}. This gives the orthogonal discriminants for the 2-modular characters 14a, 78a, 90a, 90c, 90b, 378a, and two checks coming from the 2-modular reduction of χ_{10} and χ_{18}. The 2-modular reduction of the ordinary characters of discriminant 1 have orthogonal discriminant $O+$, from which we get the orthogonal discriminants of 448a, 448b, and 832a as well as a check coming from the characters χ_{19} and χ_{20}.

4.2 The Ordinary Orthogonal Discriminants of J_2

This example illustrates how we can obtain the ordinary orthogonal discriminants of a group by only using representations over finite (prime) fields. These representations can be constructed with Richard Parker's C-meataxe by reducing permutation representations or tensor products of known representations. This way we obtain the following table.

	11	19	29	31	41	59	disc	13	17
$14a$	$O-$	$O+$	$O-$	$O+$	$O-$	$O-$	-3		
$14b$	$O-$	$O+$	$O-$	$O+$	$O-$	$O-$	-3		
$36a$	$O+$	$O+$	$O+$	$O+$	$O+$	$O+$	5	$O-$	$O-$
$70a$	$O-$	$O+$	$O-$	$O+$	$O-$	$O-$	-3		
$70b$	$O-$	$O+$	$O-$	$O+$	$O-$	$O-$	-3		
$90a$	$O+$	$O-$	$O+$	$O-$	$O-$	$O-$	-7	$O-$	$O-$
$126a$	$O-$	$O-$	$O+$	$O-$	$O+$	$O-$	-5	$O-$	$O-$
$160a$	$O+$	$O+$	$O+$	$O+$	$O+$	$O+$	1	$O+$	$O+$
$224a$	$O+$	$O+$	$O+$	$O+$	$O+$	$O+$	1		
$224b$	$O+$	$O+$	$O+$	$O+$	$O+$	$O+$	1		
$288a$	$O-$	$O-$	$O-$	$O-$	$O+$	$O+$	105	$O+$	$O-$
$300a$	$O-$	$O-$	$O-$	$O-$	$O+$	$O+$	21	$O-$	$O+$
$336a$	$O+$	$O+$	$O+$	$O+$	$O+$	$O+$	1	$O+$	$O+$

The rows of this table are named by the degrees of the ordinary even degree indicator + irreducible characters of the sporadic simple group J_2. The entries of all the columns headed by a prime are computed and in total allow us to deduce the orthogonal discriminant of the character as given in column disc. We kept the ordering of the columns as it was given in Parker's handwritten table. The character fields of the irreducible ordinary characters of J_2 are either $\mathbb{Q}$ or $\mathbb{Q}(\sqrt{5})$. This is why we first chose primes not dividing the group order for which 5 is a square, allowing us to construct the corresponding absolutely irreducible representation over the prime field. But of course this information is not enough, for instance, to decide whether the discriminant of the rational character 36a is 5 or 1. So we constructed the representations with rational characters also over $\mathbb{F}_{13}$ and $\mathbb{F}_{17}$, and computed the discriminant of an invariant quadratic form there.

As an example for our arguments, we treat the character 228a. As the character is rational of degree a multiple of 4, we know that the discriminant field is a real quadratic number field L that is unramified outside $\{2, 3, 5, 7\}$, the set of prime divisors of the group order. Our computed information yields that the primes $11, 19, 29, 31$, and 17 are inert in L and the primes $41, 59, 13$ are split in L. In other words $L = \mathbb{Q}(\sqrt{d})$ where $d \in \mathbb{N}$ is squarefree and divides $2 \cdot 3 \cdot 5 \cdot 7$. Moreover d is a square mod $41, 59$, and 13, and a non-square modulo $11, 19, 29, 31$, and 17. This yields the unique solution $d = 105 = 3 \cdot 5 \cdot 7$.

By computer, we need to solve a system of linear equations over the field $\mathbb{F}_2$: The entry (p, q) of the matrix below tells us whether p is a square modulo q (entry 0) or not (entry 1).

	11	19	29	31	41	59	13	17
−1	1	1	0	1	0	1	0	0
2	1	1	1	0	0	1	1	0
3	0	1	1	1	1	0	0	1
5	0	0	0	0	0	0	1	1
7	1	0	0	0	1	0	1	1

To compute the discriminant of the character `300a`, for example, we compute the unique linear combination of the rows of this matrix that yields $(1, 1, 1, 1, 0, 0, 1, 0)$. It is easy to see that this row is the sum of the rows of 3 and 7, so the discriminant of `300a` is 21.

5 Applications

This section lists some aspects of the computations, and implications of the results.

5.1 Which Discriminant Fields are Galois Extensions of the Rationals?

The number fields that do occur in representation theory of finite groups are usually abelian extensions of the rationals, i.e. contained in some cyclotomic fields. Also discriminant fields are very often abelian extensions of the rationals:

Theorem 15 *Let χ be an orthogonally simple ordinary character of a finite group G, and put $L := \mathrm{Disc}(\chi)$ to denote the discriminant field.*

- *If χ is not absolutely irreducible (i.e. of type $\circ$ or $-$ in Remark 7), then L is an abelian extension of $\mathbb{Q}$.*
- *If G is solvable, then L is an abelian extension of $\mathbb{Q}$ (see [Neb22a] and [Rot22])*
- *For G of type L_2, all discriminant fields are abelian extensions of the rationals (see [BN17]).*

Proposition 16 *The discriminant field is Galois over $\mathbb{Q}$ if and only if the discriminant, a square class of the character field, is stable under all Galois automorphisms of the character field.*

For the proof we need the following easy lemma in Galois theory:

Lemma 17 *Given a tower $A \subseteq B \subseteq C$ of fields such that B/A is Galois and C/B is Galois and $[C : A] < \infty$, then C/A is Galois if and only if for all $g \in \mathrm{Gal}(B/A)$, there is $f \in \mathrm{Aut}(C)$ such that $f_{|B} = g$.*

Proof Under the conditions of the lemma the sequence

$$1 \to \mathrm{Gal}(C/B) \to \mathrm{Aut}_A(C) \to \mathrm{Aut}_A(B) \to 1$$

is exact and hence $|\mathrm{Aut}_A(C)| = [C : A]$, which implies that C/A is Galois.□

Proof (of Proposition 16) Now we apply this to our situation where $F = F(\chi)$ is the character field of an ordinary orthogonally stable character χ, and $K = F[\sqrt{\delta}]$ is the discriminant field.

To prove Proposition 16, we need to show that $K/\mathbb{Q}$ is Galois if and only if $\delta(F^\times)^2$ is stable under the full Galois group of $F/\mathbb{Q}$, i.e., for all $g \in \mathrm{Gal}(F/\mathbb{Q})$ there is $k_g \in F$ such that $g(\delta) = k_g^2\delta$.

For the proof let $\alpha := \sqrt{\delta} \in K$.

Assume that $K/\mathbb{Q}$ is Galois.

Then $\langle\sigma\rangle := \mathrm{Gal}(K/F)$ is a normal subgroup of $\mathrm{Gal}(K/\mathbb{Q})$ of order 2, and hence central.

The minimal polynomial of α over F is $X^2 - \delta$ and any automorphism $f \in \mathrm{Aut}(K)$ that extends $g \in \mathrm{Gal}(F/\mathbb{Q})$ satisfies $f(\alpha)^2 = g(\delta)$ and $f(F) \subseteq F$. Now f commutes with σ so $k_g := f(\alpha)/\alpha \in \mathrm{Fix}_\sigma(K) = F$ and $k_g^2 = f(\alpha)^2/\alpha^2 = g(\delta)/\delta$, so $g(\delta) = k_g^2\delta$.

To see the opposite direction, we extend $g \in \mathrm{Gal}(F/\mathbb{Q})$ to an automorphism f of K by putting $f(a\alpha + b) := g(a)k_g\alpha + g(b)$ for all $a, b \in F$. It is easy to see that f is a field automorphism of K extending g. So Proposition 16 follows from Lemma 17. □

Remark 18 In the notation of the proof we get that the discriminant field is an abelian extension of $\mathbb{Q}$ if and only if $f(k_g)k_f = g(k_f)k_g$ for all $f, g \in \mathrm{Gal}(F/\mathbb{Q})$.

Corollary 19 *Let χ be an orthogonally stable ordinary character of G and $K := F(\chi)$ its character field. Assume that $\mathrm{Aut}(G)$ acts transitively on the Galois orbit $\chi^{\mathrm{Gal}(K/\mathbb{Q})}$. Then, $\mathrm{Disc}(\chi)$ is Galois over $\mathbb{Q}$.*

In particular all discriminant fields of the orthogonally stable characters of the alternating groups are Galois over $\mathbb{Q}$.

Example 20 Conjecture 3.9 in [Cra22] states that any absolutely irreducible character with indicator + and degree congruent to 2 (mod 4) is expected to have an orthogonal discriminant α such that $\sqrt{\alpha}$ lies in a cyclotomic field.

A counterexample is provided by the two irreducible characters of degree 169290 of the sporadic simple O'Nan group. Their orthogonal discriminants are $-53 \pm 36\sqrt{2}$, see [NP23, Remark 7.3].

So far, all non Galois discriminant fields that we are aware of occur for sporadic simple groups and their automorphism groups.

Example 21 During our computations we only found the following ordinary orthogonally simple (see Remark 7) characters of finite simple groups for which the discriminant fields $\mathbb{Q}(\sqrt{\delta})$ are not Galois over $\mathbb{Q}$:

G	χ	δ	$\mathrm{Gal}(\mathbb{Q}(\sqrt{\delta})/\mathbb{Q})$
J_1	$56ab$	$(31 + 5\sqrt{5})/2$	D_8
J_1	$120abc$	$29 - 18c_{19} - 9c_{19}^{*2}$	$C_2 \times A_4$
J_3	$1920abc$	$63 - 30y_9 - 7y_9^{*2}$	A_4
He	$21504ab$	$357 + 68\sqrt{21}$	D_8
Ru	$27000abc$	$119y_7 + 49y_7^{*2} + 170$	A_4
Ru	$34944ab$	$41 - 16\sqrt{6}$	D_8
Ru	$110592ab$	$(1015 - 185\sqrt{29})/2$	D_8
ON	$169290ab$	$-36\sqrt{2} - 53$	D_8
ON	$175616ab$	$225 + 84\sqrt{5}$	D_8
ON	$207360abc$	$-496c_{19} + 1767c_{19}^{*4} + 3472$	$C_2 \times A_4$
HN	$5103000ab$	$17 + 4\sqrt{5}$	D_8

The table lists the groups, the characters χ (full Galois orbit) in the form $\chi(1)ab...$, the orthogonal discriminant of $\chi(1)a$ in ATLAS notation (see Section 2.1.1), and the Galois group of the normal closure of the discriminant field. The characters of $G = J_3$ and $G = He$ extend to characters of $G.2$ with the same degree, character field, and orthogonal discriminant.

We can select the entries in question from the known data as follows, using the criterion from Proposition 16.

```
julia> function is_galois_discriminant_field(data)
         chi = Oscar.OrthogonalDiscriminants.character_of_entry(data)
         F, emb = character_field(chi)
         c = conductor(emb(gen(F)))
         galgens = Oscar.AbelianClosure.generators_galois_group_cyclotom⌋
         ↪  ic_field(c)
         delta = atlas_irrationality(data[:valuestring])
         return all(x -> is_square(preimage(emb, delta * x(delta))),
                    galgens)
       end;

julia> info = all_od_infos(characteristic => 0, is_simple);

julia> filter!(r -> r[:valuestring] != "?" &&
                    conductor(atlas_irrationality(r[:valuestring])) > 1,
              info);
```

```
julia> length(info)
58

julia> filter!(!is_galois_discriminant_field, info);

julia> length(info)
26

julia> println(sort!(collect(Set([r[:groupname] for r in info]))))
["HN", "He", "J1", "J3", "ON", "Ru"]
```

5.2 No Even Discriminants?

Richard Parker conjectured that orthogonal discriminants in characteristic zero are always odd (see [Neb22a, Conjecture 1.3]). This conjecture is true for characters of solvable groups (see [Neb22a, Theorem 1.5]), and it holds also for all characters of Atlas groups which we have computed so far.

```
julia> info = all_od_infos(characteristic => 0, degree => 1);

julia> all(x -> x[:valuestring] == "?" ||
                is_odd(parse(Int, x[:valuestring])),
           info)
true
```

Note that the sketch of a proof of this conjecture over the rationals given in [Cra22, p. 7] is not correct, as it assumes that there is always an even lattice of square-free discriminant.

5.3 Groups Embedding in Both Orthogonal Groups of Same Degree

The final remark in [SW91] asks whether there is a group G with irreducible orthogonal representations of the same even degree and over the same character field in characteristic two, such that one of them has orthogonal discriminant $O+$ and the other has orthogonal discriminant $O-$.

The data about Atlas groups provide exactly one such example: The simple group $G_2(3)$ has three 90-dimensional absolutely irreducible representations over the field with two elements, "90a" (the one which is invariant under the outer automorphism) has orthogonal discriminant $O+$, whereas "90b" and "90c" (which are conjugate under the outer automorphism) have orthogonal discriminant $O-$, cf. Section 4.1.

```
julia> plus = []; minus = [];
```

```
julia> for d in all_od_infos()
         if d[:valuestring] == "O+"
           push!(plus, (d[:groupname], d[:characteristic], d[:degree],
                        parse(Int, filter(isdigit, d[:charname]))))
         elseif d[:valuestring] == "O-"
           push!(minus, (d[:groupname], d[:characteristic], d[:degree],
                         parse(Int, filter(isdigit, d[:charname]))))
         end
       end

julia> both = intersect!(plus, minus);

julia> filter(x -> x[2] == 2, both)
1-element Vector{Any}:
 ("G2(3)", 2, 1, 90)

julia> length(both)
103
```

(We see that there are many examples in odd characteristic.)

5.4 Accessing the Atlas of Orthogonal Discriminants

The information about orthogonal discriminants of Atlas groups can be used in `GAP` and `OSCAR`, as follows.

The `GAP` function `Display` and the `OSCAR` function `show`, respectively, can be called with the option to extend the shown character table by a column for orthogonal discriminants. One can also access the list of known orthogonal discriminants for an ATLAS character table, via the functions `OrthogonalDiscriminants` (in `GAP`) and `orthogonal_discriminants` (in `OSCAR`), respectively.

5.5 New Findings for the Old Character Tables

The following new information has been obtained as a by-product of the computation of orthogonal discriminants.

- Listing the orthogonal discriminants of the orthogonal absolutely irreducible characters of a group requires the knowledge of the Frobenius Schur indicators of these characters (see Section 2.3). In characteristic two, this information is not known for all character tables we are interested in. Several 2-modular Frobenius Schur indicators that had been missing are now known. They have been either computed explicitly once we had the representation in question, or determined using [GW95, Lemma 1.2].

- The Brauer character tables of $L_2(49)$ mod 7, $L_2(81)$ mod 3, and $L_6(2)$ mod 2 had been missing.
- Several class fusions between Atlas character tables, which turned out to be useful for restrictions of characters to subgroups, have been added to the character table library.
- A so-called generality problem for the sporadic simple group HN and its automorphism group $HN.2$ has been solved. This problem concerns the consistency between the 11- and 19-modular character tables of these groups, as follows.
  In the ordinary character table of HN, the conjugacy classes `20A` and `20B` are distinguished only by the two algebraic conjugate irreducible characters χ_{51}, χ_{52} of degree 5 103 000. Their values on `20A` and `20B` are $1 \pm 2\sqrt{5}$.
  According to the Brauer character tables in the library of character tables up to version 1.3.4, the conjugacy class `20A` of HN was the class for which both the unique irreducible 11-modular Brauer character of degree 628 426 and the unique irreducible 19-modular Brauer character of degree 1 074 075 have the value $1 - 2r_5$. The orthogonal discriminant of χ_{51} is either $4\sqrt{5} + 17$ or $-4\sqrt{5} + 17$. In the former case, the 11-modular reduction of χ_{51} is orthogonally stable, and the 19-modular reduction is not; in the latter case, it is the other way round. However, with the above choice of the class `20A`, both the 11- and 19-modular reductions of χ_{51} are orthogonally stable (and the 11- and 19-modular reductions of χ_{52} are not). Thus we have shown that the choice of `20A` in the two character tables is not consistent. In order to make the two character tables consistent, we have changed the 11-modular table in version 1.3.5 of the table library, by swapping the columns of `20A` and `20B`.
  (As a consequence, also the 11-modular table of the automorphism group $HN.2$ of HN had to be adjusted. There are still open questions about the consistency of other conjugacy classes in Brauer character tables of HN. They are independent of the question about `20A` and `20B`, and they cannot be answered by considering orthogonal discriminants.)

Acknowledgements The authors acknowledge support by the German Research Foundation (DFG) – Project-ID 286237555 – within the SFB-TRR 195 *Symbolic Tools in Mathematics and their Applications*.

References

[BBBNP23] O. Braun, T. Braun, T. Breuer, G. Nebe, and R. Parker. *Orthogonal Discriminants of ATLAS Groups*. `https://www.math.rwth-aachen.de/~Thomas.Breuer/ctbllib/data/odresults.json`. 2023.

[BN17] O. Braun and G. Nebe. "The orthogonal character table of $SL_2(q)$". *J. Algebra* 486 (2017), pp. 64–79. DOI: 10.1016/j.jalgebra.2017.04.025.

[CCNPW85] J. H. Conway, R. T. Curtis, S. P. Norton, R. A. Parker, and R. A. Wilson. ATLAS *of finite groups*. Maximal subgroups and ordinary characters for simple groups, With computational assistance from J. G. Thackray. Oxford University Press, Eynsham, 1985, pp. xxxiv+252.

[Cra22] D. A. Craven. "An Ennola duality for subgroups of groups of Lie type". *Monatshefte für Mathematik* (2022). DOI: 10.1007/s00605-022-01676-3.

[GW95] R. Gow and W. Willems. "Methods to decide if simple self-dual modules over fields of characteristic 2 are of quadratic type". *J. Algebra* 175.3 (1995), pp. 1067–1081. DOI: 10.1006/jabr.1995.1227.

[GW97] R. Gow and W. Willems. "On the quadratic type of some simple self-dual modules over fields of characteristic two". *J. Algebra* 195.2 (1997), pp. 634–649. DOI: 10.1006/jabr.1997.7048.

[JLPW95] C. Jansen, K. Lux, R. Parker, and R. Wilson. *An atlas of Brauer characters*. Vol. 11. London Mathematical Society Monographs. New Series. Appendix 2 by T. Breuer and S. Norton, Oxford Science Publications. New York: The Clarendon Press Oxford University Press, 1995, pp. xviii+327.

[Kne02] M. Kneser. *Quadratische Formen. Neu bearbeitet und herausgegeben in Zusammenarbeit mit Rudolf Scharlau*. Berlin: Springer, 2002.

[KMRT98] M.-A. Knus, A. Merkurjev, M. Rost, and J.-P. Tignol. *The book of involutions. With a preface by J. Tits*. Vol. 44. Colloq. Publ., Am. Math. Soc. Providence, RI: American Mathematical Society, 1998.

[Mat99] A. Mathas. *Iwahori-Hecke algebras and Schur algebras of the symmetric group*. Vol. 15. University Lecture Series. American Mathematical Society, Providence, RI, 1999, pp. xiv + 188. DOI: 10.1090/ulect/015.

[Neb22a] G. Nebe. "On orthogonal discriminants of characters". *Albanian J. Math.* 16.1 (2022), pp. 41–49. DOI: 10.51286/albjm/1658730113.

[Neb22b] G. Nebe. "Orthogonal determinants of characters". *Arch. Math.* 119.1 (2022), pp. 19–26. DOI: 10.1007/s00013-022-01742-0.

[Neb99] G. Nebe. *Orthogonale Darstellungen endlicher Gruppen und Gruppenringe*. Vol. 26. Aachener Beitr. Math. Aachen: Verlag der Augustinus Buchhandlung; Aachen: RWTH Aachen (Habil.-Schr.), 1999.

[NP23] G. Nebe and R. A. Parker. "Orthogonal stability". *J. Algebra* 614 (2023), pp. 362–391. DOI: 10.1016/j.jalgebra.2022.09.017.

[PS96] W. Plesken and B. Souvignier. "Constructing rational representations of finite groups". *Exp. Math.* 5.1 (1996), pp. 39–47. DOI: 10.1080/10586458.1996.10504337.

[Rot22] M. Roth. "Ennola duality in subgroups of the classical groups". Supervised by Donna Testerman and David Craven. MA thesis. EPFL, 2022.

[Ryb90] A. J. E. Ryba. "Computer condensation of modular representations". In: vol. 9. 5-6. Computational group theory, Part 1. 1990, pp. 591–600. DOI: 10.1016/S0747-7171(08)80076-4.

[Sch85] W. Scharlau. *Quadratic and Hermitian forms.* Vol. 270. Grundlehren Math. Wiss. Springer, Cham, 1985.

[SW91] P. Sin and W. Willems. "G-invariant quadratic forms". *J. Reine Angew. Math.* 420 (1991), pp. 45–59. DOI: 10.1515/crll.1991.420.45.

[Wil+] R. A. Wilson et al. *ATLAS of Finite Group Representations.* https://atlas.math.rwth-aachen.de/Atlas/v3.

Gröbner Bases for Polynomial Ideals and Applications

Christian Eder, Rafael Mohr, and Mohab Safey El Din

Abstract We illustrate OSCAR's capabilities for computing with polynomial ideals using Gröbner bases. We give a detailed overview of the available Gröbner basis algorithms and provide some practical advice on how to use them. Based on this, we illustrate how to compute real solutions to polynomial systems in OSCAR. We then demonstrate with a range of examples from different areas of mathematics where Gröbner bases come into play when trying to answer computational questions.

Key words: Polynomial ideals, Gröbner bases, polynomial system solving

1 Introduction

Gröbner Bases of Polynomial Ideals

Algorithmically manipulating ideals in polynomial rings over fields is, as of today, heavily reliant on the notion of a *Gröbner basis*. Essentially, a Gröbner basis is a special generating set of a polynomial ideal I which allows for a membership test for I via multivariate polynomial long division. In fact, Gröbner bases let us make all sorts of ideal theoretic operations and invariants computable such as ideal intersection or the dimension of an ideal. Importantly, a Gröbner basis depends crucially on a chosen method to order monomials

Christian Eder
RPTU Kaiserslautern-Landau, Kaiserslautern, e-mail: `ederc@mathematik.uni-kl.de`

Rafael Mohr
RPTU Kaiserslautern-Landau, Kaiserslautern, and Sorbonne Université, Paris, e-mail: `rafael.mohr@lip6.fr`

Mohab Safey El Din, Sorbonne Université, Paris, e-mail: `mohab.safey@lip6.fr`

W. Decker et al. (eds.), *The Computer Algebra System OSCAR*,
Algorithms and Computation in Mathematics 32,
https://doi.org/10.1007/978-3-031-62127-7_11

(a *monomial ordering*) – we need to know which is the leading term of a polynomial when performing a polynomial long division. Some important monomial orders are closely related to different geometric operations: some allow for projections of algebraic sets, some for hyperplane cuts, some give a local and some a global picture of the polynomial ideal at hand. We refer for example to chapter 15 in [Eis95] for a comprehensive introduction to Gröbner bases, to [BW93] for a detailed and comprehensive account of the area, or to the algebraic geometry cornerstones chapter [BDS24] (cf. page 21) for some of the basics of the subject.

How Gröbner Bases Are Used in OSCAR

OSCAR's implementation of polynomial ideals and algorithms operating on them also relies on Gröbner basis computations, mainly through its interfaces to the computer algebra system Singular and to the Gröbner basis library msolve. Since Gröbner bases can be quite expensive to compute, each polynomial ideal declared in OSCAR has its own Gröbner basis cache, storing lazily a Gröbner basis per monomial ordering when needed and reusing them, if possible, for future computations involving the same polynomial ideal.

An important design goal of OSCAR's ideal theoretic capabilities is that polynomial ideals (and related objects such as quotient algebras) can be manipulated in OSCAR purely by typing in the correct ideal theoretic operation without any knowledge for how they are performed under the hood.

On the other hand, OSCAR also allows us to compute Gröbner bases explicitly, using a range of algorithms. This can allow the informed user to adapt the computation to their specific problem and hand pick which algorithm to use, for better efficiency.

To illustrate this, consider the problem of computing the Hilbert series of a homogeneous ideal, for example of the twisted cubic in $\mathbb{P}^3$. Of course we can ask OSCAR for the Hilbert series directly:

```
julia> R, (X0, X1, X2, X3) = graded_polynomial_ring(QQ, ["X$i" for i in
↪   0:3]);

julia> I = ideal(R, [X0*X2 - X1^2, X1*X3 - X2^2, X0*X3 - X1*X2]);

julia> A, _ = quo(R, I);

julia> hilbert_series(A)
(2*t^3 - 3*t^2 + 1, (-t + 1)^4)
```

However, for larger examples the user will observe that this computation slows down considerably. This is because a Gröbner basis over the rational numbers using the famous *Buchberger algorithm* [Buc65] is computed in the background, out of this Gröbner basis the Hilbert series can be extracted (via the *leading monomials* (Definition 1) of the computed Gröbner basis, see e.g. Theorem 15.26 in [Eis95]). This computation is done by Singular. To speed things up, let us first specialize the problem to a random prime field, over which arithmetic is much easier, this will make all Gröbner basis computations faster than over $\mathbb{Q}$. For a "sufficiently random" prime (i.e. a prime outside of a certain finite subset of primes) the Hilbert series that we obtain will be the same as over $\mathbb{Q}$.

```
julia> p = next_prime(rand(1:10000))
4783

julia> R, (X0, X1, X2, X3) = graded_polynomial_ring(GF(p), ["X$i" for i
↪  in 0:3]);

julia> I = ideal(R, [X0*X2 - X1^2, X1*X3 - X2^2, X0*X3 - X1*X2]);
```

Now that we are working over a prime field we can use msolve's efficient implementation of the *F4 algorithm* [Fau99] to compute the Gröbner basis that we need. F4 is generally faster than Buchberger's algorithm. Additionally, msolve's implementation of F4 is limited to finite fields and one specific monomial ordering, but this limitation allows for optimizing the implementation for these settings. For us this limitation is not a problem: We are working over a finite field and, in this case, don't care which ordering is used when computing a Gröbner basis. Here is how we get access to this F4 implementation in OSCAR:

```
julia> gb = groebner_basis_f4(forget_grading(I));
```

The Hilbert series of an ideal is the same as that of its *leading monomial* ideal (again, see 1 for a definition):

```
julia> lm_ideal = ideal(R, (leading_monomial).(gens(gb)));

julia> A, _ = quo(R, lm_ideal);

julia> hilbert_series(A)
(2*t^3 - 3*t^2 + 1, (-t + 1)^4)
```

This Chapter

This chapter has two parts: In the first part we give an overview of the available Gröbner basis algorithms in OSCAR, their use cases and the settings in which they can be applied. We show with another brief example how explicitly using the provided algorithms, rather than just asking OSCAR for a Gröbner basis, can lead to faster computations. Using an example, we also briefly describe msolve's capabilities for solving polynomial systems based on Gröbner basis computations.

In the second part we perform computations on polynomial ideals coming from a wide range of mathematical areas. We point out which Gröbner basis computations are performed in the background and where potential computational bottlenecks could lie in similar examples of larger size.

We have tried to keep this chapter as self-contained as possible. While it can be read without knowing anything about Gröbner bases, it is intended for those who have a basic understanding of the subject, like ideals in polynomial rings, want to gain some insight into how they can be computed explicitly in `OSCAR`, where and how they come up in examples of interest and where computational difficulties can lie when working with them.

2 Computing Gröbner Bases in `OSCAR`

In this section we discuss the different algorithms available in `OSCAR` for computing Gröbner bases explicitly in more detail. Let us recall the definition of a Gröbner basis, again see the algebraic geometry cornerstones chapter for more details:

Definition 1 Fix a monomial ordering $\leq$ on R, i.e. a total order on the set of monomials in R which is preserved under multiplication and extends the partial ordering given by divisibility.

1. The leading monomial of an element $f \in R$, denoted $\mathrm{lm}(f)$, is the $\leq$-largest monomial in the support of f. For a finite set $F \subset R$, denote
$$\mathrm{lm}(F) = \{\mathrm{lm}(f) \mid f \in F\}.$$
2. For an ideal $I \subset k[x_1, \ldots, x_n]$, the leading monomial ideal of I, denoted $\mathrm{lm}(I)$, is given by
$$\mathrm{lm}(I) = \langle \mathrm{lm}(f) \mid f \in I \rangle.$$
3. A Gröbner basis of I w.r.t. $\leq$ is a finite set $G \subset I$ such that
$$\langle \mathrm{lm}(G) \rangle = \mathrm{lm}(I).$$

Remark 2 Note that the restriction that a monomial ordering extends the partial ordering given by divisibility is equivalent to $1 < m$ for every monomial $m \neq 1$. In certain contexts this is called a *global* monomial ordering whereas the reverse property (i.e. $m < 1$ for every monomial $m \neq 1$) is called a *local* monomial ordering. Working with local monomial orderings is quite different, see e.g. section 10.6 in [BW93]. In this chapter, we will consider computations only with global monomial orderings.

From this definition, one can conclude that different Gröbner bases for different monomial orderings share many properties. For example, using the third item in the above definition, any Gröbner basis provides a membership test for the ideal it generates. Certain other ideal theoretic invariants may also be extracted out of any given Gröbner basis (such as, for example, the dimension or Hilbert series).

On the other hand, certain ideal theoretic operations may require Gröbner basis computations with specific monomial orderings. For example, the `eliminate` command that we will frequently use in the second part of this chapter requires the computation of a Gröbner basis w.r.t. a block ordering (see e.g. [BW93, Exercise 4.61]): we choose two monomial orderings on two disjoint blocks of variables. Then we compare two monomials by comparing lexicographically the two blocks of these monomials, corresponding to the two blocks of variables, using the two underlying monomial orderings. Doing this explicitly in OSCAR with two blocks of the `degrevlex` monomial ordering looks like this:

```
julia> R, (x, y, z) = polynomial_ring(QQ, ["x","y","z"]);

julia> o = degrevlex([x, y])*degrevlex([z]);

julia> I = ideal(R, [x^2 + y + z - 1, x + y^2 + z - 1, x + y + z^2 - 1]);

julia> groebner_basis(I, ordering = o)
Gröbner basis with elements
1 -> z^6 - 4*z^4 + 4*z^3 - z^2
2 -> 2*y*z^2 + z^4 - z^2
3 -> x + y + z^2 - 1
4 -> y^2 + x + z - 1
with respect to the ordering
degrevlex([x, y])*degrevlex([z])

julia> eliminate(I, [x,y]) # internally computes the above Gröbner basis
Ideal generated by
  z^6 - 4*z^4 + 4*z^3 - z^2
```

Through Robbiano's classification of monomial orderings as matrix orderings [Rob86], every possible monomial ordering is available in OSCAR, since we can define matrix orderings in OSCAR.

Depending on the chosen monomial ordering, Gröbner bases differ not only in their properties but also their computation with the classical Buchberger algorithm [Buc65] can wary wildly in difficulty, even for the same ideal. In many cases, computing a Gröbner basis with Buchberger's algorithm is easiest using the *degree reverse lexicographical* ordering (herein referred to as the DRL ordering, called `degrevlex` in OSCAR): Two monomials are compared by first comparing their degree, then, if these two degrees are equal, one compares the monomials lexicographically starting with the last variable but reverses the outcome of this comparison. The heuristic that Gröbner bases for the DRL ordering are usually easiest to compute is also justified in certain complexity theoretic results for computing Gröbner bases: In generic coordinates, the degrees of the elements in a DRL Gröbner basis are bounded by the so-called *regularity* of the ideal, see e.g. [BM92]. For certain ideals, this improves the well-known general doubly exponential (in the number of variables) complexity for computing a Gröbner basis to a singly exponential complexity for computing a DRL Gröbner basis.

That is why OSCAR uses DRL as a default if we don't explicitly specify another ordering. Continuing the previous example:

```
julia> groebner_basis(I)
Gröbner basis with elements
1 -> z^2 + x + y - 1
2 -> y^2 + x + z - 1
3 -> x^2 + y + z - 1
with respect to the ordering
degrevlex([x, y, z])
```

As a general rule of thumb, when a Gröbner basis is required, one should use DRL whenever possible, and avoid the use of block orderings as much as possible (although eliminating a subset of the variables using block orderings is often hard to avoid in practice, as we will see in the next section).

Let us now discuss the different algorithms for computing Gröbner bases available in OSCAR and their use cases. A popular modification of Buchberger's algorithm, Faugère's F4 algorithm [Fau99], is available in OSCAR through the C-library msolve [BES21], which also enables the computation of real solutions to polynomial equations, as we will see later. F4 works by linearizing the problem of computing Gröbner bases in a certain way, which enables the use of fast linear algebra algorithms for computing Gröbner bases. msolve itself is accessible via the Julia package AlgebraicSolving.jl, which is maintained by the authors of this chapter. This package in turn is integrated into OSCAR. At the time of writing of this chapter, *signature-based* Gröbner basis computations (see e.g. [EF17] for a survey) are already part of AlgebraicSolving.jl, and a multi-modular method for computing Gröbner bases over the rational numbers is part of msolve and soon to be integrated into AlgebraicSolving.jl.

The F4 implementation in msolve is substantially faster than the available implementation of Buchberger's algorithm, both because of the inherent efficiency of F4 over Buchberger's algorithm and because of its more specialized nature: At the time of writing of this chapter, it only accepts ideals over fields of the form $\mathbb{Z}/p\mathbb{Z}$ with $p < 2^{31}$ and computes Gröbner bases only for the DRL monomial ordering. It can also be used to eliminate the last k variables from an ideal, in which case it returns a DRL Gröbner basis of the corresponding elimination ideal.

The two other Gröbner basis algorithms (FGLM [FGLM93] and a *Hilbert driven* algorithm [Tra96]) available in OSCAR are useful if a Gröbner basis for a monomial ordering other than DRL is required. They use a "change of ordering" strategy: First compute a Gröbner basis for an "easy" ordering (usually DRL) via Buchberger's algorithm or F4, then convert that Gröbner basis into one for the desired target ordering.

The four Gröbner basis algorithms available in OSCAR, together with the restrictions on their input, are summarized in table 1.

Each implemented algorithm is available to the user in two different ways:

1. By calling the function `groebner_basis` with the appropriate `algorithm` keyword, e.g. to use the F4 algorithm to compute a Gröbner basis for a polynomial ideal `I` we may call

   ```
   groebner_basis(I, algorithm = :f4)
   ```

2. By calling directly the function corresponding to a specific algorithm e.g. to perform the same computation as in the previous item we may call

   ```
   groebner_basis_f4(I)
   ```

The second option offers more user customization options than the first one but requires more work to be done by the user. For example the Hilbert driven algorithm to compute Gröbner bases, as such, works only for ideals homogeneous w.r.t. suitable weights. With a certain homogenization-dehomogenization strategy, this algorithm becomes available also for non-graded ideals.

When we choose this algorithm by calling `groebner_basis(I, algorithm = :hilbert)`, then `OSCAR` will take care of ensuring this homogeneity-condition for us. When we call `groebner_basis_hilbert_driven` ourselves, we are the ones who have to ensure that the input ideal is given by homogeneous generators (w.r.t. suitable weights).

Depending on the target ordering a change-of-ordering strategy can be substantially faster than computing a Gröbner basis for the desired ordering directly. Let us illustrate how to do this in `OSCAR` with an example. For our example, we start by computing a Gröbner basis w.r.t. the `degrevlex` ordering. Because we are working over a suitable finite field, `groebner_basis_f4` is our fastest option. In this example we measure the time of computation with `Julia`'s `@time` macro:

```
julia> R, (x1, x2, x3, x4, x5, x6, x7, x8) = polynomial_ring(GF(65521),
↪   ["x1","x2","x3","x4","x5","x6", "x7","x8"]);

julia> I = ideal(R, [x1 + 2*x2 + 2*x3 + 2*x4 + 2*x5 + 2*x6 + 2*x7 + 2*x8 -
↪   1, x1^2 - x1 + 2*x2^2 + 2*x3^2 + 2*x4^2 + 2*x5^2 + 2*x6^2 + 2*x7^2 +
↪   2*x8^2, 2*x1*x2 + 2*x2*x3 - x2 + 2*x3*x4 + 2*x4*x5 + 2*x5*x6 +
↪   2*x6*x7 + 2*x7*x8, 2*x1*x3 + x2^2 + 2*x2*x4 + 2*x3*x5 - x3 + 2*x4*x6
↪   + 2*x5*x7 + 2*x6*x8, 2*x1*x4 + 2*x2*x3 + 2*x2*x5 + 2*x3*x6 + 2*x4*x7 -
↪   x4 + 2*x5*x8, 2*x1*x5 + 2*x2*x4 + 2*x2*x6 + x3^2 + 2*x3*x7 + 2*x4*x8 -
↪   x5, 2*x1*x6 + 2*x2*x5 + 2*x2*x7 + 2*x3*x4 + 2*x3*x8 - x6, 2*x1*x7 +
↪   2*x2*x6 + 2*x2*x8 + 2*x3*x5 + x4^2 - x7]);

julia> @time groebner_basis_f4(I);
  0.010955 seconds (57.13 k allocations: 3.081 MiB)
```

We check and discover that our ideal is zero-dimensional, so we may now use the FGLM algorithm to convert our `degrevlex` Gröbner basis into one for the `lex` ordering:

```
julia> dim(I) == 0
```

Algorithms	**Keyword**	**Restrictions**		
		base ring	monomial ordering	other
Buchberger	`:buchberger`	–	–	–
F4	`:f4`	prime field $\mathbb{Z}/p\mathbb{Z}$	DRL	$p < 2^{31}$
Hilbert Driven	`:hilbert`	field	global	–
FGLM	`:fglm`	field	global	dim 0

Table 1: Gröbner basis algorithms in OSCAR .

```
true

julia> @time fglm(I, destination_ordering = lex(R));
  0.062046 seconds (130.52 k allocations: 3.313 MiB)
```

As the reader may verify, computing a lexicographic Gröbner basis for `I` by calling `groebner_basis(I, ordering = lex(R))` takes substantially longer than the two Gröbner basis computations above combined.

Such a change-of-ordering strategy is also used by `msolve` itself to compute real solutions to polynomial systems of equations. Let us demonstrate this with an example as well.

2.1 Computing Real Solutions

For this demonstration we work with an example related to Euclidean distance problems, see [DHOST14].

Consider an algebraic map $\psi : \mathbb{R}^m \to \mathbb{R}^n$ and let $X := \overline{\mathrm{Im}(\psi)}$ be the Zariski closure of the image of ψ. For a given $u = (u_1, \dots, u_n) \in \mathbb{R}^n$ we are interested in the critical points of the function $d_u(x) = \sum_{i=1}^n (x_i - u_i)^2$ on the regular locus of X. Defining

$$D_u(t) = \sum_{i=1}^{n} (\psi_i(t) - u_i)^2$$

we have to solve the polynomial system

$$\frac{\partial D_u}{\partial t_1}(t) = \cdots = \frac{\partial D_u}{\partial t_m}(t) = 0$$

subject to the condition that the Jacobian of ψ has maximal rank. Let us set this up in OSCAR for a randomly generated $\psi : \mathbb{R}^2 \to \mathbb{R}^3$ (see also example 3.1 in [DHOST14]):

```
julia> R, (t1, t2) = polynomial_ring(QQ, ["t1", "t2"]);

julia> psi = [rand(R, 6:6, 3:5, -10:10) for _ in 1:3]
```

```
3-element Vector{QQMPolyRingElem}:
 1//2*t1^5*t2^3 - 2*t1^3*t2^4 + 2//9*t1^3*t2^3
 -3//5*t1^5*t2^5 - 8//7*t1^3*t2^4 - 17//4*t1^3*t2^3
 -23//12*t1^5*t2^5 - 3//7*t1^4*t2^3 - 13//6*t1^3*t2^5 - 9*t1^3*t2^4

julia> D = sum([(f - rand(Int))^2 for f in psi]);

julia> J = jacobian_ideal(D);
```

Next we have to add the condition that the Jacobean of ψ has maximal rank. We introduce an extra variable, and add a random linear combination of the 2×2-minors of the Jacobean of ψ as a nonzero condition, using *Rabinowitsch's trick*:

```
julia> S, (t1, t2, s) = polynomial_ring(QQ, ["t1", "t2", "s"]);

julia> phi = hom(R, S, [t1, t2]);

julia> f = sum([rand(Int) * p for p in minors(jacobian_matrix(psi), 2)]);

julia> JJ = phi(J) + ideal(S, s*phi(f) - 1);
```

Finally, we can compute the real solutions of the resulting polynomial systems:

```
julia> sols, param = real_solutions(JJ);

julia> length(sols)
8
```

We see that in this case there are 8 critical points. `real_solutions` has also computed a rational parametrization for us, containing in particular a univariate polynomial parametrizing all *complex* solutions of the polynomial system in question. In this case there are 62 complex solutions:

```
julia> degree(param.elim)
62
```

Internally, `msolve` uses a multi-modular strategy, computing DRL Gröbner bases modulo a large set of primes, converting them into *lexicographic* Gröbner bases using a particular variant of FGLM called *sparse FGLM* (see [FM17]), out of which one extracts rational parametrizations for the polynomial systems in question which are then recombined using the Chinese Remainder Theorem. The computations of the DRL Gröbner bases are optimized using *tracers* [Tra89]. The rational parametrization of our solution set then allows to compute them up to arbitrary precision.

3 Examples

Extending the algebraic geometry cornerstones chapter of this book, let us now go through a few examples of computational problems for polynomial ideals and point out where Gröbner bases come into play.

Two functions that we will frequently use are the `eliminate` and `saturation` functions. Let us recall their mathematical definition and geometric meaning. Let k be a field and let $R := k[x_1, \ldots, x_n]$ be a polynomial ring over k. Let $I \subseteq R$ be an ideal. Let $\mathbf{y} \subset \mathbf{x} := \{x_1, \ldots, x_n\}$ be a subset of the variables.

Definition 3 We define the elimination ideal of I eliminating $\mathbf{y}$ as

$$I \cap k[\mathbf{x} \setminus \mathbf{y}].$$

Geometrically, this ideal cuts out the Zariski closure of the image of the projection of $V(I)$ to the $(\mathbf{x} \setminus \mathbf{y})$-space. In OSCAR, we compute this ideal by calling `eliminate(I,y)` where `y` is a `Vector` consisting of the variables in $\mathbf{y}$. Internally, it is computed via the *elimination theorem* [CLO15, Theorem 3.1.2]: If G is a Gröbner basis for I w.r.t. a block ordering on $\mathbf{x}$, where the two blocks are $\mathbf{y}$ and $\mathbf{x} \setminus \mathbf{y}$ and the $\mathbf{y}$-block is considered lexicographically larger, then $G \cap k[\mathbf{x} \setminus \mathbf{y}]$ is a Gröbner basis for $I \cap k[\mathbf{x} \setminus \mathbf{y}]$. We want to stress again that computing Gröbner bases for block orderings is in general computationally more costly than computing DRL Gröbner bases.

Let $J \subseteq R$ be another ideal in R.

Definition 4 The saturation of I by J is defined as

$$(I : J^\infty) := \left\{p \in R \;\middle|\; pJ^k \subseteq I \text{ for some } k \in \mathbb{N}\right\}.$$

Geometrically, this ideal cuts out the Zariski closure of the locally closed set $V(I) \setminus V(J)$. In OSCAR, we compute it by calling `saturation(I,J)`. Internally, this ideal is computed using the Rabinowitsch trick, as above when we computed real solutions to polynomial systems. More precisely, if $J = \langle f \rangle$ is a principal ideal, then

$$(I : J^\infty) = (I + \langle tf - 1 \rangle) \cap k[\mathbf{x}],$$

where t is a new variable. An elimination-free algorithm for saturation has been given in the recent work [BES], this algorithm is already part of `msolve` but, as of the time of writing of this chapter, not yet ported over to OSCAR.

Knowing how to compute ideals of the form $(I : f^\infty)$ we can then compute $(I : J^\infty)$ for any J: If J is generated by $f_1, \ldots, f_r$ then

$$(I : J^\infty) = \bigcap_{i=1}^{r} (I : f_i^\infty).$$

We note that a probabilistic optimization of this, based on the prime avoidance lemma, is to instead choose a random linear combination f of the $f_1, \ldots, f_r$ and then compute $(I : f^\infty)$. Again, we already used this trick when we computed real solutions above. With high probability this ideal will be equal to $(I : J^\infty)$. It is easy to implement this trick in `OSCAR`, using `msolve`'s F4 implementation for the needed Gröbner basis computations, should the user require a quicker solution.

Computing Non-Proper Points

Suppose $f : X \to Y$ is a dominant map between algebraic sets X and Y. We are interested in computing the *non-proper* values of f:

Definition 5 A point $y \in Y$ such that there exists no neighborhood U of y with $f^{-1}(\overline{U})$ compact is called a non-proper value of f.

This notion is, for example, related to the notion of *generalized critical values* in polynomial optimization (see e.g. [KOS00; BFD21]).

In purely algebraic terms, the non-proper values are exactly the values over which f is not finite, if we are working over the complex numbers.

To compute the set of non-proper values, we follow [Sta02]. Let $\Gamma := \{(x, y) \mid f(x) = y\} \subset X \times Y$ be the graph of f and suppose that $X \subset \mathbb{A}^n$. Let $\overline{\Gamma}$ be the closure of Γ in $\mathbb{P}^n \times Y$. Then:

Proposition 6 *The set of non-proper values of f is exactly the projection of $\overline{\Gamma} \setminus \Gamma$ to Y.*

To compute this set in `OSCAR` consider the example

$$
\begin{aligned}
X &= V(xy, xz, yz, x(x-1)) \subset \mathbb{C}^3 \\
Y &= \mathbb{A}^1 \\
f &: (x, y, z) \mapsto x
\end{aligned}
$$

Let t be our coordinate on y. We set up the graph of f in `OSCAR`:

```
julia> R, (x, y, z, t) = polynomial_ring(QQ, ["x", "y", "z", "t"]);

julia> graph_id = ideal(R, [x*y, x*z, y*z, x*(x-1), x-t]);
```

Next, we have to compute the closure $\overline{\Gamma}$ of Γ in $\mathbb{P}^3 \times \mathbb{A}^1$. Just as the projective closure of an affine variety is computed by homogenizing a corresponding ideal, we compute $\overline{\Gamma}$ by homogenizing a corresponding ideal while treating the variable t as a constant (see e.g. [CLO15]) – i.e. by homogenizing w.r.t. the weighted degree given by the weight vector $(1, 1, 1, 0)$. Note that this is not as simple as homogenizing arbitrary input generators – an ideal cutting out the projective closure of an algebraic set can be computed via a

Gröbner basis computation with a suitable choice of monomial ordering, see e.g. [Eis95, Proposition 15.31]. More precisely, our monomial ordering has to extend the partial ordering on monomials given by the grading of interest. Such a Gröbner basis computation happens in the background of the following function call:

```
julia> w = [1 1 1 0]
1×4 Matrix{Int64}:
 1  1  1  0

julia> H = homogenizer(R, w, "w"); clo = H(graph_id)
Ideal generated by
  -x + t*w
  t^2 - t
  z*t
  y*t
  x*t - x
  y*z
  x*z
  x*y
  x^2 - x*w
```

Now, let π be the projection from $\overline{\Gamma}$ to $\mathbb{A}^1$. Our goal is to compute the (Zariski-closed) set $\pi\left(\overline{\Gamma} \setminus \Gamma\right)$, i.e. to project the intersection of $\overline{\Gamma}$ with the hyperplanes at infinity in $\mathbb{P}^3 \times \mathbb{A}^1$ to the t-space. This set will be the set of non-proper values. In the following code snippet we always saturate by the irrelevant ideal of $\mathbb{P}^3 \times \mathbb{A}^1$ before eliminating: this ensures that when we then algebraically eliminate the variables x, y, z, w that we indeed obtain an ideal cutting out the set $\pi\left(\overline{\Gamma} \setminus \Gamma\right)$, see p. 398 in [CLO15]:

```
julia> S = base_ring(clo)
Multivariate polynomial ring in 5 variables over QQ graded by
  x -> [1]
  y -> [1]
  z -> [1]
  t -> [0]
  w -> [1]

julia> x, y, z, t, w = gens(S);

julia> irrel_ideal = ideal(S, [x,y,z,w]);

julia> eliminate(saturation(clo + ideal(S,x), irrel_ideal), [x,y,z,w])
Ideal generated by
  t

julia> eliminate(saturation(clo + ideal(S,y), irrel_ideal), [x,y,z,w])
Ideal generated by
  t^2 - t

julia> eliminate(saturation(clo + ideal(S,z), irrel_ideal), [x,y,z,w])
Ideal generated by
  t^2 - t
```

```
julia> eliminate(saturation(clo + ideal(S,w), irrel_ideal), [x,y,z,w])
Ideal generated by
  t
```

Hence the non-proper values of f is the hypersurface defined by $t(t-1) = 0$.

Computing Conormal Spaces and Whitney Stratifications

Suppose that $X \in \mathbb{P}^n$ is a singular projective variety of dimension d. Our goal in this section is to compute a *Whitney stratification* of X, which can be understood as a certain decomposition of X into smooth pieces. For this, we will follow the algorithm presented in [HN22] and show how to perform the necessary computations in OSCAR. To introduce the problem, consider the variety cut out by the following hypersurface in $\mathbb{P}^3$:

```
julia> R, (x, y, z, w) = graded_polynomial_ring(QQ, ["x","y","z","w"]);

julia> f = y^2*w^2 + z^3*w - x^2*z^2;
```

Let us investigate the singular locus of this hypersurface $X := V(f)$ by imposing that the Jacobian of f vanishes:

```
julia> J = jacobian_matrix(f)
[           -2*x*z^2]
[            2*y*w^2]
[-2*x^2*z + 3*z^2*w]
[      2*y^2*w + z^3]

julia> I_sing = ideal(R, [f, minors(J, 1)...])
Ideal generated by
  -x^2*z^2 + y^2*w^2 + z^3*w
  -2*x*z^2
  2*y*w^2
  -2*x^2*z + 3*z^2*w
  2*y^2*w + z^3
```

We see that the singular locus of X consists of two lines. Let X_1 be the union of these two lines. A "naive" decomposition of X into smooth pieces is now given as

$$X = X \setminus X_1 \cup X_1.$$

Losing some information about the singular locus of X is an issue. We look at all associated primes of our singular locus, not just the minimal primes:

```
julia> minimal_primes(I_sing)
2-element Vector{MPolyIdeal{MPolyDecRingElem{QQFieldElem,
↪  QQMPolyRingElem}}}:
 Ideal (w, z)
 Ideal (z, y)
```

```
julia> (comp -> comp[2]).(primary_decomposition(I_sing))
6-element Vector{MPolyIdeal{MPolyDecRingElem{QQFieldElem,
↪   QQMPolyRingElem}}}:
 Ideal (w, z)
 Ideal (z, y)
 Ideal (w, z, y)
 Ideal (w, z, x)
 Ideal (z, y, x)
 Ideal (x, y, z, w)
```

In addition to our two lines, we discover the irrelevant ideal as well as three embedded points, lying on the two lines we saw previously. The fact that they appear in a primary decomposition of our singular locus suggests that they are somehow "more singular" in X than the two lines themselves. This information is lost in our decomposition above. The right geometric condition to impose additionally turns out to be the following ([HN22, section 2]):

Definition 7 Two complex manifolds $X, Y \subset \mathbb{C}^{n+1}$ with $\dim(Y) < \dim(X)$ satisfy *Whitney's condition (b)* if

- for any sequences of points $(x_i) \subset X$ and $(y_i) \subset Y$ converging to the same point p,
- such that the lines $[x_i, y_i]$ converge to some line l, and
- the tangent spaces $T_{x_i}X$ converge to some plane T

then $l \subset T$.

We refer to the introduction of [HN22] for a more detailed treatment of this example. Therein it is shown that the pair $(X \setminus X_1, X_1)$ does not satisfy Whitney's condition (b).

Whitney introduced this condition (b) in [Whi65] to define his Whitney stratifications:

Definition 8 A *Whitney stratification* of X is a filtration of X of the form

$$X = X_d \supset X_{d-1} \supset \cdots \supset X_0 = \emptyset.$$

Each connected component of $X_k \setminus X_{k-1}$, $k = 1, \ldots, d$, is called a *stratum*. Additionally, we want the filtration to satisfy the following assumptions:

- $X_k \setminus X_{k-1}$ is smooth for all $k = 1, \ldots, d$,
- every point p in X admits an open neighborhood intersecting only finitely many strata, and
- each pair of strata satisfies Whitney's condition (b).

We now would like to translate Whitney's condition (b) into an algebraic condition. Let X_{reg} be the set of smooth points of X, and X_{sing} be the set of singular points of X. At each point $p \in X_{\text{reg}}$, the projective tangent space T_pX naturally embeds as a linear space of dimension d in $\mathbb{P}^n$. Denote by $(\mathbb{P}^n)^*$ the dual projective space of dimension n, parametrizing the hyperplanes in $\mathbb{P}^n$. We define

Definition 9 The *conormal space* Conorm(X) of X is defined as the Zariski-closure of

$$U := \{(p, H) \in X_{\text{reg}} \times (\mathbb{P}^n)^* \mid T_pX \subset V(H)\}.$$

Using this conormal space, the authors of [HN22] prove the following:

Theorem 10 *Let $Y \subset X_{sing}$ be irreducible. Let κ_X : Conorm$(X) \to X$ be the canonical projection. Let*

$$I(\kappa_X^{-1}(Y)) = \bigcap_{i=1}^{r} Q_i$$

be a primary decomposition. Define

$$A := \left[\bigcup_{i \in \{1,\dots,s\}, \dim \kappa_X(V(Q_i)) < \dim Y} \kappa_X(V(Q_i)) \right] \cup Y_{sing}.$$

Then $(X_{reg}, Y \setminus A)$ satisfies Whitney's condition (b).

Our first goal is now to compute Conorm(X). Let us denote our homogeneous coordinates on $\mathbb{P}^n$ by $x_0, \dots, x_n$. We write

$$H = a_0x_0 + \cdots + a_nx_n$$

for our hyperplanes in $(\mathbb{P}^n)^*$. Let us compute this conormal space for the example $X := V(x_0^6 + x_0^4x_1x_2 + x_2^3x_3^3) \subset \mathbb{P}^3$. We start by setting up a coordinate ring for $X \times (\mathbb{P}^3)^*$ and storing the defining equation of X:

```
julia> R, (x0,x1,x2,x3,a0,a1,a2,a3) = polynomial_ring(QQ,
↪    ["x0","x1","x2","x3","a0","a1","a2","a3"]);

julia> f = x0^6 + x0^4*x1*x2 + x2^3*x3^3;
```

We compute the gradient of f, this time telling OSCAR to ignore the variables $a_0, \dots, a_3$:

```
julia> J = jacobian_matrix(f)[1:4, 1:1]
[6*x0^5 + 4*x0^3*x1*x2]
[              x0^4*x2]
[x0^4*x1 + 3*x2^2*x3^3]
[          3*x2^3*x3^2]
```

The condition $T_pX \subset V(H)$ in the definition of U is satisfied if and only if

$$\operatorname{rk} \begin{bmatrix} J(p) \\ a_0 \dots a_3 \end{bmatrix} = 1.$$

We encode this rank condition as a corresponding ideal of minors in OSCAR:

```
julia> JJ = hcat(J, R[a0; a1; a2; a3])
[6*x0^5 + 4*x0^3*x1*x2   a0]
[              x0^4*x2   a1]
[x0^4*x1 + 3*x2^2*x3^3   a2]
[          3*x2^3*x3^2   a3]

julia> id = ideal(R, minors(JJ, 2));
```

Finally, to compute Conorm(X), we add the defining equation of X to this ideal and saturate by the singular locus of X (which corresponds to taking the Zariski closure in the definition of Conorm(X)):

```
julia> id = ideal(R, minors(JJ, 2));

julia> I_sing = ideal(R, minors(J, 1));

julia> conorm = saturation(ideal(R, f) + id, I_sing);
```

To apply Theorem 10, we take $Y = X_{\text{sing}}$:

```
julia> Y = radical(ideal(R,f) + I_sing)
Ideal generated by
  x0
  x2*x3
```

Let us now look at the image of each primary component of $I(\kappa_X^{-1}(Y))$, taking the image under this projection map corresponds to eliminating the variables $a_0, \ldots, a_3$. Here we again saturate by the irrelevant ideal in the variables $x_0, \ldots, x_3$.

```
julia> id_fib_Y = saturation(conorm + Y, ideal(R, [x0,x1,x2,x3]));

julia> prim_comps = (comp -> comp[2]).(primary_decomposition(id_fib_Y));

julia> [eliminate(comp, [a0,a1,a2,a3]) for comp in prim_comps]
6-element Vector{MPolyIdeal{QQMPolyRingElem}}:
 Ideal (x2, x0)
 Ideal (x3, x2, x0)
 Ideal (x3, x0)
 Ideal (x3, x2, x0)
 Ideal (x3, x1, x0)
 Ideal (x2, x0, 4*x1^3 + 27*x3^3)
```

We see the irreducible components of Y ($\langle x_2, x_0\rangle$ and $\langle x_3, x_0\rangle$, both of dimension 1 in $\mathbb{P}^3$) and three distinct ideals corresponding to zero-dimensional varieties in $\mathbb{P}^3$ ($\langle x_3, x_2, x_0\rangle$, $\langle x_3, x_1, x_0\rangle$ and $\langle x_2, x_0, 4x_1^3 + 27x_3^3\rangle$), the union of which gives our set A in the setting of Theorem 10. In this case our Whitney stratification therefore consists of

$$X_3 := X, X_2 := Y = X_{\text{sing}}, X_1 := A \text{ and } X_0 := \emptyset.$$

We want to remark that, here, the core computational bottleneck for larger examples will be the required primary decomposition of an ideal – this is in general a very difficult problem. The algorithms for performing primary decomposition in `OSCAR` all make heavy use of elimination using Gröbner bases and polynomial factorization. We refer to [DGP97] for a survey on algorithms for primary and equidimensional decomposition available in `OSCAR`.

An Example From Combinatorics

For this example we will use elimination theory to investigate a combinatorial problem. To introduce the flavor of the problem that we will be working on, consider walks in $\mathbb{N}^2$ with steps $\{\nearrow, \searrow\}$ starting in the origin and ending at a point of the form $(x, 0)$. Let c_n be the number of such walks with exactly n steps and let $G(t) := \sum_{n=0}^{\infty} c_n t^n$ be its generating function. This function turns out to be algebraic over $\mathbb{Q}(t)$ and we wish to compute its minimal polynomial. As is easy to see, this minimal polynomial corresponds exactly to a recurrence of the sequence $(c_n)_n$. One systematic way of computing this minimal polynomial is to introduce a *catalytic variable* as follows: Consider the number $c_{n,d}$ of walks ending at a point of the form (x, d) and let $F(t, u) := \sum_{(n,d)\in\mathbb{N}^2} c_{n,d} t^n u^d$. The combinatorics of the problem allows us to derive the functional equation

$$F(t, u) = 1 + tuF(t, u) + t\frac{F(t, u) - F(t, 0)}{u}.$$

Note that $G(t) = F(t, 0)$. We define

$$P(x, z_0, t, v) = vx - t(x - z_0) - tu^2x - v,$$

so that we have the identity

$$P(F(t, u), G(t), t, u) = 0. \tag{1}$$

In [BJ06; NY23] the authors turn the equation 1 into a polynomial system which enables us to recover the algebraic equation eliminating $G(t)$ as follows: Consider

$$\begin{aligned} & \partial_u P(F(t, u), G(t), t, u) \\ = \; & \partial_u F(t, u)\partial_x P(F(t, u), G(t), t, u) + \partial_v P(F(t, u), G(t), t, u). \end{aligned}$$

If there exists a power series solution $U = U(t)$ to the equation

$$\partial_x P(F(t, U), G(t), t, U) = 0 \tag{2}$$

then also

$$\partial_v P(F(t, U), G(t), t, U) = 0. \tag{3}$$

So we have obtained a polynomial system consisting of the three equations (1), (2) and (3) in the four variables x, z_0, t, v. Eliminating the variables x and t from this polynomial system turns out to yield the desired minimal polynomial, see section 4 in [BJ06].

To work with a problem that turns out to be ideal-theoretically more interesting, consider the number of walks in $\mathbb{N}$ starting and ending at 0 with steps $\{+4, -3\}$. By again considering the endpoint as an additional parameter one obtains the functional equation

$$F(t,u) = 1 + tu^4 F(t,u) + t\frac{F(t,u) - F(t,0) - u\partial_u F(t,0) - \frac{u^2}{2}\partial_u^2 F(t,0)}{u^3}.$$

We introduce sets of variables to represent $F(t,u)$, $F(t,0)$ and its derivatives which appear in the above equation:

$$\begin{aligned} z_i &\sim \partial_u^i F(t,0), i = 0,1,2 \\ x &\sim F(t,u). \end{aligned}$$

Translating the above functional equation by plugging in these variables, the corresponding numerator yields a polynomial $P(x, z_0, z_1, z_2, t, u)$. Let us set this up in OSCAR. To make the required Gröbner basis computations easier we work modulo a large randomly chosen prime p, instead of over $\mathbb{Q}$. With high probability, this will allow us to recover the desired minimal polynomial at least modulo p. The reader can verify that working with this example over $\mathbb{Q}$ will make certain computations take a lot longer.

```
julia> F = GF(rand_bits_prime(ZZ, 20))
Prime field of characteristic 646637

julia> R, (z0, z1, z2, t, u, x) = polynomial_ring(F, ["z0","z1","z2","t",
↪  "u","x"]);

julia> P = (u^3-t*(u^7+1))*x-u^3+t*z0+t*u*z1+inv(F(2))*t*u^2*z2;
```

The polynomial system which has to be satisfied by $F(t, u(t))$ for a suitably chosen power series $u(t)$ is then:

```
julia> sys = [P,derivative(P,u),derivative(P,x)];

julia> I = saturation(ideal(R, sys), ideal(R, u));
```

Recall that our goal is now to eliminate every variable from this ideal except z_0 and t. We start by eliminating x:

```
julia> I = eliminate(ideal(R, sys), [x]);
```

Next follows the trick that will allow us to recover a unique polynomial in z_0 and t in the corresponding elimination ideal which then necessarily annihilates $F(t, 0)$. Let us look at the third equation of our system:

```
julia> sys[3]
646636*t*u^7 + 646636*t + u^3
```

This equation, when lifted to $\mathbb{Q}$, permits three distinct solutions $u(t)$ in the power series ring $\overline{\mathbb{Q}}[[t^{1/3}]]$ (see Theorem 2 in [BJ06]). As a consequence, at fixed $(z_0, z_1, z_2) \in \overline{\mathbb{Q}(t)}^3$ there exist at least three distinct solutions $u(t)$ to our polynomial system. We want to consider only such solutions. To this end we introduce duplicates of our equations with the variable u replaced by u_1 and u_2 respectively:

```
julia> S, (z0, z1, z2, t, u, u1, u2, x) = polynomial_ring(F,
↪    ["z0","z1","z2","t","u","u1","u2","x"]);

julia> phi = hom(R,S,[z0,z1,z2,t,u,x]);

julia> I = phi(I);

julia> I += ideal(S, [f(z0,z1,z2,t,u1,u1,u2,x) for f in gens(I)]);

julia> I += ideal(S, [f(z0,z1,z2,t,u2,u1,u2,x) for f in gens(I)]);
```

Then we saturate by a polynomial which vanishes if and only if at least two of our coordinates coincide. Geometrically we are taking the closure of the locus where we have at least three distinct solutions in u at fixed z_0, z_1 and z_2:

```
julia> I = saturation(I,ideal(S,(u-u1)*(u1-u2)*(u-u2)));
```

Now we can eliminate everything exceptz_0 and t:

```
julia> res = eliminate(I, [z1,z2,u,u1,u2,x]);
```

Note that we have indeed obtained a principal ideal. We verify that the generator of our polynomial ideal is irreducible to check that it is indeed the minimal polynomial of $G(t) = F(t, 0)$:

```
julia> factor(first(gens(res)))
1 * (z0^35*t^35 + 646636*z0^31*t^28 + z0^30*t^28 + 646636*z0^29*t^28 +
↪   5*z0^28*t^28 + 646636*z0^25*t^21 + z0^24*t^21 + 3*z0^23*t^21 +
↪   646633*z0^22*t^21 + 10*z0^21*t^21 + z0^19*t^14 + 646636*z0^18*t^14 +
↪   5*z0^17*t^14 + 3*z0^16*t^14 + 646631*z0^15*t^14 + 10*z0^14*t^14 +
↪   z0^13*t^7 + 646636*z0^12*t^7 + 3*z0^10*t^7 + z0^9*t^7 +
↪   646633*z0^8*t^7 + 5*z0^7*t^7 + 646636*z0 + 1)
```

Note that the trick used above to consider only fibers of cardinality at least three requires again a difficult Gröbner basis computation – we introduce a lot of extra variables only to then eliminate them. See [NY23] for a more elegant way of doing this.

Checking for Local Complete Intersections

We close with an example from pure commutative algebra that illustrates some ideal-theoretic capabilities of `OSCAR` that are more on the homological side rather than the elimination-and-saturation techniques used in the examples above. Again, let R be a polynomial ring in n variables and let I be an ideal in R.

Definition 11 I is a *local complete intersection* if the ring $(R/I)_{\mathfrak{p}}$ is a complete intersection ring for every prime $\mathfrak{p} \in \mathrm{Spec}(R/I)$.

We wish to check for a given polynomial ideal if it is a local complete intersection. For this we use (see [Vas67]):

Proposition 12 *I is a local complete intersection if and only if the conormal module I/I^2 is a projective R/I-module locally at every prime $\mathfrak{p} \in \mathrm{Spec}(R/I)$ (i.e. the module $(I/I^2)_{\mathfrak{p}}$ is projective, or equivalently free, for every prime $\mathfrak{p}$ of R/I).*

To check this condition we will use Fitting ideals (see e.g. chapter 20 in [Eis95]). For a module M over some ring S denote by $\mathrm{Fitt}_i(M)$ the ith Fitting ideal. We have (see also [Eis95])

Proposition 13 *1. If M can be generated by r elements then $\mathrm{Fitt}_r(M) = R$.*
2. $\mathrm{Fitt}_0(M) \subset \mathrm{Fitt}_1(M) \subset \dots \subset \mathrm{Fitt}_r(M) = R$.
3. for an S-algebra T

$$\mathrm{Fitt}_i(M \otimes_S T) = \mathrm{Fitt}_i(M)T.$$

4. M is projective of constant rank k if and only if $\mathrm{Fitt}_k(M) = S$ and $\mathrm{Fitt}_{k-1}(M) = 0$.

To check if a given polynomial ideal I in R is locally a complete intersection we may now proceed as follows: compute I/I^2 then compute the Fitting ideals up to index r, where r is the number of given generators of I (this suffices by the first and second item of Proposition 13). Computing Fitting ideals requires first computing syzygies for our module I/I^2. Unsurprisingly, this is again based on Gröbner bases. Out of any Gröbner basis, a Gröbner basis for the syzygy module can be extracted using Schreyer's theorem (see e.g. Theorem 15.10 in [Eis95]). We refer to the algebraic geometry cornerstones chapter for more details and to [EMSS16] for an efficient algorithm based on Schreyer's theorem available in `OSCAR`.

Once we have computed the required Fitting ideals we can check that the fourth condition in this proposition is met locally at every point $\mathfrak{p} \in \mathrm{Spec}(R/I)$. Now, the third condition lets us compute the Fitting ideals over R and then take their image in R/I. In particular, define

$$P_k(I/I^2) := \left\{\mathfrak{p} \in \mathrm{Spec}(R/I) \;\middle|\; I/I^2 \text{ is projective of rank } k\right\}.$$

By the fourth item of Proposition 13 we have

$$P_k(I/I^2) = \left[V(I) \setminus V(\mathrm{Fitt}_k(I/I^2))\right] \cap V(\mathrm{Fitt}_{k-1}(I/I^2))$$

and we may check computationally if

$$V(I) = \bigcup_{k=1}^{r} P_k(I/I^2).$$

Let us do a computational sanity check with the simple example of the coordinate lines in 3-space defined by $I = \langle xy, xz, yz \rangle$:

```
julia> R, (x, y, z) = polynomial_ring(QQ, ["x", "y", "z"]);

julia> I = ideal(R, [x*y, x*z, y*z]);
```

We set up the conormal module I/I^2 as a $\mathbb{Q}[x, y, z]$-module in OSCAR:

```
julia> conorm = subquotient(matrix(gens(I)), matrix(gens(I^2)))
Subquotient of Submodule with 3 generators
1 -> x*y*e[1]
2 -> x*z*e[1]
3 -> y*z*e[1]
by Submodule with 6 generators
1 -> x^2*y^2*e[1]
2 -> x^2*y*z*e[1]
3 -> x*y^2*z*e[1]
4 -> x^2*z^2*e[1]
5 -> x*y*z^2*e[1]
6 -> y^2*z^2*e[1]
```

The ideal I is locally defined by a complete intersection of codimension 2 everywhere in $V(I)$ except for the origin. We would therefore expect that $P_2(I/I^2) = V(I) \setminus V(x, y, z)$. Indeed this is the case:

```
julia> fitting_ideal(conorm, 1)
Ideal generated by
  y*z
  x*z
  x*y

julia> fitting_ideal(conorm, 2)
Ideal generated by
  z
  y
  x
```

Note however that our ideal I here is a *generic* complete intersection: It is unmixed and the ring $(R/I)_{\mathfrak{p}}$ is a complete intersection ring for every *minimal* prime over I. We can check for a generic complete intersection using Proposition 3.2.1 in [VGSEH97]:

Proposition 14 *Let I be unmixed of codimension c. Then I is a local complete intersection if and only if* $\operatorname{height}\operatorname{Fitt}_c(I) \geq c+1$.

Let us check this for our example. We first check, for the sake of demonstration, that I is unmixed. This can be done via equidimensional decomposition.

```
julia> equidimensional_decomposition_weak(I)
1-element Vector{MPolyIdeal{QQMPolyRingElem}}:
 Ideal (x*y, x*z, y*z)

julia> F = fitting_ideal(ideal_as_module(I), 2);

julia> codim(F) >= codim(I) + 1
true
```

Next, let us take the disjoint union of a point with a line in 3-space:

```
julia> I1 = ideal(R, [x-1, y-1]);

julia> I2 = ideal(R, [x, y, z]);

julia> I = intersect(I1, I2);

julia> conorm = subquotient(matrix(gens(I)), matrix(gens(I^2)));
```

We set up a convenience function to compute the Zariski closures of the $P_k(I/I^2)$:

```
julia> Pk(conorm, k) = saturation(I, fitting_ideal(conorm, k)) +
↪  fitting_ideal(conorm, k-1);
```

We expect now that $P_2(I/I^2)$ gives us the line defined by $\langle x-1, y-1\rangle$ and that $P_3(I/I^2)$ gives us the isolated point defined by $\langle x, y, z\rangle$:

```
julia> Pk(conorm, 2) == ideal(R, [x-1,y-1])
true

julia> Pk(conorm, 3) == ideal(R, [z,y,x])
true
```

Hence, we see in this case that I defines a local complete intersection. Note that while working with Fitting ideals can be mathematically convenient, they are more inconvenient computationally. They are *determinental ideals*, i.e. ideals generated by minors of a polynomial matrix. The degrees of the defining equations of these ideals can thus become quite large, making Gröbner basis calculations quickly become infeasible for them. Let us remark, however, that the design of Gröbner basis algorithms for determinental ideals is currently an active area of research, see e.g. [GNS23].

Acknowledgements This work has been supported by the Agence nationale de la recherche (ANR), grant agreement ANR-18-CE33-0011 (SESAME); by the EOARD-AFOSR grant agreement FA8665-20-1-7029; by the joint ANR-Austrian Science

Fund FWF grant agreements ANR-19-CE48-0015 (ECARP) and ANR-22-CE91-0007 (EAGLES); by the Deutsche Forschungsgemeinschaft (DFG, German Research Foundation) – Project-ID 286237555 – TRR 195, and by the Forschungsinitiative Rheinland-Pfalz.
The authors wish to thank Martin Helmer and Hadrien Notarantonio for providing kind advice and examples respectively for the section on Whitney stratifications and the section on combinatorics.

References

[BM92] D. Bayer and D. Mumford. "What Can Be Computed in Algebraic Geometry?" In: *In Computational Algebraic Geometry and Commutative Algebra*. University Press, 1992, pp. 1–48.

[BW93] T. Becker and V. Weispfenning. *Gröbner Bases*. Vol. 141. Graduate Texts in Mathematics. New York: Springer, 1993. DOI: 10.1007/978-1-4612-0913-3.

[BES21] J. Berthomieu, C. Eder, and M. Safey El Din. "Msolve: A Library for Solving Polynomial Systems". In: *Proceedings of the 2021 International Symposium on Symbolic and Algebraic Computation*. ISSAC'21. Saint Petersburg, Russia, 2021. DOI: 10.1145/3452143.3465545.

[BES] J. Berthomieu, C. Eder, and M. Safey El Din. *New efficient algorithms for computing Gröbner bases of saturation ideals (F4SAT) and colon ideals (Sparse-FGLM-colon)*. DOI: 10.48550/arXiv.2202.13387. arXiv: 2202.13387 [cs.SC].

[BFD21] J. Berthomieu, A. Ferguson, and M. S. E. Din. *On the Computation of Asymptotic Critical Values of Polynomial Maps and Applications*. Apr. 2021. DOI: 10.48550/arXiv.2104.00913. arXiv: 2104.00913 [cs].

[BDS24] J. Böhm, W. Decker, and F.-O. Schreyer. "Commutative Algebra and Algebraic Geometry". In: *The Computer Algebra System OSCAR. Algorithms and Examples*. Ed. by W. Decker, C. Eder, C. Fieker, M. Horn, and M. Joswig. Springer, 2024, pp. 21–79.

[BJ06] M. Bousquet-Mélou and A. Jehanne. "Polynomial Equations with One Catalytic Variable, Algebraic Series and Map Enumeration". *Journal of Combinatorial Theory, Series B* 96.5 (Sept. 2006), pp. 623–672. DOI: 10.1016/j.jctb.2005.12.003.

[Buc65] B. Buchberger. "Ein Algorithmus zum Auffinden der Basiselemente des Restklassenringes nach einem nulldimensionalen Polynomideal". PhD thesis. University of Innsbruck, Innsbruck, 1965.

[CLO15] D. A. Cox, J. Little, and D. O'Shea. *Ideals, varieties, and algorithms*. Fourth. Undergraduate Texts in Mathematics. Springer, Cham, 2015, pp. xvi+646. DOI: 10.1007/978-3-319-16721-3.

[DGP97] W. Decker, G.-M. Greuel, and G. Pfister. "Primary Decomposition: Algorithms and Comparisons". In: *Algorithmic Algebra and Number Theory*. 1997.

[DHOST14] J. Draisma, E. Horobeţ, G. Ottaviani, B. Sturmfels, and R. Thomas. "The Euclidean distance degree". In: *Proceedings of the 2014 symposium on symbolic-numeric computation, SNC 2014, Shanghai, China, July 28–31, 2014*. New York, NY: Association for Computing Machinery (ACM), 2014, pp. 9–16. DOI: 10.1145/2631948.2631951.

[EF17] C. Eder and J.-C. Faugère. "A Survey on Signature-Based Algorithms for Computing Gröbner Bases". *J. Symbolic Comput.* 80 (2017), pp. 719–784. DOI: 10.1016/j.jsc.2016.07.031.

[Eis95] D. Eisenbud. *Commutative Algebra: With a View toward Algebraic Geometry*. New York: Springer, 1995. DOI: 10.1007/978-1-4612-5350-1.

[EMSS16] B. Eröcal, O. Motsak, F.-O. Schreyer, and A. Steenpaß. "Refined Algorithms to Compute Syzygies". *J. Symbolic Comput.* 74 (May 2016), pp. 308–327. DOI: 10.1016/j.jsc.2015.07.004.

[FGLM93] J. C. Faugère, P. Gianni, D. Lazard, and T. Mora. "Efficient Computation of Zero-dimensional Gröbner Bases by Change of Ordering". *J. Symbolic Comput.* 16.4 (Oct. 1993), pp. 329–344. DOI: 10.1006/jsco.1993.1051.

[Fau99] J.-C. Faugère. "A New Efficient Algorithm for Computing Gröbner Bases (F4)". *J. Pure Appl. Algebra* 139.1 (1999), pp. 61–88. DOI: 10.1016/S0022-4049(99)00005-5.

[FM17] J.-C. Faugère and C. Mou. "Sparse FGLM Algorithms". *J. Symbolic Comput.* 80 (May 2017), pp. 538–569. DOI: 10.1016/j.jsc.2016.07.025.

[GNS23] S. Gopalakrishnan, V. Neiger, and M. Safey El Din. "Refined F5 Algorithms for Ideals of Minors of Square Matrices". In: *Proceedings of the 2023 International Symposium on Symbolic and Algebraic Computation*. ISSAC '23. New York, NY, USA: Association for Computing Machinery, July 2023, pp. 270–279. DOI: 10.1145/3597066.3597077.

[HN22] M. Helmer and V. Nanda. "Conormal Spaces and Whitney Stratifications". *Foundations of Computational Mathematics* (June 2022). DOI: 10.1007/s10208-022-09574-8. arXiv: 2106.14555 [cs, math].

[KOS00] K. Kurdyka, P. Orro, and S. Simon. "Semialgebraic Sard Theorem for Generalized Critical Values". *J. Differ. Geom.* 56.1 (Sept. 2000), pp. 67–92. DOI: 10.4310/jdg/1090347525.

[NY23] H. Notarantonio and S. Yurkevich. *Effective Algebraicity for Solutions of Systems of Functional Equations with One Catalytic Variable.* Mar. 2023. DOI: 10.48550/arXiv.2211.07298. arXiv: 2211.07298 [math].

[Rob86] L. Robbiano. "On the Theory of Graded Structures". *J. Symbolic Comput.* 2.2 (June 1986), pp. 139–170. DOI: 10.1016/S0747-7171(86)80019-0.

[Sta02] A. Stasica. "An Effective Description of the Jelonek Set". *J. Pure Appl. Algebra* 169.2 (2002), pp. 321–326. DOI: 10.1016/S0022-4049(01)00063-9.

[Tra89] C. Traverso. "Gröbner Trace Algorithms". In: *Symbolic and Algebraic Computation.* Ed. by P. Gianni. Lecture Notes in Computer Science. Berlin, Heidelberg: Springer, 1989, pp. 125–138. DOI: 10.1007/3-540-51084-2_12.

[Tra96] C. Traverso. "Hilbert Functions and the Buchberger Algorithm". *J. Symbolic Comput.* 22.4 (Oct. 1996), pp. 355–376. DOI: 10.1006/jsco.1996.0056.

[Vas67] W. V. Vasconcelos. "Ideals Generated by R-sequences". *J. Algebra* 6.3 (Aug. 1967), pp. 309–316. DOI: 10.1016/0021-8693(67)90086-5.

[VGSEH97] W. V. Vasconcelos, D. R. Grayson, M. Stillman, D. Eisenbud, and J. Herzog. *Computational Methods in Commutative Algebra and Algebraic Geometry.* Berlin, Heidelberg: Springer, 1997.

[Whi65] H. Whitney. "Tangents to an Analytic Variety". *Ann. of Math.* 81.3 (1965), pp. 496–549. DOI: 10.2307/1970400.

Invariant Theory

Wolfram Decker, Lakshmi Ramesh, and Johannes Schmitt

Abstract This is a tutorial on invariant theory in OSCAR. We begin by introducing some of the basic notions and results of invariant theory in their historical context. Then we discuss relevant algorithms and show them at work. Finally, we describe a quite recent application of invariant theory to computing the Cox ring of a linear quotient.

Key words: Invariant rings, fundamental invariants, finite groups, linearly reductive groups, tori, Cox rings

1 Introduction by Historical Remarks

One historical root of invariant theory is work in number theory by Lagrange [Lag75] and Gauss [Gau86] who studied the problem of representing integers by quadratic binary forms

$$F = ax^2 + 2bxy + cy^2,$$

Wolfram Decker
RPTU Kaiserslautern-Landau, Fachbereich Mathematik, e-mail: decker@mathematik.uni-kl.de

Lakshmi Ramesh
RPTU Kaiserslautern-Landau, Fachbereich Mathematik, e-mail: ramesh@mathematik.uni-kl.de

Johannes Schmitt
Universität Siegen, Naturwissenschaftlich-Technische Fakultät, e-mail: johannes2.schmitt@uni-siegen.de

W. Decker et al. (eds.), *The Computer Algebra System OSCAR*,
Algorithms and Computation in Mathematics 32,
https://doi.org/10.1007/978-3-031-62127-7_12

where a, b, c are integer coefficients. Gauss emphasized the importance of the *discriminant*[1] $b^2 - ac$ of F. He remarked that if another form F' is obtained from F by a substitution of type

$$x \to \alpha x + \beta y, \quad y \to \gamma x + \delta y,$$

then the corresponding discriminants differ as follows:

$$b'^2 - a'c' = (b^2 - ac)(\alpha\delta - \beta\gamma)^2. \tag{1}$$

Calling two forms F and F' *equivalent* (resp. *properly equivalent*) if they can be transformed into each other by a substitution with $\alpha\delta - \beta\gamma = \pm 1$ (resp. $= +1$), Gauss proved that equivalent forms represent the same integers.

Allowing complex coefficients, restricting ourselves to transformations with $\alpha\delta - \beta\gamma = +1$, and using the modern language of representation theory, we may interpret (1) in terms of invariant theory as follows: The special linear group $\mathrm{SL}_2(\mathbb{C})$ acts on the vector space V of quadratic binary forms over $\mathbb{C}$ by linear substitution. The induced action on the space of coordinates gives rise to an action of $\mathrm{SL}_2(\mathbb{C})$ on the graded symmetric algebra

$$\mathbb{C}[a, b, c] = \mathbb{C}[V] = S(V^*) = \bigoplus_{d \geq 0} S^d V^*.$$

The discriminant $b^2 - ac \in \mathbb{C}[a, b, c]$ is an $\mathrm{SL}_2(\mathbb{C})$-*invariant*, or simply an *invariant*, under this action.

Historically, a two-part paper of Boole [Boo42] is often regarded as the beginning of invariant theory as a discipline of its own. Motivated by the work of Lagrange and others, Boole referred to the "transformation of homogeneous functions under linear substitutions" as an "important and oft-recurring problem". Without explicit restrictions on the coefficients, he rediscovered (1) and extended it to forms of any degree in any number of variables (without giving a general proof).

Influenced by Boole, Cayley was the first to formulate the fundamental problems of invariant theory. In the second of two papers on "linear transformations" [Cay45; Cay46], Cayley proposed "to find all derivatives of any number of functions, which have the property of preserving their form unaltered after any linear transformation of the variables". He referred to "those derivatives with the property just enunciated" as *hyperdeterminants* and realized that "there remains a question to be resolved, which appears to present very great

[1] Gauss himself referred to $b^2 - ac$ as the *determinant* of F. The name *discriminant* was coined 50 years later by Sylvester [Syl51a, p. 406] "because it affords the *discrimen* or test for ascertaining whether or not equal factors enter into a function of two variables, or more generally of the existence or otherwise of multiple points in the locus represented or characterized by any algebraical function, the most obvious and first observed species of singularity in such function or locus."

difficulties, that of determining the *independent* derivatives, and the relation between these and the remaining ones".

With regard to terminology, Sylvester [Syl51b, p. 290/291] chose to rename "Mr Cayley's hyperdeterminants" generally as covariants[2], and as invariants if they are "constant": "Accordingly an invariant to a given form may be defined to be such a function of the coefficients of the form, as remains absolutely unaltered when instead of the given form any linear equivalent thereto is substituted. Of course if the determinant of the coefficients of the transformations correspondent to the respective equivalents be not taken unity as supposed in this definition, the effect will be merely to introduce as a multiplier some power of the determinant formed by the coefficients of transformation."

While Cayley's fundamental problems above were formulated from the computational point of view, they naturally raised theoretical questions[3]. Considering forms of a fixed degree in a fixed number of variables, and using more of the mathematical terminology introduced by Sylvester, these questions are:

A. Do there always exist finitely many "fundamental invariants"[4] from which all invariants can be derived as polynomials in the fundamental ones?
B. Do there always exist finitely many generators for the "syzygies"[5], that is, the polynomial relations on the fundamental invariants?

Cayley and Sylvester themselves were mainly interested in the explicit calculation of invariants and covariants of small degree, with particular focus on binary forms, and often "without the necessary theoretical underpinnings" [Par89]. Their computational methods include *Cayley's Ω-process* which finds invariants of some degree by applying a differential operator to arbitrary polynomials of that degree. An important preprocessing step was the enumeration of invariants. More precisely, having forms of a given degree in a given number of variables in mind, let a_d denote the maximal number of linearly independent invariants of degree d, for each d. Considering the counting function $d \mapsto a_d$ and its generating function $\sum_{d \geq 0} a_d t^d$, Cayley and Sylvester aimed at finding a rational function in t which represents the generating function. For example, for quadratic binary forms, the generating function is

$$\frac{1}{1-t^2} = 1 + t^2 + t^4 + \dots .$$

Equivalently, the discriminant $\theta = b^2 - ac$ is the sole fundamental invariant (up to scalar). In modern language,

---

[2] See [Stu08; Olv99] for covariants. We will not discuss covariants here.

[3] See Crilly's remarks [Cri86, p. 248/249] for Cayley's early thoughts with regard to quintic binary forms.

[4] See Sylvester [Syl51a, p. 405].

[5] See Sylvester [Syl50, p. 276].

$$\mathbb{C}[V]^{\mathrm{SL}_2(\mathbb{C})} = \mathbb{C}[\theta].$$

See [SF79] and [Syl81] for extensive lists of generating functions (with some small errors).

In Germany, Aronhold, Clebsch, and Gordan developed the *symbolic method*, which became the main tool for both computing and presenting invariants (see Dieudonné *et al.* [DC70], Kung *et al.* [KR84], Sturmfels [Stu08], Olver [Olv99], and Cröni [Crö02]). In fact, as Meyer [Mey92] puts it, the symbolic method "... [gave] birth to a new era, recognizable in that our Science [took] on a systematic development which had, from that moment on, its center in Germany."[6] The highlight of the first three decades of this new era was Gordan's finiteness theorem [Gor68] which answered question (A) for binary forms in the affirmative. In fact, Gordan gave a constructive proof based on the symbolic method which, from a practical point of view, allowed the computation of fundamental invariants in a very small number of cases.

Classical invariant theory culminated in two landmark papers of Hilbert [Hil90; Hil93] in which Hilbert introduced stunning new ideas which allowed him to extend Gordan's result to forms of any degree in any number of variables while at the same time answering question (B) in the affirmative.

The first paper consists of five parts and presents, correspondingly, five theorems. The first four theorems state general results whose importance goes far beyond their application in invariant theory. Theorem I, the starting point of it all, is Hilbert's basis theorem for polynomial rings over fields. In Part II, included as an interlude to allow for "number theoretic considerations", Hilbert proved the basis theorem for polynomial rings over the integers. Part III, connecting back to Part I, contains a proof of Hilbert's syzygy theorem, a finiteness theorem for the chain of successive syzygies on the generators of an ideal in a polynomial ring over a field (or, more generally, of a finitely generated module over such a polynomial ring). In Part IV, still working over fields, Hilbert studied the aforementioned counting function $d \mapsto a_d$ for quotients of graded polynomial rings by homogeneous ideals, and showed that this function, which is nowadays called the *Hilbert function*, is of polynomial nature (existence of the *Hilbert polynomial*). The proof is based on the syzygy theorem and allows one also to show that the corresponding generating function, nowadays called the *Hilbert series*, can indeed be always written as a rational function (*Hilbert–Serre theorem*). The final theorem, Theorem V, is the desired finiteness result for fundamental invariants of forms. Its proof relies on the basis theorem and a theoretical application of the Ω-process as a means of projecting arbitrary polynomials of some degree onto invariants of that degree.

Hilbert closed the first paper with a number of comments:

- Theorem V applies in particular to covariants and other "invariant formations" as "these formations all fall under the [...] term invariant."

[6] The English translation is taken from [Par89].

- Theorem V is also valid if we consider polynomials which are invariant under the elements of a subgroup of linear transformations rather than being invariant under all linear transformations[7] – provided there is some equivalent of the Ω-process which allows one to generate such invariants. That is, Hilbert actually worked with linearly reductive groups, although the name of this notion has only been coined later.
- As Theorems I and III apply in particular to invariants, it follows that "every finite system of invariants has only a finite number of irreducible syzygies [...] from which every other syzygy can be derived [...]."

Being criticized for his nonconstructive approach (see also Section 5), Hilbert quickly offered a constructive proof, published in the second paper. In addition to the basis theorem and the Ω-process, the proof relies on new results, namely the *Noether normalization theorem (graded case)*, *Hilbert's Nullstellensatz*, and the *Hilbert–Mumford criterion* for *Null-forms*.

Hilbert's results represent the beginning of modern algebra. However, as Dieudonné and Carell [DC70] put it: "Hilbert's success also spelled the doom of XIXth Century invariant theory, which was left with no big problems to solve and soon faded into oblivion." In the 20th century, however, "like the phoenix [invariant theory] has been again and again rising from its ashes". See Dieudonné and Carell [DC70] for remarks on post-Hilbert invariant theory, with particular emphasis on question (A) in full generality (Hilbert's 14th problem), Nagata's counter example to this problem, and Mumford's geometric invariant theory.

Nowadays, with powerful computers and new algorithms available, and with new application areas, invariant theory "is once again at the forefront of mathematics" ([KR84]). See Derksen and Kemper [DK15] for more on this.

2 The Basic Setup

In this chapter, our setting consists of a group G, a field K, a vector space V over K of finite dimension n, and a representation $\rho : G \to \mathrm{GL}(V)$ of G on V. The representation ρ gives a left action of G on V via

$$G \times V \to V,\ (\pi, v) \mapsto \rho(\pi)(v).$$

The induced right action on the dual vector space V^*,

$$V^* \times G \to V^*,\ (f, \pi) \mapsto f\,.\,\pi := f \circ \rho(\pi),$$

extends to a right action of G on the graded symmetric algebra

[7] Our main focus in this chapter is on *finite* groups of linear transformations.

$$K[V] := S(V^*) = \bigoplus_{d\geq 0} S^d V^*$$

such that the grading is preserved. The *invariants* of G are the fixed points of this action, its *invariant ring* is the graded subalgebra

$$K[V]^G := \{f \in K[V] \mid f\,.\,\pi = f \text{ for all } \pi \in G\} \subseteq K[V].$$

Remark 1 When defining the action of G on $K[V]$, one has to choose between a left and a right action. In OSCAR, group actions are by convention assumed to be right actions and we follow this convention with our definition above. However, we note that the left action given by $\pi\,.\,f := f \circ \rho(\pi^{-1})$ is quite common in the literature.

Explicitly, fixing a basis of V and its dual basis, say, $\{x_1, \ldots, x_n\}$ of V^*, we may identify $\mathrm{GL}(V) \cong \mathrm{GL}_n(K)$ and $K[V] \cong K[x_1, \ldots, x_n]$. Then the action of an element $\pi \in G$ with $\rho(\pi) = (a_{i,j})$ on a polynomial $f \in K[x_1, \ldots, x_n]$ is given as follows:

$$(f\,.\,\pi)(x_1, \ldots, x_n) = f\Big(\sum_j a_{1,j}x_j, \ldots, \sum_j a_{n,j}x_j\Big).$$

Accordingly, $K[V]^G$ may be regarded as a graded subalgebra of $K[x_1, \ldots, x_n]$:

$$K[V]^G \cong K[x_1, \ldots, x_n]^G := \{f \in K[x_1, \ldots, x_n] \mid f\,.\,\pi = f \text{ for any } \pi \in G\}.$$

Example 2 The symmetric group S_n acts on $K[x_1, \ldots, x_n] \cong K[V]$ by permuting the variables. The invariants under this action are the symmetric polynomials. The fundamental theorem on symmetric polynomials says that the invariant ring $K[V]^{S_n}$ is a polynomial algebra in the elementary symmetric polynomials. Using OSCAR, we rediscover a special case of this result:

```
julia> RS4 = invariant_ring(GF(2), symmetric_group(4))
Invariant ring
  of Sym(4)

julia> L = fundamental_invariants(RS4)
4-element Vector{MPolyDecRingElem{FqFieldElem, FqMPolyRingElem}}:
 x[1] + x[2] + x[3] + x[4]
 x[1]*x[2] + x[1]*x[3] + x[2]*x[3] + x[1]*x[4] + x[2]*x[4] + x[3]*x[4]
 x[1]*x[2]*x[3] + x[1]*x[2]*x[4] + x[1]*x[3]*x[4] + x[2]*x[3]*x[4]
 x[1]*x[2]*x[3]*x[4]

julia> is_algebraically_independent(L)
true
```

Remark 3 (The role of the ground field) From a theoretical point of view, in the context of linearly reductive groups or in the geometric context of Subsection 4.3, we will tacitly assume that our ground field is algebraically closed[8]. From a practical point of view, however, we will work over an exact field which can be handled by OSCAR. This is not a problem: The algorithms we will use rely on Gröbner bases, linear algebra, and means to compute Reynolds operators (see below). Hence they do not change the initial ground field. Hence, as long as the ground field contains all elements needed to set up the desired group representation, this field is big enough and computations over any extension of it would lead to the same results.

Remark 4 In our computational applications, we will only consider cases where $K[V]^G$ is finitely generated as a K-algebra. Then $K[V]^G$ can be represented by finitely many generators with finitely many syzygies. That is, it can be written as the quotient of a polynomial ring over K by an ideal of that polynomial ring. We then say that $K[V]^G$ is represented as an *affine algebra.*

3 Some More Terminology

3.1 Linearly Reductive Groups

The definition of a Reynolds operator below reflects the use of the Ω-process in Hilbert's papers as discussed in the introduction. In fact, Hilbert's 1890 finiteness proof applies whenever a Reynolds operator exists. Using the terminology introduced in this section, it applies to rational representations of linearly reductive groups.

Definition 5 With notation as in Section 2, a *Reynolds operator*[9] (for G acting on V) is a K-linear map $\mathcal{R} : K[V] \to K[V]$ such that:

1. $\mathcal{R} \circ \mathcal{R} = \mathcal{R}$,
2. $\mathcal{R}(f) = f$ for all $f \in K[V]^G$,
3. $\mathcal{R}(f\,.\,\pi) = \mathcal{R}(f)$ for all $f \in K[V]^G$, $\pi \in G$.

Remark 6 It easily follows from the definition that the image of $\mathcal{R}$ is $K[V]^G$. That is, we may think of a Reynolds operator as a K-linear projection onto the ring of invariants.

For a finite group G, we speak of *non-modular* invariant theory if the order $|G|$ is invertible in K, and of *modular* invariant theory otherwise. In the former case, a Reynolds operator is obtained by averaging over the group:

$$\mathcal{R} : K[V] \to K[V],\ f \mapsto \frac{1}{|G|} \sum_{\pi \in G} (f\,.\,\pi).$$

[8] Moreover, in 4.3, the characteristic of the ground field will be different from 2, 3.

[9] The name Reynolds operator was coined to honour Osbourne Reynolds who introduced a corresponding linear operator in fluid dynamics [Rey95].

Example 7 For the symmetric group S_3, we get:

```
julia> RS3 = invariant_ring(QQ, symmetric_group(3));

julia> R = polynomial_ring(RS3)
Multivariate polynomial ring in 3 variables over QQ graded by
  x[1] -> [1]
  x[2] -> [1]
  x[3] -> [1]

julia> x = gens(R);

julia> reynolds_operator(RS3, x[1]^2)
1//3*x[1]^2 + 1//3*x[2]^2 + 1//3*x[3]^2
```

On our way to define linearly reductive groups we recall that the general linear group $\mathrm{GL}_m(K)$ is an affine variety with coordinate ring

$$K[\mathrm{GL}_m(K)] \cong K[x_{ij}, \det(x_{ij})^{-1}] \cong K[x_{ij}, z]/(z \cdot \det(x_{ij}) - 1).$$

A subgroup $G \subseteq \mathrm{GL}_m(K)$ is called a *linear algebraic group* if G is closed in the Zariski topology of $\mathrm{GL}_m(K)$. Then G is itself an affine variety with coordinate ring $K[G] = K[\mathrm{GL}_m(K)]/I$, where I is the defining ideal of G in $K[\mathrm{GL}_m(K)]$.

Let G be a linear algebraic group, and let $\rho : G \to \mathrm{GL}(V) \cong \mathrm{GL}_n(K)$ be a representation of G as before. Then ρ is called *rational* if ρ is also a morphism of affine varieties.

Remark 8 The concept of invariant rings as introduced in Section 2 extends to coordinate rings of affine G-varieties (see [Spr89, Section I.3]). In particular, given a linear algebraic group G, consider the rational action of G on itself by left multiplication and the induced right action $(f, \pi) \mapsto f \,.\, \pi$ of G on $K[G]$ with $(f \,.\, \pi)(g) := f(\pi \cdot g)$ for all $g \in G$. Then the invariant ring $K[G]^G$ is isomorphic to K via the map $f \mapsto f(e_G)$.

Definition 9 A linear algebraic group G is said to be *linearly reductive* if the following holds for each rational representation $\rho : G \to \mathrm{GL}(V)$: If $W \subseteq V$ is a G-invariant subspace, then there exists a G-invariant complement, that is, a G-invariant subspace W' such that $V = W \oplus W'$.

The following result shows three of many possible characterizations of linearly reductive groups (see [Spr89, Section II.1], [DK15, Section 2.2.1]):

Theorem 10 *Let $G \subseteq \mathrm{GL}_m(K)$ be linear algebraic. Then the following are equivalent:*

1. *G is linearly reductive.*
2. *Each rational representation $\rho : G \to \mathrm{GL}(V)$ is* fully reducible, *that is, V is a direct sum of irreducible G-invariant subspaces.*
3. *For each rational representation $\rho : G \to \mathrm{GL}(V)$ there exists a unique Reynolds operator $\mathcal{R} : K[V] \to K[V]^G$.*

4. There exists a unique K-linear map $\mathcal{R}_G : K[G] \to K[G]^G \cong K$ such that $\mathcal{R}_G(1) = 1$ and $\mathcal{R}_G(f\,.\,\pi) = \mathcal{R}_G(f)$ for all $f \in K[G]^G$, $\pi \in G$.

Remark 11 Each Reynolds operator as in part 3 is graded and a $K[V]^G$-module homomorphism (see [DK15, Corollary 2.2.7]).

Remark 12 In view of Definition 5 and Remark 6, the linear map $\mathcal{R}_G$ in part 4 of Theorem 10 is called the *Reynolds operator on* $K[G]$.

Remark 13 The following groups can naturally be considered as linear algebraic groups. As such, they all are linearly reductive groups:

1. Finite groups in the non-modular case (*Maschke's theorem* [Mas99]); see also [Spr89, Section II.2.1].
2. Tori $(K^*)^m$; see, for example, [Spr89, Section II.2.2].
3. The classical groups over $K = \mathbb{C}$; see Kraft [Kra85].

Remark 14 An explicitly known Reynolds operator is a useful tool towards computing explicit generators of an invariant ring. For example, somewhat naively and inefficiently, we could proceed degree by degree, using the Reynolds operator to find K-bases of invariants of each given degree, and stop as soon as a theoretically known degree bound is reached.

Remark 15 Invariant rings arising from rational representations of linearly reductive groups are Cohen–Macaulay by a theorem of Hochster and Roberts [HR74]. That is, each such ring is a free module over any of its Noether normalizations.

3.2 Primary and Secondary Invariants

Taking up our introductory remarks on Hilbert's constructive finiteness proof, we recall that Hilbert first introduced the concept of a (graded) Noether normalization. Then he described means to compute a Noether normalization $K[p_1, \dots, p_k] \subseteq K[V]^G$ with $p_1, \dots, p_k$ homogeneous[10], and presented methods to find homogeneous generators of $K[V]^G$ as a module over $K[p_1, \dots, p_k]$. It is worth noting that the first modern algorithms for computing invariant rings of finite groups used such a two-step strategy as well. See [Stu08], [DK15], and Section 4 below for more details.

Definition 16 A system of homogeneous invariants $p_1, \dots, p_k$ as above is called a *system of primary invariants* for $K[V]^G$. A minimal system $s_0 = 1, s_1, \dots, s_l$ of homogeneous generators of $K[V]^G$ as a $K[p_1, \dots, p_k]$-module is called a *system of secondary invariants* for $K[V]^G$ (with respect to $p_1, \dots, p_k$).

[10] In other words, $p_1, \dots, p_k$ form a *homogeneous system of parameters* for $K[V]^G$.

4 Invariants of Finite Groups

That invariant rings of finite groups are finitely generated as K-algebras was shown by Emmy Noether [Noe26] with a nonconstructive proof (as outlined in the previous section, in the non-modular case, this can also be shown using the Reynolds operator). As a first step of her proof, Noether showed that if G is a finite group, then $K[V] \cong K[x_1, \ldots, x_n]$ is integral over $K[V]^G$ (see also Smith [Smi95]). For our purposes, we remark that this implies that every system of primary invariants consists of precisely n elements.

Remark 17 In the non-modular case, invariant rings of finite groups are Cohen–Macaulay by the general theorem of Hochster and Roberts (see Remark 15). For a proof directly addressing the special case considered here see Stanley [Sta79] .

In the remaining part of this section, G will always be a finite group.

4.1 Creating Invariant Rings of Finite Groups in OSCAR

The invariant theory part of OSCAR distinguishes two ways of how finite groups and their actions on $K[V] \cong K[x_1, \ldots, x_n]$ are specified:

Matrix Groups

Here, G will be explicitly given as a matrix group $G \subseteq \mathrm{GL}_n(K) \cong \mathrm{GL}(V)$ by (finitely many) generating matrices. That is, G is identified with its image under a faithful representation and the action on $K[V]$ is given as at the beginning of Section 2.

Permutation Groups

Here, G will be given as a permutation group, acting on $K[x_1, \ldots, x_n] \cong K[V]$ by permuting the variables.

The constructor for invariant rings is designed accordingly:

```
invariant_ring(G::MatrixGroup)
invariant_ring(K::Field = QQ, G::PermGroup)
```

When setting up an invariant ring as indicated above, OSCAR will automatically create an underlying polynomial ring to work with. It is often more convenient, however, to work with a user defined polynomial ring:

```
invariant_ring(R::MPolyDecRing, G::MatrixGroup))
invariant_ring(R::MPolyDecRing, G::PermGroup)
```

Example 18 We create the invariant ring of the Heisenberg group of level 3 in its Schrödinger representation:

```
julia> K, a = cyclotomic_field(3, "a");

julia> S = matrix(K, [0 0 1; 1 0 0; 0 1 0])
[0   0   1]
[1   0   0]
[0   1   0]

julia> T = matrix(K, [1 0 0; 0 a 0; 0 0 -a-1])
[1   0        0]
[0   a        0]
[0   0   -a - 1]

julia> H3 = matrix_group(S, T)
Matrix group of degree 3
  over cyclotomic field of order 3

julia> order(H3)
27

julia> exponent(H3)
3

julia> is_abelian(H3)
false

julia> R, (x, y, z) = graded_polynomial_ring(K, ["x", "y", "z"]);

julia> RH3 = invariant_ring(R, H3);
```

Example 19 The cyclic group G of order 4 acts on $\mathbb{F}_2[x_1, \ldots, x_4]$ by cyclic permutations of the variables. We create its invariant ring:

```
julia> G = permutation_group(4, [ cperm([1, 2, 3, 4]) ]);

julia> RG = invariant_ring(GF(2), G);
```

Remark 20 In OSCAR, the creation of invariant rings is lazy in the sense that no explicit computations are done until specifically invoked (for example, by the `fundamental_invariants` function).

4.2 The Molien Series

As indicated in the introduction, precomputing the Hilbert series is a crucial step in the design of algorithms in invariant theory. For finite groups, this step is based on Molien's formula [Mol97]. See also [DK15].

The computation below shows, for example, that the invariants of the Heisenberg group H3 in degree 3 admit a K-basis of two invariants.

Example 21 For the Heisenberg group H3 from Example 18, we get:

```
julia> MSH3 = molien_series(RH3)
(-t^6 + t^3 - 1)//(t^9 - 3*t^6 + 3*t^3 - 1)

julia> expand(MSH3, 10)
1 + 2*t^3 + 4*t^6 + 7*t^9 + O(t^11)
```

Example 22 We consider the regular representation of the Klein four-group in characteristic zero:

```
julia> A = matrix(QQ, [0 1 0 0; 1 0 0 0; 0 0 0 1; 0 0 1 0])
[0   1   0   0]
[1   0   0   0]
[0   0   0   1]
[0   0   1   0]

julia> B = matrix(QQ, [0 0 1 0; 0 0 0 1; 1 0 0 0; 0 1 0 0])
[0   0   1   0]
[0   0   0   1]
[1   0   0   0]
[0   1   0   0]

julia> K4 = matrix_group(A, B)
Matrix group of degree 4
  over rational field

julia> small_group_identification(K4)
(4, 2)

julia> describe(K4)
"C2 x C2"

julia> RK4 = invariant_ring(K4);

julia> MSK4 = molien_series(RK4)
(t^2 - t + 1)//(t^6 - 2*t^5 - t^4 + 4*t^3 - t^2 - 2*t + 1)

julia> expand(MSK4, 4)
1 + t + 4*t^2 + 5*t^3 + 11*t^4 + O(t^5)
```

4.3 Invariants of a Given Degree

In the non-modular case, a K-basis for the invariants of a given degree d can be computed by minimizing the generating set obtained from applying the Reynolds operator to the monomials of degree d. A method which applies in both the non-modular and modular case is based on linear algebra. In OSCAR, if no method is specified by the user, a heuristic approach to selecting one of the two methods is applied.

Example 23 As is clear from the Molien series printed in Example 21, the invariants of the Heisenberg group H3 in degree 3 have a basis formed by two invariants. Explicitly:

```
julia> P = basis(RH3, 3)
2-element Vector{MPolyDecRingElem{AbsSimpleNumFieldElem,
↪  AbstractAlgebra.Generic.MPoly{AbsSimpleNumFieldElem}}}:
 x*y*z
 x^3 + y^3 + z^3
```

From a geometric point of view, we see two curves in the projective plane, the Fermat cubic $F : x^3+y^3+z^3 = 0$ and its Hessian curve[11] $H_F : xyz = 0$. These curves generate the Hessian pencil of plane elliptic curves (see Figure 1):

$$\lambda(x^3 + y^3 + z^3) + \mu xyz = 0.$$

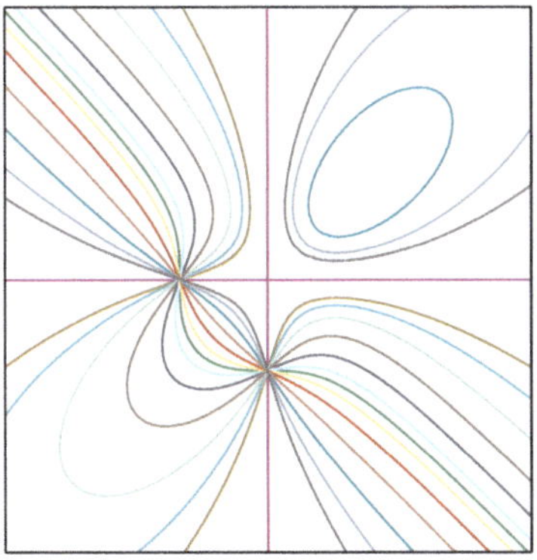

Fig. 1: Curves in the Hessian Pencil.

The Hessian pencil has 9 distinct base points:

```
julia> MP = minimal_primes(ideal(R, P));

julia> length(MP)
9

julia> [[ kernel(reduce(hcat, [ K[coeff(p, x); coeff(p, y); coeff(p, z)]
↪  for p in gens(Q) ])) for Q in MP ]]
1-element Vector{Vector{AbstractAlgebra.Generic.MatSpaceElem{AbsSimpleNu⌋
↪  mFieldElem}}}:
 [[a+1 1 0], [-a 1 0], [-1 1 0], [a+1 0 1], [-a 0 1], [-1 0 1], [0 -a 1],
 ↪  [0 a+1 1], [0 -1 1]]
```

[11] Let $C : f = 0$ be a smooth projective plane curve over an algebraically closed field, defined by a squarefree homogeneous polynomial f of degree $d \geq 3$. Then the determinant H_f of the Hessian matrix of f is a homogeneous polynomial of degree $3(d-2) \geq 3$. Suppose that the ground field is of characteristic zero or $> d$. Then $H_f \neq 0$, so that $H_C : H_f = 0$ is a curve in the projective plane, the *Hessian curve* of C. By Bezout's theorem, C and H_C have $3(d-2)d$ intersection points, counted with multiplicity. These are precisely the inflection points of C.

The 9 distinct base points are precisely the inflection points of each smooth curve E in the pencil. Indeed, since the pencil is generated by the Fermat curve and its Hessian curve, the Hessian curve of E is in the pencil, too:

```
julia> S, s = QQ["t"]; T = fraction_field(S); t = T(s);

julia> RR, (X, Y, Z) = graded_polynomial_ring(T, [ "X", "Y", "Z"]);

julia> F = X^3+Y^3+Z^3;

julia> G = t*F+hessian(F)
t*X^3 + 216*X*Y*Z + t*Y^3 + t*Z^3

julia> L = syzygy_generators([hessian(G), F, hessian(F)]);

julia> collect(coefficients(L[1]))
3-element Vector{AbstractAlgebra.Generic.FracFieldElem{QQPolyRingElem}}:
 -1
 -279936*t
 t^3 + 93312
```

Some more of the classical geometry of the Hesse pencil will be revealed by studying the group structure of H3:

```
julia> C, iC = center(H3);

julia> gens(C)[1]
[-a - 1         0         0]
[     0    -a - 1         0]
[     0         0    -a - 1]

julia> Q, pQ = quo(H3, C)
(Pc group of order 9, Hom: H3 -> Q)

julia> describe(Q)
"C3 x C3"
```

We see that H3 is a central extension of type

$$1 \to \mu_3 \to \text{ H3 } \to \mathbb{Z}/3\mathbb{Z} \times \mathbb{Z}/3\mathbb{Z} \to 0, \tag{1}$$

where $\mu_3 = \{a^k E_3 \mid k \in \mathbb{Z}/3\mathbb{Z}\}$ is a copy of the group of third roots of unity.

The four subgroups of $\mathbb{Z}/3\mathbb{Z} \times \mathbb{Z}/3\mathbb{Z}$ of order 3 give rise to twelve subgroups of H3 of order 3 divided up into four conjugacy classes:

```
julia> SSG = filter(cls -> order(representative(cls)) == 3 && length(cls)
↪   == 3, subgroup_classes(H3));

julia> length(SSG)
4
```

Using the `basis` function, we see that each of the twelve subgroups has precisely one linear invariant (up to scalar). Geometrically, this leads to twelve lines in the projective plane. Arranging these lines according to conjugacy classes, we get four triangles. For example:

```
julia> L1 = [basis(invariant_ring(R, H), 1)[1] for H in collect(SSG[1])]
3-element Vector{MPolyDecRingElem{AbsSimpleNumFieldElem,
↪  AbstractAlgebra.Generic.MPoly{AbsSimpleNumFieldElem}}}:
 x + y + z
 x + (-a - 1)*y + a*z
 x + a*y + (-a - 1)*z

julia> Triangle1 = L1[1]*L1[2]*L1[3]
x^3 - 3*x*y*z + y^3 + z^3
```

The four triangles are the singular members of the Hesse pencil, each singular member intersects each other member in the nine base points of the pencil. Altogether, we have rediscovered the celebrated $(9_4, 12_3)$ Hessian configuration of nine points and twelve lines, with four lines through each point, and three points on each line. See [AD09] and [Hul86] for more on this.

4.4 Primary Invariants

Recall from Section 3 that we call a homogeneous system of parameters of the graded algebra $K[V]^G$ a system of *primary invariants.* Primary invariants are of course not unique and neither are their degrees. However, there are certain constraints on the degrees of a system of primary invariants $p_1, \ldots, p_n \in K[V]^G$ which we can read off from the Hilbert series of $K[V]^G$. If $d_1, \ldots, d_n$ are the degrees of $p_1, \ldots, p_n$, then the Hilbert series of the polynomial algebra $K[p_1, \ldots, p_n]$ is given by

$$H(K[p_1, \ldots, p_n], t) = \frac{1}{(1 - t^{d_1}) \cdots (1 - t^{d_n})}.$$

Using the syzygy theorem as indicated in the introduction, we see that the Hilbert series of $K[V]^G$ can be written as a rational function of type

$$H(K[V]^G, t) = \frac{h(t)}{(1 - t^{d_1}) \cdots (1 - t^{d_n})},$$

for some polynomial $h(t) \in \mathbb{Z}[t]$. From this it follows that the product $d_1 \cdots d_n$ is divisible by $|G|$, see [DK15, Proposition 3.5.5].

Especially in computational applications, we are interested in finding primary invariants with "optimal" degrees. Put more precisely, we wish to minimize the degree vector $(d_1, \ldots, d_n) \in \mathbb{Z}^n_{>0}$ with respect to the ordering on $\mathbb{Z}^n_{>0}$ obtained by sorting first the products $d_1 \cdots d_n$, then sorting the sums $d_1 + \cdots + d_n$, and finally by taking the lexicographic ordering on $\mathbb{Z}^n$ as a tie breaker. This ordering ensures that the number of corresponding secondary invariants and their degrees are (at least in the non-modular case) as small as possible. The reason for this is the following. As the ring $K[V]^G$ is Cohen–Macaulay in the non-modular case, we can write the Hilbert series as

$$H(K[V]^G, t) = \frac{1 + n_1 t^{e_1} + \cdots + n_m t^{e_m}}{(1 - t^{d_1}) \cdots (1 - t^{d_n})}, \tag{2}$$

where $d_1, \ldots, d_n$ are the degrees of a fixed system of primary invariants, $e_1, \ldots, e_m \in \mathbb{Z}_{>0}$ are distinct exponents, and $n_1, \ldots, n_m \in \mathbb{Z}_{>0}$ are positive integers, see [DK15, Section 3.7.1]. We see that we need $(d_1 \cdots d_n)/|G|$ generators other than 1 for $K[V]^G$ as a $K[p_1, \ldots, p_n]$-module and that there are exactly n_i such generators in degree e_i, for each i. Hence, keeping $d_1 \cdots d_n$ small results in a lower number of secondary invariants and keeping $d_1 + \cdots + d_n$ small means that these invariants will be of lower degree.

An algorithm by Kemper [Kem99] finds primary invariants with optimal degrees. For this, the algorithm iterates, in the above ordering, over the degree vectors which fit the constraints enforced by (2), checking each time whether a system of primary invariants with the given degree vector exists. We show the OSCAR implementation of this algorithm at work by considering the Klein four-group both in the non-modular and the modular case.

Example 24 Recall from Example 22 that for the regular representation of the Klein four-group G in characteristic 0, we obtain the Hilbert series

$$H(K[V]^G, t) = \frac{t^2 - t + 1}{t^6 - 2t^5 - t^4 + 4t^3 - t^2 - 2t + 1}.$$

As the group is of order 4, the product of the degrees of the elements of any system of primary invariants is divisible by 4. According to the described ordering, we now first consider tuples $(d_1, \ldots, d_4)$ with $d_1 \cdots d_4 = 4$ and start with $(1, 1, 2, 2)$. Here we see that

$$H(K[V]^G, t) \cdot (1 - t)^2 (1 - t^2)^2 = t^2 - t + 1,$$

so we conclude with (2) that there are no primary invariants with these degrees (the coefficient of t on the right hand side is negative). An analogous argument rules out the tuple $(1, 1, 1, 4)$. We hence have to consider tuples $(d_1, \ldots, d_4)$ with $d_1 \cdots d_4 = 8$. Here, the tuple with the smallest sum of entries is $(1, 2, 2, 2)$. Checking the Hilbert series, we get

$$H(K[V]^G, t) = \frac{t^3 + 1}{(1 - t)(1 - t^2)^3}.$$

So the tuple $(1, 2, 2, 2)$ looks promising. In fact, Kemper's algorithm returns a system of primary invariants with exactly these (optimal) degrees.

```
julia> primary_invariants(RK4)
4-element Vector{MPolyDecRingElem{QQFieldElem, QQMPolyRingElem}}:
```

```
 x[1] + x[2] + x[3] + x[4]
 x[1]^2 + x[2]^2 + x[3]^2 + x[4]^2
 x[1]*x[2] + x[3]*x[4]
 x[1]*x[3] + x[2]*x[4]
```

Considering the first two polynomials above in characteristic 2, that is, in the modular case, we notice that $x_1^2+x_2^2+x_3^2+x_4^2$ is the square of $x_1+x_2+x_3+x_4$. Hence, we expect a different system of primary invariants in this characteristic. Indeed, we compute over $\mathbb{F}_2$:

```
julia> A = matrix(GF(2), [0 1 0 0; 1 0 0 0; 0 0 0 1; 0 0 1 0]);

julia> B = matrix(GF(2), [0 0 1 0; 0 0 0 1; 1 0 0 0; 0 1 0 0]);

julia> RK42 = invariant_ring(matrix_group(A, B));

julia> primary_invariants(RK42)
4-element Vector{MPolyDecRingElem{FqFieldElem, FqMPolyRingElem}}:
 x[1] + x[2] + x[3] + x[4]
 x[2]*x[3] + x[1]*x[4]
 x[1]*x[2] + x[1]*x[3] + x[2]*x[4] + x[3]*x[4]
 x[1]*x[2]*x[3]*x[4]
```

We see that now the primary invariants are of degrees $(1, 2, 2, 4)$. This means that in this case, there are no primary invariants of degrees $(1, 2, 2, 2)$.

4.5 Secondary Invariants

Fixing a system of primary invariants $p_1, \ldots, p_n \in K[V]^G$ for G, we now turn to the computation of secondary invariants. That is, we wish to find elements $s_0 = 1, s_1, \ldots, s_m$ which generate $K[V]^G$ as a $K[p_1, \ldots, p_n]$-module. With regard to algorithms, we distinguish between the modular and the non-modular case. In the latter case, from taking a theoretical point of view, the ring structure of $K[V]^G$ is better understood. Most notably, $K[V]^G$ is Cohen–Macaulay as described above. We first discuss a general algorithm for the computation of secondary invariants which applies in both cases, but is in practice only used for the modular case. We will then present a more specialized approach that can only be used in the non-modular case.

In the general case, we have almost no a priori information on the structure of $K[V]^G$ as a $K[p_1, \ldots, p_n]$-module. However, Symonds' bound [Sym11] also holds in the modular case: there exist secondary invariants whose degrees are bounded by

$$\deg(s_i) \leq \sum_{j=1}^{n} (\deg(p_j) - 1).$$

To compute a system of secondary invariants, we set $s_0 = 1$ and compute bases B_d of the K-vector spaces $K[V]_d^G$ for $d = 1, \ldots, \sum_{j=1}^{n} (\deg(p_j) - 1)$. Then, to obtain a minimal generating system as required, we iteratively choose from each basis B_d a subset which, together with the products of already selected secondary invariants of lower degree with power products of the $p_1, \dots, p_n$, generate $K[V]^G_d$ minimally.

We now turn to the non-modular case, where we make extensive use of the fact that $K[V]^G$ is Cohen–Macaulay. We may further use Molien's theorem to find the Hilbert series of $K[V]^G$. Then we can read off the number of secondary invariants and their degrees from (2). Applying the graded Nakayama lemma, we see that elements $s_0, \dots, s_m \in K[V]^G$ are secondary invariants if and only if they generate $K[V]^G / K[p_1, \dots, p_n]_+ K[V]^G$ as a K-vector space. Since we already know the number m, this is equivalent to the s_i being linearly independent modulo $K[p_1, \dots, p_n]_+ K[V]^G$. One finally sees that there is an embedding

$$K[V]^G / K[p_1, \dots, p_n]_+ K[V]^G \hookrightarrow K[V]/(p_1, \dots, p_n)K[V]$$

and concludes that the s_i need to be linearly independent modulo the ideal $(p_1, \dots, p_n) \subseteq K[V]$. In terms of the algorithm, we iterate over the degrees indicated by (2), and given such a degree d, we produce a K-basis B_d of the invariants of degree d. We then iteratively add those elements of B_d to the set of secondary invariants that result in a set of linearly independent elements modulo $(p_1, \dots, p_n)$. Note that the latter condition can be checked using Gröbner-basis techniques, and that a Gröbner basis of $(p_1, \dots, p_n)$ is already known from the computation of the primary invariants.

Both algorithms are implemented in `OSCAR` and are automatically selected depending on the characteristic of K and the group order of G. For more details on the algorithms and relevant optimizations, we refer to [DK15, Section 3.7] and [Kin07].

Example 25 For the regular representation of the Klein four-group in characteristic zero, with primary invariants as computed in Example 24, we get:

```
julia> secondary_invariants(RK4)
2-element Vector{MPolyDecRingElem{QQFieldElem, QQMPolyRingElem}}:
 1
 x[1]^3 + x[2]^3 + x[3]^3 + x[4]^3
```

The result in characteristic 2 reads as follows:

```
julia> secondary_invariants(RK42)
4-element Vector{MPolyDecRingElem{FqFieldElem, FqMPolyRingElem}}:
 1
 x[1]*x[3] + x[2]*x[4]
 x[2]^2*x[3] + x[2]*x[3]^2 + x[1]^2*x[4] + x[1]*x[4]^2
 x[1]*x[2]^2*x[3]^2 + x[1]*x[2]*x[3]^3 + x[1]^3*x[3]*x[4] +
 ↪  x[2]^3*x[3]*x[4] + x[2]^2*x[3]^2*x[4] + x[1]^2*x[2]*x[4]^2 +
 ↪  x[1]^2*x[3]*x[4]^2 + x[1]*x[2]*x[4]^3
```

Computing *module syzygies* on the secondary invariants over $K[p_1, \dots, p_n]$ as in the next example, it can be seen that the invariant ring is Cohen–Macaulay also in characteristic 2 (modular case). Note that computing module syzygies is another application of Gröbner bases, see [GP08, Section 2.8.7].

Example 26 As an example of an invariant ring which is not Cohen–Macaulay, we discuss an example of Bertin [Ber67], see also [Kem96]. In fact, we come back to Example 19 and consider the cyclic group G of order 4 acting on $\mathbb{F}_2[x_1, \ldots, x_4]$ by cyclic permutations of the variables.

```
julia> G = permutation_group(4, [ cperm([1, 2, 3, 4]) ]);

julia> RG = invariant_ring(GF(2), G);
```

We compute the primary invariants

$$p_1 = x_1 + x_2 + x_3 + x_4, \quad p_2 = x_1x_3 + x_2x_4,$$
$$p_3 = x_1x_2 + x_2x_3 + x_3x_4 + x_4x_1, \quad p_4 = x_1x_2x_3x_4 :$$

```
julia> primary_invariants(RG)
4-element Vector{MPolyDecRingElem{FqFieldElem, FqMPolyRingElem}}:
 x[1] + x[2] + x[3] + x[4]
 x[1]*x[3] + x[2]*x[4]
 x[1]*x[2] + x[2]*x[3] + x[1]*x[4] + x[3]*x[4]
 x[1]*x[2]*x[3]*x[4]
```

We then obtain five corresponding secondary invariants:

$$s_0 = 1, \quad s_1 = x_1^2x_2 + x_2^2x_3 + x_3^2x_4 + x_1x_4^2,$$
$$s_2 = x_1x_2x_3 + x_1x_2x_4 + x_1x_3x_4 + x_2x_3x_4,$$
$$s_3 = x_1^2x_2x_3 + x_2^2x_3x_4 + x_1x_3^2x_4 + x_1x_2x_4^2,$$
$$s_4 = x_1^3x_2x_3 + x_2^3x_3x_4 + x_1x_3^3x_4 + x_1x_2x_4^3.$$

```
julia> secondary_invariants(RG)
5-element Vector{MPolyDecRingElem{FqFieldElem, FqMPolyRingElem}}:
 1
 x[1]^2*x[2] + x[2]^2*x[3] + x[3]^2*x[4] + x[1]*x[4]^2
 x[1]*x[2]*x[3] + x[1]*x[2]*x[4] + x[1]*x[3]*x[4] + x[2]*x[3]*x[4]
 x[1]^2*x[2]*x[3] + x[2]^2*x[3]*x[4] + x[1]*x[3]^2*x[4] +
 ↪  x[1]*x[2]*x[4]^2
 x[1]^3*x[2]*x[3] + x[2]^3*x[3]*x[4] + x[1]*x[3]^3*x[4] +
 ↪  x[1]*x[2]*x[4]^3
```

Let us now compute a presentation of $K[V]^G$ as a $K[p_1, \ldots, p_4]$-module.

```
julia> M, MtoR, StoR = module_syzygies(RG);

julia> M
Subquotient of Submodule with 5 generators
1 -> e[1]
2 -> e[2]
3 -> e[3]
4 -> e[4]
5 -> e[5]
```

```
by Submodule with 1 generator
1 -> t2*e[2] + (t2 + t3)*e[3] + t1*e[4]
```

We see that $K[V]^G$ is not a free $K[p_1, \dots, p_4]$-module: the last line in the code above translates to the relation

$$p_2 s_1 + (p_2 + p_3) s_2 + p_1 s_3 = 0.$$

This means that $K[V]^G$ is not Cohen–Macaulay.

4.6 Fundamental Invariants and Their Syzygies

By the graded Nakayama lemma, the number and the degrees of homogeneous polynomials which form a generating system of minimal cardinality of $K[V]^G$ are uniquely determined. Following Sylvester, we refer to such a system as a *system of fundamental invariants* for $K[V]^G$.

There are two approaches to computing fundamental invariants. The first, rather natural approach starts by computing primary and secondary invariants. Then, from the combined set of these invariants, it removes elements such as $s_0 = 1$ which are superfluous as K-algebra generators. The latter step starts by computing *irreducible secondary invariants*. These form a subset of the secondary invariants which generates $K[V]^G$ as an algebra over the subalgebra generated by the primary invariants. In practice, the computation of the irreducible secondary invariants is not a separate effort, but can (and should) be done directly, via clever book-keeping, while constructing the secondary invariants. The primary invariants and the irreducible secondary invariants are then reduced further by standard Gröbner bases techniques for checking containment in a (graded) subalgebra.

The second approach for the computation of fundamental invariants is an algorithm by King [Kin13]. This algorithm is only available in the non-modular case; it computes the fundamental invariants in a direct fashion, relying on degree truncated Gröbner bases as a key computational tool.

The two algorithms are available in OSCAR. In the non-modular case, where both approaches are applicable, the first approach has the advantage of giving more information on the structure of the invariant ring. King's algorithm, however, is in general expected to be faster, which is why it is the default choice to compute fundamental invariants in OSCAR.

Once we have fundamental invariants $f_1, \dots, f_k \in K[V]^G$ on hand, we may wish to compute a full presentation of $K[V]^G$ as an affine algebra. For this, we need to construct the *syzygies* on the generators $f_1, \dots, f_k$, that is, the kernel of the surjective morphism of K-algebras

$$K[y_1, \dots, y_k] \to K[V]^G, \; y_i \mapsto f_i. \tag{3}$$

The general technique to compute such a kernel uses Gröbner bases, see for example [GP08, Section 1.8.10]. If the fundamental invariants were computed using King's algorithm, this is to our knowledge the only applicable method.

If the fundamental invariants were constructed from primary and secondary invariants, however, we can use an algorithm by Kemper and Steel [KS99] which relies on linear algebra instead of Gröbner bases. This algorithm works in the finite-dimensional K-vector spaces given by the graded components of $K[V]^G$ over the algebra generated by the primary invariants. In the modular case, one needs to take extra care with this approach as $K[V]^G$ might not be Cohen–Macaulay and one hence needs to compute the module syzygies on the secondary invariants over $K[p_1, \ldots, p_n]$.

Example 27 In coding theory, invariant polynomials appear as possible weight polynomials of self-dual codes. We consider the invariant ring of a group of order 192, which is connected to the weight polynomials of self-dual codes in characteristic 2 with weights divisible by 4, see [Gle71] for details.

We work over the field $K := \mathbb{Q}(\sqrt{2}, i)$, where i is the imaginary unit. We construct this field in OSCAR as a non-simple number field, see the chapter on number theory [FH24] (cf. page 81).

```
julia> Qt, t = QQ["t"];

julia> K, (a, b) = number_field([t^2 - 2, t^2 + 1], ["a", "b"]);
```

The group $G \leq \mathrm{GL}_2(K)$ in question is of order 192 and generated by the matrices

$$\frac{1}{\sqrt{2}}\begin{pmatrix}1 & 1\\ 1 & -1\end{pmatrix} \text{ and } \begin{pmatrix}1 & 0\\ 0 & i\end{pmatrix}.$$

```
julia> M1 = 1/a*matrix(K, [1 1; 1 -1])
[1//2*a    1//2*a]
[1//2*a   -1//2*a]

julia> M2 = matrix(K, [1 0; 0 b])
[1   0]
[0   b]

julia> G = matrix_group(M1, M2);

julia> order(G)
192
```

We compute the fundamental invariants:

```
julia> RG = invariant_ring(G);

julia> f = fundamental_invariants(RG)
2-element Vector{MPolyDecRingElem{AbsNonSimpleNumFieldElem,
↪  AbstractAlgebra.Generic.MPoly{AbsNonSimpleNumFieldElem}}}:
 x[1]^8 + 14*x[1]^4*x[2]^4 + x[2]^8
 x[1]^24 + 10626//1025*x[1]^20*x[2]^4 + 735471//1025*x[1]^16*x[2]^8 +
 ↪  2704156//1025*x[1]^12*x[2]^12 + 735471//1025*x[1]^8*x[2]^16 +
 ↪  10626//1025*x[1]^4*x[2]^20 + x[2]^24
```

So, we have one fundamental invariant of degree 8 and one of degree 24. From this, one can deduce that the invariant ring $K[x_1, x_2]^G$ is a polynomial algebra by a theorem of Kemper [Kem96] as $8 \cdot 24 = 192 = |G|$.

We can also check this using OSCAR: we now compute a presentation of $K[x_1, x_2]^G$ as an affine algebra and check that the relations are trivial[12].

```
julia> A, AtoR = affine_algebra(RG)
(Quotient of multivariate polynomial ring by ideal (0), Hom: A -> graded
↪  multivariate polynomial ring)

julia> modulus(A)
Ideal generated by
  0
```

Since we are in characteristic 0, the fact that $K[x_1, x_2]^G$ is a polynomial algebra is equivalent to G being a complex reflection group by the theorem of Chevalley–Shephard–Todd. One can check that G is the exceptional group labelled G_9 in the classification [ST54].

Example 28 In general, the invariant ring of a finite group is of course not a polynomial algebra. In the modular case, this may happen even for groups $G \subseteq \mathrm{GL}_n(K)$ generated by pseudo-reflections, that is, by elements $g \in G$ with $\mathrm{rk}(g - \mathrm{id}_V) = 1$. To demonstrate this, we use OSCAR to check an example by Nakajima [Nak79] in a special case: While the (counter-)example works for an arbitrary finite field $\mathbb{F}_q$, we do the computations for $q = 3$.

```
julia> K = GF(3);

julia> M1 = matrix(K, [1 0 1 0; 0 1 0 0; 0 0 1 0; 0 0 0 1])
[1   0   1   0]
[0   1   0   0]
[0   0   1   0]
[0   0   0   1]

julia> M2 = matrix(K, [1 0 0 0; 0 1 0 1; 0 0 1 0; 0 0 0 1])
[1   0   0   0]
[0   1   0   1]
[0   0   1   0]
[0   0   0   1]

julia> M3 = matrix(K, [1 0 1 1; 0 1 1 1; 0 0 1 0; 0 0 0 1])
[1   0   1   1]
[0   1   1   1]
[0   0   1   0]
[0   0   0   1]

julia> G = matrix_group(M1, M2, M3);
```

These matrices are indeed pseudo-reflections.

[12] Computing a kernel as in (3) via Gröbner bases is also the idea behind the function `is_algebraically_independent` used in Example (2).

```
julia> all(is_pseudo_reflection, gens(G))
true
```

We compute fundamental invariants.

```
julia> RG = invariant_ring(G);

julia> fundamental_invariants(RG)
5-element Vector{MPolyDecRingElem{FqFieldElem, FqMPolyRingElem}}:
 x[3]
 x[4]
 x[1]^3*x[3] + 2*x[1]*x[3]^3 + x[2]^3*x[4] + 2*x[2]*x[4]^3
 x[1]^9 + 2*x[1]^3*x[3]^6 + 2*x[1]^3*x[3]^4*x[4]^2 + x[1]*x[3]^6*x[4]^2 +
 ↪   2*x[1]^3*x[3]^2*x[4]^4 + x[1]*x[3]^4*x[4]^4 + 2*x[1]^3*x[4]^6 +
 ↪   x[1]*x[3]^2*x[4]^6
 x[2]^9 + 2*x[2]^3*x[3]^6 + x[1]^3*x[3]^5*x[4] + 2*x[1]*x[3]^7*x[4] +
 ↪   x[2]*x[3]^6*x[4]^2 + x[1]^3*x[3]^3*x[4]^3 + 2*x[1]*x[3]^5*x[4]^3 +
 ↪   2*x[2]^3*x[4]^6
```

We see that the invariant ring of dimension 4 is (minimally) generated by five elements. We hence expect a non-trivial relation between these generators and indeed:

```
julia> A, AtoR = affine_algebra(RG)
(Quotient of multivariate polynomial ring by ideal (y1^6*y2^2*y3 +
↪  y1^3*y4 + y2^3*y5 + 2*y3^3), Hom: A -> graded multivariate polynomial
↪  ring)

julia> modulus(A)
Ideal generated by
  y1^6*y2^2*y3 + y1^3*y4 + y2^3*y5 + 2*y3^3
```

Hence there is one relation between the fundamental invariants.

5 Invariants of Linearly Reductive Groups

In this section, we discuss the computation of invariant rings of linearly reductive groups. We will take Remark 3 on the role of the ground field into account and use the terminology introduced in Section 3: G will be a linearly reductive group, $\rho : G \to \mathrm{GL}(V)$ a rational representation of G on the n-dimensional K-vector space V, and $\mathcal{R} : K[V] \to K[V]^G$ the corresponding Reynolds operator (see Theorem 10).

We begin by presenting the details of Hilbert's 1890 finiteness proof. Let $I_{\mathcal{N}} \subseteq K[V]$ be the ideal which is generated by all homogeneous invariants of degree ≥ 1[13]. Conclude from the basis theorem that $I_{\mathcal{N}}$ is generated by finitely many such invariants:

[13] The affine subvariety of V defined by $I_{\mathcal{N}}$ is called *Hilbert's nullcone*. It plays an important role in Hilbert's 1893 paper as well as in Mumford's geometric invariant theory. See [Stu08, Section 4.6] and [DK15, Section 2.5.1].

$$I_{\mathcal{N}} = (f_1, \ldots, f_r).$$

So each homogeneous invariant f of degree ≥ 1 can be written as a linear combination $f = \sum h_i f_i$, with polynomials h_i. Here, we may assume that the h_i are homogeneous of degree $< \deg f$. Applying $\mathcal{R}$ gives a new representation $f = \sum g_i f_i$, with homogeneous invariants g_i of degree $< \deg f$ (see Remark 11). By induction, $f_1, \ldots, f_r$ generate $K[V]^G$ as a K-algebra.

The proof as it stands is nonconstructive as no recipe is given for finding $f_1, \ldots, f_r$. Considering how harshly Hilbert was criticized for this, it came as a big surprise that Derksen [Der99], more than a century later, could refine Hilbert's 1890 ideas and turn them into an algorithm. We discuss this algorithm in what follows.

5.1 Derksen's Algorithm

Derksen's point of departure is a modification of Hilbert's proof which starts from homogeneous generators of $I_{\mathcal{N}}$ that are not necessarily invariants:

Lemma 29 ([Der99]) *Suppose the homogeneous polynomials $f_1, ..., f_r \in K[V]$ generate $I_{\mathcal{N}}$. Then $\mathcal{R}(f_1), ..., \mathcal{R}(f_r)$ generate the invariant ring $K[V]^G$ as a K-algebra.*

Derksen's main result then tells us how to find $f_1, ..., f_r$ as in the lemma. To state this result, we use the following notation. Consider the map

$$\psi : G \times V \to V \times V, \ (\pi, v) \mapsto (v, \pi.v).$$

According to our assumptions, ψ is a morphism of affine varieties. Write $B \subseteq V \times V$ for the image of ψ and $\overline{B}$ for the Zariski closure of B. Identify

$$K[V \times V] \cong K[V] \otimes K[V] \cong K[x, y],$$

where $x = \{x_1, \ldots, x_n\}$ and $y = \{y_1, \ldots, y_n\}$ are the coordinate functions on the first and second copy of V, respectively. Denote the vanishing ideal of $\overline{B}$ in $K[V \times V]$ by $I_{\overline{B}}$. Then:

Theorem 30 ([Der99]) *If $f_1(x, y), ..., f_r(x, y) \in K[V \times V]$ are generators of $I_{\overline{B}}$, then $f_1(x, 0), ..., f_r(x, 0) \in K[V]$ generate $I_{\mathcal{N}}$.*

To find $I_{\overline{B}}$, we apply the general recipe for computing the ideal of the Zariski closure of a morphism of affine varieties to ψ (see [GP08, Section 1.8.3]). This amounts to a Gröbner basis computation with respect to a suitable elimination ordering. For this, it is enough to realize the linearly reductive group G by just entering its defining equations as a variety, ignoring the group structure. That is, we enter the coordinate ring $K[G]$ as an affine algebra, say $K[G] \cong K[z]/I_G$, with coordinates $z = \{z_1, \ldots, z_s\}$. For a matrix group, it is convenient to use variables with double indices.

Example 31 For the special linear group $\mathrm{SL}_m(K)$, we have:

$$K[\mathrm{SL}_m(K)] \cong K[z_{ij}]/(\det(z_{ij}) - 1).$$

A rational representation $\rho : G \to \mathrm{GL}(V) \cong \mathrm{GL}_n(K)$ can then be realized by entering an $n \times n$-matrix of polynomials in the z-variables.

Example 32 The matrix

$$\begin{pmatrix} z_{11}^2 & 2z_{11}z_{21} & z_{21}^2 \\ z_{11}z_{12} & z_{11}z_{22} + z_{12}z_{21} & z_{21}z_{22} \\ z_{12}^2 & 2z_{12}z_{22} & z_{22}^2 \end{pmatrix}$$

implements the action of $\mathrm{SL}_2(\mathbb{C})$ on the space of quadratic binary forms with respect to the basis x^2, $2xy$, y^2 as discussed in the introduction.

The above considerations allow us to find generators of $K[V]^G$ whenever an explicit Reynolds operator $\mathcal{R} : K[V] \to K[V]^G$ can be realized. The latter is possible for all rational representations of G provided there is a way of implementing the Reynolds operator $\mathcal{R}_G : K[G] \to K[G]^G \cong K$ introduced in Remark 12. Indeed, consider the map μ^* of coordinate rings corresponding to the action of G on V:

$$\mu^* : K[V] \to K[G \times V] \cong K[G] \otimes K[V].$$

Then $\mathcal{R}$ is obtained as the map below (see [DK15, Section 4.5] for details):

$$\mathcal{R} : K[V] \xrightarrow{\mu^*} K[G] \otimes K[V] \xrightarrow{\mathcal{R}_G \otimes \mathrm{id}} K[G]^G \otimes K[V] \cong K[V].$$

In the next section, we describe how for $G = \mathrm{SL}_m(K)$ we can use Cayley's Ω-process to find $\mathcal{R}_G$.

5.2 The Reynolds Operator for $G = \mathrm{SL}_m(K)$

Suppose that $\mathrm{char}(K) = 0$ and consider the coordinate ring

$$K[\mathrm{SL}_m(K)] \cong K[z_{ij}]/(\det(z_{ij}) - 1).$$

Cayley's Ω-process consists of repeated applications of the differential operator $\Omega : K[z_{ij}] \to K[z_{ij}]$ defined as follows: Set

$$\Omega = \det\left(\frac{\delta}{\delta Z}\right), \quad \text{where } Z = (z_{ij}) \text{ and } \frac{\delta}{\delta Z} = \left(\frac{\delta}{\delta z_{ij}}\right).$$

Remark 33 Let r be a positive integer. Then, as is easy to see, $\Omega^r(\det(Z)^r)$ is a non-zero constant. Moreover, by reasons of degree, $\Omega^r(f)$ is a constant for each polynomial $f \in K[z_{ij}]$ of degree rm.

Write $\mathrm{SL}_m = \mathrm{SL}_m(K)$ and let $f \in K[z_{ij}]$ be homogeneous of degree d. Set

$$\widetilde{\mathcal{R}}_{\mathrm{SL}_m}(f) = \frac{\Omega^r(f)}{\Omega^r(\det(Z)^r)} \in K$$

if $d = rm$ for some positive integer r, and set $\widetilde{\mathcal{R}}_{\mathrm{SL}_m}(f) = 0$ otherwise. Extend this K-linearly to obtain a map $\widetilde{\mathcal{R}}_{\mathrm{SL}_m} : K[z_{ij}] \to K$. As $\widetilde{\mathcal{R}}_{\mathrm{SL}_m}(\det(Z)-1) = 0$, this map factors through $K[\mathrm{SL}_m(K)]$:

$$\widetilde{\mathcal{R}}_{\mathrm{SL}_m} : K[z_{ij}] \to K[\mathrm{SL}_m(K)] \xrightarrow{\mathcal{R}'_{\mathrm{SL}_m}} K.$$

Now check that $\mathcal{R}'_{\mathrm{SL}_m}$ satisfies the properties in part 4 of Theorem 10 (see also [DK15, Section 4.5.3]). Conclude that $\mathcal{R}'_{\mathrm{SL}_m} = \mathcal{R}_{\mathrm{SL}_m}$.

5.3 Examples and Further Remarks

Example 34 (Binary Quadrics) We rediscover the discriminant as the fundamental invariant of $\mathrm{SL}_2(\mathbb{C})$ acting on quadratic binary forms:

```
julia> G = linearly_reductive_group(:SL, 2, QQ)
Reductive group SL2
  over QQ

julia> r = representation_on_forms(G, 2)
Representation of SL2
  on symmetric forms of degree 2

julia> RG = invariant_ring(r)
Invariant Ring of
graded multivariate polynomial ring in 3 variables over QQ
  under group action of SL2

julia> fundamental_invariants(RG)
1-element Vector{MPolyDecRingElem{QQFieldElem, QQMPolyRingElem}}:
 -X[1]*X[3] + X[2]^2
```

Example 35 (Ternary Cubics) It was Aronhold [Aro58] who showed that the invariant ring of $\mathrm{SL}_3(\mathbb{C})$ acting on ternary cubics[14] is generated by two fundamental invariants of degrees 4 and 6 (see also [Stu08]). We compute the degree 4 invariant, which is naturally related to the Hessian pencil (see, for example, [Kle26, Chapt. IV]):

[14] Following the classical convention, a ternary cubic in $\mathbb{C}[x_1, x_2, x_3]$ is written as a $\mathbb{C}$-linear combination of the $\mathbb{C}$-basis with elements $\binom{3}{I}x^I$.

```
julia> G = linearly_reductive_group(:SL, 3, QQ);

julia> r = representation_on_forms(G, 3);

julia> S, x = graded_polynomial_ring(QQ, "x" => 1:10);

julia> RG = invariant_ring(S, r);

julia> 75*reynolds_operator(RG, x[5]^4)
x[1]*x[4]*x[8]*x[10] - x[1]*x[4]*x[9]^2 - x[1]*x[5]*x[7]*x[10] +
↪  x[1]*x[5]*x[8]*x[9] + x[1]*x[6]*x[7]*x[9] - x[1]*x[6]*x[8]^2 -
↪  x[2]^2*x[8]*x[10] + x[2]^2*x[9]^2 + x[2]*x[3]*x[7]*x[10] -
↪  x[2]*x[3]*x[8]*x[9] + x[2]*x[4]*x[5]*x[10] - x[2]*x[4]*x[6]*x[9] -
↪  2*x[2]*x[5]^2*x[9] + 3*x[2]*x[5]*x[6]*x[8] - x[2]*x[6]^2*x[7] -
↪  x[3]^2*x[7]*x[9] + x[3]^2*x[8]^2 - x[3]*x[4]^2*x[10] +
↪  3*x[3]*x[4]*x[5]*x[9] - x[3]*x[4]*x[6]*x[8] - 2*x[3]*x[5]^2*x[8] +
↪  x[3]*x[5]*x[6]*x[7] + x[4]^2*x[6]^2 - 2*x[4]*x[5]^2*x[6] + x[5]^4
```

Example 36 (Invariants of Tori) In principle, invariants of tori can be computed using Derksen's algorithm. There is, however, a more effective specific method which, in particular, relies on algorithms from polyhedral geometry (see [DK15, Section 4.3]). We explain how to access this method in OSCAR. Let $T = (K^*)^r$ be a torus of rank r, and let an n-dimensional representation $\rho : T \to \mathrm{GL}(V)$ as in Section 2 be given. Then we may choose a basis of V and its dual basis $\{x_1, \ldots, x_n\}$ of V^* such that the induced action on $K[V] \cong K[x_1, \ldots, x_n]$ is of type

$$x_i . t = t^{\omega_i} x_i = t_1^{\omega_{i,1}} \cdots t_r^{\omega_{i,r}} x_i.$$

Here, $(\omega_{i,j})$ is an $n \times r$-matrix with integer coefficients (*weight matrix*), and $t_1, \ldots, t_r$ are the diagonal entries of the matrix representing the element $t \in T$. An example of setting all this up in OSCAR is shown below:

```
julia> T = torus_group(QQ,2)
Torus of rank 2
  over QQ

julia> r = representation_from_weights(T, [1 0; 0 1; -1 -1; -1 1]);

julia> RT = invariant_ring(r);

julia> fundamental_invariants(RT)
2-element Vector{MPolyDecRingElem{QQFieldElem, QQMPolyRingElem}}:
 X[1]*X[2]*X[3]
 X[1]^2*X[3]*X[4]
```

Remark 37 At the time of this writing, the invariant theory part of OSCAR is still under development. As some sort of motivation, note that the elimination step makes Derksen's algorithm rather inefficient for larger examples. In fact, it cannot even be used to check some of the computations done by hand in the 19th century. The same is true for Hilbert's 1893 algorithm. Here, an implementation of the symbolic method is desirable (see [Crö02]).

6 Cox Rings of Linear Quotients

Let V be a finite-dimensional $\mathbb{C}$-vector space and let $G \subseteq \mathrm{GL}(V)$ be a finite group. Then $\mathbb{C}[V]^G$ can be represented as an affine algebra.

Definition 38 The affine variety $V/G := \operatorname{Spec} \mathbb{C}[V]^G$ is called the *linear quotient* of V by G.

In the context of birational geometry, we are interested in the *Cox ring* $\operatorname{Cox}(V/G)$ of the variety V/G. The definition of this ring in the general case is a bit involved and we refer the interested reader to [ADHL15, Section 1.4]. In a nutshell, the Cox ring is a $\mathbb{C}$-algebra graded by the class group of the variety. We now present a theorem by Arzhantsev and Gaĭfullin [AG10] which says that $\operatorname{Cox}(V/G)$ is graded isomorphic to a certain invariant ring with a non-standard grading. For the precise statement we require some preparation.

Let $H \subseteq G$ be the subgroup generated by the pseudo-reflections in G. Notice that H is a normal subgroup of G and denote the factor group by $\overline{G} := G/H$. Then we have $\mathrm{Cl}(V/G) \cong \operatorname{Hom}(\overline{G}, \mathbb{C}^\times)$ for the class group by [Ben93, Theorem 3.9.2]. Here, $\operatorname{Hom}(\overline{G}, \mathbb{C}^\times)$ are the linear characters of $\overline{G}$ and by elementary character theory we see that

$$\operatorname{Hom}(\overline{G}, \mathbb{C}^\times) = \operatorname{Hom}(\operatorname{Ab}(\overline{G}), \mathbb{C}^\times),$$

where $\operatorname{Ab}(\overline{G}) := \overline{G}/[\overline{G}, \overline{G}]$ is the *abelianization* of $\overline{G}$. We abbreviate the linear characters of $\operatorname{Ab}(\overline{G})$ by $\operatorname{Ab}(\overline{G})^\vee$ in the following and note that $\operatorname{Ab}(\overline{G}) \cong \operatorname{Ab}(\overline{G})^\vee$ as groups. In conclusion, we have $\mathrm{Cl}(V/G) \cong \operatorname{Ab}(G/H)^\vee$.

Let $\Gamma := H[G, G] \subseteq G$ be the preimage of $[\overline{G}, \overline{G}]$ under the projection $G \to \overline{G}$. We endow the invariant ring $\mathbb{C}[V]^\Gamma$ with a grading by $\operatorname{Ab}(\overline{G})^\vee$. For this, notice that there is an action by $\operatorname{Ab}(\overline{G}) \cong G/\Gamma$ on $\mathbb{C}[V]^\Gamma$ induced by the action of G. This induces a grading by the group of characters $\operatorname{Ab}(\overline{G})^\vee$ where the graded component of a character $\chi \in \operatorname{Ab}(\overline{G})^\vee$ is defined to be

$$\mathbb{C}[V]^\Gamma_\chi := \{f \in \mathbb{C}[V]^\Gamma \mid f.\gamma = \chi(\gamma) f \text{ for all } \gamma \in \operatorname{Ab}(\overline{G})\}.$$

We arrive at the announced result.

Theorem 39 (Arzhantsev–Gaĭfullin) *With the above notation, there is an* $\operatorname{Ab}(\overline{G})^\vee$*-graded isomorphism* $\operatorname{Cox}(V/G) \cong \mathbb{C}[V]^\Gamma$.

See [AG10, Theorem 3.1] for further details.

We now turn to the computation of $\operatorname{Cox}(V/G)$. The computational challenge left to overcome is how to compute generators of the ring $\mathbb{C}[V]^\Gamma$ which are homogeneous with respect to the grading by $\operatorname{Ab}(\overline{G})^\vee$. In more general language, see e.g. [Sta79], we are interested in *semi-invariants* (also called *relative invariants*) of the group G. That is, we are looking for generators of the isotypic components $\mathbb{C}[V]^G_\chi$ as a module over $\mathbb{C}[V]^G$, where $\chi \in \operatorname{Hom}(G, \mathbb{C}^\times)$ is a linear character which factors through $\overline{G} = G/H$. Such semi-invariants can be computed by adapting the methods to compute secondary invariants discussed above, see [Gat96] for details.

However, in our special situation we can do better. As the action of $\mathrm{Ab}(\overline{G})$ on $\mathbb{C}[V]^{\Gamma}$ is linear, we can consider the graded components of $\mathbb{C}[V]^{\Gamma}$ with respect to the standard grading separately. Let $d \geq 0$ be an upper bound such that $\mathbb{C}[V]^{\Gamma}$ is generated by polynomials of degree at most d. For every $1 \leq k \leq d$, the action of $\mathrm{Ab}(G)$ on the graded component of $\mathbb{C}[V]^{\Gamma}$ of degree k translates to the action of pairwise commuting matrices on a finite-dimensional vector space and we can compute common eigenspaces of these matrices. The corresponding eigenvectors are then exactly the homogeneous generators of $\mathbb{C}[V]^{\Gamma}$ we are interested in, see also [DK17, Construction 2.4].

Example 40 We consider the group $G \subseteq \mathrm{GL}_3(\mathbb{C})$ generated by the matrices

$$g_1 := \begin{pmatrix} -1 & 0 & 0 \\ 0 & 1 & 0 \\ 0 & 0 & 1 \end{pmatrix} \text{ and } g_2 := \begin{pmatrix} 0 & 0 & 1 \\ 1 & 0 & 0 \\ 0 & 1 & 0 \end{pmatrix}.$$

This group is of order 24 and isomorphic to the abstract group $C_2 \times A_4$. We can clearly construct G over the field $\mathbb{Q}$. However, we are going to see soon that we require a third root of unity for our computations (this corresponds to the eigenvalues of g_2), so we directly start with the corresponding extension of $\mathbb{Q}$.

```
julia> K, z_3 = cyclotomic_field(3)
(Cyclotomic field of order 3, z_3)

julia> g1 = matrix(K, 3, 3, [ -1 0 0; 0 1 0; 0 0 1 ]);

julia> g2 = matrix(K, 3, 3, [ 0 0 1; 1 0 0; 0 1 0 ]);

julia> G = matrix_group(g1, g2)
Matrix group of degree 3
  over cyclotomic field of order 3
```

The pseudo-reflections in G are given by the matrices

$$g_1,\ g_2 g_1 g_2^{-1} = \mathrm{diag}(1, -1, 1) \text{ and } g_2^2 g_1 (g_2^2)^{-1} = \mathrm{diag}(1, 1, -1)$$

and these generate a subgroup $H \subseteq G$ of order 8. We see that $G/H = \mathrm{Ab}(G/H) \cong \mathbb{Z}/3\mathbb{Z}$ is cyclic of order 3 generated by the class $g_2 H$. The Cox ring $\mathrm{Cox}(\mathbb{C}^3/G)$ will hence have three graded components, one corresponding to the trivial character of G/H and the other two given by the characters χ_+ and χ_- defined by

$$\chi_\pm : G/H \to \mathbb{C}^\times,\ g_2 H \mapsto \zeta_3^{\pm 1},$$

where ζ_3 is a primitive 3rd root of unity.

Using the functionality for semi-invariants in OSCAR, we can compute generators of the graded component of $\mathrm{Cox}(\mathbb{C}^3/G)$ corresponding to χ_+.

```
julia> chi = character_table(G)[6];

julia> RG = invariant_ring(G)
Invariant ring
  of matrix group of degree 3 over K

julia> semi_invariants(RG, chi)
2-element Vector{MPolyDecRingElem{AbsSimpleNumFieldElem,
↪  AbstractAlgebra.Generic.MPoly{AbsSimpleNumFieldElem}}}:
 x[1]^2 + z_3*x[2]^2 + (-z_3 - 1)*x[3]^2
 x[1]^2*x[2]^2 + (-z_3 - 1)*x[1]^2*x[3]^2 + z_3*x[2]^2*x[3]^2
```

For the full Cox ring, we use the more specialized approach discussed above. For this, notice that the derived subgroup $[G, G]$ of G is contained in H as G/H is abelian in this case, so we have $\Gamma = H$ in the above notation. Let us now start by computing a presentation of the invariant ring $\mathbb{C}[V]^H$, where $V := \mathbb{C}^3$, without the grading by $\mathrm{Ab}(G/H)^\vee$.

```
julia> H, HtoG = subgroup_of_pseudo_reflections(G)
(Matrix group of degree 3 over K, Hom: H -> G)

julia> RH = invariant_ring(H)
Invariant ring
  of matrix group of degree 3 over K

julia> A, AtoR = affine_algebra(RH)
(Quotient of multivariate polynomial ring by ideal (0), Hom: A -> graded
↪  multivariate polynomial ring)

julia> map(AtoR, gens(A))
3-element Vector{MPolyDecRingElem{AbsSimpleNumFieldElem,
↪  AbstractAlgebra.Generic.MPoly{AbsSimpleNumFieldElem}}}:
 x[3]^2
 x[2]^2
 x[1]^2
```

To compute the Cox ring $\mathrm{Cox}(\mathbb{C}^3/G)$, we endow the invariant ring $\mathbb{C}[V]^H$ with the grading by $\mathrm{Ab}(G/H)^\vee$. Following the discussion above, we have to find suitable linear combinations of the generators of $\mathbb{C}[V]^H$ to get an $\mathrm{Ab}(G/H)^\vee$-homogeneous generating system. The function `cox_ring` implements this in OSCAR.

```
julia> VG = linear_quotient(G)
Linear quotient by matrix group of degree 3 over K

julia> B, BtoR = cox_ring(VG)
(Quotient of multivariate polynomial ring by ideal (), Hom: B -> graded
↪  multivariate polynomial ring)

julia> map(BtoR, gens(B))
3-element Vector{MPolyDecRingElem{AbsSimpleNumFieldElem,
↪  AbstractAlgebra.Generic.MPoly{AbsSimpleNumFieldElem}}}:
 x[1]^2 + z_3*x[2]^2 + (-z_3 - 1)*x[3]^2
```

```
x[1]^2 + x[2]^2 + x[3]^2
x[1]^2 + (-z_3 - 1)*x[2]^2 + z_3*x[3]^2
```

We see that the generators are linear combinations of the generators of $\mathbb{C}[V]^H$ we computed above. Further, if we label the generators f_1, f_2 and f_3, then g_2 acts on these by scalar multiplication, namely

$$f_1.g_2 = \zeta_3 f_1,\ f_2.g_2 = f_2 \text{ and } f_3.g_2 = \zeta_3^{-1} f_3.$$

```
julia> f1 = BtoR(B[1]); f2 = BtoR(B[2]); f3 = BtoR(B[3]);

julia> f1^G(g2) == z_3*f1
true

julia> f2^G(g2) == f2
true

julia> f3^G(g2) == z_3^(-1)*f3
true
```

This corresponds directly to the grading by $\mathrm{Ab}(G/H)^\vee$: the polynomial f_2 is in the homogeneous component of the trivial character, f_1 is homogeneous of degree χ_+ and f_3 is homogeneous of degree χ_-. In OSCAR, $\mathrm{Ab}(G/H)^\vee$ is represented as the abstract abelian group $\mathbb{Z}/3\mathbb{Z}$.

```
julia> grading_group(B)
Z/3
```

Under the choice of isomorphism $\mathrm{Ab}(G/H)^\vee \cong \mathbb{Z}/3\mathbb{Z}$, the character χ_- corresponds to 1 and χ_+ to 2.

```
julia> degree(B[1])
Abelian group element [2]

julia> degree(B[2])
Abelian group element [0]

julia> degree(B[3])
Abelian group element [1]
```

In this example, the group $\Gamma = H$ is a complex reflection group, hence $\mathbb{C}[V]^\Gamma$ is a polynomial ring by the theorem of Chevalley–Shephard–Todd.

```
julia> is_zero(modulus(B))
true
```

However, in general, the Cox ring of a linear quotient is not a polynomial algebra in contrast to the toric case.

Acknowledgements Funded by the Deutsche Forschungsgemeinschaft (DFG, German Research Foundation) - Project- ID 286237555 - TRR 195. Wolfram Decker also acknowledges support of the Potentialbereich *SymbTools – Symbolic Tools in Mathematics and their Application* of the Forschungsinitiative Rheinland-Pfalz.

References

[Aro58] S. Aronhold. "Theorie der homogenen Functionen dritten Grades von drei Veränderlichen." *J. Reine Angew. Math.* 1858.55 (1858), pp. 97–191.

[AD09] M. Artebani and I. Dolgachev. "The Hesse pencil of plane cubic curves". *Enseign. Math. (2)* 55.3–4 (2009), pp. 235–273.

[ADHL15] I. Arzhantsev, U. Derenthal, J. Hausen, and A. Laface. *Cox rings*. Vol. 144. Cambridge Studies in Advanced Mathematics. Cambridge: Cambridge University Press, 2015.

[AG10] I. V. Arzhantsev and S. A. Gaĭfullin. "Cox rings, semigroups and automorphisms of affine varieties". *Mat. Sb.* 201.1 (2010), pp. 3–24.

[Ben93] D. J. Benson. *Polynomial invariants of finite groups*. Vol. 190. London Mathematical Society Lecture Note Series. Cambridge: Cambridge University Press, 1993.

[Ber67] M.-J. Bertin. "Anneaux d'invariants d'anneaux de polynômes, en caractéristique p". *C. R. Acad. Sci., Paris, Sér. A* 264 (1967), pp. 653–656.

[Boo42] G. Boole. "Exposition of a general theory of linear transformations, Part I & II". *Cambridge Math. J.* 3 (1841, 1842), pp. 1–20, 106–119.

[Cay46] A. Cayley. "On linear transformations". *Cambridge and Dublin Math. J.* 1 (1846), pp. 104–112.

[Cay45] A. Cayley. "On the theory of linear transformations". *Cambridge Math. J.* 4 (1845), pp. 193–209.

[Cri86] T. Crilly. "The rise of Cayley's invariant theory (1841–1862)". *Hist. Math.* 13 (1986), pp. 241–254.

[Crö02] H. Cröni. "Zur Berechnung von Kovarianten von Quantiken". PhD thesis. Universität des Saarlandes, 2002.

[Der99] H. Derksen. "Computation of invariants for reductive groups". *Adv. Math.* 141.2 (1999), pp. 366–384.

[DK15] H. Derksen and G. Kemper. *Computational invariant theory*. Second enlarged edition. Vol. 130. Encyclopaedia of Mathematical Sciences. Springer, 2015.

[DC70] J. A. Dieudonné and J. B. Carrell. "Invariant theory, old and new". *Adv. Math.* 4 (1970), pp. 1–80.

[DK17] M. Donten-Bury and S. Keicher. "Computing resolutions of quotient singularities". *J. Algebra* 472 (2017), pp. 546–572.

[FH24] C. Fieker and T. Hofmann. "Number Theory". In: *The Computer Algebra System OSCAR. Algorithms and Examples*. Ed. by W. Decker, C. Eder, C. Fieker, M. Horn, and M. Joswig. Springer, 2024, pp. 81–105.

[Gat96] K. Gatermann. "Semi-invariants, equivariants and algorithms". *Appl. Algebra Engrg. Comm. Comput.* 7.2 (1996), pp. 105–124.

[Gau86] C. F. Gauss. *Disquisitiones arithmeticae.* Translated and with a preface by Arthur A. Clarke, Revised by William C. Waterhouse, Cornelius Greither and A. W. Grootendorst and with a preface by Waterhouse. Springer-Verlag, New York, 1986.

[Gle71] A. M. Gleason. “Weight polynomials of self-dual codes and the MacWilliams identities”. *Actes Congr. internat. Math. 1970* 3 (1971), pp. 211–215.

[Gor68] P. Gordan. “Beweis, dass jede Covariante und Invariante einer binären Form eine ganze Funktion mit numerischen Coeffizienten einer endlichen Anzahl solcher Formen ist”. *J. Reine Angew. Math.* 69 (1868), pp. 323–354.

[GP08] G.-M. Greuel and G. Pfister. *A Singular introduction to commutative algebra.* Second, extended edition. With contributions by Olaf Bachmann, Christoph Lossen and Hans Schönemann. Berlin: Springer, 2008.

[Hil90] D. Hilbert. “Über die Theorie der algebraischen Formen”. *Math. Ann.* 36 (1890), pp. 473–534.

[Hil93] D. Hilbert. “Über die vollen Invariantensysteme”. *Math. Ann.* 42 (1893), pp. 313–373.

[HR74] M. Hochster and J. L. Roberts. “Rings of invariants of reductive groups acting on regular rings are Cohen–Macaulay”. *Adv. Math.* 13 (1974), pp. 115–175.

[Hul86] K. Hulek. *Projective geometry of elliptic curves.* Vol. 137. Astérisque. Société Mathématique de France (SMF), Paris, 1986.

[Kem99] G. Kemper. “An algorithm to calculate optimal homogeneous systems of parameters”. *J. Symbolic Comput.* 27.2 (1999), pp. 171–184.

[Kem96] G. Kemper. “Calculating invariant rings of finite groups over arbitrary fields”. *J. Symb. Comput.* 21.3 (1996), pp. 351–366.

[KS99] G. Kemper and A. Steel. “Some algorithms in invariant theory of finite groups”. In: *Computational methods for representations of groups and algebras (Essen, 1997).* Vol. 173. Progr. Math. Basel: Birkhäuser, 1999, pp. 267–285.

[Kin07] S. King. *Fast computation of secondary invariants.* Preprint, `https://arxiv.org/abs/math/0701270`. 2007.

[Kin13] S. King. “Minimal generating sets of non-modular invariant rings of finite groups”. *J. Symbolic Comput.* 48 (2013), pp. 101–109.

[Kle26] F. Klein. *Vorlesungen über die Entwicklung der Mathematik im 19. Jahrhundert. I. Für den Druck bearbeitet von* R. Courant *und* O. Neugebauer. Ed. by R. Courant and O. Neugebauer. Vol. 24. Grundlehren Math. Wiss. Springer, Cham, 1926.

[Kra85] H. Kraft. *Geometrische Methoden in der Invariantentheorie.* 2nd ed. Vol. D1. Aspects Math. Braunschweig: Friedr. Vieweg & Sohn, 1985.

[KR84] J. P. S. Kung and G.-C. Rota. "The invariant theory of binary forms". *Bull. Am. Math. Soc., New Ser.* 10 (1984), pp. 27–85.

[Lag75] J.-L. de Lagrange. *Recherches d'arithmétique.* Vol. 3. Œuvres complètes. Hildesheim, New York: Georg Olms Verlag, 1973, 1773, 1775, pp. 695–795.

[Mas99] H. Maschke. "Beweis des Satzes, dass diejenigen endlichen linearen Substitutionsgruppen, in welchen einige durchgehends verschwindende Coefficienten auftreten, intransitiv sind". *Math. Ann.* 52 (1899), pp. 363–368.

[Mey92] F. Meyer. "Bericht über den gegenwärtigen Stand der Invariantentheorie." *Jahresber. Dtsch. Math.-Ver.* 1 (1892), pp. 79–292.

[Mol97] T. Molien. "Über die Invarianten der linearen Substitutionsgruppen". *Sitzungber. König. Preuss. Akad. Wiss.* (1897), pp. 1152–1156.

[Nak79] H. Nakajima. "Invariants of reflection groups in positive characteristics". *Proc. Japan Acad., Ser. A* 55 (1979), pp. 219–221.

[Noe26] E. Noether. "Der Endlichkeitssatz der Invarianten endlicher linearer Gruppen der Charakteristik p". *Nachr. Ges. Wiss. Göttingen, Math.-Phys. Kl.* 1926 (1926), pp. 28–35.

[Olv99] P. J. Olver. *Classical invariant theory.* Vol. 44. Lond. Math. Soc. Stud. Texts. Cambridge: Cambridge University Press, 1999.

[Par89] K. Parshall. "Toward a history of nineteenth-century invariant theory". In: *Ideas and their Reception.* Ed. by D. E. Rowe and J. McCleary. Academic Press, 1989, pp. 155–206.

[Rey95] O. Reynolds. "On the dynamical theory of incompressible viscous fluid and the determination of the criterion". *Philos. Trans. R. Soc. Lond., A* 186 (1895), pp. 123–164.

[ST54] G. C. Shephard and J. A. Todd. "Finite unitary reflection groups". *Canad. J. Math.* 6 (1954), pp. 274–304.

[Smi95] L. Smith. *Polynomial invariants of finite groups.* Vol. 6. Res. Notes Math. Wellesley, MA: A. K. Peters, 1995.

[Spr89] T. A. Springer. "Aktionen reduktiver Gruppen auf Varietäten". In: *Algebraische Transformationsgruppen und Invariantentheorie.* Vol. 13. DMV Semin. 1989, pp. 3–39.

[Sta79] R. P. Stanley. "Invariants of finite groups and their applications to combinatorics". *Bull. Amer. Math. Soc. (N.S.)* 1.3 (1979), pp. 475–511.

[Stu08] B. Sturmfels. *Algorithms in invariant theory.* Springer Science & Business Media, 2008.

[Syl51a] J. J. Sylvester. "On a remarkable discovery in the theory of canonical forms and of hyperdeterminants". *Philosophical Magazine, 4th series* 2.12 (1851), pp. 391–410.

[Syl51b] J. J. Sylvester. "On the general theory of associated algebraical forms". *Cambridge and Dublin Math. J.* 6 (1851), pp. 289–293.

[Syl50] J. J. Sylvester. "On the intersections, contacts, and other correlations of two conics expressed by indeterminate coordinates". *Cambridge and Dublin Math. J.* 5 (1850), pp. 262–282.

[Syl81] J. J. Sylvester. "Tables of the generating functions and groundforms of the binary duodecimic, with some general remarks, and tables of the irreductible syzigies of certain quantics". *Am. J. Math.* 4 (1881), pp. 41–62.

[SF79] J. J. Sylvester and F. Franklin. "Tables of the Generating Functions and Groundforms for the Binary Quantics of the First Ten Orders". *Am. J. Math.* 2.3 (1879), pp. 223–251.

[Sym11] P. Symonds. "On the Castelnuovo-Mumford regularity of rings of polynomial invariants". *Ann. of Math. (2)* 174.1 (2011), pp. 499–517.

Monomial Bases in Lie Theory

Xin Fang and Ghislain Fourier

Abstract In this survey, we present a detailed guide on using the computer algebra system `OSCAR` to compute monomial bases for simple, finite-dimensional modules of simple, complex Lie algebras. We will also demonstrate how to determine monomial bases for the homogeneous coordinate ring of a (partial) flag variety, depending on a chosen birational sequence and a monomial order. This survey will be updated to reflect any advancements in `OSCAR`'s capabilities in these areas.

Key words: Monomial bases, birational sequences, highest weight modules, toric degenerations, flag varieties

1 Introduction

For a finite-dimensional simple complex Lie algebra $\mathfrak{g}$ and its finite dimensional irreducible representation $V(\lambda)$, finding an explicit vector space basis of $V(\lambda)$ is one of the most important questions in representation theory. The study of this question dates back to the work of Young and Hodge, using Young tableaux to encode a basis of the homogeneous coordinate ring of a Grassmann variety, in order to compute its Hilbert polynomial. Later Gelfand and Tsetlin constructed such a basis for type A Lie algebras with explicit formulae for the action of Chevalley generators. Recent developments around this question include the work of Lakshmibai–Seshadri–Littelmann, Chirirì–Fang–Littelmann [LS86;

Xin Fang
RWTH Aachen University, Chair of Algebra and Representation Theory, e-mail: `fang@art.rwth-aachen.de`

Ghislain Fourier
RWTH Aachen University, Chair of Algebra and Representation Theory, e-mail: `fourier@art.rwth-aachen.de`

W. Decker et al. (eds.), *The Computer Algebra System OSCAR*,
Algorithms and Computation in Mathematics 32,
https://doi.org/10.1007/978-3-031-62127-7_13

Lit98; CFL22; CFL23] (standard monomial basis), Lusztig [Lus90] (canonical and semicanonical bases), Kashiwara [Kas93] (global crystal basis), Mirković–Vilonen [MV88] (MV basis), Feigin–Fourier–Littelmann [FFL11a; FFL11b] (FFLV basis), Gross–Hacking–Keel–Kontsevich [GHKK18] (theta basis), to name but a few.

Many such bases defined and studied in the past three decades come from different geometrizations of the universal enveloping algebra and their representations, making the computation of these bases usually a hard question. In [FFL17a] the first two authors, together with Peter Littelmann, introduced a general construction of a family of monomial bases which have both theoretical and computational convenience. The computation of these bases has been implemented in `OSCAR`.

Birational sequences are introduced in [FFL17a] as a first approximation of some of the above-mentioned bases. By the Borel–Weil–Bott theorem, such bases (or rather their dual bases) can be considered as functions on the flag varieties or their Schubert subvarieties. The key idea of the approach in *loc.cit.* is to use a birational sequence to give parameterizations of the basis elements by polynomials. By fixing a monomial order of the variables in the polynomials, we chose a leading monomial to represent the corresponding basis element. This procedure is similar to replacing an algebra by its initial algebra. Passing to the leading monomial is the first approximation of the basis element, and information will definitely get lost. What one can still read off from the leading monomial is a skeleton of the flag variety or the Schubert variety. In mathematical language, this construction produces a flat family over $\mathbb{C}$ with generic fibre the flag variety or the Schubert variety, and the special fibre is a reduced toric scheme.

The main conjecture is:

Conjecture 1 The special fibre is for any chosen birational sequence and monomial ordering an affine toric variety.

The notion of birational sequences is motivated by the Newton–Okounkov theory [KK12; LM09] and the PBW filtration [FFL17b]. In this language, the first approximation mentioned above is given by exactly the Newton–Okounkov body associated to the birational sequence and the monomial order, and the conjecture can be reformulated as: the associated Newton–Okounkov body is a rational polytope.

Birational sequences are used as a theoretical tool to unify many known toric degenerations of flag varieties and Schubert varieties. That is not all. It opens the door of computing Newton–Okounkov bodies for flag varieties and Schubert varieties, and the key idea is the notion of essential monomials, transforming the computation of the leading monomial into a down-to-earth and explicit representation theory problem. It is this notion which allows us to execute the computation in `OSCAR`, and this survey serves as a guide for both theoretical and practical aspects.

The input data are (N being the number of positive roots of $\mathfrak{g}$)

- a birational sequence S (introduced in [FFL17a]), for our purpose: a sequence of length N of (not necessarily distinct) positive roots, such that ordered monomials form a generating set of the universal enveloping algebra of the positive part of $\mathfrak{g}$,
- a fixed monomial ordering $<$ on $\mathbb{Z}^N$,
- a dominant, integral weight λ.

The OSCAR package then computes a monomial basis of $V(\lambda)$ with respect to the generating set (we consider monomials in the root vectors of the corresponding negative roots of S) and the chosen monomial order. If we denote this set of essential monomials $\mathrm{es}(S, <, \lambda)$, then $\mathrm{es}(S, <, \lambda) + \mathrm{es}(S, <, \mu) \subseteq \mathrm{es}(S, <, \lambda + \mu)$ (where the sum is the Minkowski sum of the exponents). Consequently, OSCAR computes then a minimal (with respect to a graded lexicographic ordering on weights) tuple $(\mu_1, \ldots, \mu_s)$ with $\mu_i \leq \lambda$ such that $\lambda = \mu_1 + \ldots + \mu_s$ and $\mathrm{es}(S, <, \lambda) = \sum a_i \mathrm{es}(S, <, \mu_i)$ for some $a_i \geq 0$. The "best cases" are those, where one needs fundamental weights only and obtains any set of essential monomials through Minkowski sums.

We will extend the functionality soon to Demazure modules and Schubert varieties, and on the other hand connect to the realm of OSCAR by using the capabilities provided by `polymake` to actually compute the Newton–Okounkov bodies.

2 Theoretical Background

The standard reference on basics of semisimple Lie algebras and their representation theory is [Hum72].

Motivated by the theory of Newton–Okounkov bodies and the PBW filtration, studying monomial bases of representations serves as an intermediate step towards understanding monomial bases of finite-dimensional irreducible representations of finite-dimensional complex semisimple Lie algebras. Geometrically it has a strong connection to the construction of toric degenerations of (partial) flag varieties. Roughly speaking, the idea is to replace a complicated basis element, written in a sequence of generators of the Lie algebra $\mathfrak{g}$, by a well-chosen leading term, and to express the leading term as a monomial of the fixed generators.

2.1 Representations as Functions

Let G be a simply connected semisimple algebraic group with Lie algebra $\mathfrak{g}$. Fixing a triangular decomposition $\mathfrak{g} = \mathfrak{n}^+ \oplus \mathfrak{h} \oplus \mathfrak{n}^-$ of $\mathfrak{g}$ gives a Borel subgroup B and a maximal torus T of G with Lie algebras $\mathfrak{n}^+ \oplus \mathfrak{h}$ and $\mathfrak{h}$ respectively. The complete flag variety G/B can be embedded into $\mathbb{P}(V(\lambda))$, for λ a regular dominant weight, as a highest weight orbit, from which it gets the structure of a projective variety. The homogeneous coordinate ring of G/B with respect to this embedding is then given by:

$$\mathbb{C}[G/B] \cong \bigoplus_{k\geq 0} V(k\lambda)^*,$$

where non-zero functions in $V(k\lambda)^*$ are homogeneous of degree k.

The existence of certain nice monomial bases for all $V(k\lambda)$ allows us to degenerate $\mathbb{C}[G/B]$ to its initial algebra, which is a toric algebra. In [FFL17a], a general procedure is introduced to provide a systematical way of constructing monomial bases of $V(\lambda)$. The key ingredient is the notion of a birational sequence, as will be explained in the next section.

2.2 Birational Sequences

Let U^- be the unipotent subgroup of G with Lie algebra $\mathfrak{n}^-$ of dimension, say, N. For each positive root β of $\mathfrak{g}$, we fix a non-zero root vector $f_\beta \in \mathfrak{n}^-$ of weight $-\beta$. The corresponding root subgroup is denoted by $U_{-\beta} := \{\exp(tf_\beta) \mid t \in \mathbb{C}\} \subseteq U^-$.

Definition 2 A *birational sequence* $(\beta_1, \beta_2, \ldots, \beta_N)$ is a sequence of positive roots (repetitions are allowed) of $\mathfrak{g}$ such that the multiplication map

$$U_{-\beta_1} \times \cdots \times U_{-\beta_N} \dashrightarrow U^-, \quad (u_1, \ldots, u_N) \mapsto u_1 \cdots u_N$$

is a birational map.

Example 3 Let $\underline{w}_0 = s_{i_1} s_{i_2} \cdots s_{i_N}$ be a reduced decomposition of w_0, the longest element in the Weyl group of $\mathfrak{g}$.

(1) The sequence $(\alpha_{i_1}, \alpha_{i_2}, \ldots, \alpha_{i_N})$ is birational.
(2) For $1 \leq k \leq N$, let $\beta_k := s_{i_1} \ldots s_{i_{k-1}}(\alpha_{i_k})$. The sequence $(\beta_1, \beta_2, \ldots, \beta_N)$ is birational.
(3) If we enumerate all positive roots $\gamma_1, \gamma_2, \ldots, \gamma_N$ in such a way that for $1 \leq i, j \leq N$, if $\gamma_i - \gamma_j$ is a positive root, then $i < j$. Following [FFL17b], such an enumeration is called *good*. The sequence $(\gamma_1, \gamma_2, \ldots, \gamma_N)$ is birational.

2.3 Newton–Okounkov Bodies

Given a birational sequence $S = (\beta_1, \cdots, \beta_N)$, we define a valuation on $\mathbb{C}[G/B]$.

The birational map in the definition of a birational sequence induces a birational map

$$\mathbb{C}^N \to U_{-\beta_1} \times \cdots \times U_{-\beta_N} \dashrightarrow U^- \to G/B,$$

$$(t_1, \ldots, t_N) \mapsto \exp(t_1 f_{\beta_1}) \cdots \exp(t_N f_{\beta_N}) \cdot B.$$

Passing to the field of rational functions, this gives an isomorphism

$$\mathbb{C}(t_1, \ldots, t_N) \cong \mathbb{C}(G/B).$$

Fixing a monomial order $>$ on $\mathbb{Z}^N$, one defines a valuation $\nu_> : \mathbb{C}[t_1, \ldots, t_N] \setminus \{0\} \to (\mathbb{N}^N, >)$ by sending a polynomial to its lowest exponent with respect to $>$. Extending this valuation by the rule $\nu_>(f/g) = \nu_>(f) - \nu_>(g)$ for $f, g \in \mathbb{C}[t_1, \ldots, t_N] \setminus \{0\}$ gives a valuation

$$\nu_> : \mathbb{C}(G/B) \setminus \{0\} \cong \mathbb{C}(t_1, \ldots, t_N) \setminus \{0\} \to (\mathbb{Z}^N, >).$$

By fixing a highest weight section $s_\lambda \in V(\lambda)^*$, the homogeneous coordinate ring $\mathbb{C}[G/B]$ can be embedded into $\mathbb{C}(G/B)$ by sending a homogeneous function $g \in V(k\lambda)^*$ of degree k to g/s_λ^k. The theory of Newton–Okounkov bodies associates to such a valuation

1. a monoid $\Gamma(S, >, \lambda) := \{(k, \mathbf{a}) \mid k \in \mathbb{N},\ \mathbf{a} \in \nu_>(V(k\lambda)^* \setminus \{0\})\} \subseteq \mathbb{N} \times \mathbb{Z}^N$;
2. a convex body $\Delta(S, >, \lambda) := \operatorname{cone}(\Gamma(S, >, \lambda)) \cap (\{1\} \times \mathbb{R}^N)$, called the Newton–Okounkov body associated to S and $>$.

If the monoid $\Gamma(S, >, \lambda)$ is finitely generated, there exists a flat degeneration of G/B to the toric variety associated to the monoid.

2.4 Identifications of Newton–Okounkov Bodies

When the birational sequence and the monomial order are well-chosen, one can recover well-known toric degenerations of G/B arising from canonical or global crystal bases, as well as PBW bases.

In practice, since $\mathbb{C}[G/B]$ is graded by the weight lattice with finite-dimensional graded components, the well-ordering property holds automatically; hence it suffices to consider monomial orders.

Definition 4 We consider the following four monomial orders on $\mathbb{Z}^N$: for $\mathbf{a}, \mathbf{b} \in \mathbb{Z}^N$,

1. lexicographic order: $\mathbf{a} >_{\text{lex}} \mathbf{b}$ if the first non-zero coordinate of $\mathbf{a} - \mathbf{b}$ is positive;
2. negative lexicographic order: $\mathbf{a} >_{\text{neglex}} \mathbf{b}$ if $\mathbf{b} >_{\text{lex}} \mathbf{a}$;
3. degree reverse lexicographic order: $\mathbf{a} >_{\text{degrevlex}} \mathbf{b}$ if the total degree $\deg \mathbf{a} > \deg \mathbf{b}$ or if they are equal, the first non-zero coordinate of $\mathbf{a} - \mathbf{b}$, when read from the right, is negative.
4. weighted degree reverse lexicographic order $>_{\text{wdegrevlex}}$: similar of degrevlex but the degree is given for each variable by a weight.

Theorem 5 ([FFL17a; FFL17b; FN17]) *Let $\underline{w}_0 = s_{i_1}s_{i_2}\cdots s_{i_N}$ be a reduced decomposition of w_0.*

1. *Fix the birational sequence $S = (\alpha_{i_1}, \alpha_{i_2}, \ldots, \alpha_{i_N})$.*
   (a) *The associated Newton–Okounkov body $\Delta(S, >_{\text{neglex}}, \lambda)$ is the Littelmann–Berenstein–Zelevinsky polytope.*
   (b) *The Newton–Okounkov body $\Delta(S, >_{\text{degrevlex}}, \lambda)$ is the Nakashima–Zelevinsky polytope.*
2. *Fix the birational sequence $S = (\beta_1, \beta_2, \ldots, \beta_N)$ as in Example 3 (2). The Newton–Okounkov body $\Delta(S, >_{\text{wdegrevlex}}, \lambda)$, where the weight of β is given by the height of β, is the Lusztig polytope parameterizing canonical bases.*
3. *Let $G = \mathrm{SL}_{n+1}$ or Sp_{2n}, S be a good sequence as in Example 3 (3). The Newton–Okounkov body $\Delta(S, >_{\text{neglex}}, \lambda)$ is the Feigin–Fourier–Littelmann–Vinberg polytope* $\mathrm{FFLV}(\lambda)$.

As a summary, birational sequence is a tool to unify well-known toric degenerations of flag varieties. The Conjecture 1, that the special fibre of our degeneration is in fact a toric variety, is translated into checking whether the monoid is finitely generated. All polytopes in Theorem 5 are rational, hence the monoids are finitely generated.

2.5 Essential Monomials

Computing points in a Newton–Okounkov body is a difficult task. As an advantage of the approach via birational sequences, rational points can be calculated using representation theory of Lie algebras.

We fix a birational sequence $S = (\beta_1, \beta_2, \ldots, \beta_N)$ and a monomial order on $\mathbb{N}^N$. For $\mathbf{k} = (k_1, \ldots, k_N) \in \mathbb{N}^N$, we denote $f^{\mathbf{k}} := f_{\beta_1}^{k_1} \cdots f_{\beta_N}^{k_N}$. These data allow us to define a filtration on the universal enveloping algebra $U(\mathfrak{n}^-)$ as follows: for $\mathbf{m} \in \mathbb{N}^N$,

$$U(\mathfrak{n}^-)_{\leq \mathbf{m}} := \operatorname{span}\{f^{\mathbf{k}} \mid \mathbf{k} \leq \mathbf{m}\}.$$

We set similarly $U(\mathfrak{n}^-)_{<\mathbf{m}} := \operatorname{span}\{f^{\mathbf{k}} \mid \mathbf{k} < \mathbf{m}\}$. This filtration on $U(\mathfrak{n}^-)$ induces a filtration on the module $V(\lambda)$ by assigning

$$V(\lambda)_{\leq \mathbf{m}} := U(\mathfrak{n}^-)_{\leq \mathbf{m}} \cdot \mathbf{v}_\lambda,$$

where $\mathbf{v}_\lambda$ is a highest weight vector of $V(\lambda)$. We denote $V(\lambda)_{<\mathbf{m}} := U(\mathfrak{n}^-)_{<\mathbf{m}} \cdot \mathbf{v}_\lambda$.

Definition 6 A tuple $\mathbf{m} \in \mathbb{N}^N$ is called an *essential exponent* with respect to the birational sequence S, the monomial order $>$, and the weight λ, if

$$\dim\left(V(\lambda)_{\leq \mathbf{m}}/V(\lambda)_{<\mathbf{m}}\right) = 1.$$

The set of all essential exponents will be denoted by $\mathrm{es}(S, >, \lambda)$. For $\mathbf{m} \in \mathrm{es}(S, >, \lambda)$, the monomial $f^{\mathbf{m}}$ is called an *essential monomial.*

For two dominant integral weights λ, μ, a big part of $\mathrm{es}(S, >, \lambda + \mu)$ can be computed from the following property where $+$ stands for the Minkowski sum of sets:

$$\mathrm{es}(S, >, \lambda) + \mathrm{es}(S, >, \mu) \subseteq \mathrm{es}(S, >, \lambda + \mu). \tag{1}$$

It is shown in [FFL17a] that the set $\{f^{\mathbf{m}} \cdot \mathbf{v}_\lambda \mid \mathbf{m} \in \mathrm{es}(S, >, \lambda)\}$ forms a vector space basis of $V(\lambda)$, called the essential basis of $V(\lambda)$ with respect to S and $>$. Essential bases are monomial bases for irreducible representations.

Theorem 7 ([FFL17a]) *The following holds:*

$$\bigcup_{k \in \mathbb{N}} \{k\} \times \mathrm{es}(S, >, k\lambda) = \Gamma(S, >, \lambda).$$

This theorem is the theoretic cornerstone of our algorithm for computing essential exponents. We illustrate the theorem in the following example.

Example 8 Let $G = \mathrm{SL}_3$ and $\mathfrak{g} = \mathfrak{sl}_3$ be the Lie algebra of traceless 3×3-matrices with Lie bracket the commutator of matrices. Fix $\mathfrak{n}^-$ to be the subalgebra of strictly lower triangular matrices in $\mathfrak{g}$ and $\mathfrak{h}$ to be the diagonal matrices in $\mathfrak{g}$. The simple roots are denoted by α_1, α_2, and the positive roots are $\alpha_1, \beta := \alpha_1 + \alpha_2, \alpha_2$. The simple reflections s_1 and s_2 with respect to the simple roots generate the Weyl group of $\mathfrak{g}$, which is isomorphic to the symmetric group $\mathfrak{S}_3$. We fix a reduced decomposition of the longest element $\underline{w}_0 = s_1 s_2 s_1$.

We consider the adjoint representation of $\mathfrak{g}$ on itself: it is an irreducible representation of highest weight β. Note that if we denote the fundamental weights by ϖ_1 and ϖ_2, then $\beta = \varpi_1 + \varpi_2$.

The weight space $V(\beta)_{-\beta}$ in $V(\beta)$ of weight $-\beta$ is one-dimensional. We work out in detail the essential monomial which gives a basis of $V(\beta)_{-\beta}$ for different birational sequences and different monomial orders.

1. $S = (\alpha_1, \beta, \alpha_2)$ is a birational sequence and we fix the monomial order $>_{\text{neglex}}$. All exponents $\mathbf{m} = (m_1, m_2, m_3)$ such that $f_{\alpha_1}^{m_1} f_\beta^{m_2} f_{\alpha_2}^{m_3} \cdot \mathbf{v}_\beta$ is non-zero in $V(\beta)_{-\beta}$ are: $(1, 1, 1)$ and $(0, 2, 0)$. With respect to $>_{\text{neglex}}$ we would choose $f_{\alpha_1} f_\beta f_{\alpha_2}$ as an essential monomial, and then $f_{\alpha_1} f_\beta f_{\alpha_2} \cdot \mathbf{v}_\beta$ is a basis of $V(\beta)_{-\beta}$.
2. Choosing $S = (\beta, \alpha_1, \alpha_2)$ as the birational sequence and the monomial order $>_{\text{neglex}}$. The basis of $V(\beta)_{-\beta}$ is given by the essential monomial $f_\beta^2 \cdot \mathbf{v}_\beta$ of exponent $(2, 0, 0)$.
3. Choosing $S = (\alpha_1, \alpha_2, \alpha_1)$ as the birational sequence. There exists only one tuple $\mathbf{m} = (1, 2, 1)$ such that $f^{\mathbf{m}} \cdot \mathbf{v}_\beta \in V(\beta)_{-\beta}$ is non-zero. The essential monomial giving a basis of $V(\beta)_{-\beta}$ is $f_{\alpha_1} f_{\alpha_2}^2 f_{\alpha_1} \cdot \mathbf{v}_\beta$ which is independent of the choice of the monomial order.

3 Algorithm and First Example

The computing of essential monomials is implemented in `OSCAR` with

Algorithm 5: Compute Monomial Basis

Input : Lie algebra, birational sequence, order, highest weight λ, essential monomials for fundamentals, read off essentially from `GAP`.

Output : Essential monomials for $\mathrm{es}(\lambda)$.

1 **for** *all Minkowski sums* $\lambda = \mu_1 + \mu_2$ **do**
2 Compute $\mathrm{es}(S, >, \mu_1)$ and $\mathrm{es}(S, >, \mu_2)$ recursively.
3 Compute the Minkowski sum $\mathrm{es}(S, >, \mu_1) + \mathrm{es}(S, >, \mu_2)$.
4 **end for**
5 Union all of these Minkowski sums $\mathrm{es}(S, >, \mu_1) + \mathrm{es}(S, >, \mu_2)$.
6 **if** *the total size is smaller than the expected dimension (using Weyl's dim. formula)* **then**
7 Identify weight space where monomials are missing (using the Weyl polytope, `polymake` and `GAP`).
8 Find integer solutions of linear equation system to find missing monomials.
9 **end if**

The use of equation (1) in the algorithm reduces the run time as it helps to avoid solving systems of linear equations if possible but computes Minkowski sums of lattice points only.

Note that the validation of the input is not checked in the program, the program does not check if the provided sequence is in fact birational. There will be a message, that the sequence is probably not a birational if (after a reasonable time) no such generating set can be computed.

Our goal is to provide evidence for Conjecture 1, i.e., that the essential monoid is finitely generated. To this end there is an additional return value which gives a set of highest weights that are sufficient for generating the essential monomials from smaller highest weights (component-wise when written in terms of fundamental weights). For given λ, the program gives a finite set of dominant integral weights $\{\mu_i\}$ with μ_i satisfying $\lambda = \sum a_i \mu_i$ where $a_i \in \mathbb{N}$ and

$$\mathrm{es}(S, >, \lambda) = \sum a_i \mathrm{es}(S, >, \mu_i).$$

We store a set of minimal generators with respect to the graded lexicographic ordering on the coefficients of the highest weights (along with smaller, redundant weights that were computed in the process in step **2**), that is providing a set of weights whose height is minimal and $\mathrm{es}(S, >, \lambda)$ is in the Minkowski sum of the essential monomials.

The algorithm is implemented in `OSCAR` for finite-dimensional complex simple Lie algebras of arbitrary type and rank. The main function is

```
basis_lie_highest_weight(type, rank, highest_weight,
        birational_sequence; monomial_ordering)
```

We explain the arguments of the function:

- `type` refers to the type of the Lie algebra as a julia symbol, for example `:A` or `:E`.
- `rank` refers to the rank of the Lie algebra, for example for $\mathfrak{sl}_5(\mathbb{C})$ the rank is 4.
- `highest_weight` refers to the highest weight of the module, the input has to be given as a coefficient vector in terms of the fundamental weights, for example for $\mathfrak{sl}_5(\mathbb{C})$ and the adjoint representation, the input would be `[1,0,0,1]`.
- `birational_sequence` refers to a birational sequence. The input of the roots is by their index in the output of

  `basis_lie_highest_weight_operators(type, rank)`,

  where the roots are written as linear combinations of the simple roots. For example

```
julia> basis_lie_highest_weight_operators(:A, 4)
10-element Vector{Tuple{Int64, Vector{QQFieldElem}}}:
 (1, [1, 0, 0, 0])
 (2, [0, 1, 0, 0])
 (3, [0, 0, 1, 0])
 (4, [0, 0, 0, 1])
 (5, [1, 1, 0, 0])
 (6, [0, 1, 1, 0])
 (7, [0, 0, 1, 1])
 (8, [1, 1, 1, 0])
 (9, [0, 1, 1, 1])
 (10, [1, 1, 1, 1])
```

  The birational sequence

$$(\alpha_1, \alpha_2, \alpha_3, \alpha_4, \alpha_1, \alpha_1 + \alpha_2, \alpha_1 + \alpha_2 + \alpha_3, \alpha_2, \alpha_2 + \alpha_3, \alpha_3) \tag{1}$$

  would be translated to `[1,2,3,4,1,5,8,2,6,3]`.
  Alternatively, one can provide a list of coefficient vectors with respect to simple roots. The previous birational sequence would be translated to `[[1,0,0,0], [0,1,0,0], [0,0,1,0], [0,0,0,1], [1,0,0,0]], [[1,1,0,0], [1,1,1,0], [0,1,0,0], [0,1,1,0], [0,0,1,0]]`.
  As already mentioned, it is not checked by the function that the input is indeed a birational sequence.
- `monomial_ordering` refers to a monomial ordering, default is `degrevlex`. Further monomial orderings can be found in the OSCAR documentation[1].

[1] `https://docs.oscar-system.org/v1/CommutativeAlgebra/GroebnerBases/orderings/`

Example 9 We use the function for a Lie algebra of type A, of rank 4, for the module of dominant integral weight $2\varpi_1 + \varpi_2 + 2\varpi_3 + \varpi_4$, the birational sequence from (1) and the *degrevlex* ordering.

```
julia> basis_lie_highest_weight(:A, 4, [2,1,2,1], [1,2,3,4,1,5,8,2,6,3];
↪  monomial_ordering=:degrevlex)
Monomial basis of a highest weight module
  of highest weight [2, 1, 2, 1]
  of dimension 8750
  with monomial ordering degrevlex([x1, x2, x3, x4, x5, x6, x7, x8, x9,
  ↪  x10])
over Lie algebra of type A4
  where the used birational sequence consists of the following roots
  ↪  (given as coefficients w.r.t. alpha_i):
    [1, 0, 0, 0]
    [0, 1, 0, 0]
    [0, 0, 1, 0]
    [0, 0, 0, 1]
    [1, 0, 0, 0]
    [1, 1, 0, 0]
    [1, 1, 1, 0]
    [0, 1, 0, 0]
    [0, 1, 1, 0]
    [0, 0, 1, 0]
  and the basis was generated by Minkowski sums of the bases of the
  ↪  following highest weight modules:
    [1, 0, 0, 0]
    [0, 1, 0, 0]
    [0, 0, 1, 0]
    [0, 0, 0, 1]
```

The output is the monomial basis `monomials(b)`, even more certain properties of this basis are also computed such as its length. We can read off from

```
  and the basis was generated by Minkowski sums of the bases of the
  ↪  following highest weight modules:
    [1, 0, 0, 0]
    [0, 1, 0, 0]
    [0, 0, 1, 0]
    [0, 0, 0, 1]
```

that $\mathrm{es}(S, >, 2\varpi_1 + \varpi_2 + 2\varpi_3 + \varpi_4)$ is a Minkowski sum of $\mathrm{es}(S, >, \varpi_1)$, $\mathrm{es}(S, >, \varpi_2)$, $\mathrm{es}(S, >, \varpi_3)$ and $\mathrm{es}(S, >, \varpi_4)$. In fact:

$$\begin{aligned}\mathrm{es}(S, >, 2\varpi_1 + \varpi_2 + 2\varpi_3 + \varpi_4) &= 2\mathrm{es}(S, >, \varpi_1) + \mathrm{es}(S, >, \varpi_2) \\ &\quad + 2\mathrm{es}(S, >, \varpi_3) + \mathrm{es}(S, >, \varpi_4).\end{aligned}$$

4 More Examples

There are shorthands implemented for important classes of birational sequences and monomial orderings in Theorem 5.

Example 10 Let $\mathfrak{g}$ be a finite-dimensional complex simple Lie algebra, S be the set of positive roots, ordered by their heights in a descending way and λ be a dominant integral weight. Then the essential monomials with respect to the monomial order $>_{\text{degrevlex}}$ can be computed by

```
julia> basis_lie_highest_weight_ffl(:A, 3, [1,1,1])
Monomial basis of a highest weight module
  of highest weight [1, 1, 1]
  of dimension 64
  with monomial ordering degrevlex([x1, x2, x3, x4, x5, x6])
over Lie algebra of type A3
  where the used birational sequence consists of the following roots
  ↪ (given as coefficients w.r.t. alpha_i):
    [1, 1, 1]
    [0, 1, 1]
    [1, 1, 0]
    [0, 0, 1]
    [0, 1, 0]
    [1, 0, 0]
  and the basis was generated by Minkowski sums of the bases of the
  ↪ following highest weight modules:
    [1, 0, 0]
    [0, 1, 0]
    [0, 0, 1]
```

For $\mathfrak{g} = \mathfrak{sl}_{n+1}(\mathbb{C})$ or $\mathfrak{sp}_{2n}(\mathbb{C})$, the bases are known explicitly by [FFL11a; FFL11b] and the Newton–Okounkov body $\Delta(S, >_{\text{degrevlex}}, \lambda)$ is the Feigin-Fourier-Littelmann-Vinberg polytope FFLV(λ). Again, OSCAR can be used to compute the actual FFLV polytope. Note that explicit defining inequalities are so far only known for the FFLV polytopes for Lie algebras of types A, C, G (for type G see [Gor15]). In other types, and even for some Lie superalgebras, variants of these FFLV bases are known ([BK19; BD15; Mak19; Gor19; FK21]).

Example 11 Let $\mathfrak{g}$ be a finite-dimensional complex simple Lie algebra and let $\underline{w_0} = s_{i_1} s_{i_2} \dots s_{i_N}$ be a reduced decomposition of w_0, the longest element in the Weyl group of $\mathfrak{g}$. We consider the birational sequence $(\alpha_{i_1}, \alpha_{i_2}, \dots, \alpha_{i_N})$ and fix as the monomial order $>_{\text{neglex}}$. Then the monoid Γ is finitely generated and the associated Newton–Okounkov body $\Delta(S, >_{\text{neglex}}, \lambda)$ is the Littelmann-Berenstein-Zelevinsky polytope, also known as the string polytope. This very important class of examples can be computed by the shorthand

```
julia> basis_lie_highest_weight_string(:B, 3, [1,1,1],
↪ [3,2,3,2,1,2,3,2,1])
Monomial basis of a highest weight module
  of highest weight [1, 1, 1]
  of dimension 512
```

```
  with monomial ordering neglex([x1, x2, x3, x4, x5, x6, x7, x8, x9])
over Lie algebra of type B3
  where the used birational sequence consists of the following roots
  ↪  (given as coefficients w.r.t. alpha_i):
    [0, 0, 1]
    [0, 1, 0]
    [0, 0, 1]
    [0, 1, 0]
    [1, 0, 0]
    [0, 1, 0]
    [0, 0, 1]
    [0, 1, 0]
    [1, 0, 0]
  and the basis was generated by Minkowski sums of the bases of the
  ↪  following highest weight modules:
    [1, 0, 0]
    [0, 1, 0]
    [0, 0, 1]
```

Example 12 Closely related to the string polytope are Lusztig polytopes. We fix the same notation as in the previous example. Then $S = (\beta_1, \dots, \beta_N)$ with $\beta_k = s_{i_1} \cdots s_{i_{k-1}}(\alpha_{i_k})$ is a birational sequence (Example 3 (2)). The weight of β is given by the sum of the coefficients in the linear combination of simple roots. The Newton–Okounkov body $\Delta(S, >_{\text{wdegevlex}}, \lambda)$ is the Lusztig polytope parameterizing the canonical basis of $V(\lambda)$.

```
julia> basis_lie_highest_weight_lusztig(:D, 4, [1,1,1,1],
↪  [4,3,2,4,3,2,1,2,4,3,2,1])
Monomial basis of a highest weight module
  of highest weight [1, 1, 1, 1]
  of dimension 4096
  with monomial ordering wdegrevlex([x1, x2, x3, x4, x5, x6, x7, x8, x9,
  ↪  x10, x11, x12], [1, 1, 3, 2, 2, 1, 5, 4, 3, 3, 2, 1])
over Lie algebra of type D4
  where the used birational sequence consists of the following roots
  ↪  (given as coefficients w.r.t. alpha_i):
    [0, 0, 0, 1]
    [0, 0, 1, 0]
    [0, 1, 1, 1]
    [0, 1, 1, 0]
    [0, 1, 0, 1]
    [0, 1, 0, 0]
    [1, 2, 1, 1]
    [1, 1, 1, 1]
    [1, 1, 0, 1]
    [1, 1, 1, 0]
    [1, 1, 0, 0]
    [1, 0, 0, 0]
  and the basis was generated by Minkowski sums of the bases of the
  ↪  following highest weight modules:
    [1, 0, 0, 0]
    [0, 1, 0, 0]
    [0, 0, 1, 0]
```

```
    [0, 0, 0, 1]
    [0, 0, 1, 1]
```

Example 13 A class of examples having very interesting properties is the following. We fix again the same notation as in the previous example and the birational sequence $S = (\alpha_{i_1}, \alpha_{i_2}, \dots, \alpha_{i_N})$. The Newton–Okounkov body $\Delta(S, >_{\text{degrevlex}}, \lambda)$ is the Nakashima-Zelevinsky polytope. The difference to the string polytope is the chosen order, instead of negative lexicographic one chooses the degree reverse lexicographic order.

```
julia> basis_lie_highest_weight_nz(:C, 3, [1,1,1], [3,2,3,2,1,2,3,2,1])
Monomial basis of a highest weight module
  of highest weight [1, 1, 1]
  of dimension 512
  with monomial ordering degrevlex([x1, x2, x3, x4, x5, x6, x7, x8, x9])
over Lie algebra of type C3
  where the used birational sequence consists of the following roots
  ↪   (given as coefficients w.r.t. alpha_i):
    [0, 0, 1]
    [0, 1, 0]
    [0, 0, 1]
    [0, 1, 0]
    [1, 0, 0]
    [0, 1, 0]
    [0, 0, 1]
    [0, 1, 0]
    [1, 0, 0]
  and the basis was generated by Minkowski sums of the bases of the
  ↪   following highest weight modules:
    [1, 0, 0]
    [0, 1, 0]
    [0, 0, 1]
```

The Gelfand-Tsetlin polytope for $\mathfrak{g} = \mathfrak{sl}_{n+1}$ is (up to a linear translation) an instance of a Nakashima-Zelevinsky polytope, namely for the reduced expression $(s_1 s_2 \dots s_n)(s_1 s_2 \dots s_{n-1}) \dots (s_1 s_2) s_1$.

5 Experiments on Generators of the Monoid

One of the primary motivations for this implementation is to provide evidence supporting Conjecture 1, which posits that the monoid of essential monomials $\Gamma(S, >, \lambda)$ is finitely generated for all birational sequences ([FFL17a]). According to [BZ01], this holds true for all string parameterizations (or equivalently, for all Lusztig parameterizations). However, they do not offer a set of generators or even an upper bound for the degree of elements in a minimal set of generators. In cases where the conjecture is proven true, an obvious question arises: how can one find a set of generators? In our implementation, we calculated the minimal building blocks for $\text{es}(S, >, 2\rho)$ using the string parameterization outlined in Theorem 5 for $\mathfrak{sl}_6(\mathbb{C})$ and the highest weight 2ρ, which represents the sum of all positive roots. Specifically, for every reduced expression of the longest element in S_6, we determined the essential monomials and, importantly, the necessary elements in the monoid to express $\mathrm{es}(S, >, 2\rho)$ as a Minkowski sum. To eliminate redundant cases, we utilized OSCAR to generate a list of reduced expressions, accounting for the exchange of successive orthogonal reflections, and then computed the generators for the remaining 908 cases. We organized the reduced expressions according to the necessity of generators, and here is the output, correlating each set of generators with the frequency of its occurrence. Evidently, regardless of the reduced expression, all fundamental weights appear in the list of generators. To simplify, we refer to these collectively as "fundamentals".

```
[fundamentals] => 320
[fundamentals, [1, 0, 1, 0, 0]] => 70
[fundamentals, [0, 0, 1, 0, 1]] => 70
[fundamentals, [0, 1, 0, 1, 0]] => 54
[fundamentals, [1, 0, 0, 1, 0]] => 28
[fundamentals, [0, 1, 0, 0, 1]] => 28
[fundamentals, [1, 0, 0, 1, 0], [0, 1, 0, 1, 0]] => 46
[fundamentals, [0, 1, 0, 1, 0], [0, 1, 0, 0, 1]] => 46
[fundamentals, [1, 0, 1, 0, 0], [1, 0, 0, 1, 0]] => 30
[fundamentals, [0, 1, 0, 0, 1], [0, 0, 1, 0, 1]] => 30
[fundamentals, [1, 0, 1, 0, 0], [0, 1, 0, 0, 1]] => 8
[fundamentals, [1, 0, 0, 1, 0], [0, 0, 1, 0, 1]] => 8
[fundamentals, [1, 0, 1, 0, 0], [1, 0, 0, 1, 0], [0, 1, 0, 1, 0]] => 70
[fundamentals, [0, 1, 0, 1, 0], [0, 1, 0, 0, 1], [0, 0, 1, 0, 1]] => 70
[fundamentals, [1, 0, 0, 1, 0], [0, 1, 0, 1, 0], [0, 1, 0, 0, 1]] => 10
[fundamentals, [1, 0, 1, 0, 0], [1, 0, 0, 1, 0], [0, 1, 0, 1, 0], [0, 1,
↪  0, 0, 1]] => 10
[fundamentals, [1, 0, 0, 1, 0], [0, 1, 0, 1, 0], [0, 1, 0, 0, 1], [0, 0,
↪  1, 0, 1]] => 10
```

Even in the "worst-case" scenario, no generator exceeds a height of 2 for $\mathfrak{sl}_6(\mathbb{C})$ and any arbitrary reduced expression. This observation might even be considered as evidence supporting the bold conjecture that the (global) monoid $\Gamma(S, >) := \sum_\lambda \Gamma(S, >, \lambda)$ for string polytopes is not only finitely generated but also has an upper bound on the number of generators. This global monoid parameterizes a basis of the multi-homogeneous coordinate ring. Specifically, the monoid $\Gamma(S, >)$ could be generated by all essential monomials of $V(\mu)$ for μ such that $\mu \leq 2\rho$.

6 Comparison of the Running Time Against GAP

We want to assess the performance of an OSCAR command like

```
basis_lie_highest_weight(:A, 4, [1,3,2,1]);
```

In GAP the above function call translates into the following code fragment:

```
L:= SimpleLieAlgebra("A", 4, Rationals);;
V:= HighestWeightModule(L, [1,3,2,1]);;
```

The latter function uses the GAP ordering of positive roots, ranging from simple to highest with respect to the weight, as a birational sequence, with the monomial ordering being *deglex*. Currently, GAP does not support alterations to either the birational sequence or the monomial ordering. In fact, the basis is constructed simultaneously with the module. For modules of smaller dimensions, the running time of the two algorithms is comparable. The height of the highest weight is determined by the sum of the coefficients when expressed in terms of fundamental modules. Specifically, if the height of λ is N, then $V(\lambda)$ represents the Cartan component in an N-times tensor product of fundamental representations of the Lie algebra. Our algorithm prefers to avoid computations with matrices when feasible, instead utilizing Minkowski sums of "smaller" weights. If the conjecture about the monoid's generators is true, then computing monomial bases for large highest weights simplifies to computing bases for relatively small weights and then calculating their Minkowski sum. This implies that our new algorithm is particularly efficient when dealing with large weights. We used a *Intel(R) Xeon(R) CPU E5-4617 @ 2.90GHz* for the comparison depicted in Table 1; OSCAR version v1.0 vs. GAP version v4.12.2.

Table 1: Running time comparison of type A, rank 3, in seconds.

Highest weight	Dimension of module	OSCAR	GAP
(1,3,2)	756	0.76	0.23
(3,3,3)	4096	0.77	2.64
(3,4,2)	4320	0.90	1.78
(4,3,5)	13500	0.80	15.10
(5,6,3)	34034	0.91	48.74
(5,5,5)	46656	1.18	187.91
(5,5,6)	62244	1.27	274.85
(5,6,5)	67431	1.31	296.22
(6,5,6)	82810	1.40	882.80
(6,6,6)	117649	2.70	1763.19

In the case of type A and rank 4, $\rho = (1,1,1,1)$ is a generator of the monoid, so parts of its monomial bases have to be added using the polyhedral geometry functions of OSCAR and solving systems of linear equations. We show a running time comparison with GAP on the same hardware as above in this case in Table 2.

Acknowledgements The authors would like to thank Johannes Flake and Ben Wilop for writing large parts of the code, and Lars Göttgens for providing the examples and the oscarification of the code. Ghislain Fourier has been supported by "Symbolic Tools in Mathematics and their Application" (TRR 195, project-ID 286237555).

Table 2: Running Time comparison of type A, rank 4, in seconds.

Highest weight	Dimension of module	OSCAR	GAP
(1,3,2,1)	31185	3.6	11.7
(1,2,2,3)	50400	2.4	29.9
(2,2,2,2)	59049	12.4	64.4
(1,3,3,2)	160160	5.2	98.8
(2,3,2,2)	143325	15.5	175.6
(3,2,2,2)	110565	16.5	206.5
(2,3,3,2)	332024	20.9	507.0

References

[BD15] T. Backhaus and C. Desczyk. "PBW filtration: Feigin–Fourier–Littelmann modules via Hasse diagrams". *J. Lie Theory* 25.3 (2015), pp. 815–856.

[BK19] T. Backhaus and D. Kus. "The PBW filtration and convex polytopes in type B". *J. Pure Appl. Algebra* 223.1 (2019), pp. 245–276. DOI: 10.1016/j.jpaa.2018.03.009.

[BZ01] A. Berenstein and A. Zelevinsky. "Tensor product multiplicities, canonical bases and totally positive varieties". *Invent. Math.* 143.1 (2001), pp. 77–128. DOI: 10.1007/s002220000102.

[CFL22] R. Chirivì, X. Fang, and P. Littelmann. "Seshadri stratification for Schubert varieties and standard monomial theory". *Proc. Indian Acad. Sci. Math. Sci.* 132.2 (2022), Paper No. 74, 58. DOI: 10.1007/s12044-022-00707-1.

[CFL23] R. Chirivì, X. Fang, and P. Littelmann. "Seshadri stratifications and standard monomial theory". *Inventiones mathematicae* 234 (2023), pp. 489–572. DOI: 10.1007/s00222-023-01206-4.

[FFL17a] X. Fang, G. Fourier, and P. Littelmann. "Essential bases and toric degenerations arising from birational sequences". *Adv. Math.* 312 (2017), pp. 107–149. DOI: 10.1016/j.aim.2017.03.014.

[FFL17b] E. Feigin, G. Fourier, and P. Littelmann. "Favourable modules: filtrations, polytopes, Newton–Okounkov bodies and flat degenerations". *Transform. Groups* 22.2 (2017), pp. 321–352. DOI: 10.1007/s00031-016-9389-2.

[FFL11a] E. Feigin, G. Fourier, and P. Littelmann. "PBW filtration and bases for irreducible modules in type A_n". *Transform. Groups* 16.1 (2011), pp. 71–89. DOI: 10.1007/s00031-010-9115-4.

[FFL11b] E. Feigin, G. Fourier, and P. Littelmann. "PBW filtration and bases for symplectic Lie algebras". *Int. Math. Res. Not. IMRN* 24 (2011), pp. 5760–5784. DOI: 10.1093/imrn/rnr014.

[FK21] G. Fourier and D. Kus. "PBW degenerations of Lie superalgebras and their typical representations". *J. Lie Theory* 31.2 (2021), pp. 313–334.

[FN17] N. Fujita and S. Naito. "Newton–Okounkov convex bodies of Schubert varieties and polyhedral realizations of crystal bases". *Math. Z.* 285.1-2 (2017), pp. 325–352. DOI: 10.1007/s00209-016-1709-7.

[Gor15] A. A. Gornitskiĭ. "Essential signatures and canonical bases of irreducible representations of the group G_2". *Mat. Zametki* 97.1 (2015), pp. 35–47. DOI: 10.4213/mzm10384.

[Gor19] A. A. Gornitskiĭ. "Essential signatures and monomial bases for B_n and D_n". *J. Lie Theory* 29.1 (2019), pp. 277–302.

[GHKK18] M. Gross, P. Hacking, S. Keel, and M. Kontsevich. "Canonical bases for cluster algebras". *J. Amer. Math. Soc.* 31.2 (2018), pp. 497–608. DOI: 10.1090/jams/890.

[Hum72] J. E. Humphreys. *Introduction to Lie algebras and representation theory.* Vol. Vol. 9. Graduate Texts in Mathematics. Springer-Verlag, New York-Berlin, 1972, pp. xii+169.

[Kas93] M. Kashiwara. "Global crystal bases of quantum groups". *Duke Math. J.* 69.2 (1993), pp. 455–485. DOI: 10.1215/S0012-7094-93-06920-7.

[KK12] K. Kaveh and A. G. Khovanskii. "Newton—Okounkov bodies, semigroups of integral points, graded algebras and intersection theory". *Ann. of Math. (2)* 176.2 (2012), pp. 925–978. DOI: 10.4007/annals.2012.176.2.5.

[LS86] V. Lakshmibai and C. S. Seshadri. "Geometry of G/P. V". *J. Algebra* 100.2 (1986), pp. 462–557. DOI: 10.1016/0021-8693(86)90089-X.

[LM09] R. Lazarsfeld and M. Mustaţă. "Convex bodies associated to linear series". *Ann. Sci. Éc. Norm. Supér. (4)* 42.5 (2009), pp. 783–835. DOI: 10.24033/asens.2109.

[Lit98] P. Littelmann. "Contracting modules and standard monomial theory for symmetrizable Kac-Moody algebras". *J. Amer. Math. Soc.* 11.3 (1998), pp. 551–567. DOI: 10.1090/S0894-0347-98-00268-9.

[Lus90] G. Lusztig. "Canonical bases arising from quantized enveloping algebras". *J. Amer. Math. Soc.* 3.2 (1990), pp. 447–498. DOI: 10.2307/1990961.

[Mak19] I. Makhlin. "FFLV-type monomial bases for type B". *Algebr. Comb.* 2.2 (2019), pp. 305–322. DOI: 10.5802/alco.41.

[MV88] I. Mirković and K. Vilonen. "Characteristic varieties of character sheaves". *Invent. Math.* 93.2 (1988), pp. 405–418. DOI: 10.1007/BF01394339.

Matroids

Daniel Corey, Lukas Kühne and Benjamin Schröter

Abstract: A matroid is a fundamental and actively studied object in combinatorics. Matroids generalize linear dependency in vector spaces as well as many aspects of graph theory. Moreover, matroids form a cornerstone of tropical geometry and a deep link between algebraic geometry and combinatorics. In this chapter we present `OSCAR` functions for matroids through several examples. Our focus lies in particular on computing the realization space and the Chow ring of a matroid.

Key words: Matroids, realization spaces, Hodge theory of finite geometries, Tutte polynomial

1 Introduction

A matroid is a combinatorial abstraction of independence, e.g., linear independence of vectors or spanning sets of edges in a graph. Matroids were introduced by Whitney [Whi35] and independently by Nakasawa (see [Nak09]). They are a central object in mathematics connecting multiple disciplines such as combinatorics, algebra, and geometry. Consequently, matroids have found applications in fields such as control theory, optimization, and algebraic geometry, see for

Daniel Corey
University of Nevada, Department of Mathematical Sciences, Las Vegas, USA, e-mail: `daniel.corey@unlv.edu`

Lukas Kühne
Universität Bielefeld, Fakultät für Mathematik, Germany and Institute for Andvanced Study, Princeton, USA, e-mail: `lukas.kuehne@math.uni-bielefeld.de`

Benjamin Schröter
KTH Royal Institute of Technology, Department of Mathematics, Stockholm, Sweden, e-mail: `schrot@kth.se`

W. Decker et al. (eds.), *The Computer Algebra System OSCAR*,
Algorithms and Computation in Mathematics 32,
https://doi.org/10.1007/978-3-031-62127-7_14

example [Brä11; Edm03; GGMS87] respectively. Of particular importance is the work of Adiprasito, Huh and Katz [AHK18], who demonstrate that matroids admit a Hodge theory originally stemming from algebraic geometry.

In this chapter, we demonstrate how one can use our implementation of matroids in `OSCAR` by focusing on realization spaces and Chow rings of matroids.

2 Basics

Let V be a d-dimensional vector space, E a finite set, and $(v_i)_{i\in E}$ a sequence of spanning vectors of V. There are various ways to record the linear dependencies among these vectors.

- *Independent sets*: $\mathcal{I} = \{A \subseteq E \,:\, (v_i)_{i\in A} \text{ are linearly independent}\}$.
- *Bases*: $\mathcal{B} = \{A \subseteq E \,:\, (v_i)_{i\in A} \text{ is a basis of } V\}$.
- *Rank*: $\mathrm{rk} : 2^E \to \mathbb{Z}_{\geq 0}$; $\mathrm{rk}(A) = \dim \mathrm{span}(v_i \,:\, i \in A)$.
- *Flats*: $\mathcal{F} = \{A \subseteq E \,:\, v_j \notin \mathrm{span}(v_i \,:\, i \in A) \text{ for all } j \notin A\}$.
- *Circuits*: $\mathcal{C} = \{A \subseteq E \,:\, (v_i)_{i\in A} \text{ are minimally linearly dependent}\}$.

The notions of independent sets, bases, rank, flats, and circuits may each be axiomitized, each of which leads to a definition of a matroid. We favor the description in terms of bases.

Definition 1 A *matroid* M consists of a finite set E and a non-empty collection $\mathcal{B} \subseteq 2^E$ that satisfies the *basis exchange axiom*: for each pair A, B of distinct elements of $\mathcal{B}$ and $x \in A \setminus B$, there is a $y \in B \setminus A$ such that $A \setminus \{x\} \cup \{y\}$ is in $\mathcal{B}$.

Elements of $\mathcal{B}$ are called *bases* of M. To emphasize dependence on M, especially when M is defined using a different set of axioms, we write $\mathcal{B}(\mathsf{M})$ for $\mathcal{B}$. All bases of M have the same size [Oxl11, Lemma 1.2.1]; this common size is called the *rank* of M. The remaining ways of recording dependence can be recovered from this data.

- *Independent sets*: $\mathcal{I}(\mathsf{M}) = \{A \subseteq \{1, 2, \ldots, n\} \,:\, A \subseteq B \text{ for some } B \in \mathcal{B}\}$.
- *Rank*: $\mathrm{rk}_{\mathsf{M}} : 2^E \to \mathbb{Z}_{\geq 0}$, $\mathrm{rk}(A) = \max(|I \cap A| \,:\, I \in \mathcal{I})$.
- *Flats*: $\mathcal{F}(\mathsf{M}) = \{A \subseteq E \,:\, \mathrm{rk}(A \cup e) > \mathrm{rk}(A) \text{ for all } e \in E \setminus A\}$.
- *Circuits*: $\mathcal{C}(\mathsf{M}) = \{A \subseteq E \,:\, A \notin \mathcal{I}(\mathsf{M}) \text{ and } A' \in \mathcal{I}(\mathsf{M}) \text{ for all } A' \subsetneq A\}$.

For example, consider the matroid M on $E = \{1, 2, 3, 4\}$ whose bases are $\mathcal{B}(\mathsf{M}) = \{12, 13, 14, 23, 24\}$ where we used, e.g., 12 as a short hand notation for the set $\{1, 2\}$.

```
julia> B = [[1,2],[1,3],[1,4],[2,3],[2,4]];

julia> M = matroid_from_bases(B,4)
Matroid of rank 2 on 4 elements
```

Once created, we can determine other axiomatic characterizations of M.

```
julia> independent_sets(M)
10-element Vector{Vector{Int64}}:
 []
 [1]
 [2]
 [3]
 [4]
 [1, 4]
 [2, 4]
 [1, 3]
 [2, 3]
 [1, 2]

julia> flats(M)
5-element Vector{Vector{Int64}}:
 []
 [1]
 [2]
 [3, 4]
 [1, 2, 3, 4]

julia> circuits(M)
3-element Vector{Vector{Int64}}:
 [3, 4]
 [1, 2, 4]
 [1, 2, 3]
```

Consider a vector configuration whose elements are the columns of a $r \times n$ full-rank matrix X. The matroid of this configuration, denoted $\mathsf{M}[X]$, is the matroid whose groundset is $[n] = \{1, 2, \ldots, n\}$ and its bases are the collections of r columns that are of full rank. Because multiplying a column of X by a nonzero number does not change the matroid $\mathsf{M}[X]$, we may also view the columns as points in the projective space $\mathbb{P}^{r-1}$, and so X may be viewed as a projective realization of M. This perspective is particularly useful for illustrating matroids, and projective realizations of some important matroids are depicted in Figure 1. For example, the famous *Fano matroid* F is the left-most diagram in this figure. This matroid is $\mathsf{F} = \mathsf{M}[X]$ where X is the 3×7 matrix whose columns are the 7 points of $\mathbb{P}^2(\mathbb{F}_2)$.

```
julia> X = matrix(GF(2), [1 0 1 0 1 0 1; 0 1 1 0 0 1 1; 0 0 0 1 1 1 1])
[1   0   1   0   1   0   1]
[0   1   1   0   0   1   1]
[0   0   0   1   1   1   1]


julia> F = matroid_from_matrix_columns(X)
Matroid of rank 3 on 7 elements
```

Let M_1 and M_2 be matroids with ground sets E_1 and E_2, respectively. An *isomorphism* of matroids $\varphi : \mathsf{M}_1 \to \mathsf{M}_2$ is a bijection $\varphi : E_1 \to E_2$ such that $A \in \mathcal{B}(\mathsf{M}_1)$ if and only if $\varphi(A) \in \mathcal{B}(\mathsf{M}_2)$. We may verify that $\mathsf{M}[X]$ is isomorphic to the Fano matroid.

```
julia> is_isomorphic(F, fano_matroid())
true
```

The *automorphism group* of a matroid M, denoted $\mathrm{Aut}(\mathsf{M})$, is the group of all isomorphism from M to itself. The automorphism group of the Fano matroid is $\mathrm{PSL}_3(\mathbb{F}_2)$.

```
julia> G = automorphism_group(F);

julia> describe(G)
"PSL(3,2)"
```

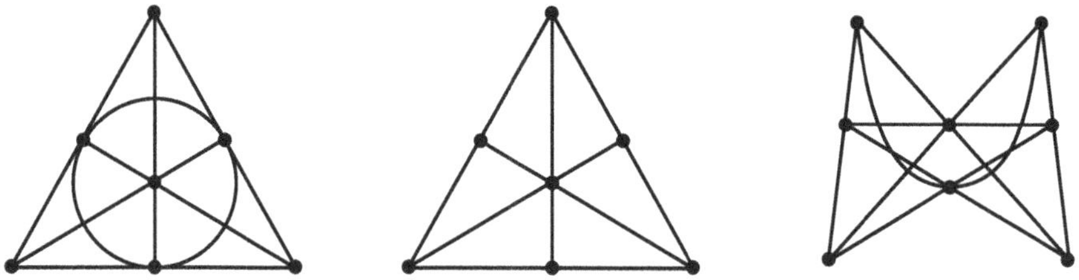

Fig. 1: Projective realizations of the Fano, non-Fano and Möbius–Kantor matroids. Points connected by an arc are interpreted as collinear.

3 Realizability

In the previous section, we saw how to obtain a matroid from a vector configuration, and when the vectors are arranged as columns of a matrix X, the resulting matroid is $\mathsf{M}[X]$. Let $\mathbb{F}$ be a field. A matroid M is $\mathbb{F}$-*realizable* if there is a matrix X with entries in the field $\mathbb{F}$ such that $\mathsf{M} \cong \mathsf{M}[X]$. If M is $\mathbb{F}$-realizable for some field $\mathbb{F}$, then M is said to be realizable.

Denote by $\binom{[n]}{r}$ the r-element subsets of $[n]$. The smallest matroid, with respect to the size of the groundset, that is not realizable over any field is the *Vámos* matroid.

```
julia> nonbases(vamos_matroid())
5-element Vector{Vector{Int64}}:
 [1, 2, 3, 4]
 [1, 2, 5, 6]
 [1, 2, 7, 8]
 [3, 4, 5, 6]
 [5, 6, 7, 8]
```

It is a rank 4 matroid on a set of size 8, and its bases (up to isomorphism) are the elements of $\binom{\{1,\ldots,8\}}{4}$ other than those listed above. We verify below with our code that it is indeed not realizable.

Determining whether a matroid is realizable is in general a hard problem. There does not exist a finite list of axioms that combinatorially characterize realizability [MNW18], which led Vámos to quip "the missing axiom of matroid theory is lost forever" [Vám78]. Nevertheless, the Vámos matroid violates Ingleton's inequality [Ing71] and hence is not realizable; see [Oxl11, §6.1, Exer. 7]. Moreover, in [Stu87], Sturmfels proves that matroid-realizability over $\mathbb{Q}$ is equivalent to the solvability of Diophantine equations over $\mathbb{Q}$. This is an extension of Hilbert's 10th problem to the rationals and it is not known whether this problem is decidable or not.

On the other hand, as a result of the discussion below, realizability of matroids over algebraically closed fields can be decided using Gröbner bases. Nevertheless, there are practical limitations in that and this algorithm becomes rather slow if the matroid is relatively large.

A closely related concept to realizability is the notion of a realization space of the matroid M. Given a field $\mathbb{F}$ and a matroid M, its *realization space* $\mathcal{R}(\mathsf{M};\mathbb{F})$ is a (possibly empty) algebraic variety (more generally, a scheme), defined over $\mathbb{F}$, whose closed points parameterize equivalence classes of point configurations in $\mathbb{P}^{r-1}_{\mathbb{F}}$ whose matroid is M, where two configurations are equivalent if one can be transformed to the other by an element of $\mathrm{PGL}_r(\mathbb{F})$. In particular, the matroid M is $\mathbb{F}$-realizable if and only if $\mathcal{R}(\mathsf{M};\mathbb{F}) \neq \emptyset$.

3.1 Algorithmic Framework

In this subsection we describe the basic algorithmic framework to compute the affine coordinate ring of a matroid realization space. We illustrate this code in the a series of examples in the next subsection. Following [Oxl11, Thm. 6.8.9] the basic algorithm does the following for a rank r matroid M on the ground set $[n]$. Assume that M has no loops (a *loop* of a matroid is an element not contained in any basis). To streamline the exposition, we consider the special case where $B_0 = \{n-r+1, \ldots, n\}$ is a basis of M. Let $A = \mathbb{Z}[x_{ij}]$ be the polynomial ring in the $r(n-r)$ variables x_{ij} for $1 \leq i \leq r$, $1 \leq j \leq n-r$. Define the matrix

$$X = \begin{bmatrix} x_{11} & x_{12} & \cdots & x_{1,n-r} & 1 & 0 & \cdots & 0 \\ x_{21} & x_{22} & \cdots & x_{2,n-r} & 0 & 1 & \cdots & 0 \\ \cdots & \cdots & \cdots & \cdots & \cdots & \cdots & \cdots & \cdots \\ x_{r1} & x_{r2} & \cdots & x_{r,n-r} & 0 & 0 & \cdots & 1 \end{bmatrix}.$$

The matroid M is realizable over a field $\mathbb{F}$ if we can replace every x_{ij} with an element of $\mathbb{F}$ such that $\mathsf{M} = \mathsf{M}[X]$. For each r-element subset N of $[n]$, write p_N for the determinant of X whose columns are indexed by the elements of N (preserving the order). To realize M, the x_{ij} must satisfy $p_N = 0$ for $N \in \binom{[n]}{r} \setminus \mathcal{B}(\mathsf{M})$ and $p_N \neq 0$ for $N \in \mathcal{B}(\mathsf{M})$. Observe that $x_{ij} = (-1)^{i+1} p_B$ for $i \in [r]$ and $j \in [n-r]$ where $B = B_0 \setminus \{n-r+i\} \cup \{j\}$. Therefore, we must have $x_{ij} = 0$ whenever $B_0 \setminus \{n-r+i\} \cup \{j\} \notin \mathcal{B}(\mathsf{M})$.

The matroid $\mathsf{M}[X]$ is invariant under row and column scaling by a nonzero scalar as this does not change the linear dependence structure among the columns. In practice, this means that we may set equal to 1 the first nonzero entry of each column, and then among the remaining entries, we may set equal to 1 the first nonzero entry of each row (if this entry exists).

Let us carefully carry out this procedure. Let us denote the row with the first nonzero entry in column $j \in [n-r]$ by $\mu(j)$ where $\mu(j) = r+1$ if there is no such entry. In other words we obtain the function $\mu : [n-r] \to [r+1]$ with

$$\mu(j) = \min(\{r+1\} \cup \{i \in [r] : B_0 \setminus \{n-r+i\} \cup \{j\} \in \mathcal{B}(\mathsf{M})\}).$$

Similarly the first nonzero (column) entry j in row $i \in [r]$ not fulfilling $\mu(j) = i$ is denoted by $\nu(i)$. That is $\nu : [r] \to [n-r+1]$ is the function

$$\nu(i) = \min(\{n-r+1\} \cup \{j \in [n-r] : B_0 \setminus \{n-r+i\} \cup \{j\} \in \mathcal{B}(\mathsf{M}) \wedge \mu(j) \neq i\}).$$

where $\nu(i) = n-r+1$ indicates that there is no such entry. Let us further set

$$A_{\mathsf{M}} = \mathbb{Z}[x_{ij} \ : \ B_0 \setminus \{n-r+i\} \cup \{j\} \in \mathcal{B}(\mathsf{M}), i \neq \mu(j), j \neq \nu(i)]$$

Define a ring homomorphism

$$\pi : A \to A_{\mathsf{M}} \quad \pi(x_{ij}) = \begin{cases} 0 & \text{if } B_0 \setminus \{n-r+i\} \cup \{j\} \notin \mathcal{B}(\mathsf{M}), \\ 1 & \text{if } i = \mu(j) \text{ or } j = \nu(i), \\ x_{ij} & \text{otherwise.} \end{cases}$$

Define the ideal I_{M} and multiplicative semigroup U_{M} by

$$I_{\mathsf{M}} = \langle \pi(p_N) \mid N \notin \mathcal{B}(\mathsf{M}),\ |N| = r \rangle, \text{ and } U_{\mathsf{M}} = \langle \pi(p_B) \mid B \in \mathcal{B}(\mathsf{M}) \rangle_{\text{smgp}}.$$

Here, the subscript smgp indicates that we take the multiplicative semigroup generated by the listed elements. The realization space $\mathcal{R}(\mathsf{M};\mathbb{F})$ is an affine scheme, and its affine coordinate ring is $S_{\mathsf{M}} \otimes_{\mathbb{Z}} \mathbb{F}$ where

$$S_{\mathsf{M}} = U_{\mathsf{M}}^{-1} A_{\mathsf{M}} / I_{\mathsf{M}},$$

and U_M^{-1} is the localization of the multiplicative semigroup. The matroid M is realizable if and only if $\mathcal{R}(\mathsf{M}, \mathbb{F})$ is nonempty, equivalently, S_{M} is not the zero ring. In practice, showing S_{M} is the zero ring amounts to showing that the saturation of I_{M} with respect to the semigroup S_{M} is the unit ideal in A_{M}. When $\mathbb{F}$ is algebraically closed, a closed point in $\mathcal{R}(\mathsf{M};\mathbb{F})$ corresponds to a realization of M via the matrix X.

The details of this implementation are also described in [BK23]. To speed up this computation we use the following techniques.

- The assumption that $B_0 = \{n - r + 1, \ldots, n\}$ is a basis of M allows us to assume that the last r columns of X form the $r \times r$ identity matrix. Of course, the above procedure may be modified so that the submatrix formed by the columns of B_0 is the identity matrix *for any* $B_0 \in \mathcal{B}(\mathsf{M})$.
- We do a short check before the realization space computation for which bases the ring A_{M} has the fewest variables and use this basis instead as this makes the following computations easier.
- We embed the space $\mathcal{R}(\mathsf{M}; \mathbb{F})$ into a smaller dimensional ambient space using techniques described in [BK23, Appendix A] and [CL23b, § 6].
- We simplify the ideal I_{M} by computing a reduced Gröbner basis and reduce the polynomials in U_{M} with respect to this basis.

3.2 Examples

We illustrate the procedure from the previous section in the following examples.

Example 2 Let's revisit the Fano matroid F introduced in the previous section:

```
julia> realization_space(F, simplify=false)
The realization space is
  [0   1    1    1    1   0   0]
  [1   0    1   x2    0   1   0]
  [1   0   x1    0   x3   0   1]
in the multivariate polynomial ring in 3 variables over ZZ
within the vanishing set of the ideal
Ideal (2, x3 + 1, x2 - 1, x1 + 1)
avoiding the zero loci of the polynomials
RingElem[x2, x3, x1, x1*x2 - x2*x3 + x3, x1 - x3 - 1, x1 + x2 - 1]
```

The ring A_{F} is the polynomial ring in 3 variables with *integer* coefficients. The output shows the matrix $\pi(X)$ which parametrizes the realizations, the ideal I_{F} and the inequations in U_{F}. The equations actually already determine the variables x_1, x_2, x_3. So we can simplify the realization space using the equations in I_{F}. We use exactly equations of this kind to embed $\mathcal{R}(\mathsf{F}; \mathbb{F})$ into a smaller dimensional ambient space as mentioned above.

```
julia> realization_space(F, simplify=true)
The realization space is
  [0   1   1   1   1   0   0]
  [1   0   1   1   0   1   0]
  [1   0   1   0   1   0   1]
in the integer ring
within the vanishing set of the ideal
2ZZ
```

Because the vanishing ideal is $\langle 2 \rangle$, the matroid is realizable only over fields of characteristic 2. Over any such field a realization of the Fano matroid is given by this matrix. We can also check directly that this matroid is not realizable over fields of other characteristics. For instance, we may check characteristic 5:

```
julia> is_realizable(F, char=5)
false
```

The default option for the parameter `simplify` is `true` so we will omit the parameter from now on.

Example 3 A matroid closely related to the Fano matroid F is the non-Fano matroid. This matroid is obtained from the F by turning one of the seven circuits of F into a basis, see Figure 1.

```
julia> realization_space(non_fano_matroid())
The realization space is
  [1   1   0   0   1   1   0]
  [0   1   1   1   1   0   0]
  [0   1   1   0   0   1   1]
in the integer ring
avoiding the zero loci of the polynomials
RingElem[2]
```

In this case, the defining ideal is trivial, but the element 2 must not vanish in the field we realize this matroid. Therefore the matroid is realizable over fields of every characteristic except 2.

Example 4 We can also confirm that the Vámos matroid is not realizable over any field:

```
julia> is_realizable(vamos_matroid())
false
```

Example 5 Let M be the *Möbius-Kantor* matroid.

```
julia> M = moebius_kantor_matroid();

julia> nonbases(M)
8-element Vector{Vector{Int64}}:
 [1, 2, 3]
 [1, 4, 5]
 [1, 7, 8]
 [2, 4, 6]
 [2, 5, 8]
 [3, 4, 7]
 [3, 6, 8]
 [5, 6, 7]
```

This is a rank-3 matroid on $\{1,\ldots,8\}$ whose bases (up to isomorphism) are the elements of $\binom{\{1,\ldots,8\}}{3}$ other than those listed above, see Figure 1 for an illustration. We may compute a simple presentation of the ring S_{M} which defines an embedding of $\mathcal{R}(\mathsf{M};\mathbb{F})$ in a smaller dimensional space:

```
julia> realization_space(M)
The realization space is
  [1   0   1   0   1     0          1     1]
  [0   1   1   0   0     1          1    x1]
  [0   0   0   1   1   -x1   -x1 + 1     1]
in the multivariate polynomial ring in 1 variable over ZZ
within the vanishing set of the ideal
Ideal (x1^2 - x1 + 1)
avoiding the zero loci of the polynomials
RingElem[x1, x1 - 1, x1^2 + 1]
```

This means that

$$S_{\mathsf{M}} \cong \mathbb{Z}[x^{\pm}]/\langle x^2 - x + 1\rangle.$$

(The polynomial $x - 1$ is already a unit in this ring.) Therefore, M is $\mathbb{F}$-realizable if and only if $x^2 - x + 1$ has a root in $\mathbb{F}$ (equivalently, $\mathbb{F}$ has a primitive 6-th root of unity). In characteristic 0, we see that M is not realizable over $\mathbb{R}$, and therefore not orientable.

We can also check whether M is realizable over a specific finite field $\mathbb{F}_q$ for some prime power q.

```
julia> is_realizable(M, q=9)
true
```

Hence M is realizable over $\mathbb{F}_9$. We can also check this systematically for all prime powers of size up to 13.

```
julia> qs = [2,3,4,5,7,8,9,11,13];
julia> println([is_realizable(M, q=a) for a in qs])
Bool[0, 1, 1, 0, 1, 0, 1, 0, 1]
```

From this list we can deduce the hypothesis that M is realizable over the finite field $\mathbb{F}_q$ if and only if $q \not\equiv 2 \mod 3$.

As mentioned in the introduction, a second motivation for matroids stems from graph theory. An undirected graph $G = (V, E)$ gives rise to a *graphic matroid* $\mathsf{M}(G)$ on the ground set E and bases the maximal subsets of edges that do not contain a cycle.

Example 6 In this example, we compute the realization space of the graphic matroid of the complete graph on four vertices K_4.

```
julia> M = cycle_matroid(complete_graph(4));
julia> realization_space(M)
The realization space is
  [1   0   1   0   1    0]
  [0   1   1   0   0    1]
  [0   0   0   1   1   -1]
in the integer ring
```

Hence there are no algebraic conditions on the field, and this matrix defines a realization of $\mathsf{M}(K_4)$ over every field (of course, $-1 = 1$ over fields of characteristic 2). That $\mathcal{R}(\mathsf{M}(K_4); \mathbb{F})$ consists of a single point for any field $\mathbb{F}$ is consistent with the fact due to White which asserts that $\mathcal{R}(\mathsf{M}; \mathbb{F})$ consists of a single point whenever M is realizable over $\mathbb{F}_2$, see [Kat16, Theorem 6.3].

Example 7 Another prominent matroid is the *Pappus* matroid as its configurations of rank two flats is a Pappus configuration. Let's begin by computing its realization space over $\mathbb{C}$:

```
julia> RS = realization_space(pappus_matroid(), char=0)
The realization space is
  [1   0   1   0   x2   x2                     x2^2    1    0]
  [0   1   1   0    1    1   -x1*x2 + x1 + x2^2    1    1]
  [0   0   0   1   x2   x1                   x1*x2   x1   x2]
in the multivariate polynomial ring in 2 variables over QQ
avoiding the zero loci of the polynomials
RingElem[x1 - x2, x2, x1, x2 - 1, x1 + x2^2 - x2, x1 - 1, x1*x2 - x1 -
↪  x2^2]
```

Thus the realization space is the affine space $\mathbb{A}^2$ over $\mathbb{C}$ with the seven specified curves removed. One can obtain a specific realization of this matroid by picking a point in that space that avoids these seven exceptional curves:

```
One realization is given by
  [1   0   1   0   2   2   4   1   0]
  [0   1   1   0   1   1   1   1   1]
  [0   0   0   1   2   3   6   3   2]
in the rational field
```

Example 8 `OSCAR` is connected to the database `polyDB` [Paf17]. To illustrate this functionality, we collect all simple matroids with rank 3 on 9 elements, up to isomorphism. [1]

```
julia> db = Polymake.Polydb.get_db();
julia> collection = db["Matroids.Small"];
julia> query = Dict("RANK"=>3,"N_ELEMENTS"=>9,"SIMPLE"=>true);
julia> results = Polymake.Polydb.find(collection, query);
julia> oscar_matroids = [Matroid(pm) for pm in results];
julia> length(oscar_matroids)
383
```

There are 383 such matroids, and only 370 of these are realizable over a field of characteristic 0.

```
julia> char_0_matroids = [M for M in oscar_matroids
         if is_realizable(M, char=0)];
julia> length(char_0_matroids)
370
```

As a consequence of Mnëv's universality theorem [Mnë85; Mnë88], realization spaces for rank-3 matroids satisfies Murphy's law in the sense of Vakil [Vak06], i.e., each singularity type appears in the realization space of a rank-3 matroid. While the modern proofs of this fact are constructive [Car15; Laf03; LV13], the ground sets of the matroids produced are large, even for the simplest singularities. In [CL23a], the authors use these `OSCAR` realization space functions to prove that realization spaces for $\mathbb{C}$–realizable, rank-3 matroids

[1] The database contains one representative for every isomorphism class of matroids.

on ground sets with fewer than 11 elements are all smooth, but there are rank-3 matroids on at least 12 elements whose realization spaces have nodal singularities.

4 Chow Rings of Geometries

To a matroid one might associate an element of an abelian group. Such a mapping is called a *matroid invariant* if it is constant for all members in an matroid isomorphism class (analogous definitions can be made for graphs). Prominent examples include the Tutte polynomial or the characteristic polynomial. For an overview on matroid invariants and the Tutte polynomial see, e.g., [FS24; EM22].

The *Tutte polynomial* of a matroid M is the bivariate polynomial

$$T_{\mathsf{M}}(x,y) = \sum_{A\subseteq E} (x-1)^{\mathrm{rk}(E)-\mathrm{rk}(A)}(y-1)^{|A|-\mathrm{rk}(A)},$$

and the *characteristic polynomial* of M is the specialization

$$\chi_{\mathsf{M}}(q) = (-1)^{\mathrm{rk}(E)} T_{\mathsf{M}}(1-q,0) = \sum_{A\subseteq E} (-1)^{|A|}\, q^{\mathrm{rk}(E)-\mathrm{rk}(A)}.$$

If M is the graphic matroid of a simple graph G with c connected components then $T_{\mathsf{M}}(x,y)$ is the Tutte polynomial of G and $q^c\chi_{\mathsf{M}}(q)$ coincides with the *chromatic polynomial* $\chi_G(q)$ of G which counts the number of node colorings using q colors such that adjacent nodes have distinct colors.

For example the complete graph K_n on n nodes has

$$\chi_{\mathsf{K}_n}(q) = q\cdot(q-1)\cdots(q-n+1)$$

q-colorings. The following lines of code reproduce this fact in `OSCAR` for the complete graph on $n=4$ nodes. They show how one can compute the Tutte and chromatic polynomial of the matroid $\mathsf{M}(\mathsf{K}_n)$.

```
julia> M = cycle_matroid(complete_graph(4))
Matroid of rank 3 on 6 elements

julia> tutte_polynomial(M)
x^3 + 3*x^2 + 4*x*y + 2*x + y^3 + 3*y^2 + 2*y

julia> char_poly = characteristic_polynomial(M)
q^3 - 6*q^2 + 11*q - 6

julia> factor(char_poly)
1 * (q - 2) * (q - 1) * (q - 3)
```

Observe that in this example the coefficients w_i of the characteristic polynomial $\chi_{\mathsf{M}}(q) = \sum_{j=0}^{\mathrm{rk(E)}} \omega_j q^j$ form a *log-concave* sequence, i.e., for all $0 < j < \mathrm{rk}(\mathsf{M})$ holds $\omega_j^2 \geq \omega_{j-1}\omega_{j+1}$. That this holds true for all matroids has been conjectured by Welsh [Wel76] and was known as the Heron–Rota–Welsh conjecture. This conjecture was famously proved by Adiprasito, Huh and Katz in [AHK18], which also contains historical background and further details about this conjecture.

To a loopfree matroid M, define the algebra $\mathrm{A}(\mathsf{M})$ by

$$\mathrm{A}(\mathsf{M}) = \mathbb{Q}[x_F \mid F \text{ is a proper flat of } \mathsf{M}]/(I+J)$$

where I and J are the ideals

$$I = \langle x_F \cdot x_G \mid F, G \text{ incomparable} \rangle \text{ and } J = \left\langle \sum_{F \ni i} x_F - \sum_{F \ni j} x_F \mid i, j \in E \right\rangle.$$

The algebra $\mathrm{A}(\mathsf{M})$ is the *Chow ring* of the matroid M. We obtain it in `OSCAR` via the following command.

```
julia> A = chow_ring(M);

julia> GR, _ = graded_polynomial_ring(QQ,symbols(base_ring(A)));

julia> AA = chow_ring(M,ring=GR);
```

This ring is graded by the polynomial degrees and generated in degree 1. In the code the ring `AA` is the graded version of the ring `A`. The component $\mathrm{A}^{\mathrm{rk}(E)-1}(\mathsf{M})$ is a one-dimensional $\mathbb{Q}$ vector space. The *volume map* is an isomorphism

$$\mathrm{vol}_{\mathsf{M}} : \mathrm{A}^{\mathrm{rk}(E)-1}(\mathsf{M}) \to \mathbb{Q}$$

normalized such that $\mathrm{vol}_{\mathsf{M}}(x_{F_1} \cdots x_{F_{\mathrm{rk}(E)-1}}) = 1$ for one (and therefore all) chains of flats $\emptyset \subsetneq F_1 \subsetneq \cdots \subsetneq F_{\mathrm{rk}(E)-1} \subsetneq E$.

We obtain the volume map in `OSCAR` via the following call.

```
julia> vol_map = volume_map(M,AA)
#3605 (generic function with 1 method)
```

The names and definitions are motivated from algebraic and toric geometry. The Chow ring of a $\mathbb{C}$-realizable matroid is the usual Chow ring of the "wonderful compactification" of the complement of a hyperplane arrangement realizing $\mathbb{C}$. See the work of de Concini and Procesi [DP95].

A degree 1 element $\sum c_F x_F \in \mathrm{A}^1(\mathsf{M})$ is called a *Lefschetz element* whenever

$$c_F + c_G > c_{F\cap G} + c_{F\cup G} \text{ for all flats } F, G \text{ of the matroid } \mathsf{M},$$

with $c_\emptyset = c_E = 0$. For example the elements

$$\alpha_i = \sum_{F \ni i} x_F \qquad \text{and} \qquad \beta_i = \sum_{F \not\ni i} x_F.$$

are limit points of Lefschetz elements as they weakly satisfy the above inequalities. Notice that neither α_i nor β_i depend on the choice of $i \in E$ as $\beta_j - \beta_i = \alpha_i - \alpha_j$, which is contained in J. Thus from now on we denote the elements α_i and β_i by α and β respectively.

To construct these two elements in `OSCAR` we pick all flats that either contain a certain element or exclude that element.

```
julia> e = matroid_groundset(M)[1];

julia> proper_flats = flats(M)[2:length(flats(M))-1];

julia> a = sum([AA[i] for i in 1:length(proper_flats) if e in
↪  proper_flats[i]])
x_{Edge(2, 1)} + x_{Edge(2, 1),Edge(3, 1),Edge(3, 2)} + x_{Edge(2,
↪  1),Edge(4, 1),Edge(4, 2)} + x_{Edge(2, 1),Edge(4, 3)}

julia> b = sum([AA[i] for i in 1:length(proper_flats) if !(e in
↪  proper_flats[i])])
x_{Edge(3, 1)} + x_{Edge(3, 2)} + x_{Edge(4, 1)} + x_{Edge(4, 2)} +
↪  x_{Edge(4, 3)} + x_{Edge(3, 1),Edge(4, 2)} + x_{Edge(3, 1),Edge(4,
↪  1),Edge(4, 3)} + x_{Edge(3, 2),Edge(4, 1)} + x_{Edge(3, 2),Edge(4,
↪  2),Edge(4, 3)}
```

A Lefschetz element gives rise to the following three properties known as the *Kähler package*.

Theorem 9 ([AHK18, Theorem 1.4 and 6.19]) *Let $\ell \in \mathrm{A}^1(\mathsf{M})$ be a Lefschetz element. Then the following three properties hold.*
Poincaré duality: For every non-negative integer $k \leq \frac{\mathrm{rk}(E)-1}{2}$, the bilinear pairing

$$\mathrm{A}^k(\mathsf{M}) \times \mathrm{A}^{\mathrm{rk}(E)-k-1}(\mathsf{M}) \to \mathbb{Q}, \qquad (\eta_1, \eta_2) \mapsto \mathrm{vol}_{\mathsf{M}}(\eta_1 \cdot \eta_2)$$

is non-degenerate.
Hard Lefschetz property: For every non-negative integer $k \leq \frac{\mathrm{rk}(E)-1}{2}$, the multiplication map

$$\mathrm{A}^k(\mathsf{M}) \to \mathrm{A}^{\mathrm{rk}(E)-k-1}(\mathsf{M}), \qquad \eta \mapsto \ell^{\mathrm{rk}(E)-2k-1} \cdot \eta$$

is an isomorphism.
Hodge–Riemann relations: For every non-negative integer $k \leq \frac{\mathrm{rk}(E)-1}{2}$, the bilinear form

$$\mathrm{A}^k(\mathsf{M}) \times \mathrm{A}^k(\mathsf{M}) \to \mathbb{Q}, \qquad (\eta_1, \eta_2) \mapsto (-1)^k \,\mathrm{vol}_{\mathsf{M}}(\eta_1 \cdot \ell^{\mathrm{rk}(E)-2k-1} \cdot \eta_2)$$

is positive definite on the kernel of the multiplication by $\ell^{\mathrm{rk}(E)-2k}$.

We now demonstrate one of many ways to verify these properties for $k=1$ and $\ell=\beta$. While β is only the limit of Lefschetz elements, it still satisfies the conclusions of the above theorem. First we generate a basis for each of the two vector spaces $\mathrm{A}^k(\mathsf{M})$ and $\mathrm{A}^{\mathrm{rk}(E)-k-1}(\mathsf{M})$. From these bases we derive the matrix Mat1 that represents the bilinear pairing of the Poincaré duality.

```
julia> k = 1
1

julia> R = base_ring(AA);

julia> g = grading_group(R)[1];

julia> PD1, mapPD1 = homogeneous_component(AA,k*g);

julia> basis_PD1 = [mapPD1(x) for x in gens(PD1)];

julia> PD2, mapPD2 = homogeneous_component(AA,(rank(M)-k-1)*g);

julia> basis_PD2 = [mapPD2(x) for x in gens(PD2)];

julia> Mat1 = matrix(QQ,[[vol_map(b1*b2) for b1 in basis_PD1] for b2 in
↪  basis_PD2])
[-1   0   0   0   0   0   0   1]
[ 0  -1   0   0   0   0   0   0]
[ 0   0  -1   0   0   0   0   1]
[ 0   0   0  -1   0   0   0   0]
[ 0   0   0   0  -1   0   0   1]
[ 0   0   0   0   0  -1   0   0]
[ 0   0   0   0   0   0  -1   0]
[ 1   0   1   0   1   0   0  -2]
```

It is easy to see that the matrix Mat1 is square and of full rank, and hence $\mathrm{A}^k(\mathsf{M}) \cong \mathrm{A}^{\mathrm{rk}(E)-k-1}(\mathsf{M})$.

Next we verify the hard Lefschetz property by looking at the bilinear form $\mathrm{A}^k\mathsf{M}) \times \mathrm{A}^k(\mathsf{M}) \to \mathbb{Q}$, $(\eta_1, \eta_2) \mapsto \mathrm{vol}_{\mathsf{M}}(\eta_1 \cdot \ell^{\mathrm{rk}(E)-2k-1} \cdot \eta_2)$ which is non-degenerate only if the hard Lefschetz map is an isomorphism as we already know $\mathrm{A}^k(\mathsf{M}) \cong \mathrm{A}^{\mathrm{rk}(E)-k-1}(\mathsf{M})$.

```
julia> Mat2 = matrix(QQ,[[vol_map(b1*b^(rank(M)-2k-1)*b2) for b1 in
↪  basis_PD1] for b2 in basis_PD1]);

julia> Mat1 == Mat2
true
```

This indeed is the case as the representing matrix Mat2 agrees with the full rank matrix Mat1.

The last property that we inspect is the Hodge–Riemann relations for which we begin by finding a basis of the kernel of the map $\eta \mapsto \ell^{\mathrm{rk}(E)-k} \cdot \eta$.

```
julia> RR, _ = graded_polynomial_ring(QQ,"y_#" => 1:length(basis_PD1));

julia> map = hom(RR,AA,basis_PD1);
```

```
julia> K = kernel(hom(RR,AA,[b^(rank(M)-2k)*b3 for b3 in basis_PD1]));

julia> basis_HR = [map(h) for h in gens(K) if degree(h).coeff==k*g.coeff]
7-element Vector{MPolyQuoRingElem{MPolyDecRingElem{QQFieldElem,
↪  QQMPolyRingElem}}}:
 x_{Edge(4, 3)}
 -x_{Edge(2, 1),Edge(3, 1),Edge(3, 2)} + x_{Edge(2, 1),Edge(4, 1),Edge(4,
 ↪  2)}
 -x_{Edge(2, 1),Edge(3, 1),Edge(3, 2)} + 2*x_{Edge(2, 1),Edge(4, 3)}
 -x_{Edge(2, 1),Edge(3, 1),Edge(3, 2)} + 2*x_{Edge(3, 1),Edge(4, 2)}
 -x_{Edge(2, 1),Edge(3, 1),Edge(3, 2)} + x_{Edge(3, 1),Edge(4, 1),Edge(4,
 ↪  3)}
 -x_{Edge(2, 1),Edge(3, 1),Edge(3, 2)} + 2*x_{Edge(3, 2),Edge(4, 1)}
 -x_{Edge(2, 1),Edge(3, 1),Edge(3, 2)} + x_{Edge(3, 2),Edge(4, 2),Edge(4,
 ↪  3)}
```

It remains to check that the bilinear form restricted to this kernel is positive definite.

```
julia> Mat3 = matrix(QQ,[[(-1)^k*vol_map(b1*b^(rank(M)-2k-1)*b2) for b1
↪  in basis_HR] for b2 in basis_HR])
[ 2   0   -2   0   -1   0   -1]
[ 0   2    1   1    1   1    1]
[-2   1    5   1    1   1    1]
[ 0   1    1   5    1   1    1]
[-1   1    1   1    2   1    1]
[ 0   1    1   1    1   5    1]
[-1   1    1   1    1   1    2]

julia> is_positive_definite(matrix(ZZ,[ZZ(i) for i in Mat3]))
true
```

Our next goal is to draw conclusions on the coefficients of the characteristic polynomial $\chi_{\mathsf{M}}(q)$ or more precisely the reduced characteristic polynomial $\overline{\chi}_{\mathsf{M}}(q) = \chi_{\mathsf{M}}(q)/(q-1) = \sum \overline{\omega}_j q^j$. The key is the fact that the j-th coefficient $\overline{\omega}_j$ of $\overline{\chi}_{\mathsf{M}}(q)$ agrees with $(-1)^j \mathrm{vol}_{\mathsf{M}}(\alpha^{\mathrm{rk}(E)-j-1}\beta^j)$ and that multiplication with α corresponds to a truncation of M, that is $\alpha \cdot \mathrm{A}(\mathsf{M}) \cong \mathrm{A}(\mathsf{M}')$ where the bases of M' are the corank-1 independent sets of M.

```
julia> reduced_characteristic_polynomial(M)
q^2 - 5*q + 6

julia> [vol_map(a^(rank(M)-j-1)*b^j) for j in range(0,rank(M)-1)]
3-element Vector{QQFieldElem}:
 1
 5
 6
```

Thus the Hodge–Riemann relations for the rank 2 truncation of M reveal that $\overline{\omega}_1^2 \geq \overline{\omega}_0\overline{\omega}_2$ as this is the determinant of the pairing with basis $\{\alpha, \beta\}$.

Iterating this argument leads then to the fact that the coefficients of $\overline{\chi}_{\mathsf{M}}(q)$ and thus $\chi_{\mathsf{M}}(q)$ do form a log-concave sequence. Of course we could have seen this for our example right away as $\chi_{\mathsf{K}_n}(q)$ has only real roots which implies log-concavity.

Acknowledgements Daniel Corey is supported by the SFB-TRR 195– 286237555 "Symbolic Tools in Mathematics and their Application". Lukas Kühne is supported by the SFB-TRR 358 – 491392403 "Integral Structures in Geometry and Representation Theory" and the SPP 2458 – 539866293 "Combinatorial Synergies". Benjamin Schröter is supported by the Swedish Research Council grant 2022-04224.

References

[AHK18] K. Adiprasito, J. Huh, and E. Katz. "Hodge theory for combinatorial geometries". *Ann. of Math. (2)* 188.2 (2018), pp. 381–452. DOI: `10.4007/annals.2018.188.2.1`.

[BK23] M. Barakat and L. Kühne. "Computing the nonfree locus of the moduli space of arrangements and Terao's freeness conjecture". *Mathematics of Computation* 92.341 (2023), pp. 1431–1452. DOI: `10.1090/mcom/3812`.

[Brä11] P. Brändén. "Obstructions to determinantal representability". *Adv. Math.* 226.2 (2011), pp. 1202–1212. DOI: `10.1016/j.aim.2010.08.003`.

[Car15] D. Cartwright. "Lifting matroid divisors on tropical curves". *Res. Math. Sci.* 2 (2015), Art. 23, 24. DOI: `10.1186/s40687-015-0041-x`.

[CL23a] D. Corey and D. Luber. "Singular matroid realization spaces". arXiv:2307.11915. 2023.

[CL23b] D. Corey and D. Luber. "The Grassmannian of 3-planes in $\mathbb{C}^8$ is schön". *Algebraic Combinatorics* 6.5 (2023), pp. 1273–1299. DOI: `10.5802/alco.302`.

[DP95] C. De Concini and C. Procesi. "Wonderful models of subspace arrangements". *Selecta Math. (N.S.)* 1.3 (1995), pp. 459–494. DOI: `10.1007/BF01589496`.

[Edm03] J. Edmonds. "Submodular functions, matroids, and certain polyhedra". In: *Combinatorial optimization—Eureka, you shrink!* Vol. 2570. Lecture Notes in Comput. Sci. Springer, Berlin, 2003, pp. 11–26. DOI: `10.1007/3-540-36478-1\{_}2`.

[EM22] J. A. Ellis-Monaghan and I. Moffatt. *Handbook of the Tutte Polynomial and Related Topics*. Chapman and Hall/CRC, 2022, p. 804. DOI: `10.1201/978-0-429-16161-2`.

[FS24] L. Ferroni and B. Schröter. "Valuative invariants for large classes of matroids". *J. Lond. Math. Soc. (2)* 110.3 (2024), Paper No. e12984. DOI: 10.1112/jlms.12984.

[GGMS87] I. M. Gelfand, R. M. Goresky, R. D. MacPherson, and V. V. Serganova. "Combinatorial geometries, convex polyhedra, and Schubert cells". *Adv. in Math.* 63.3 (1987), pp. 301–316. DOI: 10.1016/0001-8708(87)90059-4.

[Ing71] A. W. Ingleton. "Representation of matroids". In: *Combinatorial Mathematics and its Applications (Proc. Conf., Oxford, 1969)*. Academic Press, London-New York, 1971, pp. 149–167.

[Kat16] E. Katz. "Matroid theory for algebraic geometers". In: *Nonarchimedean and tropical geometry*. Simons Symposia. Springer, [Cham], 2016, pp. 435–517.

[Laf03] L. Lafforgue. *Chirurgie des grassmanniennes*. Vol. 19. CRM Monograph Series. American Mathematical Society, Providence, RI, 2003, pp. xx+170.

[LV13] S. H. Lee and R. Vakil. "Mnëv-Sturmfels universality for schemes". In: *A celebration of algebraic geometry*. Vol. 18. Clay Math. Proc. Amer. Math. Soc., Providence, RI, 2013, pp. 457–468.

[MNW18] D. Mayhew, M. Newman, and G. Whittle. "Yes, the 'missing axiom' of matroid theory is lost forever". *Trans. Amer. Math. Soc.* 370.8 (2018), pp. 5907–5929. DOI: 10.1090/tran/7408.

[Mnë88] N. E. Mnëv. "The universality theorems on the classification problem of configuration varieties and convex polytopes varieties". In: *Topology and geometry—Rohlin Seminar*. Vol. 1346. Lecture Notes in Math. Springer, Berlin, 1988, pp. 527–543. DOI: 10.1007/BFb0082792.

[Mnë85] N. E. Mnëv. "Varieties of combinatorial types of projective configurations and convex polyhedra". *Dokl. Akad. Nauk SSSR* 283.6 (1985), pp. 1312–1314.

[Nak09] T. Nakasawa. "Zur Axiomatik der linearen Abhängigkeit. I [Sci. Rep. Tokyo Bunrika Daigaku Sect. A **2** (1935), no. 43, 129–149; Zbl 0012.22001]". In: *A lost mathematician, Takeo Nakasawa*. Birkhäuser, Basel, 2009, pp. 68–88.

[Oxl11] J. Oxley. *Matroid theory*. Second. Vol. 21. Oxford Graduate Texts in Mathematics. Oxford University Press, Oxford, 2011, pp. xiv+684. DOI: 10.1093/acprof:oso/9780198566946.001.0001.

[Paf17] A. Paffenholz. "polyDB: a database for polytopes and related objects". In: *Algorithmic and Experimental Methods in Algebra, Geometry, and Number Theory*. Springer, 2017, pp. 533–547.

[Stu87] B. Sturmfels. "On the decidability of Diophantine problems in combinatorial geometry". *Bull. Amer. Math. Soc. (N.S.)* 17.1 (1987), pp. 121–124. DOI: 10.1090/S0273-0979-1987-15532-7.

[Vak06] R. Vakil. "Murphy's law in algebraic geometry: badly-behaved deformation spaces". *Invent. Math.* 164.3 (2006), pp. 569–590. DOI: 10.1007/s00222-005-0481-9.

[Vám78] P. Vámos. "The missing axiom of matroid theory is lost forever". *J. London Math. Soc. (2)* 18.3 (1978), pp. 403–408. DOI: 10.1112/jlms/s2-18.3.403.

[Wel76] D. J. A. Welsh. *Matroid theory*. L. M. S. Monographs, No. 8. Academic Press [Harcourt Brace Jovanovich, Publishers], London-New York, 1976, pp. xi+433.

[Whi35] H. Whitney. "On the Abstract Properties of Linear Dependence". *Amer. J. Math.* 57.3 (1935), pp. 509–533. DOI: 10.2307/2371182.

Generic Root Counts of Tropically Transverse Systems

An Invitation to Tropical Geometry in OSCAR

Isaac Holt and Yue Ren

Abstract The main mathematical focus of this chapter is a class of parametrised polynomial systems that we refer to as being tropically transverse. We show how their generic number of solutions can be expressed as the mixed volume of a modified system. We then provide an alternate proof of a recent result by Borovik et al, on the number of equilibria of coupled nonlinear oscillators using elementary tropical geometry. The proof draws upon a wide range of concepts across tropical geometry, which we will use as an opportunity to give an overview of various tropical features in OSCAR.

Key words: Tropical geometry, polynomial systems, root counts

1 Introduction

Tropical geometry is often described as a piecewise linear analogue of algebraic geometry. Its concepts arise naturally in many applications both inside and outside mathematics. Due to its explicit nature, algorithms and computations have always played a central role in tropical geometry. Prominent examples include: disproving strong polynomiality of certain barrier interior point methods in linear optimisation by computing tropicalisations of central paths [ABGJ18; AGV22], proving finiteness of certain central configurations in the N-body problem by computing tropical prevarieties of their algebraic

Isaac Holt
Durham University, Department of Mathematics, Durham DH1 3LE, United Kingdom
e-mail: `isaac.a.holt@durham.ac.uk`

Yue Ren
Durham University, Department of Mathematics, Durham DH1 3LE, United Kingdom
e-mail: `yue.ren2@durham.ac.uk`

W. Decker et al. (eds.), *The Computer Algebra System OSCAR*,
Algorithms and Computation in Mathematics 32,
https://doi.org/10.1007/978-3-031-62127-7_15

equations [HM06; HJ11], or breakthroughs in the geometry of moduli spaces by computing tropicalisations of linear series [FJP20].

This chapter addresses the problem of computing the generic number of solutions of a certain class of parametrized polynomial systems, which we call *tropically transverse*. These are a specialisation of the parametrised polynomial systems considered by Kaveh and Khovanskii in their work on Newton-Okounkov bodies [KK12]. Examples include the polynomials describing the stationary motion of coupled nonlinear oscillators studied by Borovik, Breiding, Del Pino, Michalek and Zilberberg [BBPMZ23]. One question of particular interest about parametrised polynomial system is that of their generic root count. It may be interesting in its own right, such as the Euclidean distance degree [DHOST14], or the maximum likelihood degree [CHKS06], but it is also a crucial piece of information for solving polynomial systems via numerical algebraic geometry [Bat+23]. We show that the generic root count of tropically transverse systems can be expressed as the mixed volume of a modified system in Theorem 30, and provide an alternate proof of [BBPMZ23, Theorem 5.1] using elementary tropical geometry in Theorem 32.

Before tackling tropically transverse systems, this paper serves as an introduction and invitation to the tropical package in `OSCAR`. While the package is still rapidly growing and evolving, it brings as part of the greater `OSCAR` and `Julia` ecosystem a plethora of powerful and sophisticated algorithms within arm's reach of the tropical user.

All algorithms pertaining to tropically transverse systems are openly available under `https://github.com/isaacholt100/generic_root_count`.

2 Tropical Functionalities in OSCAR

In this section, we give a very brief summary on some tropical features in `OSCAR`, whilst introducing some necessary concepts for the rest of the chapter. Our notation generally follows [MS15], though we may need to deviate from it for practical reasons.

2.1 Tropical Semirings

First, we begin with the most fundamental building block of tropical geometry, the tropical semiring.

Definition 1 While tropical semirings in [MS15, Section 1.1] are extensions of the real numbers, tropical semirings in `OSCAR` are either

$$\mathbb{T}_{\max} := (\mathbb{Q} \cup \{-\infty\}, \oplus, \odot) \text{ with } a \oplus b := \max(a, b), a \odot b := a + b, \text{ or}$$
$$\mathbb{T}_{\min} := (\mathbb{Q} \cup \{+\infty\}, \oplus, \odot) \text{ with } a \oplus b := \min(a, b), a \odot b := a + b \text{ (default).}$$

This allows OSCAR to avoid precision issues that arise when working with real numbers. $\mathbb{T}_{\min}$ and $\mathbb{T}_{\max}$ are referred to as the *min-plus* or *max-plus semiring*, respectively. Their elements are called *tropical numbers*.

Example 2 (Tropical Semiring and Tropical Numbers) Like [MS15], OSCAR follows the min-convention. This means that the tropical semiring is the min-plus semiring by default:

```
julia> T = tropical_semiring()
Min tropical semiring

julia> T(1)+T(2), T(1)*T(2), zero(T), one(T)
((1), (3), infty, (0))

julia> T = tropical_semiring(max)
Max tropical semiring

julia> T(1)+T(2), T(1)*T(2), zero(T), one(T)
((2), (3), -infty, (0))
```

Note:

1. Tropical numbers are printed in parenthesis to distinguish them from rational numbers.
2. Rational numbers $0, 1 \in \mathbb{Q}$ are converted to tropical numbers $0, 1 \in \mathbb{T}$, not the neutral elements of tropical addition and multiplication. Similarly, $-1 \in \mathbb{Q}$ is converted to $-1 \in \mathbb{T}$, the tropical multiplicative inverse of $1 \in \mathbb{T}$ and not the tropical additive inverse which does not exist.
3. Tropical semirings are ordered [Jos21, Section 2.7]. For two rational numbers $a < b$ we have
$$a < b \text{ in } \mathbb{T}_{\min} \quad \text{and} \quad a > b \text{ in } \mathbb{T}_{\max}.$$
Consequently, $\pm\infty$ are the largest elements of their respective semirings, and the conversion from rational to tropical numbers preserves the ordering for $\mathbb{T}_{\min}$ and reverses the ordering for $\mathbb{T}_{\max}$. The ordering is relevant for initial forms and ideals in Definition 5.

Example 3 (Tropical Matrices and Polynomials) In OSCAR, tropical semirings are concrete subtypes of `field` in which subtraction is undefined. Albeit deviating from math semantics, it allows us to access all generic features thereover, such as matrices and polynomials:

```
julia> T = tropical_semiring()
Min tropical semiring

julia> M = matrix(T,[0 1; 2 3])
[(0)   (1)]
[(2)   (3)]
```

```
julia> v = T.([-1,-1])
2-element Vector{TropicalSemiringElem{typeof(min)}}:
 (-1)
 (-1)

julia> M^2*v
2-element Vector{TropicalSemiringElem{typeof(min)}}:
 (-1)
 (1)

julia> R,(x,y) = T["x","y"];

julia> f = 1*x^2+2*y^2+0
(1)*x^2 + (2)*y^2 + (0)

julia> f^2+2*f
(2)*x^4 + (3)*x^2*y^2 + (4)*y^4 + (1)*x^2 + (2)*y^2 + (0)

julia> evaluate(f,T.([1,1]))
(0)
```

However, functions whose generic implementation requires subtraction will raise an error at runtime. In tropical mathematics, these concepts usually have subtractionless counterparts, such as the *tropical determinant* [MS15, Equation 1.2.6] [Jos21, Chapter 3.1],

$$\det(X) := \bigoplus_{\pi \in S_n} x_{1\pi(1)} \odot \cdots \odot x_{n\pi(n)} \quad \text{for } X = (x_{ij})_{i,j=1,\ldots,n} \in \mathbb{T}^{n \times n},$$

and providing implementations of specialised tropical functions will make OSCAR avoid the generic implementation when called with tropical input:

```
julia> T = tropical_semiring()
Min tropical semiring

julia> M = matrix(T,[0 1; 2 3])
[(0)   (1)]
[(2)   (3)]

julia> det(M)
(3)
```

2.2 Valued Fields

Next, we introduce some tropical notions over valued fields, which are central to many applications of tropical geometry, in particular in and around algebraic geometry.

Convention 4 For the remainder of the chapter, let K denote a field with a valuation, i.e., a field with a map $\mathrm{val}\colon K^* \to \mathbb{R}$ satisfying

1. $\mathrm{val}(a \cdot b) = \mathrm{val}(a) + \mathrm{val}(b)$,
2. $\mathrm{val}(a + b) \geq \min(\mathrm{val}(a), \mathrm{val}(b))$.

This valuation naturally induces a *ring of integers* $R := \{c \in K \mid \mathrm{val}(c) \geq 0\}$, a maximal ideal $\mathfrak{m} := \{c \in R \mid \mathrm{val}(c) > 0\} \trianglelefteq R$ and a *residue field* $\mathfrak{K} := R/\mathfrak{m}$.

We further assume that there is a fixed splitting homomorphism $\psi\colon (\mathrm{val}(K^*), +) \to (K^*, \cdot)$ such that $\mathrm{val}(\psi(r)) = r$. Such splitting homomorphisms exist for all algebraically closed fields [MS15, Lemma 2.1.15], but also for many others. For example:

- If val is trivial, i.e., $\mathrm{val}(c) = 0$ for all $c \in K^*$, then $\{0\} \to K^*, 0 \mapsto 1$ is a viable splitting homomorphism.
- If $K = \mathbb{Q}_p$ is the field of p-adic numbers and val is the usual p-adic valuation, then $\mathbb{Z} \to \mathbb{Q}_p^*, r \mapsto p^r$ is a viable splitting homomorphism.
- If $K = \mathbb{Q}(t)$ is the field of rational functions in t and val is the t-adic valuation, then $\mathbb{Z} \to \mathbb{Q}(t)^*, r \mapsto t^r$ is a viable splitting homomorphism.

Following the notation of [MS15], we abbreviate $t^r := \psi(r)$.

Definition 5 To each polynomial $f \in K[x] := K[x_1, \ldots, x_n]$, say $f = \sum_{\alpha \in \mathbb{Z}^n_{\geq 0}} c_\alpha x^\alpha$ with $c_\alpha \in K$, we can assign a tropical polynomial $\mathrm{trop}(f) \in \mathbb{T}[x]$ defined as follows:

$$\mathrm{trop}(f) := \begin{cases} \bigoplus_{c_\alpha \neq 0} \mathrm{val}(c_\alpha) \odot x^{\odot \alpha} & \text{if } \mathbb{T} = \mathbb{T}_{\min}, \\ \bigoplus_{c_\alpha \neq 0} -\mathrm{val}(c_\alpha) \odot x^{\odot \alpha} & \text{if } \mathbb{T} = \mathbb{T}_{\max}. \end{cases}$$

The *initial form* of f with respect to a weight vector $w \in \mathbb{R}^n$ is then defined as

$$\mathrm{in}_w(f) := \sum_{\substack{\mathrm{trop}(c_\alpha x^\alpha)(w) \\ = \mathrm{trop}(f)(w)}} \overline{t^{-\mathrm{val}(c_\alpha)} c_\alpha} \cdot x^\alpha \in \mathfrak{K}[x],$$

where $\overline{(\cdot)}\colon R \twoheadrightarrow \mathfrak{K}$ denotes the quotient map.

The *initial ideal* of an ideal $I \trianglelefteq K[x]$ with respect to a weight vector $w \in \mathbb{R}^n$ is given by

$$\mathrm{in}_w(I) := \langle \mathrm{in}_w(f) \mid f \in I \rangle \trianglelefteq \mathfrak{K}[x].$$

A *Gröbner basis* of an ideal $I \trianglelefteq K[x]$ with respect to a weight vector $w \in \mathbb{R}^n$ is a finite set $G \subseteq I$ such that $I = \langle g \mid g \in G \rangle$ and $\mathrm{in}_w(I) = \langle \mathrm{in}_w(g) \mid g \in G \rangle$. Note that, any Gröbner basis of I is a generating set provided I is homogeneous [MS15, Remark 2.4.4].

Example 6 In `OSCAR`, tropical features over a valued field K generally require specifying a map to a tropical semiring $\nu\colon K^* \to \mathbb{T}$ satisfying

1. $\nu(a \cdot b) = \nu(a) + \nu(b)$,
2. $\nu(a + b) \geq \min(\nu(a), \nu(b))$ (in the previously defined ordering on $\mathbb{T}$).

Most commonly, $\nu(c) = \mathrm{val}(c)$ for $\mathbb{T} = \mathbb{T}_{\min}$ and $\nu(c) = -\mathrm{val}(c)$ for $\mathbb{T} = \mathbb{T}_{\max}$. Hence, ν encodes both the valuation as well as the choice of min- or max-convention. Currently, only specific maps are supported such as:

1. trivial valuations on any field:

```
julia> nu = tropical_semiring_map(GF(3))
Map into Min tropical semiring encoding the trivial valuation on
↪  Prime field of characteristic 3

julia> nu.([0,1,2])
3-element Vector{TropicalSemiringElem{typeof(min)}}:
 infty
 (0)
 (0)
```

2. p-adic valuations on $\mathbb{Q}$:

```
julia> nu = tropical_semiring_map(QQ,3,max)
Map into Max tropical semiring encoding the 3-adic valuation on
↪  Rational field

julia> nu.([1//3,1,3])
3-element Vector{TropicalSemiringElem{typeof(max)}}:
 (1)
 (0)
 (-1)
```

3. t-adic valuations on any rational function field:

```
julia> Ft,t = rational_function_field(GF(3),"t")
(Rational function field over GF(3), t)

julia> nu = tropical_semiring_map(Ft,t,max)
Map into Max tropical semiring encoding the t-adic valuation on
↪  Rational function field over GF(3)

julia> nu.([t^(-1),1,t])
3-element Vector{TropicalSemiringElem{typeof(max)}}:
 (1)
 (0)
 (-1)
```

As in Example 2, the default codomain of tropical semiring maps is $\mathbb{T}_{\min}$. The tropical semiring map is usually passed as input, for example:

```
julia> R,(x1,x2,x3,x4) = QQ["x1","x2","x3","x4"];

julia> I = ideal([x1+2*x2-3*x3, 3*x2-4*x3+5*x4]);

julia> nu_2 = tropical_semiring_map(QQ,2);

julia> GBI = groebner_basis(I,nu_2,[0,0,0,0])
```

```
2-element Vector{QQMPolyRingElem}:
 x1 + 2*x2 - 3*x3
 3*x2 - 4*x3 + 5*x4

julia> inI = initial(I,nu_2,[0,0,0,0])
Ideal generated by
  x1 + x3
  x2 + x4

julia> GBI = groebner_basis(I,nu_2,[1,0,0,1])
2-element Vector{QQMPolyRingElem}:
 3*x2 - 4*x3 + 5*x4
 -x1 - 2*x2 + 3*x3

julia> inI = initial(I,nu_2,[1,0,0,1])
Ideal generated by
  x2
  x3

julia> coefficient_ring(inI)
Prime field of characteristic 2
```

The code above shows [MS15, Example 2.4.3], namely Gröbner bases of the ideal $I := \langle x_1 + 2x_2 - 3x_3, 3x_2 - 4x_3 + 5x_4 \rangle \trianglelefteq \mathbb{Q}[x_1, \ldots, x_4]$ with respect to the 2-adic valuation and the two weight vectors $(0,0,0,0) \in \operatorname{Trop}(V(I))$ and $(1,0,0,1) \notin \operatorname{Trop}(V(I))$. Note that the initial ideals in tropical geometry are defined over the residue field $\mathbb{F}_2$.

2.3 Tropical Varieties

Next, we recall the notion of tropical varieties. While [MS15] considers tropicalizations of very affine varieties $V(I)$, we consider tropicalizations of (Laurent) polynomial ideal I. Consequently, our tropicalizations are weighted polyhedral complexes instead of supports thereof (for more information on polyhedral complexes in OSCAR, see Chapter 1). The reason for this is two-fold:

1. From Section 3 onwards, we are interested in the number of points in $V(I)$ counted with multiplicity, so the tropical multiplicities matter.
2. For OSCAR, it is much easier to work with concrete polyhedral complexes rather than supports thereof.

Essentially, our definition of tropical varieties entails part of [MS15, Structure Theorem 3.3.5].

Definition 7 Let $I \trianglelefteq K[x]$ be a homogeneous ideal. Then the *Gröbner polyhedron* of I around a weight vector $w \in \mathbb{R}^n$ is defined to be

$$C_w(I) := \overline{\{v \in \mathbb{R}^n \mid \operatorname{in}_v(I) = \operatorname{in}_w(I)\}},$$

and the set of all Gröbner polyhedra is called the *Gröbner complex* $\Sigma(I) := \{C_w(I) \mid w \in \mathbb{R}^n\}$. As the name suggests, Gröbner polyhedra are convex polyhedra and the Gröbner complex is a finite polyhedral complex [MS15, Theorem 2.5.3].

Definition 8 The *tropical variety* of a homogeneous ideal $I \trianglelefteq K[x]$ is defined to be the following polyhedral complex consisting of all Gröbner cones whose initial ideals are monomial free:

$$\mathrm{Trop}(I) := \{C_w(I) \mid \mathrm{in}_w(I) \text{ monomial-free}\}$$

The *tropical variety* of an arbitrary ideal $I \trianglelefteq K[x]$ is defined to be

$$\mathrm{Trop}(I) := \Big\{C_w(I^h) \cap \{e_0 = 0\} \mid C_w(I^h) \in \mathrm{Trop}(I^h)\Big\},$$

where $I^h \trianglelefteq K[x_0, \ldots, x_n]$ denotes the homogenization of I in x_0, whose Gröbner polyhedra $C_w(I^h)$ lie in $\mathbb{R}^{n+1}$ with standard basis vectors $e_0, \ldots, e_n$, and we naturally identify the subspace $\{e_0 = 0\} \subseteq \mathbb{R}^{n+1}$ with $\mathbb{R}^n$. Note that both definitions coincide for homogeneous ideals.

Moreover, there is a canonical way to assign multiplicities to the maximal polyhedra of $\mathrm{Trop}(I)$ via the initial ideals [MS15, Definition 3.4.3] that make $\mathrm{Trop}(I)$ a balanced polyhedral complex if I is prime [MS15, Theorem 3.3.5]. For the sake of brevity, we omit their general definition and instead introduce them only for the special classes of tropical varieties that are of immediate interest to us below.

Remark 9 The umbrella term tropical variety extends well beyond the tropicalization of ideals covered in Definition 8. Most notably, there is an incredibly rich theory on the moduli of tropical curves, abstract graphs without a fixed embedding into some $\mathbb{R}^n$, whose significance extends well beyond the area of tropical geometry itself [Mar20].

`OSCAR` uses parameters to distinguish between embedded and abstract tropical varieties, as well as min- and max-tropical varieties. Moreover, there are four distinct types of tropical varieties, see Examples 10 and 11. This is because some functions only make sense for some types (e.g., tropical Pluecker vectors for tropical linear spaces), while other functions can be aggressively optimised depending on the type (e.g., stable intersection of tropical hypersurfaces). Embedded tropical varieties are always weighted polyhedral complexes. Abstract tropical varieties depend on the type and may not be defined for all four types.

Example 10 Objects of type `TropicalVariety` represent the most general type of tropical varieties and currently need to be embedded, abstract general tropical varieties are presently not supported. Embedded `TropicalVarietys` are weighted polyhedral complexes, that need not be balanced or pure. They can be constructed from polynomial ideals:

```
julia> K,t = rational_function_field(GF(101),"t");

julia> nu = tropical_semiring_map(K,t);

julia> R,(x,y,z) = K["x","y","z"];

julia> I = intersect(ideal([x+y+z+1,2*x+11*y+23*z+31]),
                     ideal([t^3*x*y*z-1]));

julia> TropV = tropical_variety(I,nu)
2-element Vector{TropicalVariety}:
 Min tropical variety
 Min tropical variety

julia> dim.(TropV)
2-element Vector{Int64}:
 2
 1
```

The command `tropical_variety(::MPolyIdeal)` returns an array of tropical varieties, one for each primary factor. The example above consists of a two-dimensional tropicalized binomial variety and a one-dimensional tropicalized linear space, see Figure 1. It is currently limited to special types of ideals, and an extension to general ideals is under development.

Objects of type `TropicalVariety` also arise naturally from the other types, e.g., as the (stable) intersection of two tropical hypersurfaces, see Example 15.

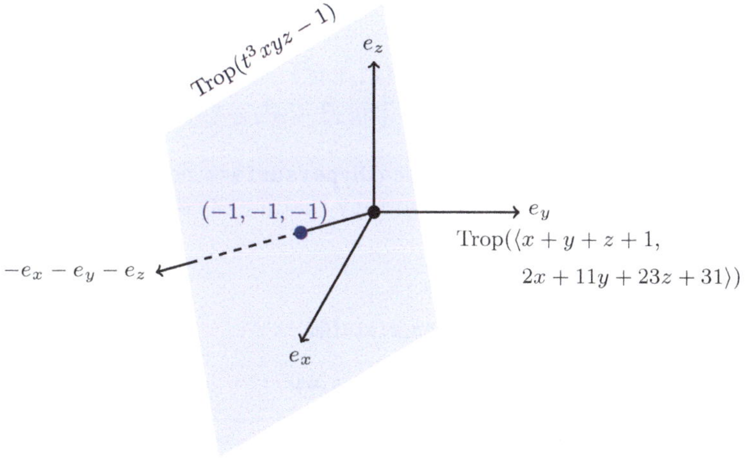

Fig. 1: The tropical variety in Example 10.

Example 11 If I is generated by a single polynomial $f=\sum_{\alpha\in\mathbb{Z}^n_{\geq 0}} c_\alpha x^\alpha$, $c_\alpha \in K$, then $\mathrm{Trop}(f) := \mathrm{Trop}(I)$ is a *tropical hypersurface* [MS15, Section 3.1].

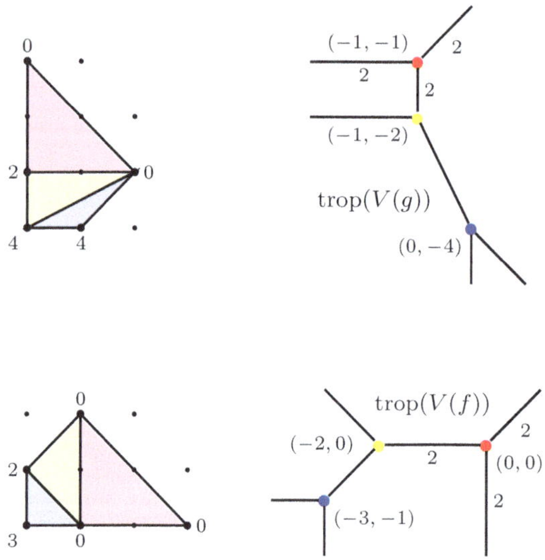

Fig. 2: Two tropical hypersurfaces and their Newton subdivisions.

Its cells are dual to the regular subdivision of its Newton polytope $\mathrm{conv}\{\alpha \mid c_\alpha \neq 0\}$ induced by the valuation of its coefficients [MS15, Lemma 3.4.6]. Its maximal cells are dual to edges and their multiplicities are the lattice lengths of the edges. Figure 2 illustrates the Newton subdivisions and max-tropical hypersurfaces of

$$\begin{aligned} f :=& t^3 + x + t^2 \cdot y + x(x^2 + y^2) \quad \text{and} \\ g :=& t^4 + t^4 \cdot x + t^2 \cdot y + y(x^2 + y^2) \in \mathbb{C}(t)[x, y]. \end{aligned}$$

In OSCAR, objects of type `TropicalHypersurface` need to be embedded and are objects of type `TropicalVariety` that are of codimension one. They can be constructed using:

1. polynomials over valued fields:

```
julia> K,t = rational_function_field(QQ,"t");

julia> nu = tropical_semiring_map(K,t,max);

julia> R,(x,y) = K["x","y"];

julia> f = t^3+x+t^2*y+x*(x^2+y^2)
x^3 + x*y^2 + x + t^2*y + t^3

julia> TropH = tropical_hypersurface(f,nu)
Max tropical hypersurface
```

```
julia> vertices_and_rays(TropH) # 1,2,3 are vertices, rest are rays
7-element SubObjectIterator{Union{PointVector{QQFieldElem},
↪  RayVector{QQFieldElem}}}:
 [0, -1]
 [-2, 0]
 [-3, -1]
 [0, 0]
 [1, 1]
 [-1, 1]
 [-1, 0]

julia> maximal_polyhedra(IncidenceMatrix,TropH)
7×7 IncidenceMatrix
[4, 5]
[1, 4]
[2, 4]
[2, 6]
[2, 3]
[1, 3]
[3, 7]
```

Note above that applying `typeof` to the seven entries of the output of `vertices_and_rays` reveals that the first three are vertices whilst the latter four are ray generators. Hence in the output of `maximal_polyhedra` the entry `[1,3]` is the cone $\mathbb{R}_{\geq 0} \cdot (0,-1) + \mathbb{R}_{\geq 0} \cdot (-3,-1)$.

2. polynomials over tropical semirings:

```
jjulia> tropf = tropical_polynomial(f,nu)
x^3 + x*y^2 + x + (-2)*y + (-3)

julia> tropical_hypersurface(tropf)
Max tropical hypersurface
```

3. subdivision of points:

```
julia> points = matrix(ZZ,collect(exponents(tropf)))
[3   0]
[1   2]
[1   0]
[0   1]
[0   0]

julia> heights = QQ.(collect(coefficients(tropf)))
5-element Vector{QQFieldElem}:
 0
 0
 0
 -2
 -3

julia> Delta = subdivision_of_points(points,heights)
Subdivision of points in ambient dimension 2
```

```
julia> tropical_hypersurface(Delta,max)
Max tropical hypersurface
```

Example 12 If I is generated by linear forms, then Trop(I) is a (realizable) *tropical linear space* [MS15, Section 4.4] [Jos21, Section 10.4 and 10.5]. It is dual to a matroid subdivision of the hypersimplex, and the support of its recession fan is the support of a Bergman fan. Its maximal cells are all of multiplicity 1. Note that OSCAR returns the same polyhedral structure as POLYMAKE, which is the one computed by Rincon's algorithm using fundamental circuits, where cyclic flats give rise to rays [Rin13].

In OSCAR, objects of type TropicalLinearSpace need to be embedded and can be thought of objects of type TropicalVariety that are of degree 1 with respect to the intersection product in Definition 14. A tropical linear space may be constructed from

1. ideals over valued fields:

```
julia> R,(w,x,y,z) = QQ["w","x","y","z"];

julia> nu = tropical_semiring_map(QQ);

julia> I = ideal([w+x+y+z,w+2*x+5*y+11*z]);

julia> TropL = tropical_linear_space(I,nu)
Min tropical linear space

julia> vertices_and_rays(TropL)
5-element SubObjectIterator{Union{PointVector{QQFieldElem},
↪   RayVector{QQFieldElem}}}:
 [0, 0, 0, 0]
 [0, -1, -1, -1]
 [0, 1, 0, 0]
 [0, 0, 1, 0]
 [0, 0, 0, 1]

julia> maximal_polyhedra(IncidenceMatrix,TropL)
4×5 IncidenceMatrix
[1, 2]
[1, 3]
[1, 4]
[1, 5]
```

2. matrices over valued fields or tropical semirings:

```
julia> A = matrix(QQ,[1 1 1 0; 1 1 0 1])
[1   1   1   0]
[1   1   0   1]

julia> TropL1 = tropical_linear_space(A, nu) # not Stiefel
Min tropical linear space

julia> tropA = nu.(A)
[(0)   (0)     (0)   infty]
```

```
[(0)   (0)   infty    (0)]

julia> TropL2 = tropical_linear_space(tropA) # not equal TropL1
Min tropical linear space
```

In the example above TropL1 is not the same as TropL2, as the tropicalised minors of a matrix need not be the minors of a tropicalised matrix. In fact, TropL1 cannot be constructed from a tropical matrix. Tropical linear spaces which can be constructed from tropical matrices are known as Stiefel topical linear spaces [RF15].

3. Plücker vectors over valued fields or tropical semirings:

```
julia> plueckerIndices = [[1,2],[1,3],[1,4],[2,3],[2,4],[3,4]];

julia> algebraicPlueckerVector = [0,-1,-1,1,1,1];

julia> TropL1 = tropical_linear_space(plueckerIndices,
                                     algebraicPlueckerVector,
                                     nu)
Min tropical linear space

julia> tropicalPlueckerVector = nu.(algebraicPlueckerVector);

julia> TropL2 = tropical_linear_space(plueckerIndices,
                                     tropicalPlueckerVector)
Min tropical linear space
```

Example 13 If $\dim(I) = 1$, then $\mathrm{Trop}(I)$ is an (embedded) tropical curve. In OSCAR, objects of type TropicalCurve are either weighted polyhedral complexes of dimension 1 if embedded or weighted graphs if abstract. Embedded TropicalCurves can be constructed by converting other tropical varieties.

```
julia> R,(x,y) = QQ["x","y"];

julia> nu = tropical_semiring_map(QQ);

julia> f = 1+x+y+x*(x^2+y^2);

julia> TropH = tropical_hypersurface(f,nu)
Min tropical hypersurface

julia> TropC = tropical_curve(TropH)
Embedded min tropical curve
```

Abstract TropicalCurve can be constructed from graphs (if no weights are specified, they are set to 1 by default).

```
julia> G = dualgraph(cube(3))
Undirected graph with 6 nodes and the following edges:
(3, 1)(3, 2)(4, 1)(4, 2)(5, 1)(5, 2)(5, 3)(5, 4)(6, 1)(6, 2)(6, 3)(6, 4)

julia> tropical_curve(G)
Abstract min tropical curve
```

2.4 Stable Intersections

Finally, we recall tropical stable intersection and tropical intersection products, which is a central concept for Section 3 onwards.

Definition 14 Let Σ_1, Σ_2 be two balanced polyhedral complexes in $\mathbb{R}^n$. Their *stable intersection* is defined to be

$$\Sigma_1 \wedge \Sigma_2 := \Big\{\sigma_1 \cap \sigma_2 \mid \sigma_1 \in \Sigma_1, \sigma_2 \in \Sigma_2, \dim(\sigma_1 + \sigma_2) = n\Big\}$$

with multiplicities

$$\mathrm{mult}_{\Sigma_1 \wedge \Sigma_2}(\sigma_1 \cap \sigma_2) := \sum_{\tau_1, \tau_2} \mathrm{mult}_{\Sigma_1}(\tau_1)\,\mathrm{mult}_{\Sigma_2}(\tau_2)[N : N_{\tau_1} + N_{\tau_2}],$$

where the sum is over all $\tau_1 \in \mathrm{star}_{\Sigma_1}(\sigma_1 \cap \sigma_2)$ and $\tau_2 \in \mathrm{star}_{\Sigma_2}(\sigma_1 \cap \sigma_2)$ with $\tau_1 \cap (\tau_2 + v) \neq \emptyset$ for some fixed generic $v \in \mathbb{R}^n$. It is either empty, or again a balanced polyhedral complex of codimension $\mathrm{codim}(\Sigma_1) + \mathrm{codim}(\Sigma_2)$ [MS15, Theorem 3.6.10]. Alternatively, its support can also be defined as the intersection under generic perturbation [MS15, Proposition 3.6.12].

If $\Sigma_1, \ldots, \Sigma_r$ are balanced polyhedral complexes in $\mathbb{R}^n$ of complementary dimension, i.e., $\sum_{i=1}^r \mathrm{codim}(\Sigma_i) = n$, then their *tropical intersection number* $\prod_{i=1}^r \Sigma_i$ is the number of points in their stable intersection $\wedge_{i=1}^r \Sigma_i$ counted with multiplicity.

Example 15 Figure 3 illustrates the stable intersection of the tropical hypersurfaces from Example 11. It can be computed using `stable_intersection`:

```
julia> K,t = rational_function_field(QQ,"t");

julia> nu = tropical_semiring_map(K,t,max);

julia> R,(x,y) = K["x","y"];

julia> f = t^3+x+t^2*y+x*(x^2+y^2)
x^3 + x*y^2 + x + t^2*y + t^3

julia> g = t^4+t^4*x+t^2*y+y*(x^2+y^2)
x^2*y + t^4*x + y^3 + t^2*y + t^4

julia> TropHf = tropical_hypersurface(f,nu)
Max tropical hypersurface

julia> TropHg = tropical_hypersurface(g,nu)
Max tropical hypersurface

julia> TropV = stable_intersection(TropHf,TropHg)
Max tropical variety

julia> vertices(TropV)
```

```
4-element SubObjectIterator{PointVector{QQFieldElem}}:
 [0, 0]
 [0, -4]
 [-3, -1]
 [-3, -2]

julia> multiplicities(TropV) # same ordering as above
4-element Vector{ZZRingElem}:
 4
 2
 2
 1
```

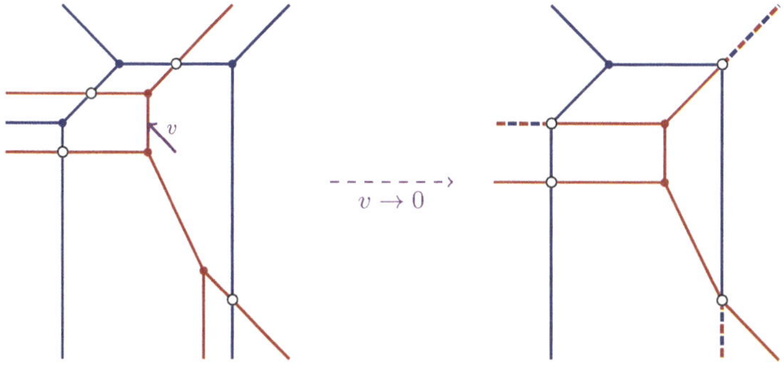

Fig. 3: Stable intersection (white vertices, right) of two tropical plane curves (red and blue, right) arising as a limit of their transverse intersection (white vertices, left) under perturbation by a generic v as $v \to 0$.

Stable intersections can be regarded as the "expected" tropicalisation when taking the sum of two ideals. This is for example captured in the following theorem:

Theorem 16 *Let $I_1, I_2 \trianglelefteq \mathbb{C}[x]$ be complete intersections. Suppose that* $\operatorname{Trop}(I_1)$ *and* $\operatorname{Trop}(I_2)$ *intersect transversely. Then*

$$\operatorname{Trop}(I_1 + I_2) = \operatorname{Trop}(I_1) \wedge \operatorname{Trop}(I_2),$$

where the equality only holds up to refinement of weighted polyhedral complexes.

Proof The statement is a consequence of [OP13, Corollary 5.1.2] and [OP13, Corollary 5.1.3] by Osserman and Payne. The first result shows the equality set-theoretically, while the second result shows that the multiplicities coincide. Both results require I_1 and I_2 to be Cohen–Macaulay, which is implied from I_1 and I_2 being complete intersections. □

Theorem 16 is a generalisation of the Transverse Intersection Theorem [MS15, Theorem 3.4.12] which relies on [OP13, Corollary 5.1.2]. From now onward, we will generally consider tropical varieties up to refinement, which does not change the tropical intersection number in Definition 14.

3 Tropical Geometry of Generic Root Counts

In this section, we recall some basic notions and a result of [HR22].

Convention 17 In addition to Convention 4, we further fix an affine space K^m, which we refer to as the *parameter space*. Let $K[a] := K[a_1, \dots, a_m]$ be its coordinate ring and $K(a) := K(a_1, \dots, a_m)$ be the field of fractions thereof.

In the following, we will refer to points $P \in K^m$ as *choices of parameters*, elements $f \in K[a][x^\pm]$ as *parametrised (Laurent) polynomials*, ideals $I \subseteq K[a][x^\pm]$ as *parametrised (Laurent) polynomial ideals*, and finite generating sets $\{f_1, \dots, f_k\} \subseteq I$ as *parametrised (Laurent) polynomial systems.*

Definition 18 Let $P \in K^m$ be a choice of parameters. For any parametrised polynomial $f \in K[a][x^\pm]$, say $f = \sum_{\alpha \in \mathbb{Z}^n} c_\alpha x^\alpha$ with $c_\alpha \in K[a]$, we define the *specialisation* of f at P to be

$$f_P = \sum_{\alpha \in \mathbb{Z}^n} c_\alpha(P) x^\alpha \in K[x^\pm].$$

For any parametrised polynomial ideal $I \trianglelefteq K[a][x^\pm]$, we define the *specialisation* of I at P to be

$$I_P = \langle f_P \mid f \in I \rangle \trianglelefteq K[x^\pm].$$

The *root count* of I at P is the vector space dimension

$$\ell_{I,P} := \dim_K(K[x^\pm]/I_P) \in \mathbb{N} \cup \{\infty\}.$$

Definition 19 The *generic specialisation* of a parametrised polynomial ideal $I \trianglelefteq K[a][x^\pm]$ is the ideal $I_{K(a)} \trianglelefteq K(a)[x^\pm]$ generated by I under the inclusion $K[a][x^\pm] \subseteq K(a)[x^\pm]$. The *generic root count* of I is the vector space dimension

$$\ell_I := \dim_{K(a)}(K(a)[x^\pm]/I_{K(a)}) \in \mathbb{N} \cup \{\infty\}.$$

We say $I_1, \dots, I_r \trianglelefteq K[a][x^\pm]$ are *generically of complementary dimension*, if

$$\operatorname{codim}(I_{1,K(a)}) + \cdots + \operatorname{codim}(I_{r,K(a)}) = n,$$

where $\operatorname{codim}(\cdot)$ denotes the Krull-codimension.

Definition 20 Recall that the torus $(K^*)^n$ is an algebraic group with group multiplication

$$(K^*)^n \times (K^*)^n \to (K^*)^n, \quad (\underbrace{(t_1,\ldots,t_n)}_{=:t}, \underbrace{(s_1,\ldots,s_n)}_{=:s}) \mapsto \underbrace{(t_1 s_1,\ldots,t_n s_n)}_{=:t\cdot s}.$$

A parametrised polynomial ideal $I \trianglelefteq K[a][x^\pm]$ is *torus-equivariant*, if there is a matching torus action on the parameter space

$$(K^*)^n \times K^m \to K^m, \quad (t,P) \mapsto t \cdot P \quad \text{such that } V(I_{t\cdot P}) = t \cdot V(I_P).$$

Moreover, several parametrised polynomial ideals $I_1,\ldots,I_r \trianglelefteq K[a][x^\pm]$ are *parametrically independent*, if their generators have distinct parameters, i.e., $I_i = \langle F_i \rangle$ with $F_i \subseteq K[a_j \mid j \in A_i][x^\pm]$ for some partition $[m] = \coprod_{i=1}^r A_i$.

Example 21 The principal ideal $I \trianglelefteq K[a_1,\ldots,a_5][x_1^\pm, x_2^\pm]$ generated by

$$f(a;x) := a_1 x_1^2 + a_2 x_1 x_2 + a_3 x_2^2 + a_4 x_1 + a_5 x_2 + 1$$

is torus equivariant under the torus action

$$(K^*)^2 \times K^5 \to K^5,$$
$$((t_1,t_2),(a_1,\ldots,a_5)) \mapsto (t_1^{-2}a_1, t_1^{-1}t_2^{-1}a_2, t_2^{-2}a_3, t_1^{-1}a_4, t_2^{-1}a_5),$$

since $t \cdot V(f(a;x)) = V(f(a;t^{-1}\cdot x)) = V(f(t\cdot a;x))$. Moreover, I is parametrically independent to the principal ideal generated by

$$g(a;x) := x_1^2 + x_1 x_2 + x_2^2 + x_1 + x_2 + 1,$$

as f and g share no parameters.

One central result for this paper is the following proposition that expresses the generic root count as a tropical intersection number:

Proposition 22 *[HR22, Proposition 4.19] Let $I_1,\ldots,I_r \trianglelefteq K[a][x^\pm]$ be parametrized polynomial ideals generically of complementary dimension, i.e., $\sum_{i=1}^r \operatorname{codim}(I_{i,K(a)}) = n$. Suppose $I_1,\ldots,I_{r-1}$ are torus equivariant and $I_1,\ldots,I_r$ are parametrically independent. Then the generic root count of their sum $I := I_1 + \cdots + I_r$ is their tropical intersection number for a generic choice of parameters:*

$$\ell_I = \prod_{i=1}^r \operatorname{Trop}(I_{i,P}) \quad \textit{for } P \in K^m \textit{ generic.}$$

4 Tropically Transverse Systems

In this section, we introduce the notion of tropically transverse systems in Definition 23 and use tropical modifications to express their generic root counts in terms of mixed volumes in Theorem 30. We briefly comment on how to check whether a system is tropically transverse and how to compute its generic root count in OSCAR.

Definition 23 We say a system $F = \{f_1, \ldots, f_n\} \subseteq K[a][x^{\pm}]$ is *horizontally parametrised* if there is a partition $[m] = \coprod_{j=1}^{n} A_j$ and a finite *support* $S = \{s_1, \ldots, s_m\} \subseteq K[x^{\pm}]$ such that

$$f_i = \sum_{j \in A_i} a_j \cdot s_j \qquad \text{for all } i = 1, \ldots, n.$$

Furthermore, we say that the support S has a *tropically transverse base*, if there is a $B = \{b_1, \ldots, b_r\} \subseteq K[x^{\pm}]$ such that

1. for each $j \in [m]$ there is a $\beta_j \in \mathbb{Z}^r$ such that $s_j = b^{\beta_j}$.
2. for any $J \subseteq [r]$ we have

$$\operatorname{codim}\Big(\bigcap_{j \in J} |\operatorname{Trop}(b_j)|\Big) = \min\Big(\sum_{j \in J} \operatorname{codim}\big(\operatorname{Trop}(b_j)\big), n\Big),$$

   where $\operatorname{codim}(\bigcap_{j\in J} |\operatorname{Trop}(b_j)|)$ is the codimension of any polyhedral complex supported on $\bigcap_{j \in J} |\operatorname{Trop}(b_j)|$.

For the sake of brevity, we will refer to a horizontally parametrised polynomial system with tropically transversal base simply as a *tropically transversal system*.

Example 24 Consider the parametrized system in [BMMT22, Equation 5.1]:

$$f_i = a_{1,i} u_i (u_i^2 + v_i^2) + a_{2,i} u_i + a_{3,i} v_i + a_{4,i} + \sum_{j \neq i} c_{j,i} v_j \quad \text{for } i = 1, \ldots, N,$$
$$g_i = b_{1,i} v_i (u_i^2 + v_i^2) + b_{2,i} u_i + b_{3,i} v_i + b_{4,i} + \sum_{j \neq i} d_{j,i} u_j \quad \text{for } i = 1, \ldots, N,$$

with parameters $a_{j,i}$, $b_{j,i}$, $c_{j,i}$, $d_{j,i}$, and variables u_i, v_i. Its support

$$S = \{1, u_i, v_i, u_i(u_i^2 + v_i^2), v_i(u_i^2 + v_i^2) \mid i = 1, \ldots, N\},$$

has a tropically transverse base $B = \{u_i, v_i, u_i^2 + v_i^2 \mid i = 1, \ldots, N\}$, as $\operatorname{Trop}(u_i) = \emptyset = \operatorname{Trop}(v_i)$ and $\operatorname{Trop}(u_i^2 + v_i^2)$ intersect transversely as their polynomials have distinct variables.

Example 25 Consider the parametrised system in [BBPMZ23, Theorem 5.1]:

$$a_0 + a_1 u + a_2 v + a_3 u(u^2 + v^2) + \cdots + a_{n+1} u (u^2 + v^2)^n = 0$$
$$b_0 + b_1 u + b_2 v + b_3 v(u^2 + v^2) + \cdots + b_{n+1} v (u^2 + v^2)^n = 0$$

with parameters a_i, b_i and variables u, v. Its support

$$S = \{1, u, v, u(u^2+v^2), \dots, u(u^2+v^2)^n, v(u^2+v^2), \dots, v(u^2+v^2)^n\},$$

has a tropically transverse base $B = \{u, v, u^2+v^2\}$.

Remark 26 It is straightforward to check whether a parametrised system $f_i = \sum_{j \in A_i} a_j \cdot s_j$ as in Definition 23 is tropically transverse. We can obtain a minimal base $B = \{b_1, \dots, b_r\}$ by computing the prime factors of all s_j. Testing whether the $\mathrm{Trop}(b_\ell)$ intersect transversely can be done by computing the regular subdivisions of the Newton polytopes $\Delta(b_1), \dots, \Delta(b_r)$ and $\Delta(\prod_{\ell=1}^r b_\ell) = \sum_{\ell=1}^r \Delta(b_\ell)$ induced by the valuation of their coefficients. Note that every cell σ of the subdivision of $\Delta(\prod_{\ell=1}^r b_\ell)$ can be written as a Minkowski sum $\sigma_1 + \dots + \sigma_r$ where σ_ℓ is a cell of the subdivisions of $\Delta(b_\ell)$ [Jos21, Section 4.2]. The $\mathrm{Trop}(b_\ell)$ intersect transversely if and only if $\sum_{\ell=1}^r \dim(\sigma_\ell) = \dim(\sigma)$ for all cells σ. Figure 4 illustrates three examples, two of which do not intersect transversely and with the defective dual cell σ highlighted.

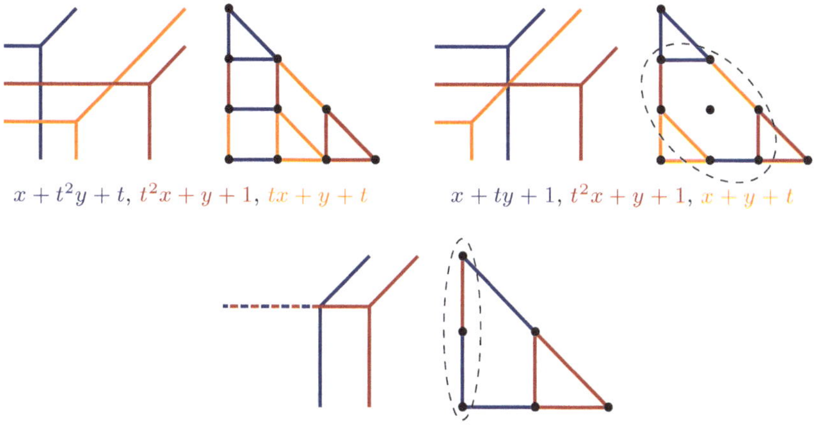

Fig. 4: Tropical transversality of hypersurfaces and its dual picture.

Definition 27 Let $F = \{f_1, \dots, f_n\} \subseteq K[a][x^\pm]$ be a tropically transverse system as in Definition 23. Its *(tropical) modification* is the parametrised system

$$\begin{aligned} \hat{f}_i &:= \sum_{j \in A_i} a_j z_j && \text{for all } i = 1, \dots, n, \\ \hat{g}_j &:= z_j - y^{\beta_j} && \text{for all } j = 1, \dots, m, \\ \hat{h}_\ell &:= y_\ell - b_\ell && \text{for all } \ell = 1, \dots, r. \end{aligned} \tag{1}$$

Here, the a_j remain parameters and the x_i, y_j, and z_ℓ are variables. We abbreviate the parametrised polynomial ring of the modification by

$$K[a][x^\pm, y^\pm, z^\pm] := K[a_1, \dots, a_m][x_1^\pm, \dots, x_n^\pm, y_1^\pm, \dots, y_r^\pm, z_1^\pm, \dots, z_m^\pm].$$

For the sake of consistency all polynomials are regarded as parametrised polynomial equations, even the g_j and h_ℓ which do not depend on the parameters. Let $\hat{I}$ be the ideal generated by the $\hat{f}_i$, $\hat{g}_j$ and $\hat{h}_\ell$. We denote the ambient space of the tropicalisations by $\mathbb{R}^{n+r+m}$ with coordinates indexed by the variables. In particular, e_{x_i}, e_{y_ℓ}, and e_{z_j} are the unit vectors in $\mathbb{R}^{n+r+m}$.

Lemma 28 *Let F be a tropically transverse system, and let $\hat{h}_\ell$ be defined as in Definition 27. Then for all $P \in K^m$*

$$\mathrm{Trop}\left(\langle \hat{h}_{\ell,P} \mid \ell = 1, \dots, r \rangle\right) = \bigwedge_{\ell=1}^{r} \mathrm{Trop}\,(\hat{h}_{\ell,P}),$$

where the equality only holds up to refinement of weighted polyhedral complexes.

Proof By Theorem 16, it suffices to show that the $\mathrm{Trop}(\hat{h}_{\ell,P})$ intersect transversely. To prove this, consider an intersection point $w \in \bigcap_{\ell=1}^{r} |\mathrm{Trop}(\hat{h}_{\ell,P})|$ and let $\sigma_\ell \in \mathrm{Trop}(\hat{h}_{\ell,P})$ be a maximal polyhedron containing w. Note that the hypersurfaces intersect transversally around w if and only if their stars around w intersect transversally around 0. We may therefore assume without loss of generality that $w = 0$ and that all σ_ℓ are cones.

Letting $(\cdot)^\perp$ denote the orthogonal complement of a vector, we then have

$$\mathrm{span}(\sigma_\ell) = \begin{cases} (e_{y_\ell} - \sum_{i=1}^{n} \alpha_i e_{x_i})^\perp & \text{for a monomial } x^\alpha \\ & \quad\text{in } b_\ell \text{ (Type Y), or} \\ (\sum_{i=1}^{n} (\beta_i - \alpha_i) e_{x_i})^\perp & \text{for two monomials } x^\alpha, x^\beta \\ & \quad\text{in } b_\ell \text{ (Type X).} \end{cases}$$

Arranging all normal vectors above as rows of a matrix, with all σ_ℓ of Type Y on top and all σ_ℓ of Type X on the bottom, with all e_{y_ℓ} on the left and all e_{x_i} on the right, yields a matrix of the form

$$\left(\begin{array}{c|c} I & * \\ \hline 0 & B \end{array}\right)$$

where I is a full-rank 0/1-submatrix in row-echelon form, and B is another full-rank submatrix due to the $\mathrm{Trop}(b_\ell)$ intersecting transversally. Hence the matrix is of full rank, and the σ_ℓ intersect each other transversally. $\square$

Lemma 29 *Let F be a tropically transverse system, and let $\hat{g}_j$, $\hat{h}_\ell$ be defined as in Definition 27. Then for all $P \in K^m$*

$$\operatorname{Trop}\Big(\langle \hat{g}_{j,P}, \hat{h}_{\ell,P} \mid j=1,\dots,m, \ell=1,\dots,r\rangle\Big) = \bigwedge_{j=1}^{m} \operatorname{Trop}(\hat{g}_{j,P}) \wedge \bigwedge_{\ell=1}^{r} \operatorname{Trop}(\hat{h}_{\ell,P}),$$

where the equality only holds up to refinement of weighted polyhedral complexes.

Proof We use a similar proof to the one of Lemma 28. By Theorem 16, it is enough to show that the hypersurfaces in the above equation intersect transversally. Let $\tau_j \in \operatorname{Trop}(\hat{g}_{j,P})$ and $\sigma_\ell \in \operatorname{Trop}(\hat{h}_{\ell,P})$ be maximal polyhedra intersecting in a point w. Again, we may assume without loss of generality that $w = 0$ and that the τ_j and σ_ℓ are polyhedral cones.

Let v_j and u_ℓ be the normal vectors on the linear span of τ_j and σ_ℓ, respectively. Arranging all normal vectors as rows of a matrix, with all v_j on top and all u_ℓ on the bottom, with all e_{z_j} on the left and all e_{y_ℓ}, e_{x_i} on the right, yields a matrix of the form

$$\left(\begin{array}{c|c} I & * \\ \hline 0 & B \end{array}\right)$$

where the submatrix I is the $m \times m$ identity matrix and the submatrix B is of full rank by Lemma 28. Hence the matrix is of full rank, and the τ_j, σ_ℓ intersect each other transversely. □

Theorem 30 *Let F be a tropically transverse system, let I be the parametrised ideal it generates, and let $\hat{I}$, $\hat{f}_i$, $\hat{g}_j$, $\hat{h}_\ell$ be as defined in Definition 27. Then the generic root count equals*

$$\begin{aligned}\ell_I &= \Big(\prod_{i=1}^{n} \operatorname{Trop}(\hat{f}_{i,P})\Big) \cdot \Big(\prod_{j=1}^{m} \operatorname{Trop}(\hat{g}_{j,P})\Big) \cdot \Big(\prod_{\ell=1}^{r} \operatorname{Trop}(\hat{h}_{\ell,P})\Big) \\ &= \operatorname{MV}\Big(\Delta(\hat{f}_{i,P}), \Delta(\hat{g}_{j,P}), \Delta(\hat{h}_{\ell,P}) \mid i \in [n], j \in [m], \ell \in [r]\Big)\end{aligned}$$

for $P \in K^m$ generic, where $\Delta(\cdot)$ denotes the Newton polytope, and $\operatorname{MV}(\cdot)$ denotes the normalized mixed volume.

Proof First note that I and $\hat{I}$ have the same generic root count. As $\langle \hat{f}_i \mid i = 1,\dots,n\rangle$ is torus equivariant and parametrically independent to $\langle \hat{g}_j, \hat{h}_\ell \mid j = 1,\dots,m, \ell = 1,\dots,r\rangle$, we have by Proposition 22:

$$\begin{aligned}\ell_I = &\operatorname{Trop}(\langle \hat{f}_{i,P} \mid i=1,\dots,n\rangle) \\ &\cdot \operatorname{Trop}(\langle \hat{g}_{j,P}, \hat{h}_{\ell,P} \mid j=1,\dots,m, \ell=1,\dots,r\rangle).\end{aligned}$$

Next, the $\operatorname{Trop}(\hat{g}_{j,P})$ and $\operatorname{Trop}(\hat{h}_{\ell,P})$ intersect each other transversely for all $P \in K^m$ by Lemma 29, hence

$$\ell_I = \mathrm{Trop}(\langle \hat{f}_{i,P} \mid i = 1, \ldots, n\rangle) \cdot \mathrm{Trop}(\langle \hat{g}_{j,P} \mid j = 1, \ldots, m\rangle) \\ \cdot \mathrm{Trop}(\langle \hat{h}_{\ell,P} \mid \ell = 1, \ldots, r\rangle).$$

The equality with the tropical intersection product then follows from the fact that $\mathrm{Trop}(\hat{f}_{i,P})$ intersect each other transversally, and the same holds for the $\mathrm{Trop}(\hat{g}_{j,P})$ and the $\mathrm{Trop}(\hat{h}_{\ell,P})$. The equality with the normalized mixed volume is from [MS15, Theorem 4.6.8]. □

Example 31 Code for computing the tropical intersection product of the modification in Theorem 30 can be found in

`https://github.com/isaacholt100/generic_root_count.`

The code currently relies on `MixedSubdivision.jl` for the computation of the tropical intersection product, it contains the functions

- `nonlinear_resonator_system(n::Int,m::Int)` for constructing System (1)
- `generic_root_count(F::Vector{<:MPolyRingElem})` for computing the tropical intersection number (or mixed volume) in Theorem 30.

The latter requires a horizontally parametrised system as a vector of polynomials with coefficients in another polynomial ring:

```
julia> include("generic_root_count/src/main.jl")

julia> A,(a0,a1,a2,a3,a4,b0,b1,b2,b3,b4) =
↪  QQ["a0","a1","a2","a3","a4","b0","b1","b2","b3","b4"];

julia> R,(u,v) = A["u","v"];

julia> f = a0+a1*u+a2*v+a3*u*(u^2+v^2)+a4*u*(u^2+v^2)^2;

julia> g = b0+b1*u+b2*v+b3*v*(u^2+v^2)+b4*v*(u^2+v^2)^2;

julia> generic_root_count([f,g])
9
```

5 Coupled Nonlinear Oscillators

This section is dedicated to an alternate proof of the generic number of equilibria of coupled nonlinear oscillators [BBPMZ23, Theorem 5.1] using elementary tropical geometry. Namely:

Theorem 32 *The following parametrised system of polynomial equations has generically* $2mn + 1$ *solutions:*

$$\begin{aligned} 0 &= a_0 + a_1x_1 + a_2x_2 + a_3x_1(x_1^m + x_2^m) + \cdots + a_{n+2}x_1(x_1^m + x_2^m)^n \\ 0 &= b_0 + b_1x_2 + b_2x_2 + b_3x_2(x_1^m + x_2^m) + \cdots + b_{n+2}x_2(x_1^m + x_2^m)^n \end{aligned} \quad (1)$$

Here, the a_i and b_i are the parameters, while x_1 and x_2 are the variables.

For that, we consider the following simplified modification of System (1):

Definition 33 Throughout the section, let:

$$\begin{aligned} f_1 &:= a_0 + a_1x_1 + a_2x_2 + a_3y_1 + \cdots + a_{n+2}y_n, \\ f_2 &:= b_0 + b_1x_1 + b_2x_2 + b_3z_1 + \cdots + b_{n+2}z_n, \\ g &:= w - (x_1^m + x_2^m), \\ p_i &:= y_i - x_1w^i \qquad \text{for } i = 1, \ldots, n, \\ q_i &:= z_i - x_2w^i \qquad \text{for } i = 1, \ldots, n, \end{aligned} \tag{2}$$

in the parametrised polynomial ring $K[a_i, b_i \mid i = 0, \ldots, n+2][x_j^{\pm}, w^{\pm}, y_i^{\pm}, z_i^{\pm} \mid j = 1, 2, i = 1, \ldots, n]$ with parameters a_i, b_i and variables x_j, w, y_i, z_i.

We will denote the ambient space of its fibrewise tropicalisations by $\mathbb{R}^{2n+3}$ and index its coordinates by the variables above, e.g., we will use e_{x_j}, e_w, e_{y_i}, and e_{z_i} to denote the unit vectors in $\mathbb{R}^{2n+3}$.

To prove Theorem 32, we explicitly compute the tropical intersection number

$$\underbrace{\mathrm{Trop}(\langle f_1, f_2\rangle)}_{=:\Sigma_{\mathrm{lin}}} \cdot \underbrace{\mathrm{Trop}(\langle g, p_i, q_i\rangle)}_{=:\Sigma_{\mathrm{nlin}}}.$$

This is done in multiple steps:

Lemma 34 Describes the three maximal cells of Σ_{nlin}: σ_0, σ_1, and σ_2.
Lemma 37 Describes the maximal cells of Σ_{lin}: $\tau_{i,j;k,l}$.
Lemma 39 Shows that $\Sigma_{\mathrm{lin}} \cap (\sigma_0 + u) = \emptyset$ for a fixed perturbation u.
Lemma 40 Shows that $\Sigma_{\mathrm{lin}} \cap (\sigma_1 + u) = \emptyset$.
Lemma 41 Describes which three $\tau_{i,j;k,l} \in \Sigma_{\mathrm{lin}}$ intersect $\sigma_2 + u$.

The section ends with the proof of Theorem 32 by computing the intersection multiplicities of the three intersection points from Lemma 41.

Lemma 34 *Let g, p_i, q_i be as defined in Definition 33. Then*

$$\Sigma_{nlin} := \mathrm{Trop}\Big(\langle g, p_i, q_i \mid i = 1, \ldots, n\rangle\Big)$$

up to refinement consists of the following three maximal cells:

$$\sigma_0 := \mathbb{R}_{\geq 0} \cdot \Big(e_w + \sum_{i=1}^{n} i(e_{y_i} + e_{z_i})\Big) + \mathbb{R} \cdot v_{nlin},$$

$$\sigma_1 := \mathbb{R}_{\geq 0} \cdot \Big(e_{x_2} + \sum_{i=1}^{n} e_{z_i}\Big) + \mathbb{R} \cdot v_{nlin},$$

$$\sigma_2 := \mathbb{R}_{\geq 0} \cdot \Big(e_{x_1} + \sum_{i=1}^{n} e_{y_i}\Big) + \mathbb{R} \cdot v_{nlin} =: \sigma_{nlin},$$

where $v_{nlin} := e_{x_1} + e_{x_2} + m \cdot e_w + \sum_{i=1}^{n}(mi+1)\cdot(e_{y_i}+e_{z_i})$ *is the generator of the lineality space of* Σ_{nlin}*. The multiplicities of* σ_1 *and* σ_2 *are* 1.

Proof The statement follows straightforwardly from Theorem 16, as the tropical hypersurfaces $\mathrm{Trop}(g)$, $\mathrm{Trop}(p_i)$ and $\mathrm{Trop}(q_i)$ all intersect transversally. Hence, we have up to refinement

$$\Sigma_{\mathrm{nlin}} = \mathrm{Trop}(g) \wedge \bigwedge_{i=1}^{n} \Big(\mathrm{Trop}(p_i) \wedge \mathrm{Trop}(q_i)\Big).$$

Note that $\mathrm{Trop}(g)$ consists of three maximal cells of codimension 1, while the intersection of the other $2n$ hypersurfaces is a linear space of codimension $2n$ in $\mathbb{R}^{2n+3}$. Together, they give rise to the three two-dimensional maximal cells σ_0, σ_1, and σ_2. The multiplicity of σ_1 and σ_2 being 1 follows directly from the definition. $\square$

Definition 35 For $\{i,j\} \in \binom{\{-2,\dots,n\}}{2}$, $i<j$, we define the following two polyhedral cones in $\mathbb{R}^{2n+3}$:

$$\begin{aligned}\theta_{i,j} &:= \{\overbrace{(v_{x_1}, v_{x_2}, v_w, v_{y_1}, \dots, v_{y_n}, v_{z_1}, \dots, v_{z_n})}^{=:v_{xwyz}} \in \mathbb{R}^{2n+3} \mid \\ &\qquad \mathfrak{y}_i = \mathfrak{y}_j \le \mathfrak{y}_k \text{ for all } k \neq i,j\},\\ \zeta_{i,j} &:= \{(v_{x_1}, v_{x_2}, v_w, v_{y_1}, \dots, v_{y_n}, v_{z_1}, \dots, v_{z_n}) \in \mathbb{R}^{2n+3} \mid \\ &\qquad \mathfrak{z}_i = \mathfrak{z}_j \le \mathfrak{z}_k \text{ for all } k \neq i,j\},\end{aligned}$$

where for $i = -2, \dots, n$

$$\mathfrak{y}_i := \begin{cases} v_{x_{-i}} & \text{if } i<0,\\ 0 & \text{if } i=0,\\ v_{y_i} & \text{if } i>0,\end{cases} \quad \text{and} \quad \mathfrak{z}_i := \begin{cases} v_{x_{-i}} & \text{if } i<0,\\ 0 & \text{if } i=0,\\ v_{z_i} & \text{if } i>0.\end{cases}$$

Moreover, for $\{k,l\} \in \binom{\{-2,\dots,n\}}{2}$, $k \in \{i,j\}$, we say $v_{xwyz} \in \mathbb{R}^{2n+3}$ *satisfies condition* (k,l) of $\theta_{i,j}$ and $\zeta_{i,j}$ to be respectively

$$\begin{cases} \mathfrak{y}_k = \mathfrak{y}_l & \text{if } \{k,l\} = \{i,j\},\\ \mathfrak{y}_k \le \mathfrak{y}_l & \text{if } l \notin \{i,j\},\end{cases} \quad \text{and} \quad \begin{cases} \mathfrak{z}_k = \mathfrak{z}_l & \text{if } \{k,l\} = \{i,j\},\\ \mathfrak{z}_k \le \mathfrak{z}_l & \text{if } l \notin \{i,j\}.\end{cases}$$

Example 36 Below are conditions for selected $\{i,j\} \in \binom{\{-2,\dots,n\}}{2}$:

(k,l)	$\theta_{-1,0}$	$\zeta_{-1,0}$
$(-1,0)$	$v_{x_1} = 0$	$v_{x_1} = 0$
$(0,-1)$	$0 = v_{x_1}$	$0 = v_{x_1}$
$(0,-2)$	$0 \le v_{x_2}$	$0 \le v_{x_2}$
$(-1,-2)$	$v_{x_1} \le v_{x_2}$	$v_{x_1} \le v_{x_2}$
$(0,s)$	$0 \le v_{y_s}$	$0 \le v_{z_s}$
$(-1,s)$	$v_{x_1} \le v_{y_s}$	$v_{x_1} \le v_{z_s}$

(k,l)	$\theta_{-1,n}$	$\zeta_{-1,n}$
$(-1,n)$	$v_{x_1} = v_{y_n}$	$v_{x_1} = v_{z_n}$
$(n,-1)$	$v_{x_1} = v_{y_n}$	$v_{x_1} = v_{z_n}$
$(-1,0)$	$v_{x_1} \le 0$	$v_{x_1} \le 0$
$(n,0)$	$v_{y_n} \le 0$	$v_{z_n} \le 0$
$(-1,-2)$	$v_{x_1} \le v_{x_2}$	$v_{x_1} \le v_{x_2}$
$(n,-2)$	$v_{y_n} \le v_{x_2}$	$v_{z_n} \le v_{x_2}$
$(-1,s)$	$v_{x_1} \le v_{y_s}$	$v_{x_1} \le v_{z_s}$
(n,s)	$v_{y_n} \le v_{y_s}$	$v_{z_n} \le v_{z_s}$

In the left table $0 < s \le n$ and in the right table $0 < s \le n-1$.

Lemma 37 *Let f_1, f_2 be as defined in Definition 33, and let $\tau_{i,j;k,l}$ be as defined in Definition 35. Then there is a Zariski dense parameter set $U_0 \subseteq K^{|a|+|b|}$ such that for all $P \in U_0$ and up to refinement the maximal cells of*

$$\Sigma_{lin} := \operatorname{Trop}\left(\langle f_{1,P}, f_{2,P}\rangle\right)$$

consists of all $\tau_{i,j;k,l} := \theta_{i,j} \cap \zeta_{k,l}$ satisfying the following condition:

$$\{i,j\} = \{k,l\} \subseteq \{-2,-1,0\} \quad \textit{does not hold.} \tag{3}$$

Proof By [MS15, Theorem 3.6.1] and torus-equivariance, there is a Zariski dense set $U_0 \subseteq K^{|a|+|b|}$ of elements with coordinate-wise valuation 0 such that for $P \in U_0$

$$\operatorname{Trop}\left(\langle f_{1,P}, f_{2,P}\rangle\right) = \operatorname{Trop}\left(f_{1,P}\right) \wedge \operatorname{Trop}\left(f_{2,P}\right).$$

Note that the $\theta_{i,j}$ and $\zeta_{k,l}$ from Definition 35 are the maximal cells of $\operatorname{Trop}(f_{1,P})$ and $\operatorname{Trop}(f_{2,P})$, respectively. Condition (3) excludes all $\theta_{i,j}$ and $\zeta_{k,l}$ that do not intersect transversally, the remaining $\tau_{i,j;k,l}$ are maximal cells of the stable intersection by definition. Each cell being of multiplicity one follows from the fact that Σ_{lin} is a tropical linear space. □

Remark 38 Note that the $\tau_{i,j;k,l}$ in Lemma 37 are not necessarily distinct. For example, $\tau_{-2,-1;-1,0} = \tau_{-2,0;-1,0}$. Important for us is only the fact that $\tau_{i,j;k,l}$ is always a maximal cell.

Lemma 39 *Let σ_0 be defined as in Lemma 34 and let Σ_{lin} be defined as in Lemma 37. Fix a perturbation*

$$u := e_{z_n} - 3n \cdot e_{x_1} - n \cdot e_{x_2}.$$

Then $\Sigma_{lin} \cap (\sigma_0 + u) = \emptyset$.

Proof Assume there is an intersection point $P \in \Sigma_{\text{lin}} \cap (\sigma_0 + u)$, say $P \in \tau_{i_0,j_0;k_0,l_0}$ for some $\tau_{i_0,j_0;k_0,l_0} \in \Sigma_{\text{lin}}$. Since $P \in \sigma_0 + u$, there are $c, d \in \mathbb{R}$, $c \ge 0$, such that

$$\begin{aligned}P = c \cdot \Big(e_w + \sum_{i=1}^{n} i(e_{y_i} + e_{z_i})\Big) \\ + d \cdot \Big(e_{x_1} + e_{x_2} + m \cdot e_w + \sum_{i=1}^{n} (mi+1) \cdot (e_{y_i} + e_{z_i})\Big) \\ + (e_{z_n} - 3n \cdot e_{x_1} - n \cdot e_{x_2}) \\ = (d-3n) \cdot e_{x_1} + (d-n) \cdot e_{x_2} + (dm+c) \cdot e_w \\ + \sum_{i=1}^{n} \Big((ci + d(mi+1)) \cdot (e_{y_i} + e_{z_i})\Big) + e_{z_n}.\end{aligned}$$

First, we show that $-2 \notin \{i_0, j_0, k_0, l_0\}$. Assume that $-2 \in \{i_0, j_0\}$ or $-2 \in \{k_0, l_0\}$, say $-2 = i_0$ or $-2 = k_0$. But then $d - n > d - 3n$ contradicts either Condition $(-2,-1)$ of θ_{-2,j_0} or Condition $(-2,-1)$ of ζ_{-2,l_0}.

Second, we prove that $0 \notin \{i_0, j_0, k_0, l_0\}$. Assume that $0 \in \{i_0, j_0\}$ or $0 \in \{k_0, l_0\}$, say $0 = i_0$ or $0 = k_0$. Then either Condition $(0,-1)$ of θ_{0,j_0} or Condition $(0,-1)$ of ζ_{0,l_0} state that $0 \leq d - 3n$ and hence $d \geq 3n > 0$, and all y_i and z_i coordinates of P are positive. Hence. $\{i_0, j_0\} = \{0,-1\} = \{k_0, l_0\}$, contradicting Condition (3).

Third, we show that $\{i_0, j_0\} = \{-1, n\}$ and $\{k_0, l_0\} \subseteq \{-1, n-1, n\}$. We have shown above that both $\{i_0, j_0\}$ and $\{k_0, l_0\}$ contain an index greater than 0. Without loss of generality, let $j_0, l_0 > 0$:

1. Condition $(j_0,-1)$ of θ_{i_0,j_0} states $cj_0 + d(mj_0+1) \leq d - 3n$, which implies $dm + c \leq -\frac{3n}{j_0} < 0$. Condition (j_0, n) of θ_{i_0,j_0} states $cj_0 + d(mj_0+1) \leq cn + d(mn+1)$, which in turn implies $(j_0 - n)(dm+c) \leq 0$. Combining both yields $j_0 = n$.
   Since $i_0 \neq j_0$, this also shows that i_0 cannot be greater than 0 (otherwise it is also n). Hence $\{i_0, j_0\} = \{-1, n\}$.
2. As the y_i and z_i coordinates of P are the same for $i = 1, \ldots, n-1$, we can use the same arguments as above to obtain $l_0 = n-1$ in the case $l_0 < n$. Hence $l_0 = n-1$ or $l_0 = n$, and therefore $\{k_0, l_0\} \subseteq \{-1, n-1, n\}$. □

Next, note that since $\{i_0, j_0\} = \{-1, n\}$, Condition $(-1, n)$ of $\theta_{-1,n}$ states

$$d - 3n = cn + d(mn+1) \quad \text{or equivalently} \quad -c - dm = 3. \tag{4}$$

Finally, we distinguish between three cases, leading each case to a contradiction:

$\mathbf{\{k_0, l_0\} = \{-1, n-1\}}$ Condition $(-1, n-1)$ of $\zeta_{-1,n-1}$ states $d - 3n = c(n-1) + d(m(n-1)+1)$. Combined with $d - 3n = cn + d(mn+1)$ from (4), we obtain $dm + c = 0$, contradicting $dm + c < 0$ established in Step (1).

$\mathbf{\{k_0, l_0\} = \{-1, n\}}$ Condition $(-1, n)$ of $\zeta_{-1,n}$ states $d - 3n = cn + d(mn+1) + 1$. This contradicts (4).

$\{\mathbf{k_0}, \mathbf{l_0}\} = \{\mathbf{n-1}, \mathbf{n}\}$ Condition $(n-1, n)$ of $\zeta_{n-1,n}$ states $c(n-1)+d(m(n-1)+1) = cn+d(mn+1)+1$ or equivalently $1 = -dm-c$. This contradicts (4).

Lemma 40 *Let σ_1 be defined as in Lemma 34 and let Σ_{lin} be defined as in Lemma 37. Fix the same perturbation as in Lemma 39:*

$$u := e_{z_n} - 3n \cdot e_{x_1} - n \cdot e_{x_2}.$$

Then $\Sigma_{lin} \cap (\sigma_1 + u) = \emptyset$.

Proof Assume there is an intersection point $P \in \Sigma_{\text{lin}} \cap (\sigma_1 + u)$, say $P \in \tau_{i_0,j_0;k_0,l_0}$ for some $\tau_{i_0,j_0;k_0,l_0} \in \Sigma_{\text{lin}}$. Since $P \in \sigma_1 + u$, there are $c, d \in \mathbb{R}$, $c \geq 0$, such that

$$\begin{aligned} P &= c \cdot \Big(e_{x_2} + \sum_{i=1}^{n} e_{z_i}\Big) + d \cdot \Big(e_{x_1} + e_{x_2} + m \cdot e_w + \sum_{i=1}^{n}(mi+1) \cdot (e_{y_i} + e_{z_i})\Big) \\ &\quad + (e_{z_n} - 3n \cdot e_{x_1} - n \cdot e_{x_2}) \\ &= (d-3n) \cdot e_{x_1} + (c+d-n) \cdot e_{x_2} + dm \cdot e_w \\ &\quad + \sum_{i=1}^{n} \Big(d(mi+1) \cdot e_{y_i} + (d \cdot (mi+1) + c) \cdot e_{z_i}\Big) + e_{z_n}. \end{aligned}$$

First, we show that $-2 \notin \{i_0, j_0, k_0, l_0\}$. Assume that $-2 \in \{i_0, j_0\}$ or $-2 \in \{k_0, l_0\}$, say $-2 = i_0$ or $-2 = k_0$. But then $c+d-n > d-3n$ contradicts either Condition $(-2, -1)$ of θ_{-2,j_0} or Condition $(-2, -1)$ of ζ_{-2,l_0}.

Second, we prove that $0 \notin \{i_0, j_0, k_0, l_0\}$. Assume that $0 \in \{i_0, j_0\}$ or $0 \in \{k_0, l_0\}$, say $0 = i_0$ or $0 = k_0$. In both cases, Condition $(0, -1)$ of θ_{0,j_0} or Condition $(0, -1)$ of ζ_{0,l_0} state that $0 \leq d - 3n$ and hence $d \geq 3n > 0$. Then all y_i and z_i coordinates of P are positive. This implies $\{i_0, j_0\} = \{0, -1\} = \{k_0, l_0\}$, which contradicts Condition (3).

Third, we show that $\{i_0, j_0\} = \{-1, n\}$ and $\{k_0, l_0\} \subseteq \{-1, n-1, n\}$. We have shown that both $\{i_0, j_0\}$ and $\{k_0, l_0\}$ contain an index greater than 0. Without loss of generality, let $j_0, l_0 > 0$:

- Condition $(j_0, -1)$ of θ_{i_0,j_0} states $d(mj_0 + 1) \leq d - 3n$, which implies $d < 0$. Condition (j_0, n) of θ_{i_0,j_0} states $d(mj_0 + 1) \leq d(mn + 1)$, which implies $j_0 = n$. Since $i_0 \neq j_0$, this also shows that i_0 cannot be greater than 0 (otherwise it is also n). Hence $\{i_0, j_0\} = \{-1, n\}$.
- Suppose $l_0 < n$. Condition $(l_0, -1)$ of ζ_{k_0,l_0} states $d(ml_0+1)+c \leq d-3n$, which implies $d < 0$. Condition $(l_0, n-1)$ states $d(ml_0 + 1) + c \leq d(m(n-1)+1)+c$. Combining both yields $l_0 = n-1$. Hence $l_0 = n-1$ or $l_0 = n$, and therefore $\{k_0, l_0\} \subseteq \{-1, n-1, n\}$. □

Next, note that since $\{i_0, j_0\} = \{-1, n\}$, Condition $(-1, n)$ of $\theta_{-1,n}$ states

$$d - 3n = d(mn+1) \quad \text{or equivalently} \quad -dm = 3. \tag{5}$$

Finally, we distinguish between three cases, leading each case to a contradiction:

$\{\mathbf{k_0}, \mathbf{l_0}\} = \{\mathbf{-1}, \mathbf{n-1}\}$ Condition $(-1, n-1)$ of $\zeta_{-1,n-1}$ states $d - 3n = d(m(n-1)+1) + c$. Combined with $d - 3n = d(mn+1)$ from (5), we obtain $c = dm$, contradicting $c \geq 0$.

$\{\mathbf{k_0}, \mathbf{l_0}\} = \{\mathbf{-1}, \mathbf{n}\}$ Condition $(-1, n)$ of $\zeta_{-1,n}$ states $d-3n = d(mn+1)+c+1$. Combined with $d - 3n = d(mn+1)$ from (5), we obtain $c = -1$, again contradicting $c \geq 0$.

$\{\mathbf{k_0}, \mathbf{l_0}\} = \{\mathbf{n-1}, \mathbf{n}\}$ Condition $(n-1, n)$ of $\zeta_{n-1,n}$ states $d(m(n-1)+1)+c = d(mn+1) + c + 1$, i.e., $-dm = 1$. This contradicts $-dm = 3$ from (5).

Lemma 41 *Let Σ_{nlin} and σ_{nlin} be defined as in Lemma 34 and let Σ_{lin} be defined as in Lemma 37. Fix again the following perturbation from Lemma 39:*

$$u := e_{z_n} - 3n \cdot e_{x_1} - n \cdot e_{x_2}.$$

Then $|\Sigma_{lin} \cap (\sigma_2 + u)| = \{P_1, P_2, P_3\}$, where

$$\begin{aligned} P_1 &:= v_2 - \frac{3}{m} \cdot v_{nlin} + u && \in \operatorname{relint}(\tau_{-1,n;-1,n}), \\ P_2 &:= 2n \cdot v_2 - \frac{n+1}{mn} \cdot v_{nlin} + u && \in \operatorname{relint}(\tau_{-2,-1;-1,n}), \\ P_3 &:= 2n \cdot v_2 + n \cdot v_{nlin} + u && \in \operatorname{relint}(\tau_{-2,-1;-1,0}) \end{aligned}$$

for $v_2 := e_{x_1} + \sum_{i=1}^n e_{y_i}$ and $v_{nlin} := e_{x_1} + e_{x_2} + m \cdot e_w + \sum_{i=1}^n (mi+1) \cdot (e_{y_i} + e_{z_i})$.

Proof It is straightforward to show that P_1, P_2, and P_3 are intersection points and contained in the relative interior of the cones specified above. It remains to show that P_1, P_2, and P_3 are the only intersection points.

Consider an intersection point $P \in \Sigma_{\text{lin}} \cap (\sigma_2 + u)$, say $P \in \tau_{i_0,j_0;k_0,l_0}$ for some $\tau_{i_0,j_0;k_0,l_0} \in \Sigma_{\text{lin}}$. We will now show that $P = P_1$, P_2, or P_3. Since $P \in \sigma_2 + u$, there are $c, d \in \mathbb{R}$, $c \geq 0$, such that

$$\begin{aligned} P &= c \cdot \Big(e_{x_1} + \sum_{i=1}^n e_{y_i}\Big) + d \cdot \Big(e_{x_1} + e_{x_2} + m \cdot e_w + \sum_{i=1}^n (mi+1) \cdot (e_{y_i} + e_{z_i})\Big) \\ &\quad + (e_{z_n} - 3n \cdot e_{x_1} - n \cdot e_{x_2}) \\ &= (c + d - 3n) \cdot e_{x_1} + (d - n) \cdot e_{x_2} + dm \cdot e_w \\ &\quad + \sum_{i=1}^n \Big((d \cdot (mi+1) + c) \cdot e_{y_i} + d(mi+1) \cdot e_{z_i}\Big) + e_{z_n}. \end{aligned}$$

Suppose $0 < j_0$. Then Condition $(j_0, -2)$ of θ_{i_0,j_0} states $d(mj_0 + 1) + c \leq d - n$, which implies $d < 0$. And Condition (j_0, n) of θ_{i_0,j_0} states $d(mj_0+1)+c \leq d(mn+1) + c$, which together imply $j_0 = n$. As the same argument holds true for i_0, we get $\{i_0, j_0\} \subseteq \{-2, -1, 0, n\}$.

Suppose $0 < l_0 < n$. Then Condition $(l_0, -2)$ of ζ_{k_0,l_0} states $d(ml_0 + 1) \leq d - n$, which implies $d < 0$. And Condition $(l_0, n-1)$ of ζ_{k_0,l_0} states $d(ml_0 + 1) \leq d(m(n-1)+1)$, which together imply $l_0 = n-1$. As the same argument holds true for k_0, we get $\{k_0, l_0\} \subseteq \{-2, -1, 0, n-1, n\}$.

Note that we always have $d \leq n$, as otherwise all coordinates of P other than x_1 are positive. This would imply $\{i_0, j_0\} = \{-1, 0\} = \{k_0, l_0\}$, contradicting Condition (3).

We now distinguish between the following cases, showing either that $P = P_1, P_2, P_3$ or deriving a contradiction:

$(\mathbf{i_0}, \mathbf{j_0}) = (\mathbf{-1}, \mathbf{0})$ First, Condition $(0, -2)$ of $\theta_{-1,0}$ states $0 \leq d - n$, which together with $d \leq n$ implies $d = n$. Second, Condition $(0, -1)$ of $\theta_{-1,0}$ states $0 = c + d - 3n$, hence $c = 2n$. This shows that $P = P_3$.

$(\mathbf{i_0}, \mathbf{j_0}) = (\mathbf{-2}, \mathbf{0})$ Condition $(0, -2)$ of $\theta_{-2,0}$ states $0 = d - n$, which implies that all y_i and z_i coordinates of P are positive. Then $P \in \zeta_{k_0,l_0}$ and Condition (3) imply $\{k_0, l_0\} = \{-2, -1\}$ or $\{k_0, l_0\} = \{-1, -0\}$. Both cases imply that the x_1 and x_2 coordinates of P are 0, i.e., $d - n = c + d - 3n = 0$. This implies $c = 2n$, showing that $P = P_3$.

$(\mathbf{i_0}, \mathbf{j_0}) = (\mathbf{0}, \mathbf{n})$ First, Condition $(0, n)$ of $\theta_{0,n}$ states $0 = d(mn+1) + c$, which implies that $d \leq 0$. Second, Condition $(0, -2)$ of $\theta_{0,n}$ states $0 \leq d - n$, which in turn implies that $n \leq d \leq 0$, contradicting $n > 0$.

$(\mathbf{i_0}, \mathbf{j_0}) = (\mathbf{-1}, \mathbf{n})$ Condition $(-1, n)$ of $\theta_{-1,n}$ states $c + d - 3n = d(mn+1) + c$ which implies $d = -\frac{3}{m}$. We then have

$$d(mn+1) + 1 = -3n - \frac{3}{m} + 1 < -3n - \frac{3}{m} + 3 = d(m(n-1)+1) < 0.$$

If $0 \in \{k_0, l_0\}$ (say $0 = k_0$), Condition $(0, n)$ of ζ_{0,l_0} states $0 \leq d(mn+1)+1$ which is a contradiction. Similarly, if $n - 1 \in \{k_0, l_0\}$ (say $n - 1 = k_0$) then Condition $(n-1, n)$ of ζ_{n-1,l_0} states $d(m(n-1)+1) \leq d(mn+1)+1$, which is a contradiction. So $0, n-1 \notin \{k_0, l_0\}$.

If $-2 \in \{k_0, l_0\}$ (say $-2 = k_0$), Condition $(-2, n)$ of ζ_{-2,l_0} states $d - n \leq d(mn+1) + 1$, which together with $d = -\frac{3}{m}$ leads to a contradiction.

The only remaining possibility is $\{k_0, l_0\} = \{-1, n\}$. Then Condition $(-1, n)$ of $\zeta_{-1,n}$ states $c + d - 3n = d(mn+1) + 1$, which together with $c + d - 3n = d(mn+1) + c$, implies that $c = 1$. Since also $d = -\frac{3}{m}$, this shows $P = P_1$.

$(\mathbf{i_0}, \mathbf{j_0}) = (\mathbf{-2}, \mathbf{n})$ Condition $(n, -1)$ of $\theta_{-2,n}$ states that $d(mn+1) + c \leq c + d - 3n$ which implies $d \leq -\frac{3}{m}$. If $0 \in \{k_0, l_0\}$ (say $0 = k_0$), then Condition $(0, n)$ of ζ_{0,l_0} states $0 \leq d(mn+1) + 1$ which leads to a contradiction to $d \leq -\frac{3}{m}$. If $-2 \in \{k_0, l_0\}$ (say $-2 = k_0$), then Condition $(-2, n)$ of ζ_{-2,l_0} states $d - n \leq d(mn+1) + 1$ which leads to a contradiction to $d \leq -\frac{3}{m}$. If $n - 1 \in \{k_0, l_0\}$ (say $n - 1 = k_0$), then Condition $(n-1, n)$ of ζ_{n-1,l_0} states $d(m(n-1)+1) \leq d(mn+1) + 1$ which leads to a contradiction to $d \leq -\frac{3}{m}$. If $\{k_0, l_0\} = \{-1, n\}$ (say $-1 = k_0$), then Condition $(-1, -2)$ of ζ_{-1,l_0} and Condition $(-2, -1)$ of $\theta_{-2,n}$ state $c + d - 3n \leq d - n$ and

$d-n \leq c+d-3n$ hence $d+c-3n = d-n$, implying $c = 2n$. Condition $(-2, n)$ of $\theta_{-2,n}$ and Condition $(-1, n)$ of $\zeta_{-1,n}$ state that $d - n = d(mn + 1) + c$ and $c + d - 3n = d(mn + 1) + 1$. Substituting $c = 2n$ then leads to a contradiction.

$(\mathbf{i_0}, \mathbf{j_0}) = (\mathbf{-2}, \mathbf{-1})$ Condition $(-2, -1)$ of $\theta_{-2,-1}$ states $c + d - 3n = d - n$ which implies $c = 2n$.
If $\{k_0, l_0\} = \{-1, 0\}$, then Condition $(-1, 0)$ of $\zeta_{-1,0}$ states $c + d - 3n = 0$ which implies $d = n$. If $\{k_0, l_0\} = \{-2, 0\}$, then Condition $(-2, 0)$ of $\zeta_{-2,0}$ states $d - n = 0$ so $d = n$. In both cases, we have $P = P_3$.
If $\{k_0, l_0\} = \{-1, n\}$, then Condition $(-1, n)$ of $\zeta_{-1,n}$ states $c + d - 3n = d(mn + 1) + 1$ which together with $c = 2n$ implies $d = -\frac{n+1}{mn}$. If $\{k_0, l_0\} = \{-2, n\}$, then Condition $(-2, n)$ of $\zeta_{-2,n}$ states $d - n = d(mn + 1) + 1$ which together with $c = 2n$ implies $d = -\frac{n+1}{mn}$. In both cases, we have $P = P_2$.
Five cases for $\{k_0, l_0\}$ remain, all of which we will lead to a contradiction:

- $(\mathbf{k_0}, \mathbf{l_0}) = (\mathbf{-2}, \mathbf{n-1})$ or $(\mathbf{k_0}, \mathbf{l_0}) = (\mathbf{-1}, \mathbf{n-1})$ In two situations, either by Condition $(-2, n-1)$ of $\zeta_{-2,n-1}$ or by combining Condition $(-1, n-1)$ of $\zeta_{-1,n-1}$ with Condition $(-2, -1)$ of $\theta_{-2,-1}$, we get $d - n = d(m(n - 1) + 1)$, which implies $-dm = \frac{n}{n-1}$. Independently, Condition $(n - 1, n)$ of ζ_{k_0,l_0} states $d(m(n-1)+1) \leq d(mn+1)+1$, which implies $-dm \leq 1$, contradicting $-dm = \frac{n}{n-1}$.
- $(\mathbf{k_0}, \mathbf{l_0}) = (\mathbf{0}, \mathbf{n-1})$ Condition $(0, n-1)$ of $\zeta_{0,n-1}$ states $0 = d(m(n-1)+1)$ which implies $d = 0$. This contradicts Condition $(0, -2)$ of $\zeta_{0,n-1}$ which is $0 \leq d - n$.
- $(\mathbf{k_0}, \mathbf{l_0}) = (\mathbf{0}, \mathbf{n})$ Condition $(0, n)$ of $\zeta_{0,n}$ states $0 = d(mn + 1) + 1$, which implies $d < 0$. This contradicts Condition $(0, -2)$ of $\zeta_{0,n}$, which states $0 \leq d - n$.
- $(\mathbf{k_0}, \mathbf{l_0}) = (\mathbf{n-1}, \mathbf{n})$ Condition $(n-1, n)$ of $\zeta_{n-1,n}$ states $d(m(n-1)+1) = d(mn + 1) + 1$, which implies $dm = -1$. This contradicts Condition $(n - 1, -2)$ of $\zeta_{n-1,n}$, which states $d(m(n - 1) + 1) \leq d - n$, implying $dm(n - 1) \leq -n$.

Proof (Proof of Theorem 32) We compute the generic root count of System (1) by computing the tropical intersection number

$$\underbrace{\mathrm{Trop}(\langle f_1, f_2\rangle)}_{=\Sigma_{\mathrm{lin}}} \cdot \underbrace{\mathrm{Trop}(\langle g, p_i, q_i\rangle)}_{=\Sigma_{\mathrm{nlin}}},$$

where f_1, f_2, g, p_i, and q_i are from the modified system in Definition 33. Lemma 41 showed that for a suitably chosen perturbation $u \in \mathbb{R}^{2n+3}$, the only intersecting cells of Σ_{lin} and $\Sigma_{\mathrm{nlin}} + u$ are:

$$\tau_{-1,n;-1,n} \cap (\sigma_2 + u), \quad \tau_{-2,-1;-1,n} \cap (\sigma_2 + u), \quad \tau_{-2,-1;-1,0} \cap (\sigma_2 + u).$$

Moreover, the intersection happens in the relative interior of the cells above, which, by Lemma 34 and Lemma 37, all have multiplicity 1. Hence, it remains to compute the sublattice indices

$$[N : N_{\tau_{-1,n;-1,n}} + N_{\sigma_2}], \quad [N : N_{\tau_{-2,-1;-1,n}} + N_{\sigma_2}], \quad [N : N_{\tau_{-2,-1;-1,0}} + N_{\sigma_2}],$$

where $N := \mathbb{Z}^{2n+3}$ and $N_\delta := N \cap \operatorname{span}(\delta)$.

Recall that

$$\begin{aligned}
\operatorname{span}(\sigma_2) &= \operatorname{span}\left(e_{x_1}+\sum_{i=1}^{n} e_{y_i}, e_{x_1}+e_{x_2}+m\cdot e_w+\sum_{i=1}^{n}(mi+1)\cdot(e_{y_i}+e_{z_i})\right),\\
\operatorname{span}(\tau_{-1,n;-1,n}) &= \operatorname{span}(e_{x_1}+e_{y_n}+e_{z_n}, e_{x_2}, e_w, e_{y_1}, \dots, e_{y_{n-1}}, e_{z_1}, \dots, e_{z_{n-1}}),\\
\operatorname{span}(\tau_{-2,-1;-1,n}) &= \operatorname{span}(e_{x_1}+e_{x_2}+e_{z_n}, e_w, e_{y_1}, \dots, e_{y_n}, e_{z_1}, \dots, e_{z_{n-1}}),\\
\operatorname{span}(\tau_{-2,-1;-1,0}) &= \operatorname{span}(e_w, e_{y_1}, \dots, e_{y_n}, e_{z_1}, \dots, e_{z_n}).
\end{aligned}$$

For the first index, note that

$$\begin{aligned}
&N_{\tau_{-1,n;-1,n}}+N_{\sigma_2}\\
&\quad= \mathbb{Z}\cdot(e_{x_1}+e_{y_n}+e_{z_n}) + \mathbb{Z}\cdot e_{x_2} + \mathbb{Z}\cdot e_w + \sum_{i=1}^{n-1}\mathbb{Z}\cdot e_{y_i} + \sum_{i=1}^{n-1}\mathbb{Z}\cdot e_{z_i}\\
&\qquad + \mathbb{Z}\cdot(e_{x_1}+e_{y_n}) + \mathbb{Z}\cdot(e_{x_1}+(mi+1)\cdot e_{y_n}+(mi+1)\cdot e_{z_n}))\\
&\quad= \mathbb{Z}\cdot e_{x_2} + \mathbb{Z}\cdot e_w + \sum_{i=1}^{n-1}\mathbb{Z}\cdot e_{y_i} + \sum_{i=1}^{n}\mathbb{Z}\cdot e_{z_i}\\
&\qquad + \mathbb{Z}\cdot(e_{x_1}+e_{y_n}) + \mathbb{Z}\cdot(mn\cdot e_{y_n}).
\end{aligned}$$

Therefore, $N/(N_{\tau_{-1,n;-1,n}} + N_{\sigma_2}) \cong \mathbb{Z}^2/\operatorname{span}_{\mathbb{Z}}((1,1),(0,mn)) \cong \mathbb{Z}/mn\mathbb{Z}$ and hence $[N : N_{\tau_{-1,n;-1,n}} + N_{\sigma_2}] = mn$.

For the second index, note that

$$\begin{aligned}
N_{\tau_{-2,-1;-1,n}}+N_{\sigma_2} &= \mathbb{Z}\cdot(e_{x_1}+e_{x_2}+e_{z_n}) + \mathbb{Z}\cdot e_w + \sum_{i=1}^{n}\mathbb{Z}\cdot e_{y_i} + \sum_{i=1}^{n-1}\mathbb{Z}\cdot e_{z_i}\\
&\quad + \mathbb{Z}\cdot e_{x_1} + \mathbb{Z}\cdot(e_{x_1}+(mi+1)\cdot e_{z_n})\\
&= \mathbb{Z}\cdot e_{x_1} + \mathbb{Z}\cdot e_w + \sum_{i=1}^{n}\mathbb{Z}\cdot e_{y_i} + \sum_{i=1}^{n-1}\mathbb{Z}\cdot e_{z_i}\\
&\quad + \mathbb{Z}\cdot(e_{x_2}+e_{z_n}) + \mathbb{Z}\cdot(mn\cdot e_{z_n}).
\end{aligned}$$

Therefore, $N/(N_{\tau_{-2,1;-1,n}} + N_{\sigma_2}) \cong \mathbb{Z}^2/\operatorname{span}_{\mathbb{Z}}((1,1),(0,mn)) \cong \mathbb{Z}/mn\mathbb{Z}$ and hence $[N : N_{\tau_{-2,1;-1,n}} + N_{\sigma_2}] = mn$.

For the third index, note that

$$N_{\tau_{-2,-1;-1,0}}+N_{\sigma_2} = \mathbb{Z}\cdot e_w + \sum_{i=1}^{n}\mathbb{Z}\cdot e_{y_i} + \sum_{i=1}^{n}\mathbb{Z}\cdot e_{z_i} + \mathbb{Z}\cdot e_{x_1} + \mathbb{Z}\cdot(e_{x_1}+e_{x_2}) = N$$

Therefore, $[N : N_{\tau_{-1,n;-1,n}} + N_{\sigma_2}] = 1$.

Summing all three indices, we obtain a generic root count of $2mn+1$. □

References

[ABGJ18] X. Allamigeon, P. Benchimol, S. Gaubert, and M. Joswig. "Log-barrier interior point methods are not strongly polynomial". *SIAM J. Appl. Algebra Geom.* 2.1 (2018), pp. 140–178. DOI: 10.1137/17M1142132.

[AGV22] X. Allamigeon, S. Gaubert, and N. Vandame. "No Self-Concordant Barrier Interior Point Method is Strongly Polynomial". In: *Proc. of the 54th Annual ACM SIGACT Symposium on Theory of Computing*. STOC 2022. Rome, Italy: Association for Computing Machinery, 2022, pp. 515–528. DOI: 10.1145/3519935.3519997.

[Bat+23] D. J. Bates et al. *Numerical Nonlinear Algebra*. 2023. eprint: arXiv:2302.08585.

[BBPMZ23] V. Borovik, P. Breiding, J. del Pino, M. Michałek, and O. Zilberberg. *Khovanskii bases for semimixed systems of polynomial equations – a case of approximating stationary nonlinear Newtonian dynamics*. 2023. eprint: arXiv:2306.07897.

[BMMT22] P. Breiding, M. Michałek, L. Monin, and S. Telen. *The Algebraic Degree of Coupled Oscillators*. 2022. eprint: arXiv:2208.08179.

[CHKS06] F. Catanese, S. Hoşten, A. Khetan, and B. Sturmfels. "The maximum likelihood degree". *Am. J. Math.* 128.3 (2006), pp. 671–697. DOI: 10.1353/ajm.2006.0019.

[DHOST14] J. Draisma, E. Horobeţ, G. Ottaviani, B. Sturmfels, and R. Thomas. "The Euclidean distance degree". In: *Proceedings of the 2014 symposium on symbolic-numeric computation, SNC 2014, Shanghai, China, July 28–31, 2014*. New York, NY: Association for Computing Machinery (ACM), 2014, pp. 9–16. DOI: 10.1145/2631948.2631951.

[FJP20] G. Farkas, D. Jensen, and S. Payne. *The Kodaira dimensions of $\overline{\mathcal{M}}_{22}$ and $\overline{\mathcal{M}}_{23}$*. 2020. eprint: arXiv:2005.00622.

[HJ11] M. Hampton and A. Jensen. "Finiteness of spatial central configurations in the five-body problem". *Celest. Mech. Dyn. Astron.* 109.4 (2011), pp. 321–332. DOI: 10.1007/s10569-010-9328-9.

[HM06] M. Hampton and R. Moeckel. "Finiteness of relative equilibria of the four-body problem". *Invent. Math.* 163.2 (2006), pp. 289–312. DOI: 10.1007/s00222-005-0461-0.

[HR22] P. A. Helminck and Y. Ren. *Generic root counts and flatness in tropical geometry.* 2022. eprint: arXiv:2206.07838.

[Jos21] M. Joswig. *Essentials of tropical combinatorics.* Vol. 219. Graduate Studies in Mathematics. Providence, RI: American Mathematical Society, 2021.

[KK12] K. Kaveh and A. G. Khovanskii. "Newton–Okounkov bodies, semigroups of integral points, graded algebras and intersection theory". *Ann. of Math. (2)* 176.2 (2012), pp. 925–978. DOI: 10.4007/annals.2012.176.2.5.

[MS15] D. Maclagan and B. Sturmfels. *Introduction to tropical geometry.* Vol. 161. Graduate Studies in Mathematics. American Mathematical Society, Providence, RI, 2015, pp. xii+363. DOI: 10.1090/gsm/161.

[Mar20] H. Markwig. "Tropical curves and covers and their moduli spaces". *Jahresber. Dtsch. Math.-Ver.* 122.3 (2020), pp. 139–166. DOI: 10.1365/s13291-020-00215-z.

[OP13] B. Osserman and S. Payne. "Lifting tropical intersections". *Doc. Math.* 18 (2013), pp. 121–175.

[Rin13] F. Rincón. "Computing tropical linear spaces". *J. Symb. Comput.* 51 (2013), pp. 86–98. DOI: 10.1016/j.jsc.2012.03.008.

[RF15] F. Rincón and A. Fink. "Stiefel tropical linear spaces". *J. Comb. Theory, Ser. A* 135 (2015), pp. 291–331. DOI: 10.1016/j.jcta.2015.06.001.

Faithful Tropicalization of Hyperelliptic Curves

Hannah Markwig, Lukas Ristau, and Victoria Schleis

Abstract We provide explicit faithful re-embeddings for all hyperelliptic curves of genus at most three and an algorithmic way to construct them. Both in the faithful tropicalization algorithm and the proofs of correctness, we showcase OSCAR-methods for commutative algebra, polyhedral and tropical geometry.

Key words: Faithful tropicalization, computational tropical geometry, hyperelliptic curves, Gröbner fans

1 Introduction

Tropical geometry allows a fruitful exchange of methods between algebraic and convex geometry, as tropicalization preserves important properties of algebraic varieties such as dimension. The question which properties can be preserved under tropicalization — and how — is of foundational importance in order to make use of the exchange of methods. In this chapter, we concretely answer this question for the case of hyperelliptic curves of genus 3.

We assume that the reader is familiar with varieties defined over the Puiseux series e.g., their tropicalizations, in particular tropical hypersurfaces and

Hannah Markwig
Fachbereich Mathematik, Universität Tübingen e-mail: `hannah@math.uni-tuebingen.de`

Lukas Ristau
Fachbereich Mathematik, Universität Tübingen e-mail: `lukas.ristau@math.uni-tuebingen.de`

Victoria Schleis
Fachbereich Mathematik, Universität Tübingen e-mail: `victoria.schleis@student.uni-tuebingen.de`

W. Decker et al. (eds.), *The Computer Algebra System OSCAR*,
Algorithms and Computation in Mathematics 32,
https://doi.org/10.1007/978-3-031-62127-7_16

duality, especially tropical plane curves and their dual Newton subdivisions, as in [MS15].

Consider a plane curve and its tropicalization. As dimension is preserved, the latter is a 1-dimensional polyhedral complex. More precisely, it is a balanced graph in $\mathbb{R}^2$ that is dual to its Newton subdivision, as described in [MS15]. Is the genus of the curve reflected in its tropicalization?

Examples of a tropical line and a tropical conic, as in [MS15], suggest that the tropicalization of a rational, i.e. genus 0, curve should be a rational graph, i.e. a tree.

Example 1 What about the elliptic curve defined by the Weierstrass equation

$$y^2 = x^3 + 4 \cdot x^2 + 8t^4 \cdot x,$$

which is of genus 1? We expect the tropicalization to be a graph containing a cycle, thus reflecting the genus. To compute its tropicalization, we use the following OSCAR commands:

```
julia> Kt,t = rational_function_field(QQ,"t")
(Rational function field over QQ, t)

julia> Kxy, (x,y) = Kt["x", "y"]
(Multivariate polynomial ring in 2 variables over rational function
↪  field, AbstractAlgebra.Generic.MPoly{AbstractAlgebra.Generic.Rationa⌋
↪  lFunctionFieldElem{QQFieldElem, QQPolyRingElem}}[x,
↪  y])

julia> f=-x^3-4*x^2+y^2+(-8*t^4)*x
-x^3 - 4*x^2 - 8*t^4*x + y^2

julia> val_t = tropical_semiring_map(Kt,t)
Map into Min tropical semiring encoding the t-adic valuation on Rational
↪  function field over QQ

julia> ftrop=tropical_polynomial(f,val_t)
x^3 + x^2 + y^2 + (4)*x

julia> Tf = tropical_hypersurface(ftrop)
Min tropical hypersurface

julia> PC = polyhedral_complex(Tf)
Polyhedral complex in ambient dimension 2

julia> visualize(PC)
```

The output of the computation is shown on the left of Figure 1. It turns out that the tropicalization is again a rational graph, i.e. the genus was somehow lost in the process of tropicalization. We should not let this discourage ourselves, as we can see that this loss happens merely because of an unfavorable choice of embedding: Let us change coordinates and define

$$g = f(x, y - 2x) = -x^3 - 4 \cdot xy + y^2 - 8t^4 \cdot x.$$

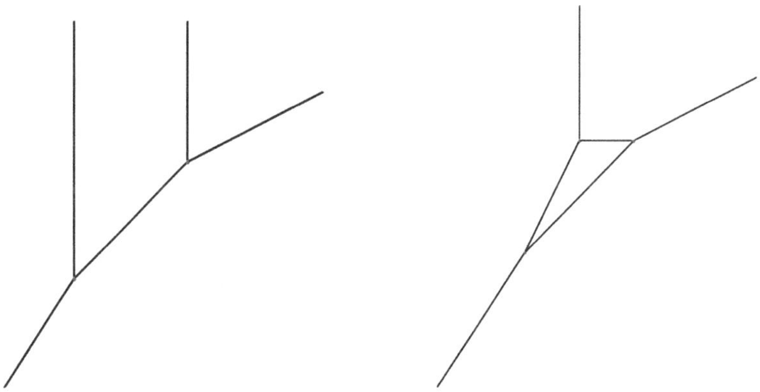

Fig. 1: The tropicalization of the elliptic curve $V(y^2 - x^3 - 4 \cdot x^2 - 8t^4 \cdot x)$ is a rational graph, depicted on the left. After the coordinate change, the tropicalization of the elliptic curve from above contains a cycle, depicted on the right.

Repeating the computation from above with g, we obtain the tropicalization depicted on the right of Figure 1, which contains a cycle as desired.

The genus is only a topological invariant: every elliptic curve is of genus 1, but their various complex structures may be distinguished using the j-invariant. Is the j-invariant of an elliptic curve reflected in the tropicalization?

As suggested by Mikhalkin in [Mik06, Example 3.15], the tropicalization of an elliptic curve defined over a field with a non-Archimedean valuation should contain a cycle whose length is equal to the valuation of the j-invariant.

Example 2 Consider the elliptic cubic curve given by the equation

$$\begin{aligned} f =&(-t^2) \cdot x^3 + t^{20} \cdot x^2 y + t^2 \cdot xy^2 + t^{14} \cdot y^3 + (-3t^3) \cdot x^2 + xy \\ &+ (t^3 + t^5 - t^6) \cdot y^2 + (-3t^4) \cdot x + (t + t^2) \cdot y + (2t^2 + t^5). \end{aligned}$$

Repeating the computation from Example 1 with f, we obtain the picture on the left in Figure 2.

After the coordinate change

$$\begin{aligned} g = f(x - t, y) = &(-t^2) \cdot x^3 + t^{20} \cdot x^2 y + t^2 \cdot xy^2 + t^{14} \cdot y^3 + (1 - 2t^{21}) \cdot xy \\ &+ (t^5 - t^6) \cdot y^2 + (t^2 + t^{22}) \cdot y + (2t^2 + 2t^5), \end{aligned}$$

we obtain the picture on the right in Figure 2.

Using the command `vertices(PC2)` for the underlying polyhedral complex `PC2` of our tropical plane curve, we obtain the coordinates of the vertices:

```
julia> vertices(PC2)
6-element SubObjectIterator{PointVector{QQFieldElem}}:
```

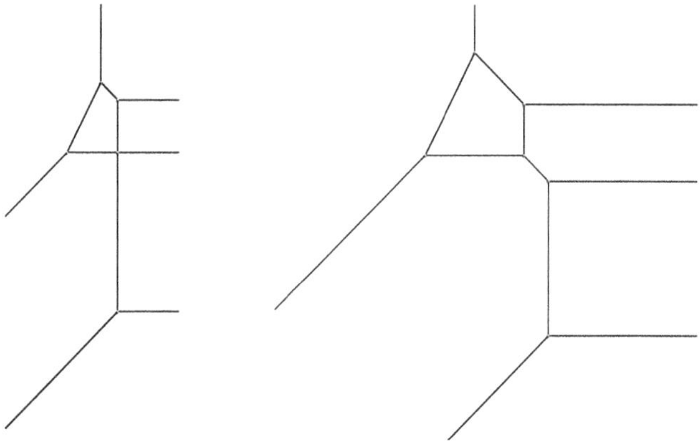

Fig. 2: The tropicalization of the elliptic curve $V(f)$ of Example 2 (depicted on the left) has a cycle, but its length does not equal the valuation of its j-invariant. After coordinate change, we obtain the expected cycle length (depicted on the right).

```
 [3, -9]
 [-2, -2]
 [2, -2]
 [3, -3]
 [2, 0]
 [0, 2]
```

Before the coordinate change, we had the vertices:

```
julia> vertices(PC1)
5-element SubObjectIterator{PointVector{QQFieldElem}}:
 [1, -11]
 [1, -2]
 [-2, -2]
 [0, 2]
 [1, 1]
```

Thus, before the coordinate change we obtained as cycle length 9, while we obtain 10 after the coordinate change.

We compute the j-invariant in OSCAR using `j_invariant_cubic(f)`, for more information see `https://book.oscar-system.org`. It turns out that 10 is the valuation of the j-invariant and thus only the second embedding, after the coordinate change, reflects the j-invariant.

Both examples illustrate that the tropicalization of a curve depends on the chosen embedding. We can see that for some, but not all, choices of embeddings, many properties of an elliptic curve are reflected in its tropicalization: dimension, genus, and j-invariant. This motivates the following definition:

Definition 3 (Faithful Tropicalizations of Elliptic Curves) An embedding of an elliptic curve E with j-invariant $j(E)$ is called *faithful* if its tropicalization is a graph containing a cycle of length $\mathrm{val}(j(E))$.

A faithful embedding of an elliptic curve can be viewed as optimal from the point of view of tropical geometry, as all important features are preserved to the zeroth order.

The fact that some, but not all, embeddings are faithful motivates the study of the limit of all tropicalizations, where tropicalization depends on the chosen embedding.

Definition 4 (Analytification as Limit of All Tropicalizations, [Pay09], Theorem 1.1) Let χ be an affine variety over K. Let e be an affine embedding of χ. Then we define the *Berkovich space* (or *analytification*) as $\chi^{\mathrm{an}} := \varprojlim \mathrm{Trop}(e(\chi))$, i.e. analytification is the limit of all tropicalizations.

Historically, the analytification is defined differently (in terms of extensions of valuations) and the statement above is a theorem. For the purpose of this Chapter however, we can take it as a definition. The statement holds analogously for projective curves after compactification, resp. by gluing affine charts.

Example 5 The tropicalization of $\mathbb{P}^1$ can be viewed as the compactification of $\mathbb{R}$. If we embed our line into the (torus of a) plane, e.g. as $V(x + y + 1)$ and tropicalize, we obtain the well-known picture of a tripod, which projects onto $\mathbb{R}$. Next, we embed our line into threespace. After tropicalization, several combinatorial types may occur. In general, we obtain a tropical curve with 2 vertices and one bounded edge as shown in the middle of Figure 3. Continuing this procedure and taking the limit of all tropicalizations, we obtain as analytification of a line an infinite tree as sketched on the right of Figure 3 which is often called the witch's broom.

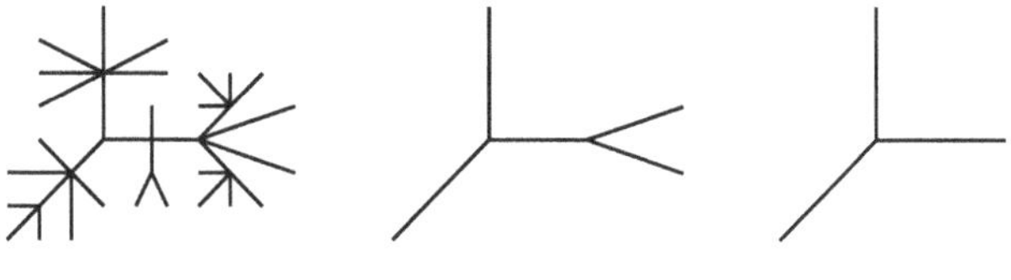

Fig. 3: Two tropicalizations and a sketch of the analytification of a line.

Analytifications satisfy many desirable properties. They are infinitely branched metric graphs, [BPR16, Section 2.4]. In some sense, analytifications can be viewed as generalizations of the j-invariant for elliptic curves: the analytification of an elliptic curve is an infinitely branched graph containing precisely one cycle of length $\mathrm{val}(j)$. More generally, the analytification of a curve χ of genus g contains a finite metric subgraph of genus g as a deformation retract, which is called the *(minimal) Berkovich skeleton* of χ. In the case of an elliptic curve, the minimal Berkovich skeleton is a loop of length $\operatorname{val}(j)$.

Using the analytification, we can extend our notion of faithful tropicalization to arbitrary curves:

Definition 6 (Faithful Tropicalization) Let χ be a projective curve and $e : \chi \to \mathbb{P}^n$ an embedding of χ into projective space. The embedding e of χ is called *faithful* on a subgraph Γ of the analytification χ^{an} if $\operatorname{Trop}(e(\chi))$ contains an isometric copy of Γ. We consider faithfulness with respect to the minimal Berkovich skeleton, unless specified otherwise.

If we fix the cycle as the subgraph Γ of the analytification of an elliptic curve, we recover Definition 3.

Faithful tropicalizations are the best tropicalizations we can hope for, in the sense that they reflect as much of the geometry of the algebraic curve as possible. Analytifications also reflect the geometry, but they are infinite objects which are hard to control. A tropicalization which is chosen arbitrarily is accessible for computations, but may not reflect the features we want to observe. A faithful tropicalization combines the advantages of both: it is a finite object which is easier to control, but it still reflects the geometric features we want to observe.

In [BPR16, Theorem 5.20], it is shown that faithful tropicalizations exist for every curve. But how can we find them concretely?

In this Chapter, we answer this question for hyperelliptic curves of genus 3. The technique we use is called re-embedding into tropical modifications and is described in detail in Section 2.

Theorem 7 *Let χ be a hyperelliptic curve of genus 3. Then Table 8 gives re-embedding polynomials such that the re-embedding into the tropical modification, induced by the polynomials, is a faithful tropicalization of χ.*

Faithful tropicalizations of curves of genus 2 have been studied in [CM19]. The concrete construction of a faithful tropicalization depends on a case-by-case analysis, considering different possibilities for equalities and inequalities among the valuations of the coefficients. As the Newton polygon of such a curve is sufficiently small, it was possible to perform this case-by-case analysis by hand for curves of genus 2.

Generalizing to hyperelliptic curves of genus 3 enlarges the amount of possibilities for inequalities among the valuations of the coefficients and the Newton subdivisions to such an extent that a case-by-case analysis by hand is out of reach. We have obtained Theorem 7 with the aid of OSCAR .

2 Techniques for Faithful Re-Embeddings

In the following, we analyze hyperelliptic curves over a valued field K of characteristic 0 given by a defining equation

$$g: \quad y^2 = x \cdot \prod_{i=1}^{8}(x_i - \alpha_i), \tag{1}$$

and we denote the valuations of coefficients by $\omega_i = \text{val}(\alpha_i)$.

In the introduction we showed an example of a coordinate change that resulted in a faithful tropicalization. In this section, we will introduce techniques to construct these coordinate changes methodically, by using *re-embeddings* into *tropical modifications*.

The faithful re-embedding construction we describe follows the divide-and-conquer principle. Given a hyperelliptic curve χ of genus $g \leq 3$, defined by an equation g as in (1), we *separate* the tropical hyperelliptic curve into subgraphs according to conditions on the initial forms of g using Table 1. For each subgraph identified this way, we change coordinates by *finding a correct re-embedding polynomial* f_i using Table 8. For all subgraphs connected by bridges, we further *reduce the dimension* of the ambient space of the re-embedding via Theorem 27.

In this section, we show how the Berkovich skeleton can be decomposed into subgraphs that can be considered independently when constructing the re-embedding. Further, we provide details on the re-embeddings themselves. Finally, we describe how to to check faithfulness on a tropical curve.

Throughout this section, we will follow the example of the elliptic curve $V(y^2 - x^3 - 4x^2 - 8t^4x)$ on the left in Figure 1, and construct its faithful tropicalization using the methods we introduce.

2.1 Classifying Hyperelliptic Curves via Their Berkovich Skeleton

Berkovich skeleta of hyperelliptic curves can be glued from a small collection of possible subgraphs, called *building blocks*. In our construction of faithful re-embeddings, these subgraphs serve a vital role: It is sufficient to determine the faithful re-embeddings of building blocks as re-embeddings can be glued with methods we present in Section 4.

Remark 8 The types of subgraphs that can occur are listed in Table 1. A subgraph at x^i is the skeleton whose leftmost vertex is dual to an edge in the Newton subdivision incident to x^i. If the subgraph consists of only a point, then x^i is the leftmost point on a length k vertical edge corresponding to the multiplicity k edge of the tropical curve adjacent to the point.

Lemma 9 *The minimal Berkovich skeleton of a hyperelliptic curve is built by gluing building blocks of the types given in Table 1, and the coefficient conditions in Table 1 hold.*

Proof The types of subgraphs that can appear in the Berkovich skeleton are a special case of [Cor21, Theorem 1.1]. The coefficient conditions for cycles, 2-thetas, points and bridges are due to (or straightforward generalizations of) the conditions given in [CM19, Table 5.1], and the correctness of the conditions for 3-thetas will be shown later in Theorem 17. □

Subgraph at x^i	**Local Berkovich skeleton**	**Local picture in the Newton subdivision**	**Coefficient conditions**
3-Theta	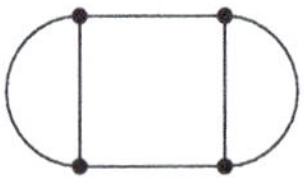	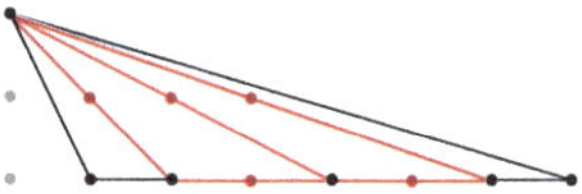	$\omega_1 < \omega_2 < \omega_3 < \omega_5$ $< \omega_7 < \omega_8$; $\omega_3 = \omega_4, \omega_5 = \omega_6$ $\text{in}(\alpha_3) = \text{in}(\alpha_4)$, $\text{in}(\alpha_5) = \text{in}(\alpha_6)$
2-Theta		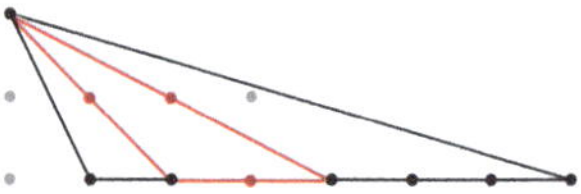	$\omega_i < \omega_{i+1} < \omega_{i+2}$ $< \omega_{i+4} < \omega_{i+5}$ i odd; $\omega_{i+2} = \omega_{i+3}$; $\text{in}(\alpha_{i+2}) = \text{in}(\alpha_{i+3})$
Cycle	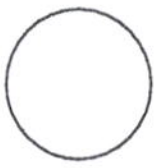		$\omega_i < \omega_{i+1} < \omega_{i+2}$ i odd
k-point	 k	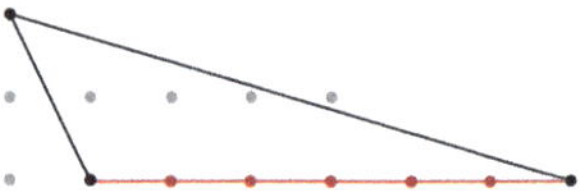	$\omega_i < \omega_{i+1} < \omega_{i+2k+1}$ $\omega_{i+1} = \cdots = \omega_{i+2k}$ $\text{in}(\alpha_l) \neq \text{in}(\alpha_j)$ for $i < l < j < 2k+2$
Bridge	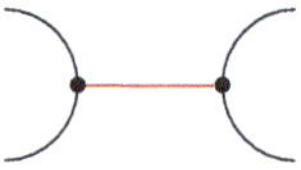	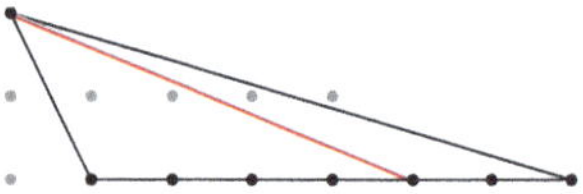	$\omega_{i-1} < \omega_i < \omega_{i+1}$ i odd
Point connector		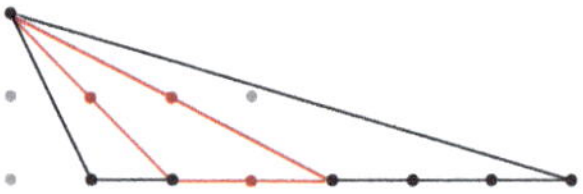	$\omega_i < \omega_{2j}$ i odd, $1 \leq j \leq k+1$ $\omega_{i+2l} = \omega_{i+2l+1}$ $\text{in}(\alpha_{i+2l}) \neq \text{in}(\alpha_{i+2l+1})$ $1 \leq l \leq k-1$

Table 1: The different building blocks appearing for hyperelliptic curves and their coefficient conditions. While point connectors and thetas have the same local Newton subdivision, their coefficient conditions and thus their behaviour under modification are distinguishable. See Remark 8 for details on notation.

2.2 Re-Embeddings

By re-embedding a plane curve, we can lift its tropicalization in $\mathbb{R}^2$ to a tropical modification of $\mathbb{R}^2$, hoping to make new features visible in the newly attached cells, and thus getting closer to the Berkovich skeleton (see, for instance, Figure 4 and Example 29). Formally, they are defined as follows:

Definition 10 (Tropical Modifications) Let F be a polynomial over the tropical semiring $\mathbb{T}$. To each cell σ of the graph of F, that does not have codimension 0, we attach a new cell ς spanned by σ and $e_{n+1} := (0, \ldots, 0, 1)$. We obtain a pure rational polyhedral complex in $\mathbb{R}^{n+1}$, the *modification* of $\mathbb{R}^n$ along F.

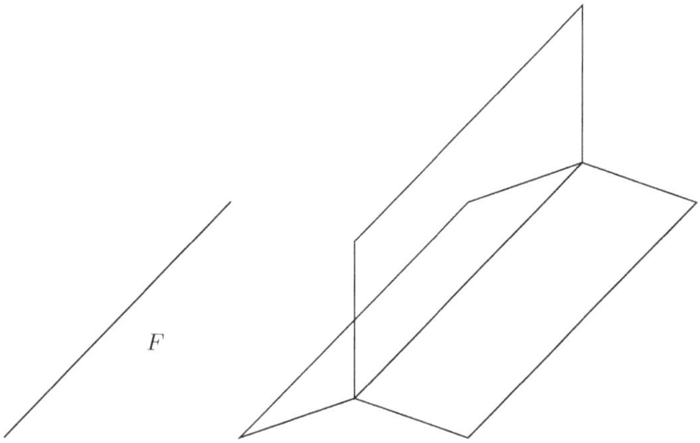

Fig. 4: The modification of $\mathbb{R}^2$ along $F = \min\{x + 2, y\}$.

We can write the modification of F as a tropicalization by finding appropriate *polynomial lifts* f of F, i.e. polynomials f whose tropicalization is F: $f(x) = \sum_{\beta \in \mathrm{supp(F)}} c_\beta x^\beta$.

Definition 11 In the remainder of the Chapter, we will use the following notation:

- χ denotes a hyperelliptic curve and g its defining equation of the form given in (1),
- uppercase letters will be used to denote objects in the tropical world. In particular, we will use F (or F_i) to denote the tropical polynomials in the tropical polynomial ring $\mathbb{T}[X, Y]$ along which we modify $\mathbb{R}^2$, and call them *modification polynomials*,
- lowercase letters will be used to denote objects over the valued field K. In particular, f (or f_i) will denote a polynomial lift of a modification polynomial. We will call these polynomials *re-embedding polynomials*.

Definition 12 ([CM19]) Let $F_1, \dots F_n \in \mathbb{T}[X,Y]$ and $f_1 \dots, f_n$ be lifts of the respective F_i. Let g be a defining equation for χ (see (1)). Then the tropicalization of the variety $V(I_{g,f})$ of the ideal $I_{g,f} = \langle g, z_1 - f_1, \dots, z_n - f_n \rangle \subset K[x^{\pm}, y^{\pm}, z_1^{\pm}, \dots, z_n^{\pm}]$ is a tropical curve in the modification of $\mathbb{R}^2$ along the F_i.

Definition 13 In the following, we will denote by $I_{g,f}$ the ideal corresponding to the re-embedding of the hyperelliptic curve χ, given by a defining equation g (see (1)), into the tropical modification induced by the re-embedding polynomials $f = (f_1, \dots, f_n)$.

2.3 Analyzing Re-Embeddings

Re-embeddings into tropical modifications only make hidden parts of the tropical curve visible if the modification curve passes through the parts we want to investigate. We can then uncover the hidden parts of the curve by investigating what happens on the new cells. As hidden cycles occur on edges with multiplicity greater than one where two edges get mapped onto each other by the embedding, we restrict our attention to re-embedding polynomials of the form $f(x,y) = y - a_1x - a_2x^2 - \cdots - a_nx^n \in K[x,y]$. Their tropicalizations pass through the multiplicity two edges of the tropical hyperelliptic curve before re-embedding, see Table 1. An example of this can be found in Figure 5.

After re-embedding the tropical curve into the modification of $\mathbb{R}^2$, we can check for faithfulness. Since re-embedding is compatible with projection, i.e. projections of curves inside a tropical modification are tropical curves under projection [CM19; Sch20], re-embeddings can be analyzed by investigating projections.

Lemma 14 ([CM19], Lemma 3.3; and [Sch20], Lemma 1.1) *Given an irreducible curve $\chi \subset (K^*)^2$, defined by polynomials $g \in K[x,y]$ and $f(x,y) = y - a_1x - a_2x^2 - \dots a_nx^n \in K[x,y]$, the tropicalization induced by the ideal $I_{g,f} = \langle g, z - f \rangle \in K[x^{\pm}, y^{\pm}, z^{\pm}]$ is completely determined by the tropical plane curves* $\mathrm{trop}(V(g))$, $\mathrm{trop}(V(I_{g,f} \cap K[x^{\pm}, z^{\pm}]))$, $\mathrm{trop}(V(I_{g,f} \cap K[y^{\pm}, z^{\pm}]))$ *and one further projection onto a rational plane P in $\mathbb{R}^3$ that is not parallel to any top-dimensional cell in the tropicalization of $z - f$, and not parallel to any of the coordinate planes.*

Finally, if we have identified a re-embedding polynomial constructed the appropriate tropical curve inside the modification, how can we be sure that the resulting tropical curve is actually faithful? The answer lies in the multiplicities of the tropical curve [MS15].

Lemma 15 ([BPR16], Theorem 5.24) *An embedded tropical curve is faithful on a subgraph Σ of the Berkovich space χ^{an} if it contains a copy of*

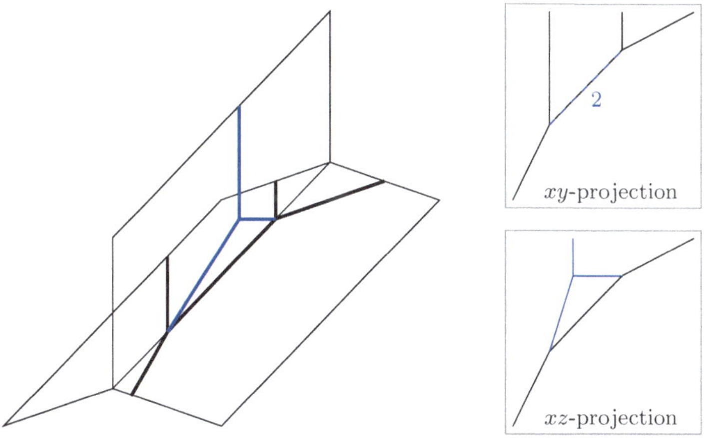

Fig. 5: The example curve $\chi = V(y^2 - x^3 - 4x^2 - 8t^4x)$ in Figure 1 inside of the modification constructed in Figure 4. We have actually already seen its xz-projection - it is $\operatorname{trop}(-x^3 - 4xy + y^2 - 8t^4x)$ on the right of Figure 1.

Σ whose edges have tropical multiplicity one and all vertices are three-valent, i.e. have three incident edges or legs.

3 Faithful Tropicalizations of Building Blocks

In this section, we find polynomials for faithful re-embeddings for curves that have genus at least one, i.e. the upper part of Table 1. We proceed by carefully analyzing the tropical curve obtained in the re-embedding by analyzing its coordinate projections.

Remark 16 (Re-Embeddings for Cycles and 2-Thetas) In [CM16], re-embedding polynomials (see Definition 11) for cycles have been constructed as $f(x, y) = y + \sqrt{-\alpha_3}x$, and in [CM19], re-embedding polynomials for 2-thetas were constructed as $f(x, y) = y - \sqrt{-\alpha_3\alpha_4\alpha_5}x + \sqrt{-\alpha_5}x^2$, where the α_i are the roots of the defining polynomial of the hyperelliptic curve.

Theorem 17 (Faithful Re-Embeddings of 3-Thetas) *Let χ be a hyperelliptic curve, given by a defining equation g satisfying the coefficient conditions of being a 3-theta as in Table 1. Then, the embedding of $\operatorname{trop}(\chi)$ into the modification of $\mathbb{R}^2$ at $y - (\sqrt{(\prod_{i=3}^{7} \alpha_i)}x + \sqrt{(\alpha_5\alpha_6\alpha_7)}x^2 - \sqrt{\alpha_7}x^3)$ given by $I_{g,f}$ is faithful on the minimal Berkovich skeleton.*

Depending on the valuations of the coefficients, several cases have to be considered. The case-by-case analysis is summed up in the following proof.

The tables, code and subprocedures providing further details of this proof will be given throughout the section.

Proof For the remainder of this section, let $\alpha_i = \beta_i^2$, for $i = 3, \dots, 6$, and $\alpha_7 = -\beta_7^2$, so that the re-embedding may be equivalently given as $y - (\prod_3^7 \beta_i x + \beta_5\beta_6\beta_7 x^2 - \beta_7 x^3)$. We have $\text{in}(\alpha_3) = \text{in}(\alpha_4)$ and $\text{in}(\alpha_5) = \text{in}(\alpha_6)$ by Table 1. Let $d_{34} = \text{val}(\alpha_3 - \alpha_4)$ and $d_{56} = \text{val}(\alpha_5 - \alpha_6)$. We control these differences by defining new variables $\beta_{34} := \beta_3 - \beta_4$ and $\beta_{56} := \beta_5 - \beta_6$, and substituting $\alpha_3 := (\beta_4 + \beta_{34})^2$ and $\alpha_5 := (\beta_6 + \beta_{56})^2$. We obtain $-\text{val}(\beta_{34}) = d_{34} + \text{val}(\alpha_4)/2 = -\text{val}(\alpha_3 - \alpha_4) + \text{val}(\beta_4) = \omega_{34}$ and $-\text{val}(\beta_{56}) = d_{56} + \text{val}(\alpha_6)/2 = -\text{val}(\alpha_5 - \alpha_6) + \text{val}(\beta_6) = \omega_{56}$.

Substituting the new variables into the Equation (1) of the hyperelliptic curve, we obtain $f(x, y) = y - \beta_4(\beta_{34} + \beta_4)\beta_6(\beta_6 + \beta_{56})\beta_7 x - \beta_6(\beta_6 + \beta_{56})\beta_7 x^2 + \beta_7 x^3$. The weight vector $\underline{u} \in \mathbb{R}^6$ encoding the valuation of the six parameters equals

$$\underline{u} = \text{val}(\beta_7, \beta_6, \beta_{56}, \beta_4, \beta_{34}, \beta_2) = \left(\frac{\omega_7}{2}, \frac{\omega_6}{2}, d_{56} - \frac{\omega_6}{2}, \frac{\omega_4}{2}, d_{34} - \frac{\omega_4}{2}, \frac{\omega_2}{2}\right).$$

Using the valuation conditions from Table 1, the valuations satisfy the following inequalities:

$$u_7 > u_6, \quad u_6 > u_{56}, \quad u_6 > u_4, \quad u_4 > u_{34}, \quad \text{and} \quad u_4 > u_2. \tag{1}$$

These inequalities span a polyhedral cone $\mathcal{C} \subset \mathbb{R}^6$. Its proper faces correspond to other types of curves. We subdivide the cone further using the data of the xz- and the yz-projections of $I_{g,f}$, computed in OSCAR via elimination. For the sake of brevity, we only supply the full computations for the maximal cones of this subdivision here. Supplementary material and further analysis of the lower dimensional cones can be found at `https://book.oscar-system.org`.

For both projections, we list the leading terms that can appear for all monomials in the projection. For the xz-projection, this gives rise to 81 cones, of which 16 are maximal, defined by the inequalities in Table 3, and give the corresponding leading terms of the xz-projection in Table 2.

Using the yz-projection, we subdivide our cones further as given in Remark 19. We do this computationally by taking the common refinement of the two fans spanned by these respective collections of cones. We list the leading terms of the yz-projection in Table 6.

We proceed by giving a precise analysis of the Newton subdivision inside of each cone in Lemma 18 for the xz-projection, and in Proposition 21 for the yz-projection. They are listed in Table 4 and Figure 8 respectively.

Combining the information of these two projections with the naive tropicalization of the curve from Section 2 and with Lemma 14, we obtain multiplicity one on all edges of the 3-theta, and thus, using Lemma 15, faithfulness on the minimal Berkovich skeleton. □

3.1 Analysis of the xz-Projection

We begin our analysis of the re-embedding $I_{g,f}$ by investigating the xz-projection of $\mathrm{Trop}(V(I_{g,f}))$. Computationally, this is done by elimination of y, using the eliminate-command in OSCAR :

```
julia> S,(b2,b34,b4,b56,b6,b7)=polynomial_ring(QQ,["b2", "b34", "b4",
↪   "b56", "b6", "b7"])
(Multivariate polynomial ring in 6 variables over QQ, QQMPolyRingElem[b2,
↪   b34, b4, b56, b6, b7])

julia> R,(x, y, z)=polynomial_ring(S,["x", "y", "z"])
(Multivariate polynomial ring in 3 variables over multivariate polynomial
↪   ring, AbstractAlgebra.Generic.MPoly{QQMPolyRingElem}[x, y, z])

julia> g = y^2-x*(x-b2^2)*(x-(b34+b4)^2)*(x-(b4)^2)*(x-(b6+b56)^2)*(x-(b ⌋
↪   6)^2)*(x+(b7)^2);

julia> f= y - b4*(b34 + b4)*b6*(b6 + b56)*b7*x + b6*(b6 + b56)*b7*x^2 -
↪   b7*x^3;

julia> Igf = ideal(R,[g,f]);

julia> eliminate(Igf,[y])
```

The resulting projection is alternatively given by the polynomial:

$$g_{xz}(x,z) := g\left(x, z + \left(\prod_{3}^{7} \beta_i\right) x + \beta_5\beta_6\beta_7 x^2 - \beta_7 x^3\right).$$

Lemma 18 *The conditions on the valuations of the coefficients for 3-thetas (see* (1)*) describe a cone $\mathcal{C}$ which is subdivided into 81 cones by inequalities of the coefficients of the xz-projection. Of these, 16 are maximal and correspond to the rows in Table 3, although only 7 distinct Newton subdivisions arise from them.*

Proof To analyze the Newton subdivision of g_{xz}, we compute all possible leading terms of all monomials in the xz-projection. Except for x^5 and x^3, all monomials admit the same leading terms for any admissible choice of ordering on the β_i using (1). Omitting constant coefficients, for x^5, the possible leading terms are β_6^4, $-\beta_{56}^2\beta_7^2$, and $-\beta_4^2\beta_7^2$. For x^3, the possible leading terms are $\beta_2^2\beta_6^4\beta_7^2$, $\beta_{34}^2\beta_6^4\beta_7^2$, and $\beta_4^4\beta_6^2\beta_7^2$. For the remaining coefficients, we refer to Table 2.

From the computed initial forms, we can determine the induced subdivision of the cone $\mathcal{C}$ by the occurring possible leading terms. The inequalities are as follows:

We can assign each of the possible leading terms of x^5 and x^3 we computed earlier to a cone: for x^5, β_6^4 leads for points in $\mathcal{C}_1$ and $\mathcal{C}_2$, $-\beta_{56}^2\beta_7^2$ for $\mathcal{C}_3$, and

x^7, z^2	x^6	x^4	x^2	x	x^3z	x^2z	xz
β_6^2	$\beta_4^2\beta_6^2\beta_7^2$	$\beta_2^2\beta_4^2\beta_6^4\beta_7^2$	$\beta_2^2\beta_4^4\beta_6^4\beta_7^2$	β_7	$\beta_6^2\beta_7$	$\beta_4^2\beta_6^2\beta_7$	

Table 2: Valuations of the xz-projection of the modification ideal $I_{g,f}$.

Cone	**Condition 1**	**Condition 2**	**Cone**	**Condition 3**	**Condition 4**
$\mathcal{C}_1$	$2\omega_6 < \omega_{56} + \omega_7$	$\omega_{56} < \omega_4$	$\mathcal{C}_A$	$2\omega_4 < \omega_{34} + \omega_6$	$\omega_{34} < \omega_2$
$\mathcal{C}_2$	$2\omega_6 < \omega_4 + \omega_7$	$\omega_4 < \omega_{56}$	$\mathcal{C}_B$	$2\omega_4 < \omega_2 + \omega_6$	$\omega_2 < \omega_{34}$
$\mathcal{C}_3$	$\omega_{56} + \omega_7 < 2\omega_6$	$\omega_{56} < \omega_4$	$\mathcal{C}_C$	$\omega_{34} + \omega_6 < 2\omega_4$	$\omega_{34} < \omega_2$
$\mathcal{C}_4$	$\omega_4 + \omega_7 < 2\omega_6$	$\omega_4 < \omega_{56}$	$\mathcal{C}_D$	$\omega_2 + \omega_6 < 2\omega_4$	$\omega_2 < \omega_{34}$

Table 3: Coefficient conditions for cones in the xz-projection.

$-\beta_4^2\beta_7^2$ for $\mathcal{C}_4$. For x^3, the leading terms are $\beta_2^2\beta_6^4\beta_7^2$ for points in $\mathcal{C}_D$, $\beta_{34}^2\beta_6^4\beta_7^2$ for $\mathcal{C}_C$, and $\beta_4^4\beta_6^2\beta_7^2$ for $\mathcal{C}_A$ and $\mathcal{C}_B$.

There are 7 possible combinatorial types of tropical curves that can arise in the xz-projection of the re-embedding, corresponding to the cones determined above. We list them and their corresponding Newton subdivisions in Table 4. All Newton subdivisions share a common part, given in Figure 6, and one that is distinct for each type, given as a subdivision of $\mathcal{P}$ in Table 4. □

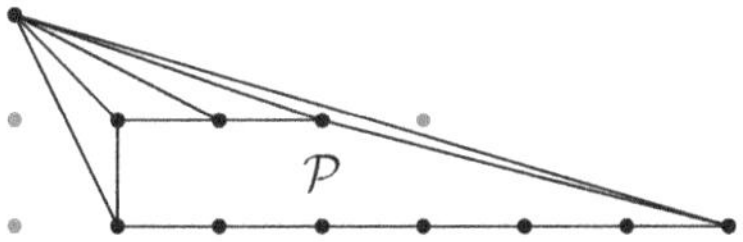

Fig. 6: General xz-subdivision.

Source codes, examples, and other material for the analysis of the remaining nonmaximal cones can be found at `https://book.oscar-system.org`. The webpage also contains an example for each maximal combinatorial type.

The following Table 4 is color-coded in accordance with the later analysis for the yz-projections, and the tropical curves are re-drawn versions of the OSCAR output to include colors.

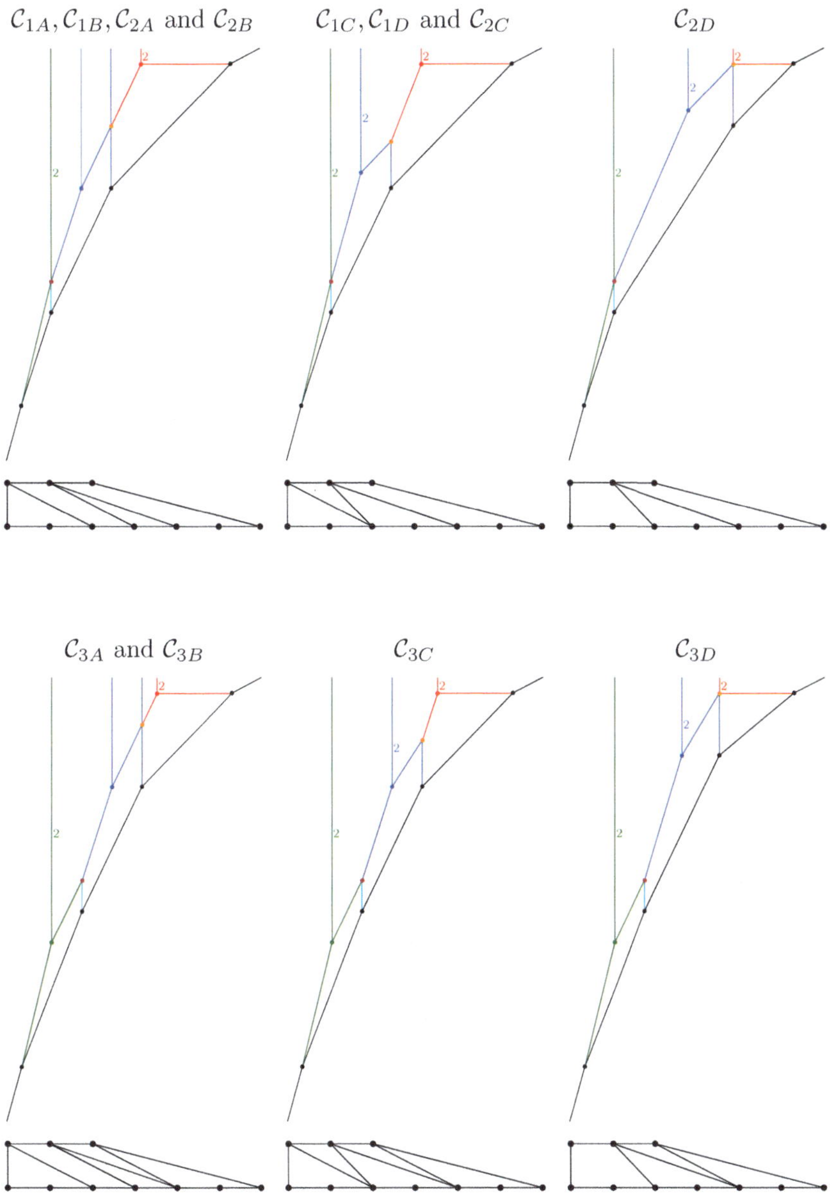
$\mathcal{C}_{1A}, \mathcal{C}_{1B}, \mathcal{C}_{2A}$ and $\mathcal{C}_{2B}$
$\mathcal{C}_{1C}, \mathcal{C}_{1D}$ and $\mathcal{C}_{2C}$
$\mathcal{C}_{2D}$
$\mathcal{C}_{3A}$ and $\mathcal{C}_{3B}$
$\mathcal{C}_{3C}$
$\mathcal{C}_{3D}$

Table 4: Tropical curves associated to cones in the xz-projection. The associated Newton subdivisions given here are subdivisions of the trapezoid $\mathcal{P}$ from Figure 6.

3.2 Analysis of the yz-Projection

Since the analogous elimination computation for the yz-projection does not terminate within reasonable time (2 weeks), we proceed as follows. First, we investigate the structure of the possible coefficients and monomials in the yz-projection.

The yz-projection is given as the variety of a single defining polynomial of degree 7 in y and z. The coefficients of the monomials $y^i z^j$ are homogeneous polynomials $p_{y^i z^j}$ in $\beta_2, \beta_{34}, \beta_4, \beta_{56}, \beta_6$ and β_7. The homogeneous degree of the $p_{y^i z^j}$ is 1 for monomials of degree 7 and $(7-(i+j)) \cdot 7$ otherwise. Using the defining equation of the cone of 3-thetas (1), we obtain the following technical *tropical dominance conditions* for coefficients:

Remark 19 (Tropical Dominance Conditions) We note that by the structure of the equations, with the exception of β_7, all exponents in leading monomials are even. If the exponent of β_7 is odd for one possible leading monomial in $p_{y^i z^j}$, they are odd in β_7 for all leading monomials. Thus, we can assume that inequalities on the valuations between leading monomials are integral. Further, there are minimal inequalities generating all inequalities, that are of the following form: $\omega_i < \omega_j$ and $\omega_i > \omega_j$ for pairs $(i, j) \in \{(2, 34), (4, 56)\}$, $2\omega_i < \omega_j + \omega_k$ and $2\omega_i > \omega_j + \omega_k$ for combinations $i, (j, k) \in \{(4, (2, 6)), (4, (34, 6)), (56, (2, 6)), (56, (34, 6)), (6, (4, 7)), (6, (56, 7))\}$. We call a cone $\mathcal{C}_i$ in the subdivision provided by Table 3 β_{56}-*dominant* if $\omega_{56} < \omega_4$. In this case, only the coefficient conditions we already determined contribute. Alternatively, we call $\mathcal{C}_i$ β_4-*dominant* (i.e. $\omega_4 < \omega_{56}$). In this case, we obtain an additional subdivision of the yz-projection, given by inequalities on ω_{56} in relation to ω_{34} and ω_2, as now, potentially $\min(\omega_{34}, \omega_2) < \omega_{56}$. This is not possible for β_{56}-dominant terms, as here, $\omega_{56} < \omega_4 < \min(\omega_{34}, \omega_2)$. In particular, in addition to the conditions considered in Table 3, the following five inequalities contribute:

$\alpha_<$	$2\beta_4 < \min(\beta_2, \beta_{34}) + \beta_{56}$
$\beta_<$	$\max(\beta_2, \beta_{34}) + \beta_{56} < \min(\beta_2, \beta_{34}) + \beta_6$
$\gamma_<$	$\min(\beta_2, \beta_{34}) + \beta_4 < \max(\beta_2, \beta_{34}) + \beta_{56}$
$\delta_<$	$\min(\beta_2, \beta_{34}) + 2\beta_6 < 2\beta_4 + \beta_7$
$\phi_<$	$2\beta_4 < \beta_{56} + \beta_6$

Of course, each of these inequalities can also be attained in the opposite direction. We indicate this by reversing the symbol, for instance, $\alpha_>$ denotes the inequality $2\beta_4 > \min(\beta_2, \beta_{34}) + \beta_{56}$.

Proposition 20 (Leading Terms) *The leading terms for the coefficients of monomials given in Table 6 are correct and complete.*

Proof We proceed in two steps. First, we consider monomials of degree ≥ 5. For monomials $y^i z^j$ of degree ≥ 5, the $p_{y^i z^j}$ are of homogeneous degree ≤ 14. Because of the structure of the equations, we then expect $p_{y^i z^j}$ to contain monomials that are only in terms of $\beta_4, \beta_{56}, \beta_6$ and β_7, and where the exponents in terms of β_4 and β_{56} sum to less than or equal 4. Using the tropical dominance conditions from Remark 19, this implies that these are indeed the only possible leading terms for the monomial satisfying the defining inequality (1). Thus, we can substitute integers for the coefficients β_2 and β_{34} and compute the elimination only with respect to the remaining coefficients. This can be computed in OSCAR as follows:

```
julia> S,(b4,b56,b6,b7)=polynomial_ring(QQ,["b4", "b56", "b6", "b7"])
(Multivariate polynomial ring in 4 variables over QQ, QQMPolyRingElem[b4,
↪  b56, b6, b7])

julia> R,(x, y, z)=polynomial_ring(S,["x", "y", "z"])
(Multivariate polynomial ring in 3 variables over multivariate polynomial
↪  ring, AbstractAlgebra.Generic.MPoly{QQMPolyRingElem}[x, y, z])

julia> g = y^2-x*(x-1000^2)*(x-(1000000+b4)^2)*(x-(b4)^2)*(x-(b6+b56)^2)⌋
↪  *(x-(b6)^2)*(x+(b7)^2);

julia> f = y - b4*(1000000 + b4)*b6*(b6 + b56)*b7*x + b6*(b6 +
↪  b56)*b7*x^2 - b7*x^3;
```

```
julia> Igf = ideal(R,[g,f]);

julia> eliminate(Igf,[x])
```

The process terminates within two hours. This procedure turns the previously homogeneous polynomials into non-homogeneous ones. The leading terms are now in the coefficients of highest degree. Here, we can identify the actual leading terms using the tropical dominance conditions from Remark 19 encoded in the following algorithm:

Algorithm 6: Leading Terms

Input : A polynomial $p_{y^i z^j}$ obtained from the elimination
Output : A list of potential leading terms lt satisfying the inequalities in (1)

$h_{b_7} \leftarrow$ maximal exponent of β_7 in $p_{y^i z^j}$
potential_monomials $\leftarrow$ list of monomials containing $\beta_7^{h_{b_7}}$
$h_{b_6} \leftarrow$ maximal exponent of β_6 in potential_monomials
while $h_{b_6} <$ *expected_degree and* $h_{b_7}, h_{b_{56}}, h_{b_4} > 0$ **do**
  curr_monomial $\leftarrow$ monomial of maximal exponent h_{b_4} of β_4 (resp. $h_{b_{56}}$ of β_{56}) among monomials containing $\beta_6^{h_{b_6}} \beta_7^{h_{b_7}}$
  if *deg(current_monomial) = expected_degree* **then**
    $lt \leftarrow lt \cup$ current_monomial
  else
    $lt \leftarrow lt \cup$ cross_verification(current_monomial)
  end if
  Adjust h_{b_7}, h_{b_6}, $h_{b_{56}}$, and h_{b_4} according to Remark 19.
end while

Now, we proceed with monomials of lower degree. Here, knowing the elimination results in coefficients $\beta_4, \beta_{56}, \beta_6$, and β_7 is not sufficient, because the leading terms contain coefficients β_2 and β_{34}.

To tackle this, we now compute the analogous eliminations, where instead of substituting integers for β_2 and β_{34}, we substitute integers for one of β_2 and β_{34} and one of β_4 and β_{56} respectively, and call the outcomes of these elimination computations *cross-verification polynomials.*

Since $\beta_4, \beta_{56}, \beta_6$, and β_7 remain the dominant coefficients, we first identify the potential monomials, again as in Algorithm 6, remembering the gap k in exponents. As we are in a maximal cone, we can assume $\omega_{34} > \omega_2$ or vice versa. Thus, in the cross-verification polynomials, we search for a term containing the correct amount of β_4 or β_{56} respectively, and the maximal exponent of β_2 or β_{34} that is less than or equal to k. We verify the existence of such a term using the remaining cross-verification polynomials.

For β_4-dominant terms, in accordance with Remark 19, we repeat the procedure and identify further leading terms by searching for terms containing maximal amounts of β_2 and β_{34} instead of β_{56}. For mixed terms, we again use the cross-verification process. This computes all leading terms in five or less variables.

For the final case of leading terms involving all six variables, we observe that the cross-verification algorithm yields two symmetric outputs of a degree smaller than the expected degree. We combine the two leading terms computed into a single leading term missing no monomials.

We supply code computing all possible leading terms at `https://book.oscar-system.org`.

Monomial	**Leading Terms**	**Cone**
$y^7, y^6z, y^5z^2,$ $y^4z^3, y^3z^4,$ y^2z^5, yz^6, z^7	1	all
y^6	$\beta_4^2\beta_6^2\beta_7^3$ $\beta_{56}^2\beta_6^2\beta_7^3$	$(2,4)$ $(1,3)$
y^5z	$\beta_4^2\beta_7^5$ $\beta_{56}^2\beta_7^5$	$(2,4)$ $(1,3)$
y^4z^2	$\beta_6^2\beta_7^5$	all
$y^3z^3, y^2z^4,$ yz^5, z^6	β_7^7	all
y^5	$\beta_{56}^2\beta_6^8\beta_7^4$ $\beta_4^6\beta_7^8$ $\beta_{56}^6\beta_7^8$	$(1,2)$ (4) (3)
y^4z	$\beta_{56}^2\beta_6^6\beta_7^6$ $\beta_4^4\beta_6^2\beta_7^8$ $\beta_{56}^4\beta_6^2\beta_7^8$	$(1,2)$ (4) (3)
$y^3z^2, y^2z^3,$ yz^4, z^5	$\beta_4^2\beta_6^4\beta_7^8$ $\beta_{56}^2\beta_6^4\beta_7^8$	$(2,4)$ $(1,3)$
y^4	$\beta_{34}^2\beta_{56}^2\beta_6^{10}\beta_7^7$ $\beta_2^2\beta_{56}^2\beta_6^{10}\beta_7^7$ $\beta_{34}^2\beta_{56}^4\beta_6^6\beta_7^9$ $\beta_2^2\beta_{56}^4\beta_6^6\beta_7^9$ $\beta_4^4\beta_{56}^2\beta_6^6\beta_7^9$	$(1A_{\delta_<},1C_{\delta_<},2A_{\delta_<},2C_{\delta_<})$ $(1B_{\delta_<},1D_{\delta_<},2B_{\delta_<},2D_{\delta_<})$ $(3A_{\alpha_>},3C_{\alpha_>},4A_{\alpha_>},4C_{\alpha_>})$ $(3B_{\alpha_>},3D_{\alpha_>},4B_{\alpha_>},4D_{\alpha_>})$ $(\alpha_<,\delta_>)$
y^3z	$\beta_4^2\beta_{56}^2\beta_6^8\beta_7^9$	all
y^2z^2, yz^3, z^4	$\beta_{56}^2\beta_6^{10}\beta_7^9$	all
y^3	$\beta_2^4\beta_{56}^2\beta_6^{12}\beta_7^{10}$ $\beta_{34}^4\beta_{56}^2\beta_6^{12}\beta_7^{10}$ $\beta_2^2\beta_4^4\beta_{56}^2\beta_6^{10}\beta_7^{10}$ $\beta_{34}^2\beta_4^4\beta_{56}^2\beta_6^{10}\beta_7^{10}$	(D) (C) (B) (A)
y^2z, yz^2, z^3	$\beta_2^2\beta_4^2\beta_{56}^2\beta_6^{12}\beta_7^{10}$ $\beta_{34}^2\beta_4^2\beta_{56}^2\beta_6^{12}\beta_7^{10}$	(B,D) (A,C)

	$\beta_2^2\beta_{34}^2\beta_4^4\beta_{56}^2\beta_6^{14}\beta_7^{11}$	$(D_{\beta_>}, C_{\beta_>})$
	$\beta_2^4\beta_4^4\beta_{56}^4\beta_6^{12}\beta_7^{11}$	$(B_{\alpha_>}, D_{\beta_<})$
y^2	$\beta_{34}^4\beta_4^4\beta_{56}^4\beta_6^{12}\beta_7^{11}$	$(A_{\alpha_>}, C_{\beta_<})$
	$\beta_2^2\beta_4^8\beta_{56}^2\beta_6^{12}\beta_7^{11}$	$(B_{\alpha_<})$
	$\beta_{34}^2\beta_4^8\beta_{56}^2\beta_6^{12}\beta_7^{11}$	$(A_{\alpha_<})$
yz, z^2	$\beta_{34}^2\beta_4^6\beta_{56}^2\beta_6^{14}\beta_7^{11}$	all
	$\beta_2^2\beta_{34}^2\beta_4^8\beta_{56}^2\beta_6^{16}\beta_7^{12}$	$(\phi_>\gamma_<)$
	$\beta_2^4\beta_4^8\beta_{56}^4\beta_6^{14}\beta_7^{12}$	$(1B, 1D, 3B, 3D)$
y, z	$\beta_{34}^4\beta_4^8\beta_{56}^4\beta_6^{14}\beta_7^{12}$	$(1A, 1C, 3A, 3C)$
	$\beta_2^2\beta_{34}^2\beta_4^{12}\beta_6^{14}\beta_7^{12}$	$(2_{\phi_<\gamma_>}, 4_{\phi_<\gamma_>})$
	$\beta_2^4\beta_4^{10}\beta_{56}^2\beta_6^{14}\beta_7^{12}$	$(2B_{\gamma_<}, 2D_{\gamma_<}, 4B_{\gamma_<}, 4D_{\gamma_<})$
	$\beta_{34}^4\beta_4^{10}\beta_{56}^2\beta_6^{14}\beta_7^{12}$	$(2A_{\gamma_<}, 2C_{\gamma_<}, 4A_{\gamma_<}, 4C_{\gamma_<})$

Table 6: Leading terms for monomials in the yz-projection.

To analyze the projections, we provide a geometric description of the re-embedding, as parallel cells need to be carefully analyzed. For the re-embedding f, given in Theorem 17, the tropical hypersurface $\mathrm{Trop}(V(z-f))$ consists of nine 2-dimensional cells, defined by the following systems of inequalities:

$\sigma_1 = \{Z = x + a \geq Y, X \leq a - b\}$,
$\sigma_2 = \{Z = 2X + b \geq Y, b - c \geq X \geq a - b\}$,
$\sigma_3 = \{Z = 3X + c \geq Y, X \geq b - c\}$,
$\sigma_4 = \{Z, Y \leq 2a - b, X = a - b\}$,
$\sigma_5 = \{Z, Y \leq 3b - 2c, X = b - c\}$,
$\sigma_6 = \{Y = X + a \geq Z, X \leq a - b\}$,
$\sigma_7 = \{Y = 2X + b \geq Z, a - b \leq X \leq b - c\}$,
$\sigma_8 = \{Y = 3X + c \geq Z, b - c \leq X\}$,
$\sigma_9 = \{Z = Y \geq X + a, 2X + b, 3X + c\}$.

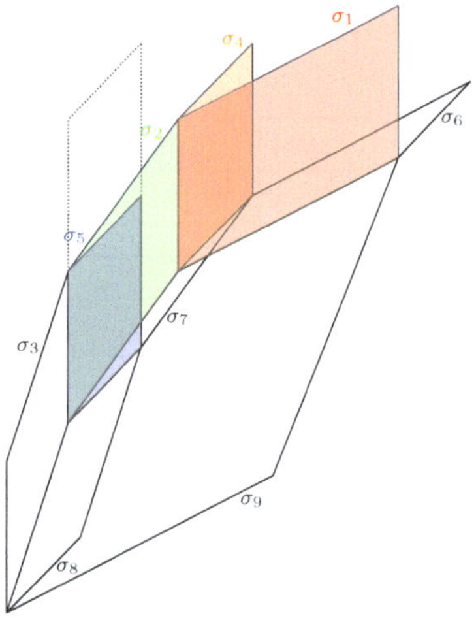

Fig. 7: The generic re-embedding for 3-thetas.

Proposition 21 *Using the re-embedding of Theorem 17, all edges of the re-embedded tropical curve* $\mathrm{Trop}(g)$ *have tropical multiplicity one, i.e. the re-embedding yields a faithful tropicalization.*

Proof We observe that in the yz-projection, the pairs of cells σ_1 and σ_4, σ_1 and σ_5, σ_2 and σ_5, and σ_4 and σ_5 overlap in parts. Further, the behaviour on these cells cannot be directly observed on the xz-projection. Thus, we expect some of the edges in the projections to have higher multiplicities, and need to carefully analyze them to understand the multiplicities and structure of the re-embedded curve. Overall, there are four types of such irregularities that can occur:

(i) Some multiplicity one edges or legs inherit higher multiplicities in the yz-tropicalization due to the push-forward formula for multiplicities, see [Tev07]. Multiplicities inherited in this way are the black multiplicities in Figure 8.

On the images of the overlapping cells σ_1, σ_2, σ_4 or σ_5, the following further behaviour can occur:

(ii) Parallelograms inside the Newton subdivision may indicate either a vertex (red in the subdivision) or not (marked in blue), depending on whether two edges or legs of $I_{g,f}$ that intersect in the yz-projection intersect in $I_{g,f}$ or not.

(iii) Two edges or legs overlap in the yz-tropicalization and their multiplicities get added accordingly, indicated by dashed edges and sums of multiplicities on the curve in Figure 8.

(iv) Vertices lie in the relative interiors of edges or on another vertex. The vertices for which this can happen are the purple and the orange vertices in Table 4 and Figure 8.

We resolve these issues by investigating the corresponding xz-projection for each case and combining the information from both projections. A color-coded analysis of each individual representative is given in Table 4 for the xz-projection and at `https://book.oscar-system.org` for the yz-projection. Edges that are the same color in the projections correspond to each other. See Example 22, where a three dimensional reconstruction is carried out in full for the cone $\mathcal{C}_0$. □

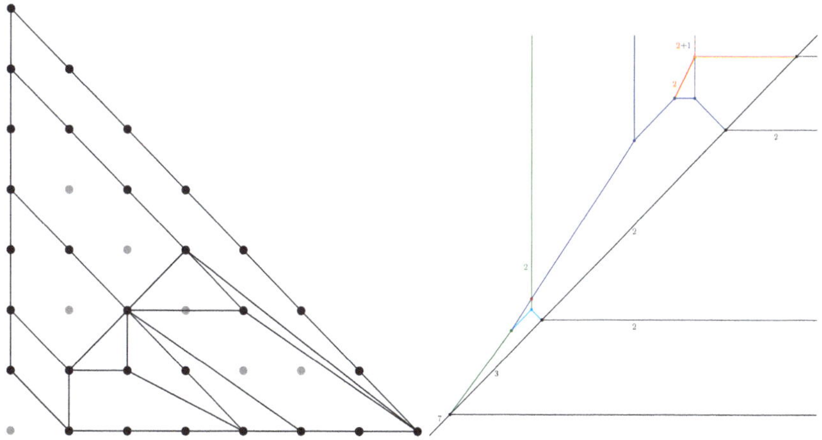

Fig. 8: The yz-projection in Example 22. The tropical curve is a re-drawing of the OSCAR output to include colors.

Example 22 We give an example of a re-embedded curve in Cone $\mathcal{C}_{1A_{\alpha_{<\delta>}}}$ in Figure 8. The colors of edges and vertices coincide with the colors in the xz- and yz-projections in Table 4 and Figure 8.

4 Gluing Re-Embeddings

In the previous section we have seen how to locally modify the tropical hyperelliptic curve according to specific building blocks of the associated Berkovich skeleton. This allows us to faithfully re-embed in the case of 3-thetas and 3-points. But what happens if the Berkovich skeleton consists of multiple building blocks?

In this section, we show how to glue the constructed re-embeddings of building blocks to a re-embedding that works for the whole curve.

Remark 23 First, let us assume that we are considering two building blocks B_1 and B_2 connected by a point connector p, and that we have a re-embedding that is faithful on both B_1 and B_2. Then, the re-embedding is already faithful on the minimal Berkovich skeleton. Now, we assume that both building blocks are connected by a bridge e_i. The coefficient condition of bridges gives rise to a dual edge of length one, as can be observed in Table 1. Therefore, e_i has tropical multiplicity one and is faithful on every subgraph of the Berkovich skeleton that contains it. Thus, if the tropicalizations on both sides of the e_i are faithful on respective subgraphs of the Berkovich skeleton connected by e_i in the Berkovich skeleton, the whole curve is faithful on the union.

The second piece of the gluing puzzle is the ability to shift re-embeddings: The modifications given in Section 3 work for the first monomial in the building block being x, but multiple building blocks will necessarily start at different monomials. The next proposition solves just that, i.e. multiplying the modification polynomial by the correct monomial at which the building block starts preserves faithfulness.

Lemma 24 *Let χ be an irreducible hyperelliptic curve given by the defining equation $g(x,y) := y^2 - h(x) = 0$, and let $\tilde{\chi}$ be another curve given by the defining equation $\tilde{g}(x,y) := y^2 - x^{2l+1}h(x) = 0$. Further, let $f \in K[x,y]$ be a modification polynomial such that the tropicalization induced by the ideal $I_{g,f} = \langle g, z-f\rangle \in K[x^{\pm}, y^{\pm}, z^{\pm}]$ is faithful. Then, for the modification polynomial $\tilde{f}(x,y) = y - x^{l-1}f \in K[x,y]$, the tropicalization induced by the ideal $I_{\tilde{g},\tilde{f}} = \langle \tilde{g}, z-\tilde{f}\rangle \in K[x^{\pm}, y^{\pm}, z^{\pm}]$ is faithful.*

Proof Let $\mathcal{P}$ be the Newton polytope of f, with the negative valuation of every monomial corresponding to a point attached as a last coordinate. Let $\tilde{\mathcal{P}}$ be the extended Newton polytope of $\tilde{f}$. Then, for every vertex $\tilde{v}$ in $\tilde{\mathcal{P}}$ corresponding to the monomial m there is a vertex v in $\mathcal{P}$ corresponding to

mx^{-l} with the same valuation. Thus, $\mathcal{P}$ and $\tilde{\mathcal{P}}$ yield congruent subdivisions on their respective Newton polytopes $\mathcal{N}$ and $\tilde{\mathcal{N}}$ under the coordinate shift $x^i \mapsto x^{i+l}$. Analogous computations hold for the projections of their respective re-embeddings. Further, lattice lengths stay constant under this shift. Thus, by construction, $\tilde{g}$ only has multiplicity one edges in the re-embedding along $\tilde{f}$ and is thus faithful by Lemma 15. □

Combining the results from Section 3 and the shifting from Lemma 24, we obtain the re-embedding polynomials for building blocks in Table 8.

Subgraph at x^i	**Local Berkovich skeleton**	**Re-embedding polynomial**
3-Theta	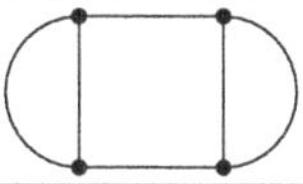	$y - (\sqrt{(\prod_{i=3}^{7} \alpha_i)}x + \sqrt{(\alpha_5\alpha_6\alpha_7)}x^2 - \sqrt{\alpha_7}x^3)$
2-Theta		$y - \sqrt{x^{i-1}} \cdot (\sqrt{-\alpha_3\alpha_4\alpha_5}x - \sqrt{-\alpha_5}x^2)$
Cycle	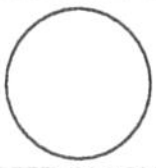	$y - \sqrt{x^{i-1}} \cdot \sqrt{-\alpha_3}x$

Table 8: Re-embedding polynomials for the building blocks. The α_i are the coefficients in the defining equation of the hyperelliptic curve χ, see (1), and the i is the exponent of the first monomial contributing to the cycle or 2-theta, as in Table 1.

Finally, we can combine all the pieces constructed in the last Sections to obtain our main result, Theorem 7.

Theorem 25 *Let χ be an irreducible hyperelliptic curve of genus 3 defined by an equation g as in Equation (1). The embedding of χ in the ideal given by modifications along hyperplanes generated by $z_i - f_i(x, y)$, where $f_i(x, y) = y - x^l(a_1x - a_2x^2 - \cdots - a_{k_i}x^{k_i})$ for suitable l and a_i given in Table 8, by considering $I_{g,f} = \langle g, z_1 - f_1, \ldots, z_m - f_m \rangle \in K[x^\pm, y^\pm, z_1^\pm, \ldots, z_m^\pm]$, is faithful.*

Proof First, we separate the defining equation g into its different components that identify the different (possibly shifted) building blocks using the coefficient conditions in Table 1. In Section 3, we showed that for the first building block of the Berkovich skeleton of χ, this is a suitable faithful re-embedding regardless of the type of the building block. By applying Proposition 24, we see that this is true for all building blocks contained in the tropical curve. For two connected adjacent building blocks carrying genus, we proceed pairwise by elimination of variables.

We denote by I_{g,f_i,k_1} the re-embedding along f_i, by I_{g,f_{i+1},k_2} the re-embedding along f_{i+1} and by $I_{g,f_i,f_{i+1},k_1,k_2}$ the re-embedding along both f_i and f_{i+1}. Since the two sets of variables of the re-embedding are disjoint, eliminating further the variables $z_1, \ldots, z_m$, we obtain $I_{g,f_i,f_{i+1},k_1,k_2} \cap K[x^{\pm}, y^{\pm}, w_1^{\pm}, \ldots, w_{\tilde{m}}^{\pm}] = I_{g,f_{i+1},k_2}$. Analogously, we obtain the equality $I_{g,f_i,f_{i+1},k_1,k_2} \cap K[x^{\pm}, y^{\pm}, z_1^{\pm}, \ldots, z_m^{\pm}] = I_{g,f_i,k_1}$. Since $I_{g,f_i,k_1} \subset I_{g,f_i,f_{i+1},k_1,k_2}$ and $I_{g,f_{i+1},k_2} \subset I_{g,f_i,f_{i+1},k_1,k_2}$, all edges that appear in I_{g,f_i,k_1} and I_{g,f_{i+1},k_2} can be identified with an edge of the same length in $I_{g,f_i,f_{i+1},k_1,k_2}$. Thus, as I_{g,f_i,k_1} and I_{g,f_{i+1},k_2} contain isometric copies of their respective building blocks, $I_{g,f_i,f_{i+1},k_1,k_2}$ does as well.

By assumption, all multiplicities of edges in I_{g,f_i,k_1} and I_{g,f_{i+1},k_2} corresponding to isometric copies of the building blocks of genus k_1 and k_2 are one. Thus, all multiplicities of edges in $I_{g,f_i,f_{i+1},k_1,k_2}$ corresponding to isometric copies of the building blocks of genus k_1 and k_2 are one. Then, by Remark 23, the tropicalization induced by $I_{g,f_i,f_{i+1},k_1,k_2}$ is faithful on both building blocks and their connector. Thus, we obtain a faithful tropicalization of χ.□

Example 26 (Partial Modification of Two Cycles) Consider the curve $g(x, y) = y^2 - x(x - t^2)(x - t^4)(x - t^6)(x - t^8)$. All valuations of the monomials in x are distinct. Using the coefficient conditions in Table 1, we obtain that g consists of two cycles, separated by a bridge.

First, we modify at the lower edge in direction z with $f = y - \sqrt{t^2 t^4 t^6} x = y - t^6 x$. Then, we modify at the upper edge using $\tilde{f} = y - \sqrt{t^2} x^2 = y - t x^2$, and obtain a faithful tropicalization. The resulting modification is depicted on the left in Example 29.

4.1 Reducing Dimensions

We have a faithful re-embedding now, but we can do better. For curves whose Berkovich skeleton contains bridges, we can systematically cut down ambient dimension by combining multiple re-embedding polynomials into one.

Theorem 27 *Let χ be an irreducible hyperelliptic curve given by a defining equation g, as in Equation (1), consisting of two k-thetas of genus k_1 and k_2 respectively, connected by a bridge. Let $I_{g,\hat{f},k_1} = \langle g, z_1 - \hat{f} \rangle \in K[x^{\pm}, y^{\pm}, z_1^{\pm}]$ and $I_{g,\tilde{f},k_2} = \langle g, w_1 - \tilde{f}_1 \rangle \in K[x^{\pm}, y^{\pm}, w_1^{\pm}]$ be the ideals determined in Section 3 for which the embedding is faithful, where $\hat{f} = y - \hat{h}(x)$ and $\tilde{f} = y - x^{k_1+1} \tilde{h}(x)$, obtained from Proposition 24. Let $f = y - \hat{h}(x) - x^{k_1+1} \tilde{h}(x)$.*

Then, the ideal $I_{g,f} = \langle g, z_1 - f \rangle \in K[x^{\pm}, y^{\pm}, z_1^{\pm}]$ induces a faithful tropicalization.

Proof By Theorem 25, $I_{g,f,k_1,k_2} = \langle g, z_1 - \hat{f}, w_1 - \tilde{f} \rangle \in K[x^{\pm}, y^{\pm}, z_1^{\pm}, w_1^{\pm}]$ induces a faithful tropicalization on the whole Berkovich skeleton.

Let $F = \text{trop}(f)$. As $\hat{F} = \text{trop}(\hat{f})$ and $\tilde{F} = \text{trop}(\tilde{f})$ pass through the respective multiplicity two edges of the thetas in the standard xy-tropicalization of g by Section 2, so does F by its construction.

Let C be the additional cell in F that is not contained in any cell of $\hat{F}$ or $\tilde{F}$. (For an example, see Example 29)

Outside of C, the modification agrees with $I_{g,\hat{f},k_1}$ before the bridge and with $I_{g,\tilde{f},k_2}$ after the bridge respectively. They are thus faithful on the two thetas as they lie on distinct cells. Hence, by Remark 23, the modification is faithful on the minimal Berkovich skeleton. By construction, there is no edge or leg on C. Further, all legs and edges of the projected curve around C have multiplicity one by Remark 23. By Lemma 14, this extends to the re-embedding. Thus, the modification is faithful on C as well. □

Remark 28 If the Berkovich skeleton of χ consists only of thetas and bridges (i.e. χ_{trop} is in a *top-dimensional cone of the moduli space* $M_{3,n}^{trop}$), the tropicalization of χ can be faithfully modified in dimension 3 by re-embedding $\text{Trop}(\chi)$ into a tropical hypersurface obtained as the tropicalization of $V(z - (y - a_1x - a_2x^2 - \cdots + a_7x^7) \in K[x, y]$.

Example 29 (Global Tropical Modification of Two Cycles) Consider the curve $g(x, y) = y^2 - x(x - t^2)(x - t^4)(x - t^6)(x - t^8)$ from Example 26. By reducing dimension using Remark 28, we can further simplify Example 26 while preserving faithfulness. By modifying $\mathbb{R}^2$ at $f = y - t^6x - tx^2$, we obtain a re-embedding that is faithful on the tropical curve, see Figure 9.

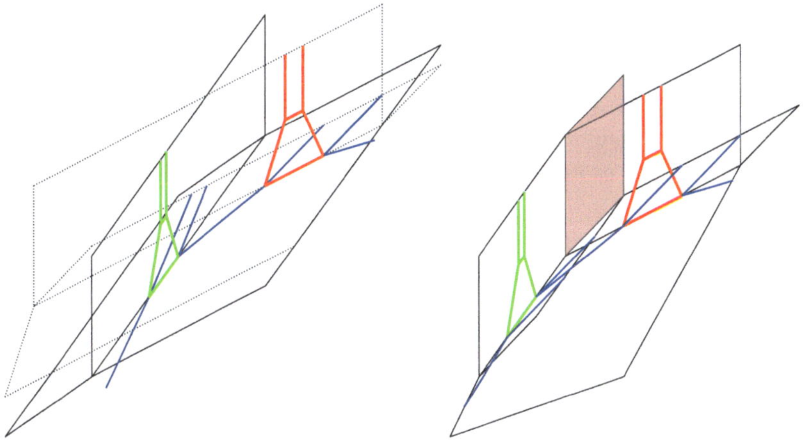

Fig. 9: The picture on the left shows the modification of Example 26, and the picture on the right shows the combined modification. The purple cell on the right is C referred to in the proof of Theorem 27.

Acknowledgements All three authors were supported by the Deutsche Forschungsgemeinschaft (DFG, German Research Foundation), Project-ID 286237555, TRR 195.

References

[BPR16] M. Baker, S. Payne, and J. Rabinoff. "Nonarchimedean geometry, tropicalization, and metrics on curves". *Algebr. Geom* 3.1 (2016), pp. 63–105.

[Cor21] D. Corey. "Tropical curves of hyperelliptic type". *J. Algebraic Combin.* 53.4 (2021), pp. 1215–1229. DOI: 10.1007/s10801-020-00959-y.

[CM16] M. A. Cueto and H. Markwig. "How to repair tropicalizations of plane curves using modifications". *Experimental Mathematics* 25.2 (2016), pp. 130–164.

[CM19] M. A. Cueto and H. Markwig. "Tropical geometry of genus two curves". *J. Algebra* 517 (2019), pp. 457–512.

[MS15] D. Maclagan and B. Sturmfels. *Introduction to tropical geometry.* Vol. 161. Graduate Studies in Mathematics. American Mathematical Society, Providence, RI, 2015, pp. xii+363. DOI: 10.1090/gsm/161.

[Mik06] G. Mikhalkin. "Tropical geometry and its applications". In: *International Congress of Mathematicians. Vol. II.* Eur. Math. Soc., Zürich, 2006, pp. 827–852.

[Pay09] S. Payne. "Analytification is the limit of all tropicalizations". *Mathematical Research Letters* 16.3 (2009), pp. 543–556.

[Sch20] V. M. Schleis. *Faithful tropicalizations of hyperelliptic curves of genus three.* Bachelor thesis, TU Kaiserslautern, 2020.

[Tev07] J. Tevelev. "Compactifications of subvarieties of tori". *American Journal of Mathematics* 129.4 (2007), pp. 1087–1104.

Tropical Implicitization Revisited

Kemal Rose, Bernd Sturmfels, and Simon Telen

Abstract Tropical implicitization means computing the tropicalization of a unirational variety from its parametrization. In the case of a hypersurface, this amounts to finding the Newton polytope of the implicit equation, without computing its coefficients. We present a new implementation of this classical procedure in `Oscar.jl`. It solves challenging instances, and can be used for classical implicitization as well. We discuss implicitization in higher codimension via Chow forms and pose several open questions.

Key words: Implicitization, tropical geometry, Newton polytope, Chow polytope, discriminant, Horn uniformization

1 Introduction

Let $X \subset \mathbb{C}^n$ be a d-dimensional affine variety defined as the closure of the image of a map

$$f : \mathbb{C}^d \dashrightarrow \mathbb{C}^n, \quad t \longmapsto \big(f_1(t), f_2(t), \ldots, f_n(t)\big). \tag{1}$$

Kemal Rose
Max-Planck-Institut für Mathematik in den Naturwissenschaften, Leipzig, Germany, e-mail: `krose@mis.mpg.de`

Bernd Sturmfels
Max-Planck-Institut für Mathematik in den Naturwissenschaften, Leipzig, Germany, e-mail: `bernd@mis.mpg.de`

Simon Telen
Max-Planck-Institut für Mathematik in den Naturwissenschaften, Leipzig, Germany, e-mail: `simon.telen@mis.mpg.de`

W. Decker et al. (eds.), *The Computer Algebra System OSCAR*,
Algorithms and Computation in Mathematics 32,
https://doi.org/10.1007/978-3-031-62127-7_17

Here $t = (t_1, \ldots, t_d)$, and $f_1, f_2, \ldots, f_n \in \mathbb{C}(t)$ are rational functions. The problem of *implicitization* asks for the defining polynomial equations of X in the coordinates $x = (x_1, \ldots, x_n)$ on $\mathbb{C}^n$. When $f_1, \ldots, f_n \in \mathbb{Q}(t)$ have rational coefficients, these equations can be computed via *symbolic elimination* [CLO97, Chapter 3, §3]. More precisely, one eliminates $t_1, \ldots, t_d$ from

$$x_1 - f_1(t) = x_2 - f_2(t) = \cdots = x_n - f_n(t) = 0 \tag{2}$$

using Gröbner basis or resultant techniques. Unfortunately, these methods prove to be inefficient for larger instances. This has motivated the question: whether we can obtain interesting partial information in cases where computing the ideal of X is out of reach?

Tropical geometry [MS15] replaces an algebraic variety by a polyhedral complex which encodes many of its geometric properties. A commonly used slogan is that this complex serves as a *combinatorial shadow* of the original variety. The *tropicalization* $\mathrm{trop}(X)$ of X is a pure d-dimensional polyhedral fan in $\mathbb{R}^n$, satisfying a balancing condition. The task of *tropical implicitization* [ST08; STY06] is to compute $\mathrm{trop}(X)$ from the data in (1). This was the goal of the paper [SY08], which includes an implementation called `TrIm`. Theorem 6.4 in [SY08] suggests the following two-step procedure for performing tropical implicitization:

1. Compute the tropicalization of the graph Γ_f of f, given by the n equations in (2). The result is the d-dimensional balanced fan $\mathrm{trop}(\Gamma_f)$ in the product space $\mathbb{R}^d \times \mathbb{R}^n$.
2. Project this fan to $\mathbb{R}^n$ and assign appropriate multiplicities to each cone in the image.

For an illustration of this procedure, see Figure 2. Computing $\mathrm{trop}(\Gamma_f)$ in step 1 can be complicated in general. It involves the computation of a *tropical basis* for the ideal generated by $x_1 - f_1(t), \ldots, x_n - f_n(t)$.

In this article, we consider two different assumptions on the map f. Both assumptions circumvent the tropical basis computation, and are relevant in practice. In Section 2, we assume that the functions f_i are Laurent polynomials which are generic with respect to their Newton polytopes. This is the assumption in [ST08; STY06; SY08]. It reduces step 1 above to computing the stable intersection of n codimension one fans in $\mathbb{R}^d \times \mathbb{R}^n$. In Section 3, we assume that f is the composition of a linear map $\lambda : \mathbb{C}^d \to \mathbb{C}^\ell$ followed by a Laurent monomial map $\mu : \mathbb{C}^\ell \dashrightarrow \mathbb{C}^n$. In symbols, we have $f = \mu \circ \lambda$. This allows to compute $\mathrm{trop}(X)$ as the linear projection of a tropical linear space in $\mathbb{R}^\ell$, for details see [MS15, Section 5.5]. An important special case arises from the computation of *tropical A-discriminants* [DFS07].

Tropical implicitization is a first step towards classical implicitization. Let $X = V(F)$ be the hypersurface defined by a polynomial $F \in \mathbb{C}[x]$. Then $\mathrm{trop}(X)$ is the union of the $(n-1)$-dimensional cones in the normal fan of the Newton polytope $\mathcal{N}(F)$, decorated with multiplicities. From $\mathrm{trop}(X)$ we can recover $\mathcal{N}(F)$. The key ingredient is a *vertex oracle* which, for a generic weight vector $w \in \mathbb{R}^n$, returns the vertex v of $\mathcal{N}(F)$ that minimizes the dot product with w on $\mathcal{N}(F)$. The algorithm realizing the oracle is suggested by [DFS07, Theorem 2.2]. We provide an implementation in `Oscar.jl` and use it to recover $\mathcal{N}(F)$ via the algorithm in [Hug06]. Once we have the Newton polytope $\mathcal{N}(F)$, we can find F via (numerical) linear algebra. The task is to compute the unique kernel vector of a matrix constructed via *numerical integration* [CGKW00] or *sampling* [BKSW18; EKKB13]. Sampling is preferred when the f_i have rational coefficients. We can then use the parametrization to find rational points on X, and compute F using exact arithmetic over $\mathbb{Q}$. However, the size of the matrix is the number of lattice points in $\mathcal{N}(F)$, and we may have to resort to floating point arithmetic when this number is too large. An alternative is rational reconstruction from linear algebra over finite fields. We discuss these techniques in Section 4 and solve instances for which elimination via Gröbner bases does not terminate within reasonable time.

If the f_i are Laurent polynomials which are generic with respect to their Newton polytopes, as in Section 2, then $\mathcal{N}(F)$ is a *mixed fiber polytope* [EKP07; EK08; STY06]. Our implementation in `Oscar.jl` for computing $\mathcal{N}(F)$ gives a practical way of computing mixed fiber polytopes.

In Section 5, we present a new way of finding its implicit equations from $\mathrm{trop}(X)$, when $\dim X < n-1$. The idea is to pass through the *Chow form* $\mathrm{Ch}(X)$ of X [DS95]. The polytope we compute is the *Chow polytope* $\mathcal{C}(X)$, which is a linear projection of the Newton polytope $\mathcal{N}(\mathrm{Ch}(X))$. This computation rests on a result by Fink [Fin13], which describes the (weighted) normal fan of $\mathcal{C}(X)$ in terms of $\mathrm{trop}(X)$. We explain how to recover $\mathrm{Ch}(X)$ from $\mathcal{C}(X)$, using the parameterizing functions and an appropriate ansatz. Defining equations for X are obtained from $\mathrm{Ch}(X)$ in the standard manner [DS95, Proposition 3.1].

The implementation of the algorithms supporting this work have benefited from the flexibility provided by `Oscar.jl`. The possibility to combine polyhedral computations with symbolic linear and nonlinear algebra in the same environment has greatly simplified the task. This feature has been our incentive to revisit tropical implicitization. Throughout the article, we include several open problems and computational challenges which we hope will inspire the reader to join this effort. Our software and data are made available in the `MathRepo` collection at MPI-MiS via `https://mathrepo.mis.mpg.de/TropicalImplicitization`.

2 Generic Tropical Implicitization

In this section, we start with n Laurent polynomials in n variables:

$$f_i = \sum_{a \in A_i} c_{i,a}\, t^a \in \mathbb{C}[t_1^{\pm 1}, \ldots, t_d^{\pm 1}] \quad \text{for} \quad i = 1, 2, \ldots, n.$$

We use these Laurent polynomials in (1). The tuple $f = (f_1, \ldots, f_n)$ gives a map $(\mathbb{C}^*)^d \to (\mathbb{C}^*)^n$. Let $X \subset (\mathbb{C}^*)^n$ be the closure of the image of f. Our first task is to find its tropicalization $\operatorname{trop}(X)$. In Section 4, we use $\operatorname{trop}(X)$ for classical implicitization. As a set,

$$\operatorname{trop}(X) \;=\; \{w \in \mathbb{R}^n \;:\; \operatorname{in}_w(I(X)) \text{ does not contain a monomial } \} \;\subset \mathbb{R}^n.$$

Here $I(X) \subset \mathbb{C}[x_1^{\pm 1}, \ldots, x_n^{\pm 1}]$ is the vanishing ideal of X, and in_w takes the initial ideal with respect to the weight vector w. It is well known that $\operatorname{trop}(X)$ is the support of a fan Σ of dimension $\dim(X)$. This fan is not unique, but for the purposes of this text we can choose any fan Σ with support $\operatorname{trop}(X)$. Assigning a *multiplicity* m_σ to each top dimensional cone $\sigma \in \Sigma$ in the appropriate way [MS15, Definition 3.4.3], the fan Σ is *balanced* [MS15, Theorem 3.4.14]. We will see that these multiplicities are crucial when using $\operatorname{trop}(X)$ for implicitization.

Classically, the variety $X \subset (\mathbb{C}^*)^n$ is the closure of the projection $\Gamma_f \to (\mathbb{C}^*)^n$ of the graph

$$\Gamma_f \;=\; \{(x,t) \in (\mathbb{C}^*)^n \times (\mathbb{C}^*)^d \;:\; x_1 - f_1(t) = 0, \ldots, x_n - f_n(t) = 0\}$$

onto the n x-coordinates. It turns out this has an easy tropical analog.

Theorem 1 *Let* $X \;=\; \overline{\operatorname{im} f} \;\subset\; (\mathbb{C}^*)^n$. *The tropical variety* $\operatorname{trop}(X)$ *is the image of the projection* $\operatorname{trop}(\Gamma_f) \to \mathbb{R}^n$*, where* $\operatorname{trop}(\Gamma_f) \subset \mathbb{R}^n \times \mathbb{R}^d$ *is the tropicalization of the graph of* f.

This is an instance of [STY06, Theorem 2.1]. See also [SY08, Theorem 6.4]. We can thus obtain $\operatorname{trop}(X)$ from $\operatorname{trop}(\Gamma_f)$ via a simple projection. However, Theorem 1 is only useful in practice when $\operatorname{trop}(\Gamma_f)$ is easy to compute. Our next theorem describes $\operatorname{trop}(\Gamma_f)$ under the assumption that the f_i are *generic* with respect to their Newton polytopes $\mathcal{N}(f_i)$. It uses the following notation: for a polytope $P \subset \mathbb{R}^k$ and a vector $w \in (\mathbb{R}^k)^*$, we write $P_w = \{p \in P : w \cdot p \leq w \cdot q \text{ for all } q \in P\}$. In words, P_w is the face of P supported by w.

Theorem 2 *Suppose* f_i *is generic with respect to* $\mathcal{N}(f_i)$*, and let* $P_i = \mathcal{N}(x_i - f_i(t)) \subset \mathbb{R}^n \times \mathbb{R}^d$ *for* $i = 1, \ldots, n$. *The tropical variety* $\operatorname{trop}(\Gamma_f)$ *is the support of a* d*-dimensional subfan of the normal fan of* $P = P_1 + \cdots + P_n$. *It consists of the normal cones* σ *of* P *for which the face polytopes* $(P_1)_w, \ldots, (P_n)_w$ *have positive mixed volume* MV_w *in the affine lattice of* P_w*, for each* $w \in \operatorname{int}(\sigma)$. *Moreover, the multiplicity* m_σ *of* σ *in* $\operatorname{trop}(\Gamma_f)$ *equals* MV_w*, see [ST08, Theorem 4.3].*

We illustrate this for a parametric curve.

Example 3 Consider the parametrization $(f_1, f_2) : \mathbb{C}^* \longrightarrow (\mathbb{C}^*)^2$ given by

$$f_1 = 11\,t^2 + 5\,t^3 - t^4 \quad \text{and} \quad f_2 = 11 + 11\,t + 7\,t^8.$$

The image is the plane curve $C = \overline{\operatorname{im} f}$ given by the implicit equation $F(x,y) = 0$, with

$$F = 2401\,x^8 - 1372\,x^6 y - 422576\,x^5 y + \cdots + y^4 + \cdots + 1247565503668. \quad (1)$$

This has 25 terms, one for each lattice point of $\mathcal{N}(F)$, shown on the right side of Figure 1. The Newton polytopes of $x - f_1$ and $y - f_2$ of Γ_f are the triangles seen on the left side.

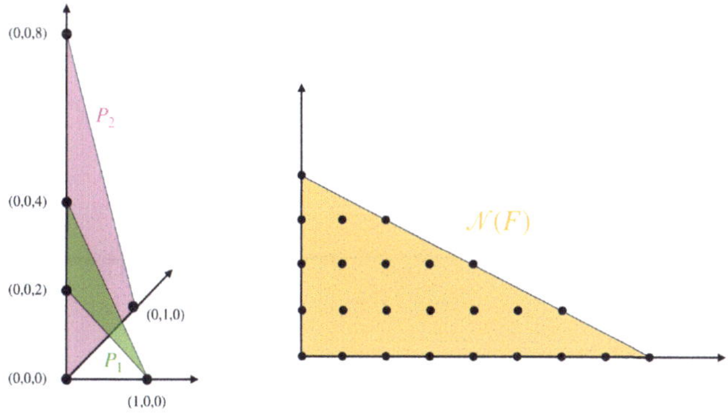

Fig. 1: Newton polytopes P_1, P_2 and $\mathcal{N}(F)$ from Example 3.

The tropical curve $\operatorname{trop}(\Gamma_f)$ can be constructed according to Theorem 2. It is shown in blue on the right of Figure 2. The result is a balanced, one-dimensional fan with four rays:

$$\operatorname{trop}(\Gamma_f) \;=\; \mathbb{R}_+ \cdot (1,0,0) \,\cup\, \mathbb{R}_+ \cdot (-4,-8,-1) \,\cup\, \mathbb{R}_+ \cdot (0,1,0) \,\cup\, \mathbb{R}_+ \cdot (2,0,1),$$

with respective multiplicities $2, 1, 8$, and 1. We demonstrate how to obtain these multiplicities.

Consider the primitive ray generator $w = (2,0,1)$, revealing the faces

$$(P_1)_w \;=\; \operatorname{conv}((0,0,2),(1,0,0)) \quad \text{and} \quad (P_2)_w \;=\; \operatorname{conv}((0,0,0),(0,1,0)).$$

The multiplicity of $\mathbb{R}_+ \cdot (2,0,1)$ in $\operatorname{trop}(\Gamma_f)$ can be computed as the mixed volume of the line segments $(P_1)_w$ and $(P_2)_w$ inside the 2-dimensional lattice defined by the affine hull of their Minkowski sum. This is the mixed volume MV_w, and we find that it is equal to 1.

Now that we know $\operatorname{trop}(\Gamma_f)$ and its multiplicities (when the Laurent polynomials f_i are generic), and we know that $\operatorname{trop}(X)$ is obtained from its projection, it remains to determine the multiplicities of $\operatorname{trop}(X)$ from those of $\operatorname{trop}(\Gamma_f)$. The answer is given by [ST08, Theorem 1.1], which is the second part of [SY08, Theorem 6.4]. In order to recall the formula, we introduce

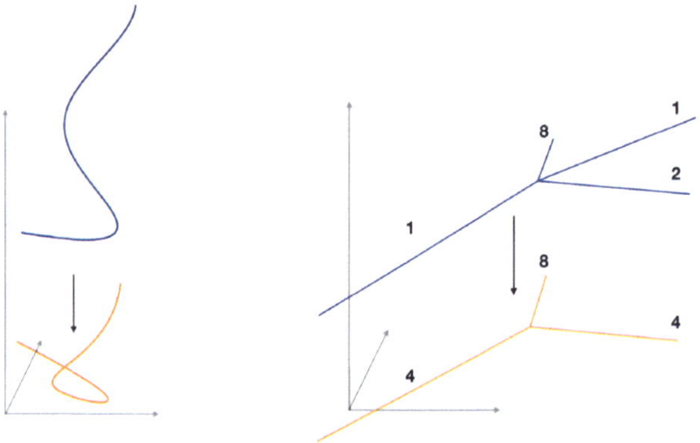

Fig. 2: Classical (left) and tropical (right) implicitization of a parametric plane curve.

some more notation. Let Σ_X be a fan in $\mathbb{R}^n$ whose support is $\mathrm{trop}(X)$, and Σ_{Γ_f} a fan in $\mathbb{R}^n \times \mathbb{R}^d$ whose support is $\mathrm{trop}(\Gamma_f)$. Let v be a point in the interior of a top dimensional cone $\sigma_v \in \Sigma_X$. We write $\mathbb{L}_v$ for the linear span of a small open neighborhood of v in $\mathrm{trop}(X)$. Similarly, $w \in \mathrm{int}(\sigma_w)$ for a top dimensional cone $\sigma_w \in \Sigma_{\Gamma_f}$ defines a linear space $\mathbb{L}_w$. If the projection $\Gamma_f \to X$ is generically finite of degree δ, then the multiplicity of $\sigma_v \in \Sigma_X$ is

$$m_{\sigma_v} \;=\; \frac{1}{\delta} \sum_{w \in \pi^{-1}(v)} m_{\sigma_w} \cdot \mathrm{index}\big(\, \mathbb{L}_v \cap \mathbb{Z}^n : \pi(\mathbb{L}_w \cap \mathbb{Z}^{n+d})\,\big). \tag{2}$$

Here π is the projection $\mathbb{R}^n \times \mathbb{R}^d \to \mathbb{R}^n$ and the sum is over all points w in the pre-image of v under the map $\pi_{|\mathrm{trop}(\Gamma_f)} : \mathrm{trop}(\Gamma_f) \to \mathrm{trop}(X)$. It is assumed that there are only finitely many such points, and each of them lies in the interior of a top dimensional cone of Σ_{Γ_f}.

With the choice of weights (2), the image fan is balanced. This is a non-trivial fact, derived in a more general setting in [MS15, Lemma 3.6.3]. See also [MS15, Theorem 6.5.16] for a textbook discussion of tropical implicitization in the context of geometric tropicalization.

Example 4 According to Theorem 1, the tropical curve $\mathrm{trop}(\Gamma_f)$ projects to $\mathrm{trop}(C)$. This is displayed on the right of Figure 2, where $\mathrm{trop}(C)$ is shown in orange as the fan

$$\mathrm{trop}(C) \;=\; \mathbb{R}_+ \cdot (1,0) \;\cup\; \mathbb{R}_+ \cdot (-1,-2) \;\cup\; \mathbb{R}_+ \cdot (0,1).$$

This fan is balanced with ray multiplicities $4,4,8$, in that order. We demonstrate the computation of $m_\rho = 4$ for the first ray $\rho = \mathbb{R}_+ \cdot (1,0)$ using (2). The ray $\hat{\rho} = \mathbb{R}_+ \cdot (2,0,1)$ of $\mathrm{Trop}(\Gamma_f)$ projects to ρ. Its primitive ray generator $(2,0,1)$ projects to the imprimitive lattice vector $(2,0)$. The contribution of

$\hat{\rho} = \mathbb{R}_+ \cdot (2,0,1)$ to the multiplicity m_ρ is the product of two numbers: its intrinsic multiplicity $m_{\hat{\rho}} = 1$, and the lattice index 2. The ray $\mathbb{R}_+ \cdot (1,0,0)$ also projects to ρ, which leads to a total of $m_\rho = 1 \cdot 2 + 2 \cdot 1$. The tropical curve $\text{trop}(C)$ equals the normal fan of the Newton polytope $\mathcal{N}(F)$, displayed on the right of Figure 1.

The discussion above leads to Algorithm 7, which makes the results in this section effective. It takes the Newton polytopes $Q_i = \mathcal{N}(f_i)$ as an input, and returns the tropicalization of $X = \overline{\text{im} f}$. Here the Laurent polynomials f_i are assumed to be generic with respect to their Newton polytopes Q_i. The output is a set of pairs (m_τ, τ), where $\tau \subset \mathbb{R}^n$ is a cone, and m_τ is a positive integer. The tropical hypersurface $\text{trop}(X)$ is the union of all these cones τ, and the multiplicity of $\text{trop}(X)$ at a generic point x is the sum $\sum_{x \in \tau} m_\tau$.

We warn the reader that, although the union of all cones τ forms the support of a fan, the collection of cones itself is generally not a fan. This representation of a tropical variety is unconventional. However, it is easy to compute, and convenient for our algorithmic purposes.

Algorithm 7: Generic Tropical Implicitization

Input : Polytopes $Q_1, \ldots, Q_n$ in $\mathbb{R}^{n-1}$.
Output: A tropical cycle $\text{trop}(X)$ in $\mathbb{R}^n$.

1 **for** $i \in \{1, \ldots, n\}$ **do**
2 $\quad P_i \leftarrow \text{conv}(e_i \cup Q_i)$
3 $P \leftarrow P_1 + \cdots + P_n$
4 $\Sigma \leftarrow$ normal fan of P
5 $\text{trop}(X) \leftarrow \emptyset$
6 **for** $\sigma \in \Sigma$ **do**
7 $\quad m_\sigma \leftarrow \text{MV}((P_1)_\sigma, \ldots, (P_n)_\sigma)$ **if** $m_\sigma > 0$ **then**
8 $\quad\quad \tau \leftarrow \pi(\sigma)$
9 $\quad\quad m_{\text{lattice}} \leftarrow \text{index}(\mathbb{L}_\tau \cap \mathbb{Z}^n : \pi(\mathbb{L}_\sigma \cap \mathbb{Z}^{n+d}))$
$\quad\quad \text{trop}(X) \leftarrow \text{trop}(X) \cup \{(m_{\text{lattice}} \cdot m_\sigma, \tau)\}$

We now explain Algorithm 7. The polytopes $P_1, \ldots, P_n$ in line 2 are the Newton polytopes of the equations $x_1 - f_1, \ldots, x_n - f_n$ of Γ_f. The standard basis $e_1, e_2, \ldots$ of $\mathbb{R}^{n+d}$ is indexed by the variables $x_1, \ldots, x_n, t_1, \ldots, t_d$ in that order. Following Theorem 2, Algorithm 7 selects all cones σ in the normal fan of $P_1 + \cdots + P_n$ that contribute to the tropicalization $\text{trop}(\Gamma_f)$. Line 7 computes the mixed volume $m_\sigma = \text{MV}((P_1)_\sigma, \ldots, (P_n)_\sigma)$, where $(P_i)_\sigma = (P_i)_w$ for any $w \in \text{int}(\sigma)$. We denote by $\pi(\sigma)$ the projection of $\sigma \subset \mathbb{R}^{n+d}$ to the first n coordinates. The multiplicity with which $\pi(\sigma)$ contributes to $\text{trop}(X)$ is computed in lines 7 and 9. Based on (2), it is the product of m_σ with the index of the lattice $\pi(\mathbb{L}_\sigma \cap \mathbb{Z}^{n+d})$ in the lattice $\mathbb{L}_\tau \cap \mathbb{Z}^n$. Here $\mathbb{L}_\sigma$ is the linear span of σ and $\mathbb{L}_\tau = \pi(\mathbb{L}_\sigma)$.

Example 5 We show how to apply our `Julia` implementation of Algorithm 7, to Example 3:

```
julia> using TropicalImplicitization, Oscar;
julia> R, (t,) = polynomial_ring(QQ,["t"]);
julia> f1 = 11*t^2 + 5*t^3 - 1*t^4;
julia> f2 = 11 + 11*t + 7*t^8;
julia> Q1 = newton_polytope(f1);
julia> Q2 = newton_polytope(f2);
julia> newton_pols = [Q1, Q2];
julia> cone_list, weight_list = get_tropical_cycle(newton_pols);
```

The lists `cone_list` and `weight_list` returned by our program have four elements each. The first list contains the planar cones

$$\mathbb{R}_+ \cdot (1,0)\,, \quad \mathbb{R}_+ \cdot (1,0)\,, \quad \mathbb{R}_+ \cdot (0,1)\,, \quad \mathbb{R}_+ \cdot (-1,-2),$$

and the second list consists of their respective multiplicities $(2,2,8,4)$. Notice that $\mathbb{R}_+ \cdot (1,0)$ appears twice, and its multiplicity is split up as $4 = 2+2$, like in Example 4.

Problem 6 Suppose the coefficients of $f_1, \ldots, f_n$ lie in a field with a nontrivial valuation, such as the p-adic numbers $\mathbb{Q}_p$ or the Puiseux series $\mathbb{C}\{\{t\}\}$. While the theory of tropical implicitization generalizes nicely to this setting, with balanced fans replaced by balanced polyhedral complexes, useful algorithms and their implementations are yet to be developed.

3 *A*-Discriminants

We fix a $d \times n$ integer matrix A of rank d which has the vector $(1,1,\ldots,1)$ in its row span. The associated $(d-1)$-dimensional projective toric variety X_A is the closure in $\mathbb{P}^{n-1}$ of the set

$$\big\{ \, (t^{a_1} : t^{a_2} : \cdots : t^{a_n}) \in \mathbb{P}^{n-1} \; : \; t = (t_1, \ldots, t_d) \in (\mathbb{C}^*)^d \, \big\}. \tag{1}$$

Here a_i denotes the ith column of the matrix A. We are interested in the dual variety X_A^*, which parametrizes hyperplanes that are tangent to X_A at some points. Equivalently, X_A^* is the closure in $\mathbb{P}^{n-1}$ of the set of points $x = (x_1 : x_2 : \cdots : x_n)$ such that the hypersurface

$$\big\{ \, t \in (\mathbb{C}^*)^d \; : \; \sum_{i=1}^{n} x_i \, t^{a_i} \; = \; 0 \, \big\} \tag{2}$$

has a singular point. The variety X_A^* is irreducible, and it is usually a hypersurface. The *A-discriminant* Δ_A is the unique (up to scaling) irreducible polynomial vanishing on X_A^*.

In this section, we address the following computational problem: given the matrix A, compute its A-discriminant Δ_A. Along the way, we will discover whether or not X_A^* is a hypersurface. In this event, we turn to Section 5, and we compute its Chow form instead.

Our algorithm is based on the *Horn uniformization*, which writes X_A^* as the image of a map whose coordinates are products of linear forms. We follow the exposition given in [DFS07]. For additional information, see the book references in [GKZ08, Section 9.3.F] and [MS15, Section 5.5]. Given two vectors u and v in $(\mathbb{C}^*)^n$, we define $u \star v = (u_1v_1 : u_2v_2 : \cdots : u_nv_n) \in \mathbb{P}^{n-1}$. If U and V are varieties in $\mathbb{P}^{n-1}$, neither contained in a coordinate hyperplane, then their *Hadamard product* $U \star V$ is the closure of all such points $u \star v$, where $u \in U$ and $v \in V$.

Theorem 7 (Horn Uniformization) *The dual variety X_A^* of X_A is the Hadamard product in $\mathbb{P}^{n-1}$ of the $(d-1)$-dimensional toric variety X_A with an $(n-d-1)$-dimensional linear space:*

$$X_A^* \quad = \quad X_A \star \text{kernel}(A). \tag{3}$$

We illustrate this theorem with several examples. In each of them, we refer to the $(d-1)$-dimensional polytope $Q = \text{conv}(a_1, a_2, \ldots, a_n)$, and we fix an $(n-d) \times n$-matrix B whose rows span the kernel of A. In polytope language, B is a Gale transform of the polytope Q. For (3), we introduce unknowns $u = (u_1, \ldots, u_{n-d})$ and we write uB for vectors in kernel(A).

Example 8 (Determinant) Fix $n = k^2$ and $d = 2k-1$, for some integer $k \geq 2$, and let A represent the linear map that extracts the row sums and column sums of a $k \times k$ matrix. Naively, this matrix has $2k$ rows, but only $2k-1$ of them are linearly independent. Here $Q = \Delta_{k-1} \times \Delta_{k-1}$ is the product of two $(k-1)$-simplices. The toric variety X_A consists of $k \times k$ matrices of rank 1 and X_A^* consists of $k \times k$ matrices of rank $\leq k-1$. We parametrize X_A^* by the Hadamard product of a rank 1 matrix with a matrix whose row and columns are zero. For example for $k = 3$, the Horn uniformization writes all singular 3×3 matrices as follows:

$$\begin{bmatrix} t_1t_4\,u_1 & t_1t_5(u_2-u_1) & t_1t_6(-u_2) \\ t_2t_4(u_3-u_1) & t_2t_5(u_1-u_2-u_3+u_4) & t_2t_6(u_2-u_4) \\ t_3t_4(-u_3) & t_3t_5(u_3-u_4) & t_3t_6\,u_4 \end{bmatrix}. \tag{4}$$

This matrix has $(t_4^{-1}, t_5^{-1}, t_6^{-1})^t$ in its right kernel and $(t_1^{-1}, t_2^{-1}, t_3^{-1})$ in its left kernel. The A-discriminant Δ_A is the determinant of a square matrix, which obviously vanishes on (4).

Example 9 (Resultant) The resultant of a square system of homogeneous polynomials is the A-discriminant where A is the Cayley configuration of the given monomial supports. We examine the Sylvester resultant Δ_A of two binary quadrics $(d = 3, n = 6)$. We set

$$A = \begin{bmatrix} 1 & 1 & 1 & 0 & 0 & 0 \\ 0 & 0 & 0 & 1 & 1 & 1 \\ 0 & 1 & 2 & 0 & 1 & 2 \end{bmatrix} \quad \text{and} \quad B = \begin{bmatrix} 1 & -2 & 1 & 0 & 0 & 0 \\ 0 & 0 & 0 & 1 & -2 & 1 \\ 1 & -1 & 0 & -1 & 1 & 0 \end{bmatrix}.$$

This yields the following parametrization for pairs of univariate quadrics with a common zero:

$$\begin{aligned} x_1 + x_2 z + x_3 z^2 &= t_1(t_3 u_1 z - u_1 - u_3)(t_3 z - 1), \\ x_4 + x_5 z + x_6 z^2 &= t_2(t_3 u_2 z - u_2 - u_3)(t_3 z - 1). \end{aligned}$$

These Horn uniformizations exist for resultants of polynomials in any number of variables.

Example 10 (Hyperdeterminant) The hyperdeterminant of a multidimensional tensor vanishes whenever the hypersurface defined by the associated multilinear form is singular. In our notation, this is the A-discriminant Δ_A where the columns of A are the vertices of a product of simplices. As an illustration, we here present the Horn uniformization for the hyperdeterminant of format $2 \times 2 \times 2$. Here $n = 8$ and our configuration is the regular 3-cube:

$$A = \begin{bmatrix} 1&1&1&1&1&1&1&1 \\ 0&0&0&0&1&1&1&1 \\ 0&0&1&1&0&0&1&1 \\ 0&1&0&1&0&1&0&1 \end{bmatrix} \quad \text{and} \quad B = \begin{bmatrix} 1&-1&-1&1&0&0&0&0 \\ 0&0&0&0&1&-1&-1&1 \\ 1&-1&0&0&-1&1&0&0 \\ 1&0&-1&0&-1&0&1&0 \end{bmatrix}.$$

These two matrices yield the following map from $\mathbb{C}^8$ into the space of $2 \times 2 \times 2$ tensors

$$\begin{aligned} &x_{000} = t_1(u_1 + u_3 + u_4), \quad x_{001} = t_1 t_4(-u_1 - u_3), \quad x_{010} = t_1 t_3(-u_1 - u_4), \\ &x_{011} = t_1 t_3 t_4 u_1, \qquad x_{100} = t_1 t_2(u_2 - u_3 - u_4), \; x_{101} = t_1 t_2 t_4(u_3 - u_2), \\ &x_{110} = t_1 t_2 t_3(u_4 - u_2), \qquad x_{111} = t_1 t_2 t_3 t_4 u_2. \end{aligned}$$

Implicitization of this parametrization gives us the hyperdeterminant:

$$\begin{aligned} \Delta_A \;=\; & x_{000}^2 x_{111}^2 + x_{001}^2 x_{110}^2 + x_{011}^2 x_{100}^2 \\ & + x_{010}^2 x_{101}^2 + 4 x_{000} x_{011} x_{101} x_{110} + 4 x_{001} x_{010} x_{100} x_{111} \\ & - 2 x_{000} x_{001} x_{110} x_{111} - 2 x_{000} x_{010} x_{101} x_{111} - 2 x_{000} x_{011} x_{100} x_{111} \\ & - 2 x_{001} x_{010} x_{101} x_{110} - 2 x_{001} x_{011} x_{100} x_{110} - 2 x_{010} x_{011} x_{100} x_{101}. \end{aligned}$$

We now return to tropical implicitization. Our aim is to compute the tropical variety $\mathrm{trop}(X_A^*)$ directly from A. Here we identify X_A^* with its affine cone in $(\mathbb{C}^*)^n$. If X_A^* has codimension 1 then $\mathrm{trop}(X_A^*)$ is an $(n-1)$-dimensional balanced fan in $\mathbb{R}^n$, with a one-dimensional lineality space. This is the normal fan of the Newton polytope of the A-discriminant Δ_A. We recover the polytope using Algorithm 9 below; see also [MS15, Remark 3.3.11].

The Horn uniformization of Theorem 7 gives a convenient way of computing $\mathrm{trop}(X_A^*)$. It is an instance of parametrizations given by monomials in linear forms. These admit an elegant solution to the tropical implicitization problem;

see [MS15, Section 5.5]. Let U and V be integer matrices of size $r \times m$ and $s \times r$ respectively. The rows of V are $v_1, \ldots, v_s \in \mathbb{Z}^r$. We denote by λ_U the map defined by U, and by μ_V the monomial map specified by V:

$$\begin{aligned} \lambda_U : (\mathbb{C}^*)^m &\dashrightarrow (\mathbb{C}^*)^r & \qquad \mu_V : (\mathbb{C}^*)^r &\longrightarrow (\mathbb{C}^*)^s \\ v &\longmapsto U\,v & x &\longmapsto (x^{v_1}, \ldots, x^{v_s}). \end{aligned}$$

The composition of these maps gives the unirational variety $Y_{U,V} = \overline{\operatorname{im}(\mu_V \circ \lambda_U)}$ in $(\mathbb{C}^*)^s$. Its tropicalization $\operatorname{trop}(Y_{U,V})$ is obtained by tropicalizing the map $\mu_V \circ \lambda_U$. The tropical linear space $\operatorname{trop}(\operatorname{im} \lambda_U)$ is computed purely combinatorially, as the *Bergman fan* of the matroid of U, see [MS15, Section 4.2]. The monomial map μ_V tropicalizes to the linear map $V : \mathbb{R}^r \to \mathbb{R}^s$. The following result is from [DFS07, Theorem 3.1] and [MS15, Theorem 5.5.1].

Theorem 11 *The tropical variety* $\operatorname{trop}(Y_{U,V})$ *is the image, as a balanced fan via [MS15, Lemma 3.6.3], of the Bergman fan* $\operatorname{trop}(\operatorname{im} \lambda_U)$ *under the linear map* $\mathbb{R}^r \to \mathbb{R}^s$ *given by* V.

By Theorem 7, the affine cone over the A-discriminant in $(\mathbb{C}^*)^n$ is the variety $Y_{U,V}$ with

$$U = \begin{pmatrix} B^t & 0 \\ 0 & I_d \end{pmatrix}, \quad V = \begin{pmatrix} I_n & A^t \end{pmatrix}. \tag{5}$$

Here $m = s = n$ and $r = n + d$. This leads to Algorithm 8 for computing $\operatorname{trop}(X_A^*)$.

Algorithm 8: Compute the Tropical A-Discriminant

Input : An integer matrix $A \in \mathbb{Z}^{d \times n}$, of full rank containing the all ones vector $(1, 1, \ldots, 1)$ in its row span.
Output: A tropical cycle $\operatorname{trop}(X_A^*) \subseteq \mathbb{R}^n$

1 $B \leftarrow$ Gale dual of A
2 $U \leftarrow \begin{pmatrix} B^t & 0 \\ 0 & I_d \end{pmatrix}$
3 $V \leftarrow \begin{pmatrix} I_n & A^t \end{pmatrix}$
4 $M \leftarrow$ matroid of U
5 $\operatorname{trop}(\operatorname{im} \lambda_U) \leftarrow$ Bergman fan of M
6 $\operatorname{trop}(X_A^*) \leftarrow \emptyset$
7 **for** $(m_\sigma, \sigma) \in \operatorname{trop}(\operatorname{im} \lambda_U)$ **do**
8     $\tau \leftarrow V\sigma$
9     $m_{\text{lattice}} \leftarrow \operatorname{index}(\mathbb{L}_\tau \cap \mathbb{Z}^n : V(\mathbb{L}_\sigma \cap \mathbb{Z}^{n+d}))$
10     $\operatorname{trop}(X_A^*) \leftarrow \operatorname{trop}(X_A^*) \cup \{(m_\sigma \cdot m_{\text{lattice}}, \tau)\}$

The matrix B in line 1 is Gale dual to A. Using the symbolic linear algebra functionality provided by `Oscar.jl`, we find this with the command `nullspace(A)`. Lines 4 and 5 compute the tropicalization $\operatorname{trop}(\operatorname{im} \lambda_U)$

of the column span of U. They are based on the `Oscar.jl` commands `Oscar.Polymake.matroid.Matroid(VECTORS = U)` and `Oscar.Polymake.tropical.matroid_fan{min}(matroid)`. From line 7 on, the algorithm computes a projection of the Bergman fan trop(im λ_U). This is analogous to Algorithm 7.

Example 12 We compute the tropicalized $2 \times 2 \times 2$ hyperdeterminant from Example 10:

```
julia> A = [1 1 1 1 1 1 1 1; 0 0 0 0 1 1 1 1;
            0 0 1 1 0 0 1 1; 0 1 0 1 0 1 0 1];

julia> cone_list, weight_list = get_trop_A_disc(A);
```

The result consists of 32 7-dimensional cones and a list of their multiplicities, constituting the weighted normal fan of the Newton polytope $\mathcal{N}(\Delta_A)$. The following code uses an implementation of Algorithm 9 below. It computes $\mathcal{N}(\Delta_A)$, its lattice points, and its f-vector.

```
julia> Delta = get_polytope_from_cycle(cone_list, weight_list);
julia> f_vec, lattice_pts = f_vector(Delta), lattice_points(Delta);
julia> print(f_vec);
ZZRingElem[6, 14, 16, 8]
```

Then `lattice_pts` contains the 12 exponents of Δ_A.

Mixed discriminants [CCDDS13] are special cases of A-discriminants. We discuss a non-trivial one.

Example 13 We revisit [DFS07, Example 5.1]. Here, $d = 4$ and $n = 8$, and we fix the matrix

$$A = \begin{bmatrix} 1 & 1 & 1 & 1 & 0 & 0 & 0 & 0 \\ 0 & 0 & 0 & 0 & 1 & 1 & 1 & 1 \\ 2 & 3 & 5 & 7 & 11 & 13 & 17 & 19 \\ 19 & 17 & 13 & 11 & 7 & 5 & 3 & 2 \end{bmatrix}.$$

This represents the following sparse system of two polynomial equations in two variables:

$$\begin{aligned} & x_1 s^2 t^{19} + x_2 s^3 t^{17} + x_3 s^5 t^{13} + x_4 s^7 t^{11} \\ = \; & x_5 s^{11} t^7 + x_6 s^{13} t^5 + x_7 s^{17} t^3 + x_8 s^{19} t^2 \; = \; 0. \end{aligned}$$

The mixed volume of the two Newton polygons equals 39, so we expect 39 common solutions in $(\mathbb{C}^*)^2$. The A-discriminant Δ_A is the condition for two of these solutions to come together. We know from [DFS07, Example 5.1] that Δ_A is a polynomial of degree 126 in $x_1, \ldots, x_8$. Our computations show that $\mathcal{N}(\Delta_A)$ has 43400 lattice points, and f-vector $(45, 92, 63, 16)$.

Problem 14 Computing the coefficients of Δ_A in Example 13 amounts to solving a linear system of equations with 43400 unknowns. This is one of our topics in Section 4. Solving that system is hard for at least two reasons. First, systems of this size are beyond the reach of symbolic black box solvers on most personal computers at present. Second, the large condition number and the unbalanced nature of the coefficients of the implicit equation, as in (1), hinder the naive use of numerical linear algebra. It is a challenge to develop symbolic or mixed symbolic-numerical techniques for solving such problems.

4 Polytope Reconstruction and Interpolation

Suppose that X is an irreducible hypersurface in $(\mathbb{C}^*)^n$, given by its parametrization (1) or (3). Using Algorithms 7 and 8, we have computed the tropical variety $\Sigma_X = \operatorname{trop}(X)$. Thus, Σ_X is a balanced fan of dimension $d = n-1$ in $\mathbb{R}^n$, represented by a collection of weighted cones. Our aim in this section is to compute the polynomial $F \in \mathbb{C}[x_1, \ldots, x_n]$ that defines the hypersurface X. We identify X with its closure in $\mathbb{C}^n$. This makes F uniquely defined up to scaling. In particular, the Newton polytope $P = \mathcal{N}(F)$ is uniquely specified.

Before spelling out the details, we summarize our approach. First, we compute the Newton polytope P from the balanced fan Σ_X. This relies on Theorem 15 below. Second, we find the polynomial F from the parametrization by interpolation. Here we use the ansatz

$$F(x) \quad = \sum_{a \in \mathcal{N}(F) \cap \mathbb{Z}^n} c_a \, x^a, \tag{1}$$

and we determine the unknown coefficients c_a by evaluating (1) at many points x on X.

We start with computing $P = \mathcal{N}(F)$. The fan Σ_X is dual to the Newton polytope P, namely, it is the $(n-1)$-skeleton of the normal fan of P. Taking into account the multiplicities of all maximal cones of Σ_X, we can go back and forth between P and Σ_X. Obtaining Σ_X and its multiplicities from P is straightforward. The multiplicity of an $(n-1)$-dimensional cone in Σ_X is the lattice length of the corresponding edge of P. The other direction is more interesting to us, that is, we want to compute P from the output of Algorithm 7. This is discussed in [MS15, Remark 3.3.11]. The main tool is a *vertex oracle*, provided by [DFS07, Theorem 2.2].

Theorem 15 *Let $X = V(F) \subset \mathbb{C}^n$ be a hypersurface, whose tropicalization* $\operatorname{trop}(X) \subset \mathbb{R}^n$ *is the support of a fan Σ_X. For a generic weight vector $w \in \mathbb{R}^n$, the vertex $\mathcal{N}(F)_w$ is*

$$\sum_{i=1}^{n} \left(\sum_{\sigma \in \Sigma_X} m_\sigma \cdot \operatorname{IM}(w + \mathbb{R}_+ \cdot e_i, \sigma) \right) \cdot e_i.$$

Here e_i is a standard basis vector, and the inner sum is over all maximal cones of Σ_X.

The intersection multiplicity $\mathrm{IM}(w + \mathbb{R}_+ \cdot e_i, \sigma)$ is the *lattice multiplicity* of the intersection of the ray $\mathbb{R} \cdot e_i$ with the hyperplane $\mathbb{L}_\sigma = \mathbb{R} \cdot \sigma$. This is the absolute value of the determinant of any $n \times n$ matrix whose columns are e_i and a lattice basis for $\mathbb{L}_\sigma \cap \mathbb{Z}^n$. Algorithm 9 implements Theorem 15, that finds the vertex $\mathcal{N}(F)_w$ from the output of Algorithm 7 or 8. Theorem 15 and Algorithm 9 are illustrated in Figure 3 for the curve in Example 4.

Algorithm 9: Compute Vertex Oracle From a Tropical Hypersurface

Input : A tropical hypersurface $\mathrm{trop}(X)$
Output: The vertex $v = \mathcal{N}(F)_w$

$v \leftarrow 0$
for $i \in \{1, \ldots, n\}$ **do**
  for $(m_\sigma, \sigma) \in \mathrm{trop}(X)$ **do**
    $v \leftarrow v + \mathrm{IM}(w + \mathbb{R}_+ \cdot e_i, \sigma) \cdot m_\sigma \cdot e_i$

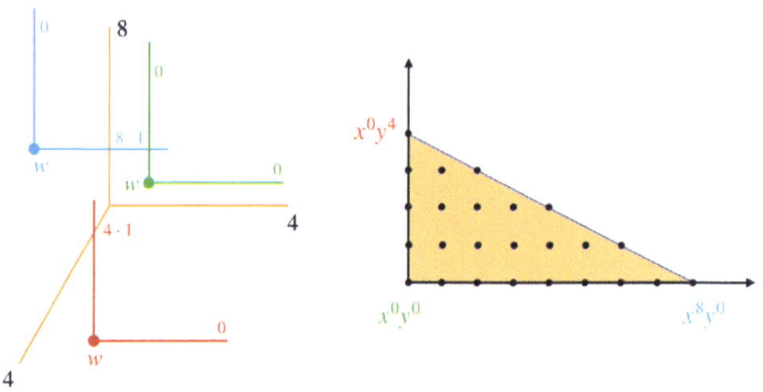

Fig. 3: Computing vertices of $\mathcal{N}(F)$ by intersecting $\mathrm{trop}(X)$ with $w + \mathbb{R}_{\geq 0} \cdot e_i$.

Algorithm 9 can be used to compute *all* vertices of $\mathcal{N}(F)$. A naive approach applies the vertex oracle to many random vectors w. However, it is not clear how many w would be needed, and which stop criterion to use. A way of constructing $\mathcal{N}(F)$ using a vertex oracle like Algorithm 9 was proposed by Huggins [Hug06]. Our implementation uses that but is not deterministic. A deterministic alternative is algorithm 2.2.3 in the dissertation [Bry20].

Example 16 The following `Julia` code computes a polytope from a tropical hypersurface:

```
julia> Delta = get_polytope_from_cycle(cone_list, weight_list);
```

If the variables `cone_list`, `weight_list` are carried over from Example 5, then `Delta` is the yellow polytope shown in Figure 3. For `cone_list`, `weight_list` from Example 12, the polytope `Delta` is the Newton polytope of the hyperdeterminant Δ_A from Example 10.

Remark 17 Under the assumptions of Section 2, i.e., the f_i are Laurent polynomials which are generic with respect to their Newton polytopes, $\mathcal{N}(F)$ is a *mixed fiber polytope.* This was discovered independently by several authors [EKP07; EK08; STY06]. For instance, in Figure 1, $\mathcal{N}(F)$ is the mixed fiber polytope of P_1 and P_2. Our implementation of Huggins' algorithm [Hug06] combined with Algorithm 9 provides a practical way of computing mixed fiber polytopes. This includes the computation of fiber polytopes and secondary polytopes [SY08, Section 3].

Once the Newton polytope $\mathcal{N}(F)$ of the defining equation $F = 0$ of X is known, we can obtain its coefficients c_a in (1) using linear algebra. The set $\mathcal{B} = \mathcal{N}(F) \cap \mathbb{Z}^n$ is a superset of the monomial support $\operatorname{supp}(F)$ of F. It can be computed in `Oscar.jl` via the command `lattice_points`. The interpolation method is most efficient when $\mathcal{B}$ is not much larger than $\operatorname{supp}(F)$, that is, few of the c_a in (1) are zero. The inclusion $\mathcal{B} \supseteq \operatorname{supp}(F)$ can be strict:

Example 18 Consider the map $f : \mathbb{C} \to \mathbb{C}^2$ given by $f_1(t) = a_1 t^4 + a_2 t$ and $f_2(t) = a_3 t^2 + a_4 t$ for generic complex numbers a_1, a_2, a_3, a_4. Here the implicit polynomial $F(x, y)$ equals

$$a_1^2 y^4 - 2a_1 a_3^2 x y^2 + a_3^4 x^2 - 4a_1 a_3 a_4^2 x y$$
$$+3a_1 a_2 a_3 a_4 y^2 - a_4(a_1 a_4^3 - a_2 a_3^3)x + a_2(a_1 a_4^3 - a_2 a_3^3)y.$$

Note that the term y^3 does not appear in spite of it being in $\mathcal{N}(F)$. This shows that some lattice points in a predicted Newton polytope may never appear with a nonzero coefficient.

Problem 19 We propose to refine the observation in Remark 17 by predicting the monomial support of F from the monomial support of $f_1, \ldots, f_n$. That is, which lattice points in the mixed fiber polytope, other than its vertices, contribute to the implicit equation?

For simplicity, we work with the superset $\mathcal{B} \supseteq \operatorname{supp}(F)$ and allow some coefficients to be zero. We identify a set $\mathcal{P}$ of m points in X, so that the interpolation conditions $F(p) = 0$ for $p \in \mathcal{P}$ uniquely determine F (up to a constant factor). We obtain $\mathcal{P} \subset X$ by sending random points in $\mathbb{C}^d$ through the parametrization (1). The unknown coefficients c_a in (1) form a vector $c = (c_a)_{a \in \mathcal{B}} \in \mathbb{C}^{\mathcal{B}}$. For each point $p \in \mathbb{C}^n$, let $p^{\mathcal{B}} = (p^a)_{a \in \mathcal{B}}$ be the vector of monomials corresponding to $\mathcal{B}$, evaluated at p. We interpret $p^{\mathcal{B}}$ as an element of the dual vector space $(\mathbb{C}^{\mathcal{B}})^*$. With this setup, c is the unique vector (up to scaling) satisfying

$$p^{\mathcal{B}} \cdot c = 0 \qquad \text{for all } p \in X.$$

If the sample points $\mathcal{P} \subset X$ are sufficiently random and $m \geq |\mathcal{B}| - 1$, this is equivalent to

$$p^{\mathcal{B}} \cdot c = 0, \qquad \text{for all } p \in \mathcal{P}.$$

The *Vandermonde matrix* $M(\mathcal{B},\mathcal{P})$ has the vectors $p^{\mathcal{B}}$ for its rows, where $p \in \mathcal{P}$. It has size $m \times |\mathcal{B}|$ and by the above discussion, the kernel of $M(\mathcal{B},\mathcal{P}) : \mathbb{C}^{\mathcal{B}} \to \mathbb{C}^m$ is spanned by c.

Our problem is now reduced to the computation of the one-dimensional kernel of a Vandermonde matrix $M(\mathcal{B},\mathcal{P})$. Below, we will fix $\mathcal{B}$ and $\mathcal{P}$ and use the simpler notation $M = M(\mathcal{B},\mathcal{P})$, where there is no danger of confusion. When the parametrizing functions f_i have coefficients in $\mathbb{Q}$, like in the case of A-discriminants in Section 4, we can use $\mathcal{P} \subset \mathbb{Q}^n$. In particular, M has rational entries, and its kernel can be computed in exact arithmetic.

Example 20 We now demonstrate our implementation of the above discussion by computing the implicit equation F from Example 3. The following code computes the Vandermonde matrix $M(\mathcal{B},\mathcal{P})$ of size 24×25 with rational entries. This is done by plugging 24 random rational numbers into the parametrization (1). The functions `f1`, `f2` are taken from Example 5, and the Newton polytope `Delta` $= \mathcal{N}(F)$ was computed in Example 16.

```
julia> Delta = get_polytope_from_cycle(cone_list, weight_list);
julia> B = lattice_points(Delta);
julia> n_samples = length(B)-1;
julia> P = sample([f1,f2], n_samples);
julia> M_BP = get_vandermonde_matrix(B,P);
julia> coeffs_F = nullspace(M_BP)[2];
```

Often, in practical computations, the points $p \in \mathcal{P}$ are approximations of points on X, so the entries of M are finite precision floating point numbers. In that case, the task of computing $\ker M$ is one of *numerical linear algebra*. This is not supported in the current version of `Oscar.jl`. The standard way to proceed using the numerical linear algebra functionality in `Julia`, is via the *singular value decomposition* (SVD) of M. Alternatives include *QR factorization with optimal pivoting* and *iterative eigenvalue methods*. We refer to [BKSW18, Section 5] for such numerical considerations and pointers to the relevant literature.

When f is defined over $\mathbb{Q}$, one might still want to use floating point computations for speed. Let c_a be a nonzero entry of a generator c for $\ker M$. The vector $c_a^{-1}c$ has rational entries. Its numerical approximation $\tilde{c}_a^{-1}\tilde{c}$ is contaminated by rounding errors. We approximate the entries of $\tilde{c}_a^{-1}\tilde{c}$ by rational numbers using the built in function `rationalize` in `Julia`. This has an optional input `tol`, so that `rationalize(a,tol = e)` returns a rational number `q` which satisfies $|$`q` $-$ `a`$| \leq$ `e`. A sensible choice for `tol` is $100 \cdot \tilde{c}_a^{-1} \cdot \varepsilon \cdot \sigma_1/\sigma_{|\mathcal{B}|-1}$.

If symbolic computation is preferable to numerical methods, then one might solve the linear equations over various finite fields and recover rational solutions via the Chinese remainder theorem. This can be done in a computer algebra system. Sometimes, one is only interested in a fixed finite field. We illustrate the finite field computation in `Oscar.jl`.

Example 21 We seek the A-discriminant for a matrix with large entries:

$$A = \begin{bmatrix} 1 & 1 & 1 & 1 & 1 & 1 \\ 2 & 3 & 5 & 7 & 11 & 13 \\ 13 & 8 & 5 & 3 & 2 & 1 \end{bmatrix}$$

The following code finds that the Newton polytope of Δ_A over $\mathbb{Q}$ has dimension 3 and f-vector $(12, 18, 8)$. It terminated on a MacBook Pro with a 3,3 GHz Intel Core i5 processor within 120 seconds. The number of lattice points equals 2295. In order to compute the coefficients of the A-discriminant, we must solve a linear system of 2294 equations with large integer coefficients. We solve this over the field with 101 elements instead:

```
julia> A = [1 1 1 1 1 1; 2 3 5 7 11 13; 13 8 5 3 2 1];
julia> cone_list, weight_list = get_trop_A_disc(A);
julia> Delta = get_polytope_from_cycle(cone_list, weight_list);
julia> @time mons, coeffs = compute_A_discriminant(A,Delta,GF(101));
Collect all lattice points.
Sample 2754.0 points from Discriminant
Construct Vandermonde matrix.
Compute coefficients of the Discriminant.
 97.247222 seconds (210.17 M allocations: 17.505 GiB, 6.51% gc time,
 ↪  0.00% compilation time)
```

For the same computation over the rational numbers, the computation failed due to the increasing memory cost imposed by exact arithmetic.

We close this section with a combinatorics problem arising from Remark 17.

Problem 22 Let $P_1, \ldots, P_n$ be polytopes in $\mathbb{R}^{n-1}$ having $v_1, \ldots, v_n$ vertices. Give a sharp upper bound in terms of $v_1, \ldots, v_n$ for the number of vertices of their mixed fiber polytope. In other words, prove an Upper Bound Theorem for f-vectors arising in tropical implicitization.

Example 23 We illustrate Problem 22 for three triangles ($n = v_1 = v_2 = v_3 = 3$). After many runs for different random configurations, the following example is our current winner:

```
julia> verts1 = [898 -614; -570 817; 892 -594];
julia> verts2 = [-603 -481; -623 -127; -36 732];
julia> verts3 = [-548 -864; -151 873; 800 -861];
julia> pols = convex_hull.([verts1, verts2, verts3]);
julia> Delta = get_polytope_from_cycle(get_tropical_cycle(pols)...);
julia> print(f_vector(Delta))
ZZRingElem[25, 49, 26]
```

This code computes a mixed fiber polytope that has 25 vertices, 49 edges and 26 facets. Can you find three triangles in $\mathbb{R}^2$ whose mixed fiber polytope has more than 25 vertices?

5 Higher Codimension

In this section, we address the implicitization problem for varieties X that are not hypersurfaces. The role of the Newton polytope $\mathcal{N}(F)$, of a polynomial F, will now be played by the Chow polytope $\mathcal{C}(X)$. We begin by reviewing some definitions from [DS95] and [GKZ08, Chapter 6].

Let X be an irreducible projective variety of dimension d in complex projective space $\mathbb{P}^n$. Suppose we are given the tropical variety trop(X), a balanced fan of dimension d in $\mathbb{R}^{n+1}/\mathbb{R}\mathbf{1}$. Our goal is to compute the *Chow form* Ch(X), which is a hypersurface in the Grassmannian Gr$(n-d-1, \mathbb{P}^n)$. Its points are the linear subspaces of dimension $n-d-1$ whose intersection with X is non-empty. We identify Ch(X) with its defining polynomial of degree deg(X) in primal Plücker coordinates $p_{i_0 i_1 \cdots i_d}$, where $1 \leq i_0 < i_1 < \cdots i_d \leq n$. The $p_{i_0 i_1 \cdots i_d}$ are the maximal minors of any $(d+1) \times (n+1)$ matrix whose kernel is the subspace to be represented. The Chow form Ch(X) is only well-defined up to the Plücker relations that vanish on Gr$(n-d-1, \mathbb{P}^n)$. By [Stu08, Theorem 3.1.7], Ch(X) is a unique linear combination of standard tableaux. In our computations, we always use that standard representation for Chow forms.

The weight of the Plücker coordinate $p_{i_0 i_1 \cdots i_d}$ is the vector $e_{i_0} + e_{i_1} + \cdots + e_{i_d}$ in $\mathbb{Z}^n$, and the weight of a Plücker monomial is the sum of the weights of its variables, with multiplicity. By definition, the Chow polytope $\mathcal{C}(X)$ is the convex hull of the weights occurring in Ch(X).

Fink [Fin13] gave a combinatorial recipe for constructing the weighted normal fan of the Chow polytope $\mathcal{C}(X)$ from the tropical variety trop(X). Let L_{n-d-1} denote the standard tropical linear space of dimension $n-d-1$ in $\mathbb{R}^{n+1}/\mathbb{R}\mathbf{1}$. Its maximal cones are the orthants spanned by $(n-d-1)$-tuples of unit vectors. It is proved in [Fin13, Theorem 4.8] that the weighted normal fan of $\mathcal{C}(X)$ is the *stable sum* of trop(X) with the negated linear space $-L_{n-d-1}$. The stable sum is a dual operation to the stable intersection. It always produces a balanced fan of expected dimension. Hence trop$(X) - L_{n-d-1}$ is a balanced fan of codimension 1 in $\mathbb{R}^{n+1}/\mathbb{R}\mathbf{1}$. Fink's result states that this is the *outer* normal fan of $\mathcal{C}(X)$.

We can compute $\mathcal{C}(X)$ from trop$(X) - L_{n-d-1}$ by the algorithm for building Newton polytopes in Section 4, *up to an integer translation.* Indeed, the normal fan of $\mathcal{C}(X)$ and $\mathcal{C}(X) + \mathbf{t}$ are identical, for any $\mathbf{t} \in \mathbb{Z}^n$. Algorithm 9 finds vertices of $\mathcal{C}(X) + \mathbf{t}$, where $\mathbf{t}$ shifts $\mathcal{C}(X)$ so that it touches each coordinate hyperplane. In previous examples, we had $\mathbf{t} = 0$. Indeed, if F is irreducible, then the polytope $\mathcal{N}(F)$ touches all coordinate hyperplanes. This is not true for the Chow polytope, as illustrated by the example below. Finding the correct $\mathbf{t}$ is an interesting combinatorial problem which we plan to investigate in a future project.

Example 24 ($d = 1, n = 3$) Let X be the curve in $\mathbb{C}^3$ which is given by the parametrization

$$x_1 = t(t-1)(t+1), \quad x_2 = t^2(t+1), \quad x_3 = t^3(t-1).$$

The tropical curve is determined by the orders of the coordinate functions at all zeros and poles. Hence $\operatorname{trop}(X)$ is the fan with four rays $(1,2,3)$, $(1,1,0)$, $(1,0,1)$, and $(-3,-3,-4)$. We identify X with its projective closure in $\mathbb{P}^3$, obtained by adding an extra coordinate x_0. The tropical line L_1 is spanned by e_0, e_1, e_2, e_3, and we form the sum of $\operatorname{trop}(X)$ with the negated line $-L_1$. This 2-dimensional fan is the normal fan of the Chow polytope $\mathcal{C}(X)$. We implemented the stable sum using `Oscar.jl`, and obtain this fan as follows.

```
julia> cone_list = positive_hull.([[1, 1, 0],[1, 2, 3], [1,0,1], [-1, -1,
↪  -4//3]]);
julia> weight_list = ones(Int64, 4);
julia> cone_list, weight_list = get_chow_fan(cone_list,weight_list);
```

The output consists of 16 2-dimensional cones and their multiplicities. A translated version of the Chow polytope is obtained from this output as in the previous section:

```
julia> C_translated = get_polytope_from_cycle(cone_list,weight_list);
```

This is a three-dimensional polytope touching all coordinate hyperplanes. The following is a list of its 12 vertices:

$$(0,2,3,1), (0,3,1,2), (0,4,1,1), (1,0,4,1), (1,2,3,0), (1,3,0,2),$$
$$(1,4,0,1), (1,4,1,0), (2,0,1,3), (2,0,4,0), (2,4,0,0), (3,0,0,3).$$

To identify the shift $\mathbf{t}$, we compare this to the Chow polytope. We obtain $\mathcal{C}(X)$ as the convex hull of the weights of the Plücker monomials in the Chow form $\operatorname{Ch}(X)$ of our curve:

$$\begin{aligned}
&p_{03}^4 - p_{01}^3p_{13} - 3p_{01}^2p_{02}p_{13} - 3p_{01}p_{02}^2p_{13} - p_{02}^3p_{13} + 3p_{01}^2p_{03}p_{13} + 9p_{01}p_{02}p_{03}p_{13} + 6p_{02}^2p_{03}p_{13}\\
&+p_{01}p_{03}^2p_{13} - 5p_{02}p_{03}^2p_{13} + 2p_{01}^2p_{12}p_{13} + p_{01}p_{02}p_{12}p_{13} + 2p_{01}^2p_{13}^2 - 2p_{01}p_{02}p_{13}^2 + 4p_{02}^2p_{13}^2\\
&+p_{01}p_{03}p_{13}^2 - 4p_{01}p_{12}p_{13}^2 - p_{01}^3p_{23} - 3p_{01}^2p_{02}p_{23} - 3p_{01}p_{02}^2p_{23} - p_{02}^3p_{23} + 4p_{01}^2p_{03}p_{23}\\
&+11p_{01}p_{02}p_{03}p_{23} + 7p_{02}^2p_{03}p_{23} - 2p_{01}p_{03}^2p_{23} - 10p_{02}p_{03}^2p_{23} + 2p_{03}^3p_{23} + 2p_{01}^2p_{12}p_{23}\\
&+p_{01}p_{02}p_{12}p_{23} + 9p_{01}^2p_{13}p_{23} - p_{01}p_{02}p_{13}p_{23} + 6p_{02}^2p_{13}p_{23} + 2p_{01}p_{03}p_{13}p_{23} - 2p_{02}p_{03}p_{13}p_{23}\\
&-6p_{01}p_{12}p_{13}p_{23} + 2p_{01}p_{13}^2p_{23} + 9p_{01}^2p_{23}^2 + 2p_{01}p_{02}p_{23}^2 + 2p_{02}^2p_{23}^2 - 4p_{01}p_{03}p_{23}^2 - 2p_{01}p_{12}p_{23}^2.
\end{aligned}$$

We find that $\mathcal{C}(X)$ is the 3-dimensional polytope with the following vertices:

$$(1,2,3,2), (1,3,1,3), (1,4,1,2), (2,0,4,2), (2,2,3,1), (2,3,0,3),$$
$$(2,4,0,2), (2,4,1,1), (3,0,1,4), (3,0,4,1), (3,4,0,1), (4,0,0,4).$$

We conclude that $\mathbf{t} = (-1,0,0,-1)$. For now, we apply this shift manually.

After listing all lattice points in $\mathcal{C}(X)$, we can compute the Chow form $\operatorname{Ch}(X)$ by interpolation. This is done as follows: for each lattice point u in $\mathcal{C}(X)$, we list all standard Plücker monomials of weight u, and form their linear combination with unknown coefficients. Our ansatz is the sum of these $\mathbb{Z}^n$-homogeneous Plücker polynomials, with distinct unknown coefficients. We generate random points on the Chow hypersurface as follows: pick a random point in X and a random linear space of dimension $n-d-1$ through that point. We read off the Plücker coordinates of that linear space and substitute them into the ansatz. Repeating this process many times gives the desired linear system of equations in the unknown coefficients. Up to scaling, this system has a unique solution, namely the Chow form $\mathrm{Ch}(X)$.

Example 25 We use this strategy to recover the Chow form $\mathrm{Ch}(X)$ from Example 24. For each lattice point u in the polytope $\mathcal{C}(X)$, we form the general linear combination of standard Plücker monomials of weight u. For instance, for $u=(2,2,2,2)$ this linear combination is

$$\gamma_{u,1}\cdot p_{01}^2 p_{23}^2 \,+\, \gamma_{u,2}\cdot p_{01}p_{02}p_{13}p_{23} \,+\, \gamma_{u,3}\cdot p_{02}^2 p_{13}^2.$$

Our ansatz for the Chow form $\mathrm{Ch}(X)$ is the sum of these expressions over all $u\in\mathcal{C}(X)\cap\mathbb{Z}^4$.

We sample from the Chow hypersurface by picking matrices of the form

$$\begin{bmatrix} \alpha_0 & \alpha_1 & \alpha_2 & \alpha_3 \\ 1 & t(t-1)(t+1) & t^2(t+1) & t^3(t-1) \end{bmatrix}.$$

The 2×2 minors of this matrix are the dual Plücker coordinates of the corresponding sample point in $\mathrm{Ch}(X)\subset \mathrm{Gr}(1,\mathbb{P}^3)$. We read off its primal Plücker coordinates as follows:

$$\begin{aligned} p_{01} &= (t^4-t^3)\alpha_2+(t^3+t^2)\alpha_3, & p_{02} &= (t^3-t^4)\alpha_1+(t^3-t)\alpha_3,\\ p_{03} &= (t^3+t^2)\alpha_1+(t-t^3)\alpha_2, & p_{12} &= (t^4-t^3)\alpha_0-\alpha_3,\\ p_{13} &= (t^3+t^2)\alpha_0+\alpha_2, & p_{23} &= (t^3-t)\alpha_0-\alpha_1. \end{aligned}$$

We substitute many such sample points into the ansatz, and we solve the resulting system of linear equations for the unknown coefficients $\gamma_{u,i}$. The output is the desired Chow form. This yields defining equations for X by setting $p_{ij}=\alpha_i x_j-\alpha_j x_i$ for any $\alpha_0,\ldots,\alpha_3\in\mathbb{Q}$.

In this article, we discussed an application of tropical geometry to computer algebra, namely implicitization with tropical preprocessing. It was shown that `Oscar.jl` provides excellent capabilities for performing tropical implicitization in practice. Our implementation in `Oscar.jl` realizes the vision in [SY08] and fulfills the promise made by `TrIm`. In Section 5, we ventured into a setting where the desired hypersurface is not in an affine or projective space, but inside a Grassmannian. This suggests yet one more problem for future research.

Problem 26 Many applications lead to interesting subvarieties of Grassmannians. For instance, in computer vision certain cameras are represented by curves and surfaces in $\mathrm{Gr}(1,\mathbb{P}^3)$. Their tropicalizations lie inside the tropical Grassmannian, and their cohomology classes are computed by Schubert calculus. It would be desirable to develop tropical implicitization in the setting when the ambient spaces are Grassmannians, or even flag varieties.

References

[BKSW18] P. Breiding, S. Kališnik, B. Sturmfels, and M. Weinstein. "Learning algebraic varieties from samples". *Revista Matemática Complutense* 31 (2018), pp. 545–593.

[Bry20] T. Brysiewicz. "Newton polytopes and numerical algebraic geometry". 2020. arXiv: 2004.12177.

[CCDDS13] E. Cattani, M. A. Cueto, A. Dickenstein, S. Di Rocco, and B. Sturmfels. "Mixed discriminants". *Math. Z.* 274.3-4 (2013), pp. 761–778.

[CGKW00] R. M. Corless, M. W. Giesbrecht, I. S. Kotsireas, and S. M. Watt. "Numerical implicitization of parametric hypersurfaces with linear algebra". In: *International Conference on Artificial Intelligence and Symbolic Computation*. Springer. 2000, pp. 174–183.

[CLO97] D. A. Cox, J. Little, and D. O'Shea. "Ideals, Varieties, and Algorithms". *Undergraduate Texts in Mathematics* (1997).

[DS95] J. Dalbec and B. Sturmfels. "Introduction to Chow forms". In: *Invariant Methods in Discrete and Computational Geometry: Proceedings of the Curaçao Conference, 13–17 June, 1994*. Springer. 1995, pp. 37–58.

[DFS07] A. Dickenstein, E. Feichtner, and B. Sturmfels. "Tropical discriminants". *J. Amer. Math. Soc.* 20.4 (2007), pp. 1111–1133.

[EKKB13] I. Z. Emiris, T. Kalinka, C. Konaxis, and T. L. Ba. "Implicitization of curves and (hyper) surfaces using predicted support". *Theoretical Computer Science* 479 (2013), pp. 81–98.

[EKP07] I. Z. Emiris, C. Konaxis, and L. Palios. "Computing the Newton polytope of specialized resultants". In: *Presented at MEGA (Effective Methods in Algebraic Geometry)*. 2007.

[EK08] A. Esterov and A. Khovanskii. "Elimination theory and Newton polytopes". *Functional Analysis and Other Mathematics* 2.1 (2008), pp. 45–71.

[Fin13] A. Fink. "Tropical cycles and Chow polytopes". *Beiträge zur Algebra und Geometrie/Contributions to Algebra and Geometry* 54 (2013), pp. 13–40.

[GKZ08] I. M. Gelfand, M. M. Kapranov, and A. V. Zelevinsky. *Discriminants, resultants and multidimensional determinants*. Modern Birkhäuser Classics. Reprint of the 1994 edition. Birkhäuser Boston, Inc., Boston, MA, 2008, pp. x+523.

[Hug06] P. Huggins. "iB4e: A software framework for parametrizing specialized LP problems". In: *Mathematical Software-ICMS 2006: Second International Congress on Mathematical Software, Castro Urdiales, Spain, September 1-3, 2006. Proceedings 2*. Springer. 2006, pp. 245–247.

[MS15] D. Maclagan and B. Sturmfels. *Introduction to tropical geometry*. Vol. 161. Graduate Studies in Mathematics. American Mathematical Society, Providence, RI, 2015, pp. xii+363. DOI: 10.1090/gsm/161.

[Stu08] B. Sturmfels. *Algorithms in invariant theory*. Springer Science & Business Media, 2008.

[ST08] B. Sturmfels and J. Tevelev. "Elimination Theory for Tropical Varieties". *Mathematical Research Letters* 15.3 (2008), pp. 543–562. DOI: 10.4310/MRL.2008.v15.n3.a14.

[STY06] B. Sturmfels, J. Tevelev, and J. Yu. "The Newton Polytope of the Implicit Equation". *Computing Research Repository - CORR* 7 (July 2006). DOI: 10.17323/1609-4514-2007-7-2-327-346.

[SY08] B. Sturmfels and J. Yu. "Tropical implicitization and mixed fiber polytopes". In: *Software for algebraic geometry*. Vol. 148. IMA Vol. Math. Appl. Springer, New York, 2008, pp. 111–131. DOI: 10.1007/978-0-387-78133-4_7.

F-Theory Applications

Martin Bies and Andrew P. Turner

Abstract String theory is a prominent contender for offering a comprehensive explanation of all the physical phenomena we observe. This encompasses a wide spectrum, ranging from the quantum interactions among subatomic particles to the gravitational forces that steer the motion of colossal astronomical objects across galactic expanses. A significant challenge with string theory arises from its immense set of solutions, known as the string landscape. Many of these solutions describe universes with markedly different laws of physics from what we observe in experiments. The pursuit of solutions that align with experimental observations becomes exceedingly demanding, especially as we seek greater consistency between string theory predictions and experimental results. The quest to discover a solution that harmonizes perfectly with all existing experimental observations—while potentially making predictions for future experiments—stands as a revered goal in the realm of string phenomenology.

Within this framework, F-theory emerges as a formulation of string theory that manages to encapsulate a substantial portion of the information regarding the various solutions within the string landscape through its geometric aspects. Mathematically, solutions to F-theory correspond to singular elliptically fibered Calabi–Yau manifolds. Given the lack of direct methods of analysis, the F-theory community has adopted the standard approach of first resolving the singularities, and subsequently deriving the underlying physics from the resolved space. The computational process of finding these resolutions is both demanding and occasionally intricate. This chapter illustrates how the application of `OSCAR` can simplify the computation of resolutions and thereby facilitate the exploration of F-theory solutions.

Martin Bies
RPTU Kaiserslautern-Landau, Kaiserslautern, Germany, e-mail: `martin.bies@rptu.de`

Andrew P. Turner
University of Pennsylvania, Philadelphia, USA, e-mail: `turnerap@sas.upenn.edu`

W. Decker et al. (eds.), *The Computer Algebra System OSCAR*,
Algorithms and Computation in Mathematics 32,
https://doi.org/10.1007/978-3-031-62127-7_18

Key words: String theory, F-theory, elliptic fibration, Weierstrass model, Tate model, Kodaira classification, resolution of singularities, (toric) blow-ups, intersection theory, Mordell-Weil group, Chow ring, literature model database

1 Introduction

1.1 String Phenomenology

String theory aims to describe all phenomena of inanimate nature—ranging from quantum physics at submicroscopic scales to gravitational effects at astronomically large scales—in a unified framework. Today, it is arguably one of the leading candidates for a theory of quantum gravity. For consistency reasons, string theory requires a ten-dimensional spacetime. In making contact with our observed four-dimensional world, it is standard to assume that six of these ten real dimensions are very small. In other words, in the simplest case, the ten-dimensional spacetime is a direct product of a four-dimensional space $\mathcal{E}_4$ resembling our observed world and a "small" six-dimensional Riemannian manifold B_3. It turns out that the geometry of B_3 has a significant impact on the physics observed in the four-dimensional space $\mathcal{E}_4$. There are many candidates for the choice of B_3, leading to a vast array of solutions to string theory, known as the *string landscape*. It is natural to wonder which—if any—choice of B_3 leads to perfect agreement between the four-dimensional physics predicted by string theory and our experimental observations.

String phenomenology is the branch of string theory that tries to tackle this very question. As such, one employs techniques from neighboring fields (typically mathematics, but more recently also machine learning and data science) to identify "good" geometries. Finding a geometry that exactly reproduces the physics of our four-dimensional everyday experience is considered the holy grail of string phenomenology. Attempts to find this holy grail date back as far as 1985 [CHSW85]. In fact, there exist so many works in this regard, that any attempt to provide a complete list is doomed to fail. Let us therefore apologize to all authors whose important works we fail to list; the relevant works conducted in the context of perturbative string theory include [CHSW85; GKMR86; BDL96; AFIRU01a; AFIRU01b; IMR01; BKLO01; CSU01b; CSU01a; BHOP05; BD06; GLS07; BCLS05; BCD06; BD08; AGHL10; AGLP11; AGLP12].

Often, these string theory constructions are severely constrained by global consistency conditions. For instance, [BD06] was the first work to solve the subtle slope-stability conditions in heterotic string theory and thereby managed to give an explicit geometric construction of the minimal supersymmetric standard model in the context of heterotic string theory. The interested reader can find details on these subtle conditions in [BCD06].

While these global consistency conditions are often highly nontrivial to satisfy, for some string theory constructions they are almost automatically satisfied. Specifically, this is the case for F-theory [Vaf96; MV96a; MV96b]. F-theory solutions comprise a certain nonperturbative regime of string theory. We will introduce F-theory in more detail momentarily. Let it suffice for now to state that its ease of solving global consistency conditions has led many people to investigate standard model constructions in F-theory. Such works include [GH12; KMW12a; KMW12b; BGK13; CGKP14; BMPW14; CK-POR15; LMTW16; LW16; BMW17; Bie18; CLLO18; TT20; RTT20; JTT23b]. A recent such set of constructions of interest is the Quadrillion F-theory Standard Models (F-theory QSMs) [CHLLT19], which have subsequently been investigated in more detail in [BCDLO21; BCL21; BCDO22] by means of root bundles.[1] See [BCDO23] for a recent overview of these efforts. For a broad overview of model building in F-theory we refer the interested reader to [Wei18].

1.2 F-Theory

F-theory is a special domain in string theory that employs a powerful dictionary between physics and geometry. Specifically, the physics is encoded in the geometry of an elliptic fibration $\pi \colon Y_4 \twoheadrightarrow B_3$. For nontrivial physics, Y_4 must be singular, and so the study of singularities in elliptic fibrations is a central focus of F-theory. We will provide more details in the following sections. The interested reader can find more background information in [Wei18] and the references contained therein.

While we can glean some information about the physics directly from the singularities in the original singular model [AHK14; CS15a; CS15b; CGSV16; AHKS17], the situation is more easily understood by resolving these singularities. Due to constraints coming from the physics, we must focus on *crepant* resolutions. While algorithms for general desingularization do exist, there are none known for general crepant desingularization. A related challenge is that a crepant resolution can fail to render a singular space smooth; specifically, the presence of $\mathbb{Q}$-factorial terminal singularities precludes the possibility of a full crepant resolution. The presence of such singularities in F-theory models has physical significance; however, for the sake of simplicity we will focus in this chapter on models that can be fully crepantly resolved.

In addition to the appearance of $\mathbb{Q}$-factorial terminal singularities, there can be non-minimal singularities that lead to non-flat fibrations after resolution. This means that the dimension of the fiber jumps, i.e., the fiber over some loci could have complex dimension two. While this phenomenon is of interest

[1] Of note, OSCAR has already seen use in these works, namely in performing an involved triangulation task in [BCDO22].

to the physics community [ALM19; Oeh21], we again exclude this scenario from the discussion in this article for the sake of simplicity.

There are yet more ways in which the field of study for F-theory can be enlarged. For instance, one can consider not only elliptic fibrations but, more generally, torus fibrations [Wit96; BM14; AGGK14; KMOPR15; GGK14; MPTW14; MPTW15; MT16; CDKPP15; LMTW16; Kim17]. Also, based on physical motivation, one can study F-theory constructions with different base dimension. Bases of complex dimension two are of interest for the engineering of superconformal field theories, as discussed for instance in [HR19; ALLSW19; ALLSW20a; ALLSW20b], while dimension three is typically employed in attempts to construct F-theory solutions that directly resemble observed physics.

In summary, the exposition here assumes a three-dimensional base B_3 and focuses on singular elliptic fibrations Y_4 that admit at least one smooth, flat, crepant resolution $\widehat{\pi}\colon \widehat{Y}_4 \twoheadrightarrow B_3$. This article will focus on resolutions via a sequence of ordinary blowups, though other techniques such as weighted blowups have been investigated [AJO23].

Finally, let us mention that beyond the elliptic fibration $\widehat{\pi}\colon \widehat{Y}_4 \twoheadrightarrow B_3$, additional data are required to fully specify an F-theory solution. Among others, this includes the G_4-flux, with $G_4 \in H^{2,2}_{\mathbb{Z}}(\widehat{Y}_4)$. There exists an extensive toolbox with established methodology to handle these G_4-fluxes [GH12; KMW12b; BGK13; CGKP14; CKPOR15; LMTW16; LW16; JTT23a]. For brevity of exposition, we will not go into much detail about this flux. Future planned extensions of the suite of F-theory tools within OSCAR will strongly depend on this flux and yet finer information, for instance Deligne cohomology [BMPW14; BMW17; Bie18] or root bundles [BCDLO21; BCL21; BCDO22; Bie24; BCDO23].

1.3 Article Navigation

Section 2 introduces F-theory, which should give the reader a first glimpse at the required computations that we aim to automate. OSCAR is presented as a tool for this purpose. Section 3 demonstrates the use of OSCAR for one such calculation. It is worth noting that this suite of tools stems from an initially independent Julia package, `FTheoryTools` [BT22], that has since been incorporated as a module of OSCAR. In Section 4, we propose creating a comprehensive database of F-theory constructions. One often revisits established constructions with new mathematical techniques, necessitating recomputation of complex data; such a database would bypass this redundancy, enhancing research velocity by collecting both fundamental and advanced geometric data from existing literature. Final thoughts on future directions are provided in Section 5.

2 Technology for F-Theory and Elliptic Fibrations

F-theory can be thought of as a particular nonperturbative formulation of Type IIB string theory that allows for the analysis of systems containing extended objects called 7-branes. The presence of 7-branes generates a varying profile for the *axio-dilaton field*,

$$\tau\colon \mathcal{E}_4 \times B_3 \to \mathbb{C}\,, \quad x^\mu \mapsto C_0(x^\mu) + ie^{-\phi(x^\mu)}\,. \tag{1}$$

Recall that $\mathcal{E}_4$ here refers to the four-dimensional extended spacetime that resembles the world of our everyday experience, while B_3 denotes the "internal" three-dimensional complex space. The variation induced by the 7-branes necessitates the existence of spacetime regions where the string coupling $g_s = e^\phi$ becomes large, precluding analysis using the standard toolkit of perturbative string theory.

This issue is resolved by utilizing the nonperturbative $\mathrm{SL}(2,\mathbb{Z})$ invariance of Type IIB string theory, which allows us to interpret the value of the axio-dilaton $\tau(x)$ as the complex structure modulus of an elliptic curve $\mathbb{C}_{1,\tau(x)}$ for each $x \in B_3$. The physics of F-theory dictates that the axio-dilaton remain constant on $\mathcal{E}_4$ but vary over B_3; in fact, $\tau(x)$ must diverge at specific loci of B_3 for the model to describe nontrivial physics. Wherever $\tau(x)$ diverges, the elliptic curve $\mathbb{C}_{1,\tau(x)}$ becomes singular. Consequently, a singular elliptic fibration $\pi\colon Y_4 \twoheadrightarrow B_3$ with $\pi^{-1}(x) = \mathbb{C}_{1,\tau(x)}$ can serve as geometric book-keeping device to allow us to keep track of the behavior of the axio-dilaton. This is the basis of F-theory.

As alluded to before, we assume that Y_4 admits a smooth, flat, crepant resolution $\widehat{\pi}\colon \widehat{Y}_4 \twoheadrightarrow B_3$. Typically, this resolution is performed by a sequence of (non-weighted) blowups.

2.1 Weierstrass Models

Weierstrass models are one of the most common presentations of singular elliptic fibrations over a base manifold (or variety) B_3. We consider a line bundle $\mathcal{L} \in \mathrm{Pic}(B_3)$ as well as the projective bundle[2] $\rho\colon \mathbb{P}^{2,3,1}[\mathcal{L}^2 \oplus \mathcal{L}^3 \oplus \mathcal{O}_{B_3}] \twoheadrightarrow B_3$. Let $X = \mathbb{P}^{2,3,1}[\mathcal{L}^2 \oplus \mathcal{L}^3 \oplus \mathcal{O}_{B_3}]$ be the total space of this projective bundle. Then a section of the bundle $\mathcal{O}_X(6) \otimes \rho^* \mathcal{L}^6$ cuts out an elliptic fibration as a hypersurface in X. We can describe such a section explicitly. To this end,

[2] Note that we use the classical convention for the projectivization of a vector bundle, so that the fibers are the lines passing through the origin rather than the hyperplanes. This is the opposite of the convention in Hartshorne's book on algebraic geometry. Also, note that the usual projection symbol π is reserved in this text for the projection of an elliptic fibration, i.e., $\pi\colon Y_4 \twoheadrightarrow B_3$. To avoid confusion, we thus use the symbol ρ for the projection of the projective bundle at hand.

let us denote the homogeneous coordinates of the ambient $\mathbb{P}^{2,3,1}$-bundle as $[x : y : z]$. We see that x, y, z are sections of $\mathcal{O}_X(2) \otimes \rho^*\mathcal{L}^2$, $\mathcal{O}_X(3) \otimes \rho^*\mathcal{L}^3$, and $\mathcal{O}_X(1)$, respectively. Over each point of the base B_3, the fiber is the elliptic curve cut out by the equation

$$y^2 = x^3 + fxz^4 + gz^6 \,, \tag{2}$$

where f and g are respectively sections of the line bundles $\rho^*\mathcal{L}^4$ and $\rho^*\mathcal{L}^6$. For the purposes of F-theory constructions, we are particularly interested in elliptic fibrations in which the total space is Calabi–Yau. Note, however, that the base B_3 is only required to be Kähler and indeed cannot be Calabi–Yau for a nontrivial theory. The Calabi–Yau constraint sets

$$c_1(\mathcal{L}) = c_1(B_3) = \bar{K}_{B_3} \,, \tag{3}$$

with K_{B_3} the canonical class of the base, so $\mathcal{L}$ is the anticanonical bundle.

There are more general definitions of elliptic fibrations that are also relevant for F-theory, for instance where the elliptic fibration is not defined as a hypersurface in some ambient space but rather a complete intersection, but we will not discuss them here for the sake of space.

2.2 The Kodaira Classification

Of particular physical importance is the singularity structure of a Weierstrass model. Singularities occur first at complex codimension one in the base B_3, and can enhance at higher codimension. The singular locus is called the *discriminant locus*, as it is the zero locus $\mathbb{V}(\Delta)$ of the discriminant $\Delta = 4f^3 + 27g^2$. We picture the situation in Figure 1.

All possible singularity types supported at codimension one have been classified by Kodaira [Kod63a; Kod63b] and Néron [Nér64] in what is commonly referred to as the *Kodaira classification*. A singularity supported at a codimension-one locus is classified according to the orders of vanishing of f, g, and the discriminant Δ at this locus. In actuality, Kodaira classified codimension-one singularities in elliptically fibered surfaces, which have no higher-codimension loci. In the F-theory context, where the base can be of complex dimension higher than one, there is an additional scenario that can arise related to monodromy when moving around within the codimension-one singular locus, known as Tate monodromy. Including these data along with the Kodaira classification yields the *refined Tate fiber type* of each codimension-one singularity. To each refined Tate fiber type, there is an associated compact simple Lie algebra; the Lie algebra associated to a given codimension-one singularity corresponds precisely to a factor of the gauge algebra in the physical theory. This classification of the singularities is compactly collected

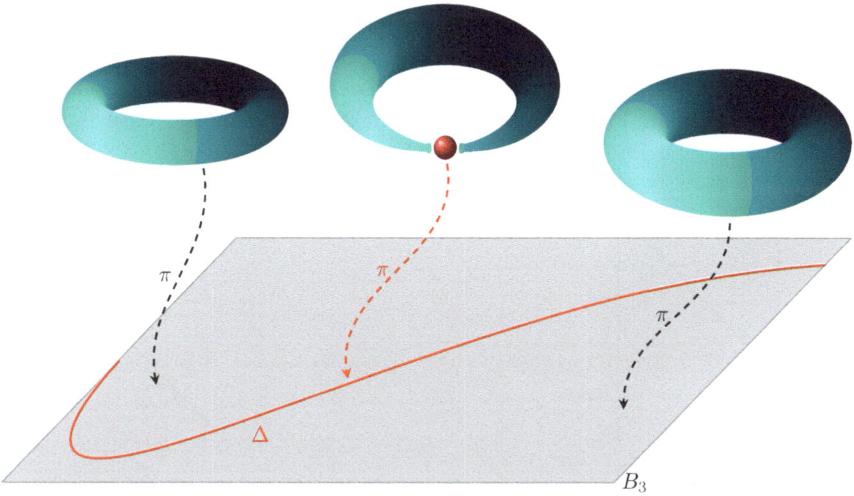

Fig. 1: A cartoon depiction of a singular elliptic fibration. The fiber over a generic point is a smooth elliptic curve. Over the discriminant locus $\mathbb{V}(\Delta)$—which is codimension one in the base B_3—the fiber is a singular elliptic curve. This picture is taken from [Bie18].

in the Kodaira table, which is available in the OSCAR documentation[3] (see also [Wei18]):

https://docs.oscar-system.org/stable/Experimental/FTheoryTools/weierstrass/

The singularity type may enhance at codimension-two loci where codimension-one singular loci collide with one another or with the residual discriminant locus. We may heuristically associate a Kodaira type to such codimension-two singular loci based on the orders of vanishing of f, g, Δ, but in reality the classification of such singularities is more subtle. Under certain sets of assumptions, these singularities have been classified by Miranda [Mir83] and Szydlo [Szy98]. In the physical theory, codimension-two loci where the singularity type enhances support localized matter charged under the gauge algebras associated to the colliding codimension-one singular loci. Such codimension-two loci in the base are thus termed *matter curves* when the base has dimension three. The type of the codimension-two singularity determines the representation in which the charged matter transforms. We picture a typical situation in Figure 2.

[3] Please note that FTheoryTools is currently subject to much development and therefore an experimental module of OSCAR. Once FTheoryTools joins the OSCAR core, the location of the documentation will change. Consequently, the provided URL will eventually cease to function. In this case, the interested reader is encouraged to visit https://docs.oscar-system.org/stable/ and use the search function to identify the relevant sections in the documentation.

In a similar manner, the singularity type may enhance yet further at codimension-three loci, which is associated in the physical theory with Yukawa couplings among the charged matter localized at the codimension-two loci. Such enhancement loci are called *Yukawa points* when the base is three-dimensional.

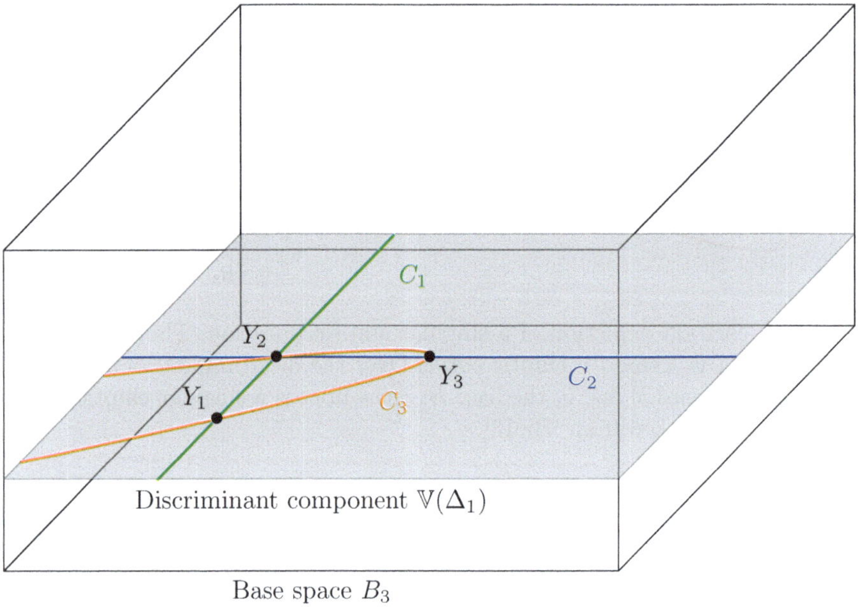

Fig. 2: A cartoon depiction of an irreducible component $\mathbb{V}(\Delta_1)$ of the discriminant locus $\mathbb{V}(\Delta_0\Delta_1\cdots)$ in a base B_3. This discriminant component contains the matter curves C_1, C_2, and C_3. At the intersections of these matter curves, we find the Yukawa points Y_1, Y_2, Y_3, respectively. This picture is modified from [Bie18].

2.3 The Geometry of the Resolved Elliptic Fibration

Next, we wish to discuss the geometry of the smooth, flat, crepant resolution $\widehat{\pi}\colon \widehat{Y}_4 \twoheadrightarrow B_3$. This geometry differs from Y_4 in the fiber structure over the discriminant locus $\mathbb{V}(\Delta)$, over which the elliptic fibers in the original fibration are singular. Consequently, let us focus on the fiber geometry over $\mathbb{V}(\Delta)$. The typical situation over an irreducible component of the discriminant locus is pictured in Figure 3.

Over a generic point of a codimension-one singular locus, the fiber of the resolved space will consist of some number of $\mathbb{P}^1$s intersecting one another in the structure of a simply laced affine Dynkin diagram, depicted in grey in Figure 3. In the case of trivial Tate monodromy, this affine Dynkin diagram is the one corresponding to the simply laced Lie algebra associated with the refined Tate fiber type of the singularity; in the case of nontrivial Tate monodromy, the Dynkin diagram of the Lie algebra associated with the refined Tate fiber type is a particular folding of the Dynkin diagram arising from the intersection structure of the local fiber. In either case, the associated Lie algebra appears as a factor of the gauge algebra in the physical theory.

It is important to emphasize that the physics requires a crepant resolution, i.e., because $c_1(T_{Y_4})$ is trivial, then it must also be trivial for the resolved space. No algorithm for general crepant resolution is known. Furthermore, it is well-known that certain types of singularities cannot be resolved crepantly. Such singularities are known as $\mathbb{Q}$-factorial terminal singularities. Another interesting "exotic" situation is that of non-minimal singularities, whose resolution leads to non-flat fibrations for which the dimension of the fiber increases over certain loci. A typical example is that over certain loci of B_3 the fiber is a complex surface, whereas for generic loci of B_3 the fiber is the union of a collection of curves. As alluded to in the introduction, we will not discuss these situations here.

At higher codimension, the singularity type of the singular fibers in the elliptic fibration $\pi\colon Y_4 \twoheadrightarrow B_3$ may enhance further. Typically, this means that the corresponding fibers in the resolved fibration $\widehat{\pi}\colon \widehat{Y}_4 \twoheadrightarrow B_3$ consist of more $\mathbb{P}^1$s. This is depicted in Figure 3, where there are five green $\mathbb{P}^1$s over the matter curve C_1 and five red $\mathbb{P}^1$s over C_2. Similarly, there are yet more $\mathbb{P}^1$s over the Yukawa point Y. One can relate the generic (grey) $\mathbb{P}^1$s to the $\mathbb{P}^1$s over the matter curves, which in turn relate to the $\mathbb{P}^1$s over the Yukawa point. These relations, together with the intersection structure of the $\mathbb{P}^1$s—as reflected in the Dynkin diagrams—encode important information about the physics. Specifically, the codimension-two fiber structure determines the representations of charged matter, while the codimension-three fiber structure determines the types of interactions between these matter representations. The extraction of these intersection properties will be a key step in the future development of `FTheoryTools`. However, for the sake of space, we will not discuss this matter in more detail. The interested reader can find background information in [Wei18] and the references therein.

2.4 Global Tate Models

As explained above, the singularities of the elliptic fibers over the discriminant locus $\mathbb{V}(\Delta)$ are of great importance to the physics. One often wishes to engineer the singularities over the irreducible components of $\mathbb{V}(\Delta)$ to be of particular

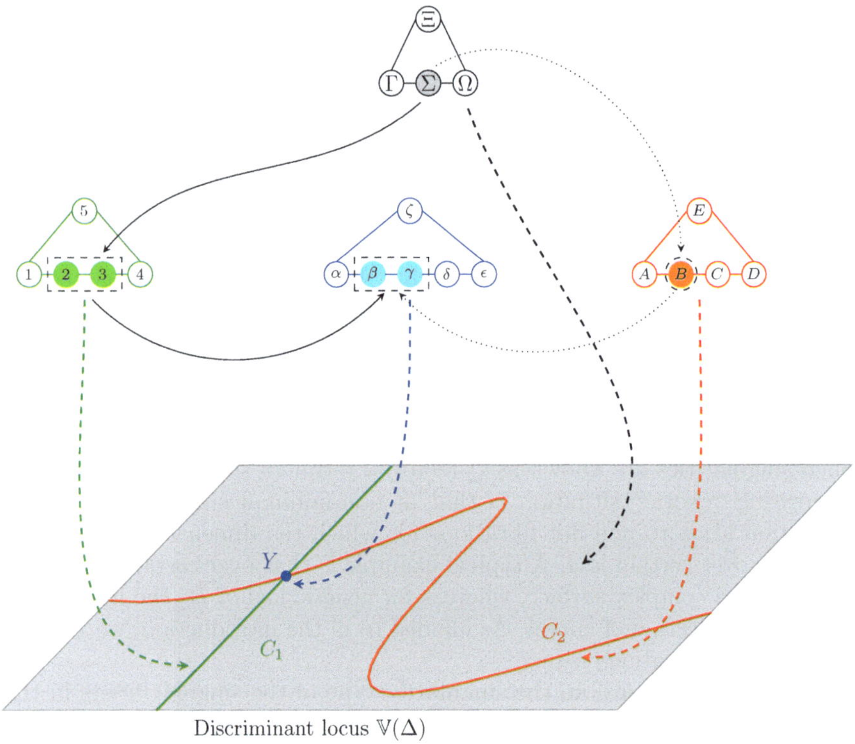

Fig. 3: A cartoon depiction of a resolved elliptic fibration. Note that the singularity type can enhance with increasing codimension in the discriminant locus $\mathbb{V}(\Delta)$. Consequently, the number of $\mathbb{P}^1$s in the resolved fibers typically increases as the codimension in the discriminant increases. Upon restriction, certain $\mathbb{P}^1$s in the fiber at low codimension may split into collections of $\mathbb{P}^1$s in the fiber at higher codimension. This picture is modified from [Bie18].

Kodaira types. One systematic approach to engineering these singularities is to use what are known as Tate models.

Let us consider a point $p \in B_3$. It was pointed out in [Ber+96] that there is always an open neighborhood $p \in U \subset B_3$ such that for every $q \in U$, the elliptic fiber $\mathbb{C}_{1,\tau(q)}$ can described as the hypersurface cut out by

$$y^2 + a_1(q)xyz + a_3(q)yz^3 = x^3 + a_2(q)x^2z^2 + a_4(q)xz^4 + a_6(q)z^6 \,. \quad (4)$$

In this expression, the a_i are (local) sections of $\overline{K}_{B_3}^{\otimes i}$ over U.[4] This description is referred to as a Tate model or long Weierstrass form. Similarly to the

[4] In an abuse of notation, we use the symbol $\overline{K}_{B_3}$ to refer to both the anticanonical class and the anticanonical bundle of B_3.

Kodaira classification, the singularities of such a model can be classified by the vanishing orders of the sections a_i, but no monodromy action needs to be taken into account (except for in a small number of special cases). The corresponding Tate table is available in the OSCAR documentation[5] (see also [Ber+96, Table 2], [Wei18], and [HT19]):

https://docs.oscar-system.org/stable/Experimental/FTheoryTools/tate/

In practice, the design of singularities over $\mathbb{V}(\Delta)$ is easier with a Tate model than a Weierstrass model. Note however that a Weierstrass model can in general only locally be brought into Tate form. To circumvent this hurdle, one can instead cut out the entire elliptic fibration from a single Tate polynomial of the form in Equation 4. Such constructions, known as a *global* Tate models, are less general than Weierstrass models, but are useful for model building and are widely used in the F-theory literature. A particular instance, which we will discuss in the next section with the help of computations in OSCAR, are the geometries discussed in [KMW12a].

3 Resolution of a Global Tate Model

As an example, let us use OSCAR to construct and resolve an $\mathfrak{su}(5) \oplus \mathfrak{u}(1)$ singular global Tate model that was analyzed in detail in [KMW12a]. For the sake of simplicity, we will construct this model over the toric base $\mathbb{P}^3$:

```
julia> B3 = projective_space(NormalToricVariety, 3)
Normal toric variety

julia> cox_ring(B3)
Multivariate polynomial ring in 4 variables over QQ graded by
  x1 -> [1]
  x2 -> [1]
  x3 -> [1]
  x4 -> [1]
```

Next, recall that a global Tate model is realized as hypersurface cut out by the polynomial

$$y^2 + a_1 xyz + a_3 yz^3 = x^3 + a_2 x^2 z^2 + a_4 xz^4 + a_6 z^6 \,, \tag{1}$$

where $a_i \in H^0(B_3, \overline{K}_{B_3}^{\otimes i})$. Following [KMW12a], we now wish to specialize (or *tune*) the model to enforce a certain type of singularity over a particular divisor of B_3, which will appear as a factor of the discriminant locus Δ of the elliptic fibration. We opt for $\mathbb{V}(x_1) \subset B_3$, with x_1 the first variable in the Cox ring of B_3 as listed above. The singularity that we wish to tune is a

[5] This link will cease to function once FTheoryTools joins the OSCAR core. The interested reader should then visit https://docs.oscar-system.org/stable/ and use the search function to identify the relevant sections in the documentation.

split (indicating the Tate monodromy type) I_5 fiber type. This corresponds to an $\mathfrak{su}(5)$ factor in the F-theory gauge algebra. We achieve this singularity structure by factoring the a_i as follows:

$$a_1 = x_1^0 a_{1,0}\,, \quad a_2 = x_1^1 a_{2,1}\,, \quad a_3 = x_1^2 a_{3,2}\,, \quad a_4 = x_1^3 a_{4,3}\,, \quad a_6 = x_1^5 a_{6,5}\,. \tag{2}$$

Additionally, we wish to tune the model so that the gauge algebra contains a $\mathfrak{u}(1)$ factor, which is achieved in this case by setting $a_{6,5} = 0$.

To construct this model in OSCAR, we first compute the sections $a_{i,j}$:

```
julia> Kbar = anticanonical_bundle(B3)
Toric line bundle on a normal toric variety

julia> W = toric_line_bundle(torusinvariant_prime_divisors(B3)[1])
Toric line bundle on a normal toric variety

julia> x1 = basis_of_global_sections(W)[4]
x1

julia> a10 = sum([rand(Int) * b for b in
↪   basis_of_global_sections(Kbar)]);

julia> a21 = sum([rand(Int) * b for b in basis_of_global_sections(Kbar^2
↪   * W^(-1))]);

julia> a32 = sum([rand(Int) * b for b in basis_of_global_sections(Kbar^3
↪   * W^(-2))]);

julia> a43 = sum([rand(Int) * b for b in basis_of_global_sections(Kbar^4
↪   * W^(-3))]);

julia> a65 = 0;
```

Next, we construct the global Tate model and compute its singular loci, in order to confirm that our construction indeed works:

```
julia> t = global_tate_model(B3, [a10, a21 * x1, a32 * x1^2, a43 * x1^3,
↪   a65 * x1^5], completeness_check = false)
Global Tate model over a concrete base

julia> sing = singular_loci(t);

julia> length(sing)
2

julia> singular_loci(t)[2]
(Ideal (x1), (0, 0, 5), "Split I_5")

julia> singular_loci(t)[1][2:3]
((0, 0, 1), "I_1")
```

This indeed confirms the presence of a split I_5 singularity at the codimension-one locus $\mathbb{V}(x_1)$ as desired. There is additionally a codimension-one locus of B_3 supporting an I_1 singularity; while the fiber is singular over this base locus, the total space of the fibration remains smooth. Such singularities do not support nontrivial gauge algebras.

The resolution of this singular model can be achieved by a sequence of toric blowups of the toric ambient space of this global Tate model. We begin by analyzing the ambient space:

```
julia> amb = ambient_space(t)
Normal toric variety

julia> cox_ring(amb)
Multivariate polynomial ring in 7 variables over QQ graded by
  x1 -> [1 0]
  x2 -> [1 0]
  x3 -> [1 0]
  x4 -> [1 0]
  x -> [8 2]
  y -> [12 3]
  z -> [0 1]
```

We follow the first non-weighted resolution sequence detailed in [KMW12a], entailing a total of five toric blowups. The first step is to blow up the locus $\mathbb{V}(x, y, x_1)$ in the toric ambient space. In keeping with the original work, we denote the resulting exceptional divisor coordinate by e_1. This is achieved with OSCAR as follows:

```
julia> t1 = blow_up(t, ["x", "y", "x1"]; coordinate_name = "e1")
Partially resolved global Tate model over a concrete base

julia> amb1 = ambient_space(t1)
Normal toric variety

julia> cox_ring(amb1)
Multivariate polynomial ring in 8 variables over QQ graded by
  x1 -> [1 0 0]
  x2 -> [0 1 0]
  x3 -> [0 1 0]
  x4 -> [0 1 0]
  x -> [1 1 2]
  y -> [1 2 3]
  z -> [0 -3 1]
  e1 -> [-1 1 0]
```

We now repeat this process for the subsequent four blowups:

```
julia> t2 = blow_up(t1, ["y", "e1"]; coordinate_name = "e4")
Partially resolved global Tate model over a concrete base

julia> t3 = blow_up(t2, ["x", "e4"]; coordinate_name = "e2")
Partially resolved global Tate model over a concrete base

julia> t4 = blow_up(t3, ["y", "e2"]; coordinate_name = "e3")
Partially resolved global Tate model over a concrete base

julia> t5 = blow_up(t4, ["x", "y"]; coordinate_name = "s")
Partially resolved global Tate model over a concrete base
```

The resulting model, `t5`, is fully resolved. We can now, for instance, return the strict transform of the Tate polynomial via `tate_polynomial(t5)`, though we omit the output here as it is rather long in this case.

4 Literature Models

Owing to nature of string phenomenology, it is common to revisit previous string theory constructions with the latest mathematical tools. Sadly, this process is often cumbersome, as one has to repeat many of the computations that were conducted in earlier papers. This also makes it more difficult for new people to enter the field. In order to simplify this task, we are building a database of well-established F-theory constructions within OSCAR. This database will not only include the construction of the model, but also load in known results such as topological data or known crepant resolutions. While this area is currently under rapid development, let us still exemplify how the above $\mathfrak{su}(5) \oplus \mathfrak{u}(1)$ model could be read in from the literature model database and then resolved:

```
julia> B3 = projective_space(NormalToricVariety, 3)
Normal toric variety

julia> W = torusinvariant_prime_divisors(B3)[1]
Torus-invariant, prime divisor on a normal toric variety

julia> t = literature_model(arxiv_id = "1109.3454", equation = "3.1",
↪  base_space = B3, model_sections = Dict("w" => W), completeness_check
↪  = false)
Construction over concrete base may lead to singularity enhancement.
↪  Consider computing singular_loci. However, this may take time!

Global Tate model over a concrete base -- SU(5)xU(1) restricted Tate
↪  model based on arXiv paper 1109.3454 Eq. (3.1)
```

This constructs the model in question by using the arXiv ID of the paper [KMW12a]. Furthermore, we provide the equation in which the defining Tate sections a_i were specified. With this information, we retrieve the defining data of this model from our database and construct it.

Among other data, we store different crepant resolutions of this model in our database. Here, we choose the first non-weighted resolution to resolve this model. This resolution uses the exact sequence of toric blowups that we demonstrated in the previous section. However, this computation is now executed with a single command:

```
julia> t5 = resolve(t, 1)
Partially resolved global Tate model over a concrete base -- SU(5)xU(1)
↪  restricted Tate model based on arXiv paper 1109.3454 Eq. (3.1)
```

This clearly takes much less effort than writing and executing the code from the previous section. Hopefully, this should exemplify the kind of boost in productivity that one can hope to gain from employing the literature model database for such investigations in the future.

5 Outlook

In the future, we hope to extend the current functionality to cover more areas of mathematics and their applications in F-theory. We will outline these concepts, starting with the simplest and gradually moving towards increasingly complex future prospects.

As alluded to briefly in Section 1.2, there is a G_4-flux that determines, among other things, how many matter particle families exist in a given compactification. This is a key feature to compare a given F-theory solution to the observed particle physics in the real world. We hope to extend our technology so that we can read off chiral matter (or enumerate fluxes for flux breaking of gauge algebras). More ambitiously, we also hope to include tools for the Deligne cohomology and root bundles that are being used in the study of vector-like matter, e.g., in telling if an F-theory solution includes zero, one, or multiple copies of the famous Higgs boson [BMPW14; BMW17; Bie18; BCDLO21; BCL21; BCDO22; Bie24; BCDO23]. This is another important feature in making contact between F-theory solutions and observed particle physics.

Additionally, we hope to incorporate more general approaches to resolution, including weighted blowups [AJO23]. Weighted blowups give more flexibility, allowing one to find new resolutions for existing crepantly resolvable singularities and in some cases crepantly resolving singularities that cannot be crepantly resolved with ordinary blowups. Another important tool we intend to incorporate is the pushforward formulae, which relate intersection theory computations in the fully-resolved 4-fold to intersection theory computations in the base [JT23]. This simplifies the involved computations, and often allows one to determine important physical quantities entirely in terms of the intersection data of the base.

Recall from Section 2 that nonabelian gauge factors in the physical theory arise from codimension-one singular loci. However, $\mathfrak{u}(1)$ gauge factors are also crucial for F-theory model building. For instance, they are key in replicating selection rules of particle physics, which makes the computation of $\mathfrak{u}(1)$-charges of matter fields in F-theory a pivotal task in many F-theory constructions [Rag18; RT18; RT22]. Unlike nonabelian gauge factors, abelian gauge factors in F-theory models are intrinsically global and are associated with the torsion-free subgroup of the Mordell–Weil group of the elliptic fibration (see [Wei18] and references therein). This is the group of infinite-order rational sections of the fibration, with group law given by fiberwise elliptic curve addition. Thus, the rank of the abelian part of the gauge algebra is exactly the Mordell–Weil rank. It should be noted that the computation of the Mordell–Weil group is extremely ambitious. Still, we hope to make tools available that support the study of abelian gauge factors and the computation of $\mathfrak{u}(1)$-charges, e.g., based on the results in [RT22].

The torsion subgroup of the Mordell–Weil rank is also relevant to the physics; namely, this subgroup is associated with the global structure of the gauge group in the physical theory. Specifically, to each codimension-one singularity, the refined Tate fiber type associates a nonabelian Lie algebra $\mathfrak{g}$, and this has a corresponding simply connected compact Lie group G. In the gauge group, this group G should be quotiented out by an action of each of the factors of the torsion subgroup of the Mordell–Weil group; the particular action is determined by the way in which the torsional sections intersect the $\mathbb{P}^1$ components of the resolved fiber over the codimension-one discriminant locus.

Last but not least, recall from Section 1.2 that one can also investigate the physics of F-theory on genus-one fibrations. Such geometries are torus fibrations that do not have a zero section. In these cases, there is always a k-section; locally, this k-section maps a point in the base to k points in the fiber, but globally these k points are exchanged by monodromy so that only the collection of all k points together, when fibered over the base, produce a well-defined divisor. Mathematically, this is related to the Weil–Châtelet group of the fibration, and physically, to the appearance of a discrete factor $\mathbb{Z}_k$ in the gauge group of the associated F-theory model. This is a domain that we also hope to eventually cover with `FTheoryTools`.

Acknowledgements Martin Bies and Andrew Turner express their gratitude and appreciation for the support provided by the `OSCAR` team, led by Claus Fieker, Max Horn, Michael Joswig, and Wolfram Decker. Martin Bies acknowledges financial support from the Forschungsinitiative des Landes Rheinland-Pfalz through the project *SymbTools – Symbolic Tools in Mathematics and their Application*. Andrew Turner acknowledges funding from *DOE (HEP) Award DE-SC0013528* of the United States Department of Energy (DOE). This work was supported by the SFB-TRR 195 *Symbolic Tools in Mathematics and their Application* of the German Research Foundation (DFG).

References

[AFIRU01a] G. Aldazabal, S. Franco, L. E. Ibáñez, R. Rabadan, and A. M. Uranga. "D=4 chiral string compactifications from intersecting branes". *Journal of Mathematical Physics* 42 (July 2001), pp. 3103–3126. DOI: 10.1063/1.1376157.

[AFIRU01b] G. Aldazabal, S. Franco, L. E. Ibáñez, R. Rabadan, and A. M. Uranga. "Intersecting brane worlds". *JHEP* 02 (2001), p. 047. DOI: 10.1088/1126-6708/2001/02/047.

[AGGK14] L. B. Anderson, I. García-Etxebarria, T. W. Grimm, and J. Keitel. "Physics of F-theory compactifications without section". *JHEP* 12 (2014), p. 156. DOI: 10.1007/JHEP12(2014)156.

[AGHL10] L. B. Anderson, J. Gray, Y.-H. He, and A. Lukas. "Exploring Positive Monad Bundles And A New Heterotic Standard Model". *JHEP* 02 (2010), p. 054. DOI: 10.1007/JHEP02(2010)054.

[AGLP12] L. B. Anderson, J. Gray, A. Lukas, and E. Palti. "Heterotic line bundle standard models". *JHEP* 2012 (June 2012). DOI: 10.1007/jhep06(2012)113.

[AGLP11] L. B. Anderson, J. Gray, A. Lukas, and E. Palti. "Two Hundred Heterotic Standard Models on Smooth Calabi–Yau Threefolds". *Phys. Rev.* D84 (2011), p. 106005. DOI: 10.1103/PhysRevD.84.106005.

[AHK14] L. B. Anderson, J. J. Heckman, and S. Katz. "T-Branes and Geometry". *JHEP* 05 (2014), p. 080. DOI: 10.1007/JHEP05(2014)080.

[AHKS17] L. B. Anderson, J. J. Heckman, S. Katz, and L. Schaposnik. "T-Branes at the Limits of Geometry". *JHEP* 58 (2017). DOI: 10.1007/JHEP10(2017)058.

[ALLSW20a] F. Apruzzi, C. Lawrie, L. Lin, S. Schäfer-Nameki, and Y.-N. Wang. "5d Superconformal Field Theories and Graphs". *Phys. Lett. B* 800 (2020), p. 135077. DOI: 10.1016/j.physletb.2019.135077.

[ALLSW19] F. Apruzzi, C. Lawrie, L. Lin, S. Schäfer-Nameki, and Y.-N. Wang. "Fibers add Flavor, Part I: Classification of 5d SCFTs, Flavor Symmetries and BPS States". *JHEP* 11 (2019), p. 068. DOI: 10.1007/JHEP11(2019)068.

[ALLSW20b] F. Apruzzi, C. Lawrie, L. Lin, S. Schäfer-Nameki, and Y.-N. Wang. "Fibers add Flavor, Part II: 5d SCFTs, Gauge Theories, and Dualities". *JHEP* 03 (2020), p. 052. DOI: 10.1007/JHEP03(2020)052.

[ALM19] F. Apruzzi, L. Lin, and C. Mayrhofer. "Phases of 5d SCFTs from M-/F-theory on Non-Flat Fibrations". *JHEP* 05 (2019), p. 187. DOI: 10.1007/JHEP05(2019)187.

[AJO23] V. Arena, P. Jefferson, and S. Obinna. "Intersection Theory on Weighted Blowups of F-theory Vacua" (Apr. 2023). arXiv: 2305.00297.

[BDL96] M. Berkooz, M. R. Douglas, and R. G. Leigh. "Branes intersecting at angles". *Nuclear Physics B* 480.1-2 (Nov. 1996), pp. 265–278. DOI: 10.1016/s0550-3213(96)00452-x.

[Ber+96] M. Bershadsky et al. "Geometric singularities and enhanced gauge symmetries". *Nucl. Phys.* B481 (1996), pp. 215–252. DOI: 10.1016/S0550-3213(96)90131-5.

[Bie18] M. Bies. "Cohomologies of coherent sheaves and massless spectra in F-theory". PhD thesis. Heidelberg U., Feb. 2018. DOI: 10.11588/heidok.00024045.

[Bie24] M. Bies. "Root bundles: applications to F-theory standard models." In: *Proceedings of the Symposia in Pure Mathematics, String-Math 2022.* Vol. 107. 2024, pp. 17–44. arXiv: 2303.08144.

[BCDLO21] M. Bies, M. Cvetič, R. Donagi, M. Liu, and M. Ong. "Root bundles and towards exact matter spectra of F-theory MSSMs". *JHEP* 2021.9 (Sept. 2021). DOI: 10.1007/jhep09(2021)076.

[BCDO22] M. Bies, M. Cvetič, R. Donagi, and M. Ong. "Brill-Noether-general limit root bundles: absence of vector-like exotics in F-theory Standard Models". *JHEP* 11 (2022), p. 004. DOI: 10.1007/JHEP11(2022)004.

[BCDO23] M. Bies, M. Cvetič, R. Donagi, and M. Ong. "Improved statistics for F-theory standard models" (July 2023). arXiv: 2307.02535.

[BCL21] M. Bies, M. Cvetič, and M. Liu. "Statistics of limit root bundles relevant for exact matter spectra of F-theory MSSMs". *Physical Review D* 104 (Sept. 2021). DOI: 10.1103/physrevd.104.1061903.

[BMPW14] M. Bies, C. Mayrhofer, C. Pehle, and T. Weigand. "Chow groups, Deligne cohomology and massless matter in F-theory" (Feb. 2014). arXiv: 1402.5144.

[BMW17] M. Bies, C. Mayrhofer, and T. Weigand. "Gauge Backgrounds and Zero-Mode Counting in F-Theory". *JHEP* 11 (2017), p. 081. DOI: 10.1007/JHEP11(2017)081.

[BT22] M. Bies and A. P. Turner. *FTheoryTools: Julia tools for F-Theory compactifications.* https://github.com/Julia-meets-String-Theory/FTheoryTools.jl. 2022.

[BCLS05] R. Blumenhagen, M. Cvetič, P. Langacker, and G. Shiu. "Toward realistic intersecting D-brane models". *Annual Review of Nuclear and Particle Science* 55.1 (Dec. 2005), pp. 71–139. DOI: 10.1146/annurev.nucl.55.090704.151541.

[BKLO01] R. Blumenhagen, B. Kors, D. Lüst, and T. Ott. "The Standard Model from stable intersecting brane world orbifolds". *Nuclear Physics B* 616.1-2 (Nov. 2001), pp. 3–33. DOI: 10.1016/s0550-3213(01)00423-0.

[BCD06] V. Bouchard, M. Cvetič, and R. Donagi. "Tri-linear couplings in an heterotic minimal supersymmetric Standard Model". *Nuclear Physics B* 745.1-2 (June 2006), pp. 62–83. DOI: 10.1016/j.nuclphysb.2006.03.032.

[BD06] V. Bouchard and R. Donagi. "An SU(5) heterotic standard model". *Physics Letters B* 633.6 (Feb. 2006), pp. 783–791. DOI: 10.1016/j.physletb.2005.12.042.

[BD08] V. Bouchard and R. Donagi. "On heterotic model constraints". *JHEP* 2008.8 (Aug. 2008), pp. 060–060. DOI: 10.1088/1126-6708/2008/08/060.

[BGK13] V. Braun, T. W. Grimm, and J. Keitel. "Geometric Engineering in Toric F-Theory and GUTs with U(1) Gauge Factors". *JHEP* 12 (2013), p. 069. DOI: 10.1007/JHEP12(2013)069.

[BHOP05] V. Braun, Y.-H. He, B. A. Ovrut, and T. Pantev. "A heterotic standard model". *Physics Letters B* 618.1-4 (July 2005), pp. 252–258. DOI: 10.1016/j.physletb.2005.05.007.

[BM14] V. Braun and D. R. Morrison. "F-theory on Genus-One Fibrations". *JHEP* 08 (2014), p. 132. DOI: 10.1007/JHEP08(2014)132.

[CHSW85] P. Candelas, G. T. Horowitz, A. Strominger, and E. Witten. "Vacuum configurations for superstrings". *Nuclear Physics B* 258 (1985), pp. 46–74. DOI: 10.1016/0550-3213(85)90602-9.

[CGSV16] A. Collinucci, S. Giacomelli, R. Savelli, and R. Valandro. "T-branes through 3d mirror symmetry". *JHEP* 07 (2016), p. 093. DOI: 10.1007/JHEP07(2016)093.

[CS15a] A. Collinucci and R. Savelli. "F-theory on singular spaces". *JHEP* 09 (2015), p. 100. DOI: 10.1007/JHEP09(2015)100.

[CS15b] A. Collinucci and R. Savelli. "T-branes as branes within branes". *JHEP* 09 (2015), p. 161. DOI: 10.1007/JHEP09(2015)161.

[CDKPP15] M. Cvetič, R. Donagi, D. Klevers, H. Piragua, and M. Poretschkin. "F-theory vacua with $\mathbb{Z}_3$ gauge symmetry". *Nucl. Phys.* B898 (2015), pp. 736–750. DOI: 10.1016/j.nuclphysb.2015.07.011.

[CGKP14] M. Cvetič, A. Grassi, D. Klevers, and H. Piragua. "Chiral four-dimensional F-theory compactifications with SU(5) and multiple U(1)-factors". *JHEP* 04 (2014), p. 010. DOI: 10.1007/JHEP04(2014)010.

[CHLLT19] M. Cvetič, J. Halverson, L. Lin, M. Liu, and J. Tian. "Quadrillion *F*-Theory Compactifications with the Exact Chiral Spectrum of the Standard Model". *Phys. Rev. Lett.* 123 (2019), p. 101601. DOI: 10.1103/PhysRevLett.123.101601.

[CKPOR15] M. Cvetič, D. Klevers, D. K. M. Peña, P.-K. Oehlmann, and J. Reuter. "Three-Family Particle Physics Models from Global F-theory Compactifications". *JHEP* 08 (2015), p. 087. DOI: 10.1007/JHEP08(2015)087.

[CLLO18] M. Cvetič, L. Lin, M. Liu, and P.-K. Oehlmann. "An F-theory Realization of the Chiral MSSM with $\mathbb{Z}_2$-Parity". *JHEP* 09 (2018), p. 089. DOI: 10.1007/JHEP09(2018)089.

[CSU01a] M. Cvetič, G. Shiu, and A. M. Uranga. "Chiral four-dimensional N=1 supersymmetric type IIA orientifolds from intersecting D6-branes". *Nuclear Physics B* 615.1-3 (Nov. 2001), pp. 3–32. DOI: 10.1016/s0550-3213(01)00427-8.

[CSU01b] M. Cvetič, G. Shiu, and A. M. Uranga. "Three family supersymmetric standard-like models from intersecting brane worlds". *Phys. Rev. Lett.* 87 (2001), p. 201801. DOI: 10.1103/PhysRevLett.87.201801.

[GGK14] I. García-Etxebarria, T. W. Grimm, and J. Keitel. "Yukawas and discrete symmetries in F-theory compactifications without section". *JHEP* 11 (2014), p. 125. DOI: 10.1007/JHEP11(2014)125.

[GLS07] T. L. Gómez, S. Lukic, and I. Sols. "Constraining the Kähler Moduli in the Heterotic Standard Model". *Communications in Mathematical Physics* 276.1 (Sept. 2007), pp. 1–21. DOI: 10.1007/s00220-007-0338-8.

[GKMR86] B. R. Greene, K. H. Kirklin, P. J. Miron, and G. G. Ross. "A superstring-inspired standard model". *Physics Letters B* 180.1 (1986), pp. 69–76. DOI: 10.1016/0370-2693(86)90137-1.

[GH12] T. W. Grimm and H. Hayashi. "F-theory fluxes, chirality and Chern-Simons theories". *JHEP* 2012 (Mar. 2012). DOI: 10.1007/jhep03(2012)027.

[HR19] J. J. Heckman and T. Rudelius. "Top Down Approach to 6D SCFTs". *J. Phys. A* 52.9 (2019), p. 093001. DOI: 10.1088/1751-8121/aafc81.

[HT19] Y.-C. Huang and W. Taylor. "Comparing elliptic and toric hypersurface Calabi–Yau threefolds at large Hodge numbers". *JHEP* 02 (2019), p. 087. DOI: 10.1007/JHEP02(2019)087.

[IMR01] L. E. Ibáñez, F. Marchesano, and R. Rabadan. "Getting just the standard model at intersecting branes". *JHEP* 2001.11 (Nov. 2001). DOI: 10.1088/1126-6708/2001/11/002.

[JTT23a] P. Jefferson, W. Taylor, and A. P. Turner. "Chiral Matter Multiplicities and Resolution-Independent Structure in 4D F-Theory Models". *Commun. Math. Phys.* 404.3 (2023), pp. 1361–1461. DOI: 10.1007/s00220-023-04860-0.

[JTT23b] P. Jefferson, W. Taylor, and A. P. Turner. "Chiral spectrum of the universal tuned $(SU(3) \times SU(2) \times U(1))/\mathbb{Z}_6$ 4D F-theory model". *JHEP* 02 (2023), p. 254. DOI: 10.1007/JHEP02(2023)254.

[JT23] P. Jefferson and A. P. Turner. "Generating functions for intersection products of divisors in resolved F-theory models". *Nucl. Phys. B* 991 (2023), p. 116177. DOI: 10.1016/j.nuclphysb.2023.116177.

[Kim17] Y. Kimura. "Discrete Gauge Groups in F-theory Models on Genus-One Fibered Calabi–Yau 4-folds without Section". *JHEP* 04 (2017), p. 168. DOI: 10.1007/JHEP04(2017)168.

[KMOPR15] D. Klevers, D. K. Mayorga Pena, P.-K. Oehlmann, H. Piragua, and J. Reuter. "F-Theory on all Toric Hypersurface Fibrations and its Higgs Branches". *JHEP* 01 (2015), p. 142. DOI: 10.1007/JHEP01(2015)142.

[Kod63a] K. Kodaira. "On Compact Analytic Surfaces: II". *Ann. of Math.* 77.3 (1963), pp. 563–626.

[Kod63b] K. Kodaira. "On Compact Analytic Surfaces: III". *Ann. of Math.* 78.1 (1963), pp. 1–40.

[KMW12a] S. Krause, C. Mayrhofer, and T. Weigand. "G_4-flux, chiral matter and singularity resolution in F-theory compactifications". *Nuclear Physics B* 858.1 (May 2012), pp. 1–47. DOI: 10.1016/j.nuclphysb.2011.12.013.

[KMW12b] S. Krause, C. Mayrhofer, and T. Weigand. "Gauge Fluxes in F-theory and Type IIB Orientifolds". *JHEP* 2012 (Aug. 2012). DOI: 10.1007/jhep08(2012)119.

[LMTW16] L. Lin, C. Mayrhofer, O. Till, and T. Weigand. "Fluxes in F-theory Compactifications on Genus-One Fibrations". *JHEP* 01 (2016), p. 098. DOI: 10.1007/JHEP01(2016)098.

[LW16] L. Lin and T. Weigand. "G4-flux and standard model vacua in F-theory". *Nuclear Physics B* 913 (Dec. 2016), pp. 209–247. DOI: 10.1016/j.nuclphysb.2016.09.008.

[MPTW14] C. Mayrhofer, E. Palti, O. Till, and T. Weigand. "Discrete Gauge Symmetries by Higgsing in four-dimensional F-Theory Compactifications". *JHEP* 12 (2014), p. 068. DOI: 10.1007/JHEP12(2014)068.

[MPTW15] C. Mayrhofer, E. Palti, O. Till, and T. Weigand. "On Discrete Symmetries and Torsion Homology in F-Theory". *JHEP* 06 (2015), p. 029. DOI: 10.1007/JHEP06(2015)029.

[Mir83] R. Miranda. "Smooth models for elliptic threefolds". In: *The Birational geometry of degenerations (Cambridge, Mass., 1981).* Ed. by R. Friedman, D. R. Morrison, H. University, and S. A. G. S. H. U. (1981). Vol. 29. Progress in mathematics ix. Boston, MA: Birkhäuser, 1983, pp. 85–133.

[MT16] D. R. Morrison and W. Taylor. "Sections, multisections, and U(1) fields in F-theory". *Journal of Singularities* 15 (2016), pp. 126–149. DOI: 10.5427/jsing.2016.15g.

[MV96a] D. R. Morrison and C. Vafa. "Compactifications of F-theory on Calabi–Yau threefolds (II)". *Nuclear Physics B* 476.3 (Sept. 1996), pp. 437–469. DOI: 10.1016/0550-3213(96)00369-0.

[MV96b] D. R. Morrison and C. Vafa. "Compactifications of F-theory on Calabi–Yau threefolds. (I)". *Nuclear Physics B* 473.1-2 (Aug. 1996), pp. 74–92. DOI: 10.1016/0550-3213(96)00242-8.

[Nér64] A. Néron. "Modèles minimaux des variétés abéliennes sur les corps locaux et globaux". *Publications Mathématiques de l'Institut des Hautes Études Scientifiques* 21.1 (1964), pp. 5–125. DOI: 10.1007/BF02684271.

[Oeh21] P.-K. Oehlmann. "Non-flat elliptic four-folds, three-form cohomology and strongly coupled theories in four dimensions". *JHEP* 08 (2021), p. 097. DOI: 10.1007/JHEP08(2021)097.

[Rag18] N. Raghuram. "Abelian F-theory Models with Charge-3 and Charge-4 Matter". *JHEP* 05 (2018), p. 050. DOI: 10.1007/JHEP05(2018)050.

[RT18] N. Raghuram and W. Taylor. "Large U(1) charges in F-theory". *JHEP* 10 (2018), p. 182. DOI: 10.1007/JHEP10(2018)182.

[RTT20] N. Raghuram, W. Taylor, and A. P. Turner. "General F-theory models with tuned $(SU(3) \times SU(2) \times U(1))/\mathbb{Z}_6$ symmetry". *JHEP* 04 (2020), p. 008. DOI: 10.1007/JHEP04(2020)008.

[RT22] N. Raghuram and A. P. Turner. "Orders of vanishing and U(1) charges in F-theory". *JHEP* 03 (2022), p. 051. DOI: 10.1007/JHEP03(2022)051.

[Szy98] M. Szydlo. "Flat Regular Models of Elliptic Schemes". PhD thesis. Harvard University, Dec. 1998.

[TT20] W. Taylor and A. P. Turner. "Generic Construction of the Standard Model Gauge Group and Matter Representations in F-theory". *Fortsch. Phys.* 68.5 (2020), p. 2000009. DOI: 10.1002/prop.202000009.

[Vaf96] C. Vafa. "Evidence for F-theory". *Nuclear Physics B* 469.3 (June 1996), pp. 403–415. DOI: 10.1016/0550-3213(96)00172-1.

[Wei18] T. Weigand. "F-theory". *PoS* TASI2017 (2018), p. 016. arXiv: 1806.01854.

[Wit96] E. Witten. "Nonperturbative superpotentials in string theory". *Nucl.Phys.* B474 (1996), pp. 343–360. DOI: 10.1016/0550-3213(96)00283-0.

Confirmable Workflows

Michael Joswig, Lars Kastner and Benjamin Lorenz

Abstract We discuss what is special about the reproducibility of workflows in computer algebra. It is emphasized how `Julia`'s and `OSCAR`'s features support such a reproducibility, and how users can benefit for their own work. The combined code examples for this book serve as one running example.

Key words: Methodology of mathematics, software, workflows

1 Introduction

There is a current trend to rethink the role of and the use of data in all areas of scientific research. Some aspects are common to all fields of science, such as questions concerning, e.g., where to store the data to make it long-term accessible to the scientific community, or the role of metadata, which should help finding specific data sets. But some other aspects also vary drastically from field to field. Most importantly, the fundamental question about which kind of data is worth storing and how. The primary goals are described in the FAIR-principles, which mean that data should be Findable, Accessible, Interoperable and Reproducible. Here we look at research data in mathematics,

Michael Joswig
TU Berlin, Chair of Discrete Mathematics/Geometry, Berlin, Germany; Max-Planck-Institut für Mathematik in den Naturwissenschaften, Leipzig, Germany, e-mail: `joswig@math.tu-berlin.de`

Lars Kastner
TU Berlin, Chair of Discrete Mathematics/Geometry, Berlin, Germany, e-mail: `kastner@math.tu-berlin.de`

Benjamin Lorenz
TU Berlin, Chair of Discrete Mathematics/Geometry, Berlin, Germany, e-mail: `lorenz@math.tu-berlin.de`

W. Decker et al. (eds.), *The Computer Algebra System OSCAR*,
Algorithms and Computation in Mathematics 32,
https://doi.org/10.1007/978-3-031-62127-7_19

and computer algebra in particular, from the angle of reproducibility. On the way we will also talk about data interoperability, but only to a lesser degree. Throughout we will give examples based on computations with the computer algebra system `OSCAR`, which is written in `Julia`. These examples are chosen to highlight the specific features which the pair `Julia/OSCAR` has to offer toward the reproducibility of computer algebra experiments.

Computer experiments have formed an invaluable resource in mathematical research since the advent of computers. Large calculations that are infeasible for humans with pen and paper can be completed in seconds. A hypothesis can be tested on a million examples. However these advantages can only unfold their true potential if the calculation in the computer is not regarded as a black box. Instead a computer experiment needs to be accompanied by documentation and supplementary material such that everyone reading these details is convinced of the correctness of the calculation. Here we are restricting our attention to computer algebra rather than, say, numerical analysis. This distinction means that, in our scenario, computations are always exact, and questions concerning the precision or the quality of an approximation, do not arise. Furthermore, data sets in computer algebra often tend to be smaller than their counter-parts in numerical analysis. In contrast, computer algebra data tends to be more varied, and equipped with more intricate semantics.

It is a main goal of this chapter to sketch guidelines for the proper documentation of a computer experiment with `OSCAR`, such that the results can easily be confirmed by other mathematicians, with `OSCAR` or without. This confirmation has two steps: First there is the confirmation that re-running the experiment leads to an equivalent result. Second there is the mathematical soundness that the experiment provides the mathematical insight claimed. Along the way we discuss design decisions and other choices.

1.1 Related Work

There is an ongoing effort to discuss the reliability and the reproducibility of research results in all areas of science. In Germany this led to setting up a large scientific network under the name *Nationale Forschungsdateninitiative (NFDI)* (National Research Data Initiative). The *Mathematical Research Data Initiative (MaRDI)* is the single consortium within the NFDI which addresses research data in the mathematical sciences. Recently, MaRDI published a white paper [Con23] on research data management in mathematics.

The MaRDI recommendations are derived from a substantial body of previous work. For example, Stodden et al. [SKB18] published a survey on 306 articles from the Journal of Computational Physics. They found that more than half of the results were impossible to reproduce. The authors emphasized that none of the results could be reproduced with minimal effort. One of their main conclusions was that many problems could have been avoided if the principles of [BBS16] had been followed.

A more recent example is a study of Samuel and Mietchen [SM24] of about 27 000 `Jupyter` notebooks from biomedical publications. That study focuses on `Python` notebooks. In the end, less than 8% of the notebooks could be rerun successfully. Even fewer produced the same results as originally.

An article by Riedel et al. [Rie+22] "analyzes the reproducibility of 108 publications from an interdisciplinary Collaborative Research Center on applied mathematics in various scientific fields." To summarize the conclusion: Only about 40% of the publications considered gave enough data for attempting replication at all. Out of these, only 4 were considered "fully reproducible".

A comprehensive study of reproducibility in computer algebra seems to be lacking so far. However, we invite the reader to try to pick a random computer algebra publication from 10 years ago and to reproduce these results. Our own experiments suggest that the reader will most likely experience some problems. The project `MathRepo` [FG22] seeks to improve this situation.

2 The FAIR Principles

The Paris Call on Research Assessment[1] establishes the following principles for main conditions concerning research data in general:

> Ensure independence and transparency of the data, infrastructure and criteria necessary for research assessment and for determining research impacts; in particular by clear and transparent data collection, algorithms and indicators, by ensuring control and ownership by the research community over critical infrastructures and tools, and by allowing those assessed to have access to the data, analyses and criteria used.

These principles are also reflected in the San Francisco Declaration on Research Assessment[2]. Requirements for data to be useful to other scientists is often translated into the *FAIR principles* [Wil+16]. Here F stands for "findable", A for "accessible", I for "interoperable", and R for "reusable". This means that a researcher who is working in the field that the data belongs to will find it without prior knowledge, she can actually get at the data, and use it interoperable in other contexts. Reusability entails that the data can be reproduced. Only the latter property turns a data set a scientifically meaningful contribution.

That being said, it is not entirely obvious what it requires to call a data set in computer algebra FAIR. Here we focus on the reusability/reproducibility and briefly mention interoperability in computer algebra. That is to say, we largely ignore the important aspects of findability and accessibility. As we see

[1] `https://osec2022.eu/paris-call/`

[2] `https://sfdora.org/`

it, reproducibility is the outcome of an entire workflow, which is confirmable, i.e., adhering to the FAIR principles, at any point. This has a wide range of implications. For instance, this requires that the software code producing the data needs to be FAIR itself. This clearly invites the use of open source software. Conversely, closed software is maybe not ruled out entirely, but the reproducibility of computations with current commercial software systems would require a level of service by the manufacturers of these systems, which seems to be hard to find.

It is worth noting that the reproducibility of proofs in mathematics may be a challenge, independent from computers. A prominent example is the classification of the finite simple groups, which was accepted as accomplished by many around 1980, but some questions remained. Consequently, Gorenstein, Lyons and Solomon initiated a revised classification in a sequence of books [GLS94], which is still growing. A major outcome of this classification and, at the same time, also an indispensable tool along the way, is the *Atlas Of Finite Groups* [CCNPW85]. The question of its reproducibility was raised by Jean-Pierre Serre, when he asked for proofs of the Atlas tables during a talk in 2015[3]. In [BMO17] Breuer, Malle and O'Brien made an effort towards the verification of the Atlas, with `GAP` and `Magma`. So this is an example where mathematical software has been used for checking proofs which had been obtained by more traditional ways of doing mathematics.

Ultimately, any issues concerning the correctness of mathematical proofs, computational or not, should be resolved by theorem provers like `Coq`[4] or `LEAN`[5]. Despite tremendous progress, these provers are still far from capable of certifying large computer algebra computations. Some algorithms can be certified this way. However, typically neither an actual implementation nor its results (considered coming from a black box) can be verified with `Coq`, `LEAN` or their friends. This discrepancy leaves a gap, which needs to be filled with a computational workflow which can be confirmed.

3 Computer Algebra Experiments

A confirmable workflow essentially entails two things: First, the entire workflow must be reproducible. Second, its theoretical claims need to be verifiable by other researchers. A typical scenario is the review process of an article which contains claims which are proved computationally. In the discussion of related work, we gave references to existing principles and guidelines. The purpose of this section is to put these guidelines in the context of computer algebra.

[3] Serre's presentation can be viewed on YouTube at `https://youtu.be/MZ6_JKYdKog`; the Atlas is mentioned after 35min, approximately.

[4] `https://coq.inria.fr/`

[5] `https://leanprover-community.github.io/`

One particularity of computer algebra is its fragmentation. There are many niche software packages solving only very special problems in a highly optimized fashion. Examples include `cohomCalg`[6] [BJRR10] for computing the cohomology of toric varieties, `TOPCOM` [Ram02] and `mptopcom` [JJK18] for enumerating triangulations of finite point sets, or `Arb` [Joh17] for arbitrary precision interval arithmetic. This list could be continued nearly indefinitely. If one needs many of these, the resulting workflow may become complicated. Consequently, it is impossible to provide a scheme for a confirmable workflow in computer algebra which is too rigid. The discussion of general purpose computer algebra systems will be deferred to Section 5. Since the developers of these systems provide a coherent view on the components used, the situation becomes somewhat easier for users of general purpose computer algebra systems. However, even in this case, it is necessary to take into account that some libraries may be linked dynamically, which thus depend on the local installation.

In the rest of this section we give explicit recommendations toward confirmability.

3.1 Connect Your Experiment to the Mathematical Theory

Especially in mathematics and in computer algebra, objects and the theory behind them may be quite complicated. At the same time the computation may seem easy, like simply enumerating lattice points in a polytope or computing the kernel of a matrix. It is essential to explain how the computation fits into the theory. The detailed steps are the items 1, 2 and 3 in [BBS16, Section 9.2.1].

Linking computations with the underlying theory can be achieved in several ways: A very simple approach suggests to add code snippets to a written article or, conversely, references to that article as comments in the code. It is a drawback that this method tends to get messy fast, and it is hard to maintain in large projects. A second approach suggests using a `Jupyter` notebook or similar as an intermediate layer between the code and the paper. Another key part is the proper documentation of the code, especially if it more than just a script, e.g., a `Julia` package.

The same remarks hold for any data, too. Often data looks deceptively simple, like matrices with integer or rational entries or a mere collection of finite sets. For this data to be usable in the future it needs to be thoroughly connected to the mathematical context. For instance, if a matrix is supposed to describe some polytope, then it needs to be specified explicitly if the rows or the columns are supposed to form the generators of an interior description, or if the rows or the columns should be interpreted as linear or affine inequalities.

[6] `http://wwwth.mppmu.mpg.de/members/blumenha/cohomcalg/`

3.2 Publish Your Data

To be able to replicate your experiment any reader needs to know the input and output of your program or code, see item 10 in [BBS16, Section 9.2.1]. In particular, it is not enough to publish the code only and then let the reader figure out the input for your examples. If your experiment produces intermediate results, it makes sense to store these intermediate results too. This is especially important when the intermediate result is the output of one software which is passed on as the input to another software component. Intermediate results provide a level of redundancy which is desirable. They help to understand the full workflow, and they also help to detect the precise point of failure, when things should go wrong.

It is a subtle point that most current computer algebra systems do not provide a high-level view on the data produced with them. For instance, `SageMath` makes use of `Python`'s *pickle* system, which is fragile because it relies on the memory layout of the data during runtime. `SageMath` sometimes compensates such pickles by also storing code snippets which can ideally be used for the reproduction. This setup poses a challenge for the long-term maintenance of data sets. As detailed below, `OSCAR` follows a different strategy, by employing the new `mrdi` file format [DJL24].

Data formats in computer algebra are still being established and probably will not be unified in the near future. Thus documentation will need to be done very carefully and diligently. As mentioned in the previous section, the data stored may actually be simple in structure, e.g. consisting of matrices or vectors. However the mathematical object encoded can be much more complex. In order for the data to be FAIR, see Section 2, every dataset needs to be published with backlinks to the mathematical theory.

An important side aspect is item 13 of [BBS16, Section 9.2.1], choosing an appropriate license. There are tools and questionnaires for finding a suitable license[7]. So let us just highlight the following: In most areas of the world, data without a license cannot be used for further research, legally.

3.3 Publish Your Code

Your code is the key ingredient to rerunning your experiment. There are different ways to publish it, for example on the arXiv you can upload your code as an auxiliary file. This approach is only suitable in case your code is very small, say a `Jupyter` notebook or a simple script file. In most cases though hosting your code in a public git repository is the gold standard. This can be on GitHub or a similar platform. Many universities offer GitLab

[7] `https://opendefinition.org/licenses/`

instances for publishing code. To guarantee long term availability of the code, sign it up for mirroring by software heritage[8].

3.4 Write Down All Software Versions

Usually one will not write the entire algorithm from scratch, but instead rely on algorithms implemented by other mathematicians, experts in their respective fields.

Not every setup will produce the same results. Here by "the same" we mean really the exact same digital data, bit by bit the same. Depending on your setup, data might be represented differently internally. Then algorithms that run in parallel might not produce output in the same order. Finally the mathematical definition of things being equal often differs from the much more strict notion of "digital equality". Besides recording these mathematical details carefully, one should also document details about the setup used. Not all aspects apply to every experiment, so apply common sense, and when in doubt, record whichever thing was causing you doubts.

Changes to the underlying software may change the behavior of your code. The best case is of course that your code still does the same. The second best case is that you get an error message, because some function has been renamed or some other part of your code now fails. The worst case is a silent change of behavior, undetected by your tests, but nevertheless eliminating mathematical soundness.

Here is one small example. The following is valid code in all versions of `Julia`:

```
x = [1,2]
x[2], x[1] = x
```

Yet, the behavior in `Julia` version 1.6 is very different from the behavior in version 1.7 and onward. To analyze and understand such a small example is not difficult, but issues like the above can make it difficult to analyze larger chunks of code. Such challenges, which arise from semantic changes during the evolution of a programming language, are quite common.

We conclude this section with another pitfall which is sometimes overlooked. Most executables (on a Linux system anyway) are dynamically linked to system libraries. They may be moving parts, too. The following command in a Linux shell shows the dependencies of the `Julia` interpreter itself.

```
[lars@nomurai ~]$ ldd `which julia-1.9.3`
        linux-vdso.so.1 (0x00007ffd14f34000)
        libdl.so.2 => /usr/lib/libdl.so.2 (0x00007f12c07f0000)
        libpthread.so.0 => /usr/lib/libpthread.so.0 (0x00007f12c07eb000)
        libc.so.6 => /usr/lib/libc.so.6 (0x00007f12c0609000)
```

[8] `https://www.softwareheritage.org`

```
/lib64/ld-linux-x86-64.so.2 => /usr/lib64/ld-linux-x86-64.so.2
↪ (0x00007f12c0836000)
libjulia.so.1 => not found
```

3.5 Document Your Hardware and Operating System

This section is especially important when measuring the performance of algorithms, but it can also help explain to some reader, why your experiment was running out of memory on their machine, but not on yours. Additionally different architectures may influence the results in subtle ways, meaning that you may get different results on a Mac vs a Windows machine, e.g., if the relevant algorithms employ randomization. In that case you should of course make sure that these results remain the same mathematically speaking.

On Linux, a tool like `inxi` can help collect information about CPU, memory and storage.

```
[lars@nomurai ~]$ inxi -a
CPU: quad core 11th Gen Intel Core i7-1165G7 (-MT MCP-)
speed/min/max: 745/400/4700 MHz Kernel: 5.16.20-200.fc35.x86_64 x86_64
Up: 17d 16h 41m Mem: 11804.1/31819.6 MiB (37.1%)
Storage: 1.86 TiB (29.2% used) Procs: 373 Shell: Bash 5.1.8 inxi: 3.3.14
```

You can get basic information about your operating system using `uname`:

```
[lars@nomurai ~]$ uname -a
Linux nomurai 5.16.20-200.fc35.x86_64 #1 SMP PREEMPT Wed Apr 13 22:09:20
↪ UTC 2022 x86_64 x86_64 x86_64 GNU/Linux
```

For Linux systems it is useful to also add information about the Linux distribution.

4 Julia and Her Friends

OSCAR relies on Julia as a programming language to combine the cornerstones into a unified computer algebra system. In addition to that, the Julia environment provides several key features that support confirmable workflows.

4.1 The Package Manager

The package manager `Pkg.jl` is integrated as a core package into the `Julia` distribution and can be accessed directly from the `Julia` prompt. This allows for quick installations and does not require any extra permissions.

All installed packages are stored in a content-addressed manner in a user-specific folder. This way multiple versions of the same package can be installed at the same time even though only one may be loaded in a single session. But this allows users to work on different projects which might require different dependencies. Content-addressed refers to the fact that each version of a package is stored using an identifier that is based on a *tree hash*, a combined hash of all files and folders in that package directory.

From a developer perspective, the packaging system is also very easy to use and well integrated into GitHub to allow easy registration of new packages and version updates into the *general registry*, the main package database that `Pkg.jl` uses. It is open to anyone, provided the package is covered by an open license, follows reasonable naming guidelines and semantic versioning. The tree hash is also used as an identifier to mirror all registered package versions from their repository to the `Julia` package servers. This makes sure that the packages stay available independent of their original repositories.

4.2 Project Environments

A key feature for confirmable workflows is that the package manager is built around project environments, these consist of two TOML files to describe all packages that are used for a given project. TOML is short for "Tom's Obvious Minimal Language"[9].

`Project.toml` The first one is the project file which lists all direct dependencies for the project, together with optional compatibility bounds. Any `Julia` package contains such a project file, in this case it will also contain the name, version, and a unique identifier for that package. But it may also be used for an unnamed project which only lists other packages that are in use by this project. See page 7 for `OSCAR`'s own `Project.toml`.

`Manifest.toml` The second file describes a concrete instantiation of such a project file and contains the full list of all direct and indirect dependencies together with their exact version numbers and tree hashes.

Given a project file, the package manager can be used to create or update the manifest to allow installation of all packages for a given project. But from a reproducibility standpoint the key feature is that given a project file together with a manifest, the package manager can also be used to reinstall the precise versions that were used when this manifest was created. This can be extremely useful to recreate software experiments.

The currently active manifest can be inspected in the `Pkg.jl` REPL. Even for a `Julia` core package like `Downloads.jl` the manifest already contains 12 entries.

---

[9] `https://toml.io/en/`

```
(project) pkg> status --manifest
Status `~/project/Manifest.toml`
  [0dad84c5] ArgTools v1.1.1
  [56f22d72] Artifacts
  [f43a241f] Downloads v1.6.0
  [7b1f6079] FileWatching
  [b27032c2] LibCURL v0.6.4
  [8f399da3] Libdl
  [ca575930] NetworkOptions v1.2.0
  [deac9b47] LibCURL_jll v8.4.0+0
  [29816b5a] LibSSH2_jll v1.11.0+1
  [c8ffd9c3] MbedTLS_jll v2.28.2+0
  [14a3606d] MozillaCACerts_jll v2022.10.11
  [83775a58] Zlib_jll v1.2.13+0
  [8e850ede] nghttp2_jll v1.52.0+1
```

An established approach to document and store computer experiments in computer algebra are notebooks, and here `Jupyter` is the open source standard. While a `Jupyter` notebook in general does not suffice to fully document the setup [SM24], it can be augmented with `Julia`'s project environments to make sure the correct dependencies are installed automatically.

The combination of the two TOML files can be seen as `Julia`'s contribution to a predefined software environment, meaning a software environment that can reproduce the exact same setting a computer experiment was previously run in. A computer experiment with a more intricate setup can be made reproducible with a suitable *MaPS*[10] runtime environment or other containerization solution, like Docker[11] or even a VM. *MaPS* provides a dedicated runtime

`org.oscar_system.oscar/x86_64/1.0.0`

for using `OSCAR` version 1.0, in particular, for running the examples from this book.

4.3 Binary Packages

In addition to `Julia` packages `Pkg.jl` also manages binary packages for non-julia software that may be used by `Julia` packages. This works by having a small `Julia` package in the general registry that refers to a set of pre-compiled binaries for all supported host platforms.

In particular, this means that the project files, including the manifest, not only fix specific versions for `Julia` packages, but these will also be used to install all required binary packages in the correct versions.

---

[10] `https://github.com/aaruni96/maps`

[11] `https://www.docker.com/`

All binary packages for `Julia` are built using `BinaryBuilder.jl`[12] and the build recipes are managed in the Yggdrasil repository[13]. The build system also includes several mechanisms to allow the binaries to be reproducible, i.e. rebuilding a package from the same recipe will in many cases produce bit-identical output.

4.4 GitHub Integration

The package managing process for `Julia` is well integrated into GitHub, we highlight just a few important features provided by various `Julia` packages:

- Automatic registration of new packages and new package versions with a simple comment. This allows for a quick turnaround for bug fixes and new features. And it also ensures that all versions are properly registered, together with a tree hash, and stay available for download.
- Each package version can automatically generate a new GitHub release which can then be imported to Zenodo to create a citable DOI for the package.
- There are several helpers for the free GitHub Actions continuous integration system which makes it very easy to automatically run tests during the development, even on different platforms that might not be available to the developers.

5 OSCAR

In this section we discuss features of OSCAR which support a confirmable workflow. First, we give hints to how users can organize their own OSCAR experiments such that they become reproducible by other researchers. Second, we address code documentation and maintenance of OSCAR itself as well as the serialization of data. Knowing about these more technical aspects will help to further improve the documentation of user experiments.

5.1 Documenting Computational Experiments

`Julia` and OSCAR offer a variety of tools to aid in documentation of computer experiments of varying size. Here we will discuss how these tools support the user in the tasks of sections 3.1 to 3.4.

[12] `https://binarybuilder.org/`

[13] `https://github.com/JuliaPackaging/Yggdrasil`

As described in section 3.1, the code should be connected to the surrounding mathematical theory. This can be done via a `Jupyter` notebook for small projects, or via some custom documentation using `Documenter.jl` as described in section 5.2.

Data can be published according to section 3.2 using `OSCAR`'s serializer [DJL24], see section 5.3 for more details. One vision of the MaRDI project is to establish this file format in the community such that other computer algebra systems can read data from `OSCAR` and vice versa, thereby increasing interoperability.

Publishing the code as in section 3.3 can be done using any suitable repository, see for example [Con23] for guidance. Again, for smaller projects a notebook may be enough, for a larger project with intended reuse it makes sense to think about creating a separate `Julia` module and adding it to the `Julia` package manager, see section 4.1.

If the project results in code that are interesting to a part of the `OSCAR` community there is a dedicated process for adding new functionality to `OSCAR`[14].

Lastly, both `OSCAR` and `Julia` provide convenient `versioninfo` commands for collecting the relevant software versions of section 3.4. The Oscar variant also takes several additional parameters to give an even fuller view of the versions installed:

```
julia> Oscar.versioninfo(full=true)
OSCAR version 1.0.0
  combining:
    AbstractAlgebra.jl   v0.40.1
    GAP.jl               v0.10.3
    Hecke.jl             v0.30.2
    Nemo.jl              v0.43.1
    Polymake.jl          v0.11.14
    Singular.jl          v0.22.4
  building on:
    Antic_jll               v0.201.500+0
    Arb_jll                 v200.2300.0+0
    Calcium_jll             v0.401.100+0
    FLINT_jll               v200.900.9+0
    GAP_jll                 v400.1200.200+9
    Singular_jll            v403.214.1400+0
    libpolymake_julia_jll   v0.11.4+0
    libsingular_julia_jll   v0.43.0+0
    polymake_jll            v400.1100.1+0
See `]st -m` for a full list of dependencies.

Julia Version 1.10.2
Commit bd47eca2c8a (2024-03-01 10:14 UTC)
Build Info:
  Official https://julialang.org/ release
Platform Info:
```

[14] https://docs.oscar-system.org/dev/Experimental/intro/

```
  OS: Linux (x86_64-linux-gnu)
  CPU: 8 × 11th Gen Intel(R) Core(TM) i7-1165G7 @ 2.80GHz
  WORD_SIZE: 64
  LIBM: libopenlibm
  LLVM: libLLVM-15.0.7 (ORCJIT, tigerlake)
Threads: 1 default, 0 interactive, 1 GC (on 8 virtual cores)
Official https://julialang.org/ release
```

You can see that this also entails information on the OS and the CPU. Furthermore it has information on the dependencies and even their version numbers. This version data is also stored in the TOML files mentioned in Section 4.2.

5.2 User Manual

The entire OSCAR user manual resides next to the actual code, in the OSCAR source tree. In fact, it comprises separate sections dedicated to specific areas of mathematics. Each such area starts with an introductory page which explains the basic concepts and definitions. In each section we list standard mathematical textbooks which are used as the basis for all functions in that section.

For instance, toric varieties are algebraic varieties which are defined via a polyhedral fan associated with a lattice polytope. In the literature, two competing conventions exist, which are dual to one another. Either the fan is the normal fan of the relevant polytope, or the face fan. The textbook [CLS11] by Cox, Little and Schenck is chosen as OSCAR's standard reference, and so toric varieties in OSCAR are defined via normal fans.

OSCAR's functions are documented in the `Julia` code, and we rely on `Documenter.jl` to compile the various pieces of the documentation into HTML. Those HTML pages are linked on OSCAR's GitHub page. They exist in two flavors: one for the latest stable release[15] and one for the current developer's version[16]. Further versions, e.g., for git branches, can be built locally by calling `Oscar.build_doc()`. The code examples that accompany the documentation are verified via `Documenter.jldoctest()` as part of the continuous integration process; see Section 5.5.

5.3 Data Serialization

Serialization in computer science is the process of taking a chunk of data and putting it into a form such that it can be written to a file and read back

[15] `https://docs.oscar-system.org/stable/`

[16] `https://docs.oscar-system.org/dev/`

identically. Sometimes writing to a file is replaced by network or interprocess communication. Between these tasks there are differences that matter on a technical level, but there is no difference between the communication over a network and writing to and reading from a file on a logical level. Principally, `Serialization.jl` takes care of the serialization of objects in `Julia`, and here is the beginning of the documentation of its main function `serialize`:

> Write an arbitrary value to a stream in an opaque format, such that it can be read back by deserialize. The read-back value will be as identical as possible to the original, but note that Ptr values are serialized as all-zero bit patterns (NULL).

The little caveat at the end of that text fragment gives a first hint that serialization is sometimes not as straightforward as it might seem. This is particularly true in semantically rich environments like in computer algebra.

The traditional way of storing data resulting from computer algebra computations is via notebooks, which combine code, computational results and text. This idea was pioneered by `Mathematica` and later picked up by `Maple` and others. `Julia` and thus `OSCAR` is supported by `Jupyter`, and so such notebooks exist for `OSCAR`, too. The notebook concept comes with serious limitations, however. Namely, it is difficult or impossible to store intermediate results without the need to recompute them when they are needed for a subsequent new computation. In this way `Mathematica`, `Maple` and `Jupyter` behave almost the same.

`OSCAR` employs the new `mrdi` file format [DJL24] for the serialization of its data. The design goals are different from `Serialization.jl` or from notebooks. In `OSCAR`, the user should be able to start a computation on one machine, store the resulting output in a file, read that file on a different machine and continue that computation, without the need of recomputation. Of course, it is trivial to print output and to store it in a text file. For computer algebra this typically does not suffice, for several reasons. First, parsing that text before continuing the computation may be tedious. Second, the same may be even impossible since the usual text output sometimes suppresses crucial information. Third, what if between the two steps of the computation several years have passed, and the specific way of printing a certain output has changed? Ideally, such a serialization could even pave the road to a certain degree of interoperability among several computer algebra systems. However, this is a challenge of its own. We refer to the article [DJL24] for answers to these questions and more technical details.

To conclude this section we give one example, to show what `OSCAR` can do, and which would be impossible with `Serialization.jl`. We create a multivariate polynomial ring with coefficients in the finite field of order 49.

```
julia> F, o = finite_field(7,2)
(Finite field of degree 2 and characteristic 7, o)

julia> R, (y,z) = F["y","z"]
```

```
(Multivariate polynomial ring in 2 variables over GF(7, 2),
↪  FqMPolyRingElem[y, z])
```

Then we save two polynomials that are loaded back and added.

```
julia> save("p.mrdi", 2*y^3*z^4 + (o + 3)*z^2 + 5*o*y + 1)

julia> save("q.mrdi", y^17*z^3 + (o + 4)*z + 2*o*y)

julia> print( load("p.mrdi") + load("q.mrdi") )
y^17*z^3 + 2*y^3*z^4 + (o + 3)*z^2 + (o + 4)*z + 1
```

The key point here is that the semantics of `OSCAR` allows for adding two polynomials only if they live in the same ring; it is not enough that the two polynomials are both bivariate over the same domain of coefficients. Those coefficients pose a further problem; since 49 is not a prime number there is no canonical encoding for the coefficients. Again `OSCAR` semantics requires the coefficients to live in the same field. To sketch the range of applications that are possible with the `mrdi` file format, let us imagine the following scenario. After storing the polynomials to the files `p.mrdi` and `q.mrdi`, the user, Alice, could send those files via email to her friends Bob and Charlie. Then Bob computes the square of `p.mrdi` on his computer, while Charlie computes the third power `q.mrdi`. Both store their results in new files. Afterwards they send their files back to Alice, who is able to compute the sum of that square and that third power. Bob and Charlie do not even need to know about each other.

5.4 Confirmable `OSCAR` Book

As a demonstration we have prepared a workflow that runs all the examples from the different chapters of this book. This is done via a specialized package called `OscarBookExamples.jl`[17]. Let us summarize the steps to do so:

1. For every chapter, we read the LaTeX code and extract the lines including code files. Then we read those code files in order and compose them into markdown files.
2. These markdown files are then added to the documentation part of `OscarBookExamples.jl` and the `Julia` package `Documenter.jl` is run on this documentation. We set it to automatically correct the output in the examples which are comprised of the examples from the `OSCAR` book.
3. The corrected examples are extracted from the markdown files and compared to the original examples. In case there are differences, these are reported and can then be fixed by the authors.

[17] `https://github.com/oscar-system/OscarBookExamples.jl`

We use several mechanisms from `Documenter.jl`. All examples are wrapped in `jldoctest` environments. For every chapter a custom label is introduced and all `jldoctests` are equipped with this label. This ensures that these examples are run in the same environment and that variables may be reused.

Some examples use code that is not in `OSCAR`, this code has been placed separate directory `auxiliary_code`.

Already during the writing phase, the continuous testing of the code included exhibited many shortcomings of the examples:

1. Typos in the code.
2. Examples use code that has not been merged to master yet.
3. Examples use external code that is not up to `OSCAR` standards.

We collected the examples from this book in the `OscarBookExamples.jl` repository.

5.5 Continuous Integration

`OSCAR` utilizes a multitude of software components written in several languages, including `C`, `C++`, `Perl`, `Python` and, of course, `Julia`. The largest contributions come from the four *cornerstone systems*: `ANTIC`, `GAP`, `polymake` and `Singular`. These in turn interface other software; e.g., `polymake` will often dispatch convex hull computations to `cddlib`, `lrslib`, or `Normaliz`. What makes `OSCAR` special is that the development team of `OSCAR` comprises the developers of the four cornerstones. This results in an unusually high degree of integration of each cornerstone into `OSCAR`, leading to a coherent semantics and fast computations.

This peculiar setup requires a comprehensive approach to testing and maintaining the code, to avoid any regression, with respect to correctness and performance, during the development process. To this end `OSCAR` employs the continuous integration (CI) workflow of GitHub, including code coverage analysis. This aids the developer in properly checking all new contributions. Additionally, the cornerstones continue to exist as separate software systems, which thus come with their own CI features and setups. It is useful that GitHub's CI workflow directly allows for continuous delivery (CD). Such a setup for CI/CD is is considered quite standard by now and similarly implemented for many other computer algebra systems. These tools are crucial for guaranteeing robustness in a complex system like `OSCAR` with many moving parts, while keeping development speed high.

For instance, duplicate implementations of standard algorithms can be used for cross-certification. Here is a Smith normal form computation:

```
julia> m = matrix(ZZ, [1 2; 3 4])
[1   2]
[3   4]
```

```
julia> snf(m)
[1  0]
[0  2]
```

OSCAR attaches multiple implementations of Smith normal forms to the same function, for instance an optimized version for integer matrices, which is the one used above. A second one from AbstractAlgebra works for arbitrary Euclidean rings.

```
julia> methods(snf, Tuple{MatrixElem{ZZRingElem}})
# 2 methods for generic function "snf" from AbstractAlgebra:
 [1] snf(x::ZZMatrix)
     @ Nemo ~/.julia/packages/Nemo/xwSoX/src/flint/fmpz_mat.jl:1096
 [2] snf(a::MatrixElem{T}) where T<:RingElement
     @ ~/.julia/packages/AbstractAlgebra/dsta0/src/Matrix.jl:5431

julia> invoke(snf, Tuple{MatrixElem{ZZRingElem}}, m) == snf(m)
true
```

OSCAR uses sensible defaults for the user through Julia's multiple dispatch system which is why invoking the more general AbstractAlgebra implementation requires a syntactic effort.

6 Conclusion and Outlook

OSCAR was created with the FAIR principles in mind. It allows various users to document their computer experiments, and in particular, saving intermediate results. Already now OSCAR is able to interact with the database polyDB [Paf17], which contains objects from polyhedral and tropical geometry. We plan to setup further databases of a similar kind. Modern document based databases support JSON, which underlies the mrdi file format. In this way our approach to serialization naturally works with existing database solutions.

Ideally, computer algebra data would last forever, much like theorems in mathematics. This is in contrast with existing software that needs to evolve and adapt to new standards and hardware. Similarly, data files need to evolve, too. In OSCAR we rely on versioning and automatic upgrade scripts, based on two decades of experience with the polymake file format [GHJ16], see also [JP02].

We envision interacting with computer algebra systems other than OSCAR through the mrdi file format and namespaces.

We integrated the collected examples of this book into the OSCAR CI to continuously monitor whether the examples still work with the most recent OSCAR version. This will help us to migrate the existing example code to future versions of OSCAR.

Acknowledgements Michael Joswig and Lars Kastner were supported by MaRDI (Mathematical Research Data Initiative), funded by the Deutsche Forschungsgemeinschaft (DFG), project number 460135501, NFDI 29/1 "MaRDI – Mathematische Forschungsdateninitiative"[18]. The authors wish to thank Thomas Breuer for pointing out and detailing the pitfalls and struggles while verifying the Atlas tables. Michael Joswig is supported by the SFB-TRR 195–286237555 "Symbolic Tools in Mathematics and their Application".

References

[BBS16] D. H. Bailey, J. M. Borwein, and V. Stodden. "Facilitating Reproducibility in Scientific Computing: Principles and Practice". In: *Reproducibility*. John Wiley & Sons, Ltd, 2016. Chap. 9, pp. 205–231. DOI: 10.1002/9781118865064.ch9.

[BJRR10] R. Blumenhagen, B. Jurke, T. Rahn, and H. Roschy. "Cohomology of line bundles: A computational algorithm". *Journal of Mathematical Physics* 51.10 (Oct. 2010), p. 103525. DOI: 10.1063/1.3501132.

[BMO17] T. Breuer, G. Malle, and E. A. O'Brien. "Reliability and reproducibility of *Atlas* information". In: *Finite simple groups: thirty years of the atlas and beyond*. Vol. 694. Contemp. Math. Amer. Math. Soc., Providence, RI, 2017, pp. 21–31. DOI: 10.1090/conm/694.

[Con23] T. M. Consortium. *Research Data Management Planning in Mathematics*. Oct. 2023. DOI: 10.5281/zenodo.10018246.

[CCNPW85] J. H. Conway, R. T. Curtis, S. P. Norton, R. A. Parker, and R. A. Wilson. $\mathbb{ATLAS}$ *of finite groups*. Maximal subgroups and ordinary characters for simple groups, With computational assistance from J. G. Thackray. Oxford University Press, Eynsham, 1985, pp. xxxiv+252.

[CLS11] D. A. Cox, J. B. Little, and H. K. Schenck. *Toric varieties*. Vol. 124. Graduate Studies in Mathematics. American Mathematical Society, Providence, RI, 2011. DOI: 10.1090/gsm/124.

[DJL24] A. Della Vecchia, M. Joswig, and B. Lorenz. "A FAIR file format for mathematical software". In: *Mathematical software—ICMS 2024*. Vol. 14749. Lecture Notes in Computer Science. Springer, Cham, 2024, pp. 234–244. DOI: 10.1007/978-3-031-64529-7_25.

[FG22] C. Fevola and C. Görgen. "The Mathematical Research-Data Repository MathRepo". *Der Computeralgebra-Rundbrief* 70 (Mar. 2022). arXiv: 2202.04022.

[18] https://www.mardi4nfdi.de/

[GHJ16] E. Gawrilow, S. Hampe, and M. Joswig. "The polymake XML file format". In: *Mathematical software – ICMS 2016. 5th international congress, Berlin, Germany, July 11–14, 2016. Proceedings.* Berlin: Springer, 2016, pp. 403–410. DOI: 10.1007/978-3-319-42432-3_50.

[GLS94] D. Gorenstein, R. Lyons, and R. Solomon. *The classification of the finite simple groups.* Vol. 40.1. Mathematical Surveys and Monographs. American Mathematical Society, Providence, RI, 1994, pp. xiv+165. DOI: 10.1090/surv/040.1.

[Joh17] F. Johansson. "Arb: efficient arbitrary-precision midpoint-radius interval arithmetic". *IEEE Transactions on Computers* 66 (8 2017), pp. 1281–1292. DOI: 10.1109/TC.2017.2690633.

[JJK18] C. Jordan, M. Joswig, and L. Kastner. "Parallel enumeration of triangulations". *Electron. J. Combin.* 25.3 (2018), Paper 3.6, 27. DOI: 10.37236/7318.

[JP02] M. Joswig and K. Polthier. "EG-Models — a New Journal for Digital Geometry Models". In: *Multimedia Tools for Communicating Mathematics.* Ed. by J. Borwein, M. H. Morales, K. Polthier, and J. F. Rodrigues. Springer, 2002, pp. 165–190. DOI: 10.1007/978-3-642-56240-2_10.

[Paf17] A. Paffenholz. "polyDB: a database for polytopes and related objects". In: *Algorithmic and Experimental Methods in Algebra, Geometry, and Number Theory.* Springer, 2017, pp. 533–547.

[Ram02] J. Rambau. "TOPCOM: Triangulations of point configurations and oriented matroids." In: *Mathematical software. Proceedings of the 1st international congress, Beijing, China, August 17–19, 2002.* Singapore: World Scientific, 2002, pp. 330–340.

[Rie+22] C. Riedel et al. "Including Data Management in Research Culture Increases the Reproducibility of Scientific Results". In: *GI-Jahrestagung.* 2022.

[SM24] S. Samuel and D. Mietchen. "Computational reproducibility of Jupyter notebooks from biomedical publications". *GigaScience* 13 (Jan. 2024), giad113. DOI: 10.1093/gigascience/giad113.

[SKB18] V. Stodden, M. S. Krafczyk, and A. Bhaskar. "Enabling the Verification of Computational Results: An Empirical Evaluation of Computational Reproducibility". In: *Proceedings of the First International Workshop on Practical Reproducible Evaluation of Computer Systems.* P-RECS'18. Tempe, AZ, USA: Association for Computing Machinery, 2018. DOI: 10.1145/3214239.3214242.

[Wil+16] M. D. Wilkinson et al. "The FAIR Guiding Principles for scientific data management and stewardship". *Scientific Data* 3 (2016). DOI: 10.1038/sdata.2016.18.

The manufacturer's authorised representative in the EU is Springer Nature Customer Service Centre GmbH, Europaplatz 3, 69115 Heidelberg, Germany. If you have any concerns regarding our products, please contact ProductSafety@springernature.com

Printed and bound by CPI Group (UK) Ltd, Croydon, CR0 4YY
01/10/2026
02225914-0016